SERIES ENTOMOLOGICA

EDITOR

E. SCHIMITSCHEK

Wien

VOLUMEN 2

Springer-Science+Business Media, B.V. 1967

A SYSTEMATIC CATALOGUE OF THE GENUS *ZYGAENA* FABRICIUS (LEPIDOPTERA: *ZYGAENIDAE*)

by

HUGO REISS
and
W. GERALD TREMEWAN

Springer-Science+Business Media, B.V. 1967

ISBN 978-94-017-5857-4 ISBN 978-94-017-6333-2 (eBook)
DOI 10.1007/978-94-017-6333-2

Ursprünglich erschienen bei Dr W. Junk Publisher, The Hague 1967
Softcover reprint of the hardcover 1st edition 1967

INTRODUCTION

It is forty years since Burgeff published, in 1926, the first comprehensive catalogue of the genus *Zygaena* Fabricius, forming part 33 of the Lepidopterorum Catalogus. Following the pattern and general layout of Burgeff's work, we have attempted to produce a catalogue in which all names in the genus *Zygaena* are included, with references to the literature where these names were originally published. Additional references are included when these refer to illustrations of a species, subspecies or form, or to a taxonomic change, e.g., a change in status. References to misidentifications are generally omitted unless a new species has been described at a later date.

In compiling this work we have adhered to the International Code of Zoological Nomenclature as adopted by the XV International Congress of Zoology. However, although the International Commission on Zoological Nomenclature recognises the necessity of names of lower rank than subspecies, they do not at present deal with such names. The provisions of the Code do not apply to them and, therefore, such names have no nomenclatural status.

Every subspecies is given equal status in the catalogue although their relative value is not always the same. Certain authors have very often separated subspecies on minute differences and a subsequent examination of further material, taken over a number of years, has shown that the differences are not always constant. In many cases, however, we have been unable to verify the status of each subspecies.

The names of forms and aberrations are included in the catalogue and are, in our opinion, a necessity to geneticists and collectors. Long descriptions are thereby avoided by the use of such names. This also applies to the names of hybrids for the same reasons.

The hybrids which were described by Stauder from specimens captured in the field are listed in the catalogue as aberrations or synonyms, since an examination of many of his types has shown that the genitalia are identical with those of normal specimens. It has been shown that the genitalia of hybrids, bred in captivity from a cross-pairing between two different species, exhibit intermediate characters.

It has often been customary to employ the same name when describing similar aberrations occurring in two or more species of the genus *Zygaena*. For example, the name *flava* has been applied by many authors to yellow aberrations occurring in different species. This seems to be a more suitable method than employing many different names when describing aberrations analogous for colour. The name *flava* immediately suggests that the aberration is yellow. There should be no confusion if the name is cited with its respective specific and subspecific names. If the Law of Homonymy were applied to the names of aberrations, a large percentage would immediately become junior primary homonyms. The species of the genus *Zygaena* are extremely variable and many hundreds of names of aberrations can be found in the existing literature. When describing an aberration, some authors have used a name that has already been employed for a similar aberration occurring in the same species. For example, Rühl (1896, Ent.

Z., 12: 117) applied the name *grossmanni* to a yellow aberration of *Z. purpuralis fatrensis* Reiss. Later, Burgeff (1914, Mitt. münch. ent. Ges., 5: 43) employed the same name for the yellow aberration occurring in *Z. purpuralis nubigena* Lederer, as follows: "*purpuralis nubigena* ab. *grossmanni* Rühl (n. em.)". This was quoted in later publications as ab. *grossmanni* (Rühl) Burgeff, 1914. This method was originally introduced by Burgeff (loc. cit., 5: 37) who employed the term "n. em." (= nomen emendatum). This is not truly an emendation, which is either the correction of an incorrect spelling of a name or an incorrect subsequent spelling of a name. We do not condone Burgeff's method of naming aberrations and, in the catalogue, those described in this manner are included and date from the publication of the second author to whom the name is also attributed. Such aberrations are considered separate entities and the descriptions are regarded as new. This should not be confused with misidentifications.

With the exception of *Z. ephialtes* Linné, little is known of the genetics of the various forms and aberrations of the *Zygaena* species. Dryja (1959, Badania nad Polimorfizmem Kraśnika Zmiennego) has made an extensive study of *ephialtes*, his work being the result of thirty-two years of breeding experiments. It has been shown by Dryja that the two basic forms (peucedanoid and ephialtoid) each depends on a pair of independently inherited allelomorphic genes and that peucedanoid forms are dominant to ephialtoid forms. The peucedanoid and ephialtoid forms can be either red or yellow, the former colour being dominant to the latter. Each of these colours is dependent on a second pair of independently inherited alleles. These characters showed, in the F_2, F_3 and back-crosses, typical Mendelian segregation. The peucedanoid and ephialtoid forms can be five- or six-spotted. It was found that five-spotted forms are dominant to six-spotted forms and that each depends on a third pair of independently inherited alleles. The same results have been obtained by Burgeff and Bovey and the references to their publications may be found in this catalogue.

Some species of the genus *Zygaena* are closely allied and the genitalia do not exhibit good characters. The concept of a species varies according to the author. If cross-breeding between closely related species were undertaken, it would show whether the resulting hybrids (F_1) could interbreed freely and produce a further generation (F_2). If breeding experiments produced such results it would suggest that the separation of these species is not justified. The closely allied species *trifolii* Esper and *lonicerae* Scheven have almost identical genitalia and the minute genital differences are not always constant. When these two species are crossed the resulting hybrids can interbreed quite freely and produce fertile progeny. However, the larvae of the two species are different in coloration and setal structure and, in our opinion, such differences are of considerable taxonomic value. Closely allied species, such as *trifolii* and *lonicerae*, have a wide geographical distribution which overlaps. A large number of well defined subspecies have evolved but each of these subspecies has, however, the basic characters of either *trifolii* or *lonicerae*. The grouping of such species into one species would, in our opinion, be quite unjustifiable.

Although twenty-two subgenera have been erected within the genus *Zygaena*, the species are grouped in the catalogue under three subgenera only, viz., *Mesem-*

brynus Hübner, *Agrumenia* Hübner and *Zygaena* Fabricius. These three subgenera are divided into sections or species groups, many of which were formerly placed as subgenera. The remaining subgeneric names are now placed in synonymy. This arrangement follows, in general, the classification of the senior author (Reiss, 1958, Z. wien. ent. Ges., 43: 140 et seq.) who based his conclusions on the morphology, external characters, such as wing pattern and coloration, and the biology of the species. The species in subgenus *Mesembrynus* are considered to be the most primitive. This subgenus is followed by the subgenus *Agrumenia* which in turn is followed by subgenus *Zygaena*, the latter containing what are considered to be the most recently evolved species. The coloration and spot-formation of the forewings give support to this theory. In what are considered to be the most primitive species of *Mesembrynus* the forewing spots are unicolorous cream, yellow, orange, light red or carmine. Most of the species in *Agrumenia* have the forewing spots encircled with whitish cream, cream or light yellow coloration. It is thought that the red coloration in the forewing spots in *Agrumenia* began from the centre in the more primitive species. The yellowish or whitish ring then appeared as the remainder of the former colour of the forewing spots which, as already stated above, are in the more primitive species of *Mesembrynus*, yellow, orange, light red or carmine. The closely related species *Z. banghaasi* Burgeff and *cocandica* Erschoff are good examples. Only in the more recently evolved species of *Agrumenia*, e.g., *exulans* Reiner & Hohenwarth and *loti* Denis & Schiffermüller, are the rings normally absent around the forewing spots. When present, the rings are found only in aberrant specimens. It follows that the most recently evolved species grouped in *Zygaena* have lost the cream or whitish rings surrounding the forewing spots. It is interesting to note that occasionally, specimens of *filipendulae* Linné and *transalpina* Esper have been captured with traces of whitish rings around the forewing spots. Such forms occur only as extremely rare aberrations in Italy and suggest a possible link with the more primitive *Agrumenia* species.

A red collar and abdominal ring are present in most species of the subgenera *Mesembrynus* and *Agrumenia*. The abdominal ring also occurs in the more primitive species of subgenus *Zygaena*. In the more recently evolved species in *Zygaena*, however, the abdominal ring occurs only in aberrant specimens and has not been detected at all in *ramburii* Herrich-Schäffer, *trifolii* and *lonicerae*, which are considered to be the most recently evolved species of the whole genus *Zygaena*.

The larvae of the species in *Mesembrynus* feed on plants of the families *Umbelliferae*, *Compositae* and *Labiatae*. The larvae of the species in *Agrumenia* feed mostly on hard-leaved *Papilionaceae* while the larvae of the species in *Zygaena* feed mostly on soft-leaved *Papilionaceae*. A study of the larvae, pupae and cocoons, as far as they are known, has provided valuable characters for placing the species into species-groups or sections.

The habits of the species by day and their resting place at night are also worth noting. The more recently evolved species, when not flying or feeding, sit about on flowers and rest singly but more often in groups during wet weather and at night. The more primitive species fly for short periods in the sunshine and, after feeding, rest singly on stems of plants and bushes.

It is thought that the species in subgenus *Mesembrynus*, whose larvae feed on *Umbelliferae*, *Compositae* and *Labiatae*, originated in the Tertiary (possibly the middle Miocene) period. It is assumed that the subgenus *Agrumenia* originated rather early, possibly at the end of the Miocene period and after the transition of the larval feeding habits to the *Papilionaceae* had taken place. The most recently evolved subgenus *Zygaena* probably originated from the middle of the Pliocene period. It is not possible to conceive whether a second transition took place of the larval feeding habits of the umbelliferous and composite feeders to the *Papilionaceae*.

Many species of *Mesembrynus*, *Agrumenia* and *Zygaena* probably had a far wider distribution than they have in present times. The present, more localised and discontinuous distribution was probably caused by the influence of the Ice Age. It is thought that the more primitive species originated in and around the Mediterranean region, spreading from here to the western and northern parts of Europe, to Africa north of the Sahara, and north-east from Asia Minor to Central Asia and Siberia. As far as it is known, only one species, *niphona* Butler, has reached Japan. The distribution of the genus is palaearctic with the exception of two species, viz., *rubricollis* Hampson and *transpamirina* Koch from Chitral.

In this catalogue the subspecies of each species are arranged according to the geographical distribution, approximately from south to north and east to west. The type locality of each subspecies is provided as it was originally quoted by the author (to avoid errors in translation), and is placed in the righthand column opposite the subspecific name. The names of forms and aberrations are arranged under the subspecies to which they belong. The subspecific names are prefixed with the term "ssp." and the names of aberrations are prefixed with the term "ab.". We have refrained from using the term "f." (= forma) for the aberrations, since a form is considered to be of higher status than an aberration as, for example, seasonal forms (f.t. = forma tempestatis). The term "f." is used, however, to denote the constantly recurring forms of the polymorphic species *ephialtes*. The term "f. loc." (= forma alicuius loci) is used to denote a local population which differs sufficiently from the subspecies to merit a name. This terminology is based on that of Rothschild and Jordan (1903, Novit. zool., 9, Supplement). The term "var.", which has been used quite indiscriminately by many authors to denote both subspecies or geographical races and aberrations, is here discarded as it has no status in nomenclature. Junior synonyms are placed in italics and are listed under their respective senior synonyms. Junior homonyms are also given in italics and are placed under an available, existing name.

We have experienced some difficulty in determining the dates of publication of certain names of Esper, Hübner and Herrich-Schäffer. Esper's "Die Schmetterlinge" was issued in several parts. For the dates of publication of these various parts we have referred to the paper by Sherborn & Woodward (1901, Ann. Mag. nat. Hist., (7) 7: 137-140). The dates of publication of the various parts of Hübner's "Sammlung europäischer Schmetterlinge" and "Verzeichniss bekannter Schmettlinge [sic]" are taken from Hemming's excellent work "Hübner", published in two volumes by the Royal Entomological Society of London in 1937. The dates of names published in Herrich-Schäffer's "Systematische Bearbeitung der Schmetterlinge von Europa" are also taken from Hemming's work. References to periodicals are abbreviated and, where possible, follow the "World List of Scientific Periodicals, 1900-1950". Titles of books and other publications are quoted in full. All references and the spelling of every name have been compared with the original literature for verification.

We are most grateful to Sylvia M. Tremewan, the wife of the junior author, for her help in compiling the index of the catalogue, a formidable task, comprising over three thousand four hundred names.

HUGO REISS,
7 Stuttgart 1,
Traubenstrasse 15B[1],
West Germany.

W. GERALD TREMEWAN,
56, Hart Road,
Byfleet,
Surrey, England.

10th September, 1966.

EIN SYSTEMATISCHER KATALOG DER GATTUNG *ZYGAENA* FABRICIUS (LEPIDOPTERA: *ZYGAENIDAE*)

von

HUGO REISS
und
W. GERALD TREMEWAN

EINFÜHRUNG

40 Jahre ist es her, dass Burgeff im Jahre 1926 den ersten umfassenden Katalog über die Gattung *Zygaena* Fabricius als Pars 33 des Lepidopterorum Catalogus veröffentlichte. Wir sind der allgemeinen Anlage von Burgeff's Katalog gefolgt und haben versucht, einen Katalog zu schaffen, in dem alle Namen der Gattung *Zygaena* mit genauen Literaturangaben enthalten sind. Zusätzliche Literaturangaben sind erfolgt, wenn diese sich auf Abbildungen einer Art (Species), Unterart (Subspecies) oder Form oder auf eine taxonomische Änderung, z. B. eine Änderung im Status, beziehen. Literaturangaben über Falschbestimmungen wurden weggelassen, es sei denn, dass später eine neue Art beschrieben wurde.

Bei der Aufstellung des Katalogs haben wir uns an den internationalen Code für zoologische Nomenklatur gehalten, wie er vom XV. internationalen Kongress für Zoologie angenommen wurde. Obgleich die internationale Kommission für zoologische Nomenklatur die Notwendigkeit von Namen von geringerem Rang als die Subspecies erkennt, befasst sie sich gegenwärtig nicht mit solchen Benennungen. Der Code wird auf sie nicht angewendet. Aus diesem Grunde haben solche Namen keinen nomenklatorischen Status.

Im Katalog wurde jeder Unterart (Subspecies) der gleiche Status gegeben, obgleich ihr relativer Wert nicht immer der gleiche ist. Verschiedene Autoren haben sehr oft Unterarten mit winzigen Unterschieden abgetrennt, bei denen eine anschliessende Prüfung von weiterem Material, das in späteren Jahren gesammelt wurde, zeigte, dass die Unterschiede nicht immer konstant sind. In vielen Fällen ist es uns indessen nicht möglich, den Status von jeder Unterart nachzuprüfen.

Die Namen von Formen und Aberrationen sind im Katalog enthalten, weil sie nach unserer Meinung für die Genetiker und für die Sammler notwendig sind. Lange Beschreibungen können vermieden werden, wenn man die gegebenen Formen- und Aberrationsnamen verwendet. Dieselben Gründe sind auch für die Anwendung von Namen von Hybriden massgebend.

Die Hybriden, welche Stauder nach gefangenen Stücken beschrieben hat, sind im Katalog als Aberrationen oder Synonyme angeführt, weil eine Nachprüfung von vielen seiner Typen gezeigt hat, dass die Genitalien mit denen normaler Stücke identisch sind. Es hat sich herausgestellt, dass die Genitalien von in Gefangenschaft erzogenen Hybriden aus Kreuzungen verschiedener Arten intermediären Charakter zeigen.

Es ist üblich geworden, den gleichen Namen bei der Beschreibung von ähnlichen Aberrationen, die in 2 oder mehr Arten der Gattung *Zygaena* vorkommen, zu verwenden. Zum Beispiel wurde der Name *flava* von vielen Autoren für gelbe Aberrationen bei verschiedenen Arten gegeben. Dies scheint eine passendere Methode zu sein als die Anwendung von vielen verschiedenen Namen bei der Beschreibung von Aberrationen mit gleicher Farbe. Der Name *flava* deutet sofort an, dass die Aberration gelb ist. Es dürfte keine Verwirrung geben, wenn der Name zusammen mit dem betreffenden Art- oder Unterart-namen zitiert wird. Wenn das Gesetz der Homonymie auf die Aberrationsnamen angewendet würde, würde ein grosser Prozentsatz derselben zu jüngeren primären Homonymen werden. Die

Arten der Gattung *Zygaena* sind meistens sehr variabel und viele hundert Namen von Aberrationen wurden in der vorhandenen Literatur gefunden.

Bei der Beschreibung einer Aberration haben einige Autoren einen Namen verwendet, welcher schon für eine ähnliche Aberration in der gleichen Art angewendet wurde. Zum Beispiel, Rühl (1896, Ent. Z., 12: 117) gab einer gelben Aberration von *Z. purpuralis fatrensis* Reiss den Namen *grossmanni*. Später verwendete Burgeff (1914, Mitt. münch. ent. Ges., 5: 43) den gleichen Namen für die gelbe Aberration bei *Z. purpuralis nubigena* Lederer mit nachstehender Bezeichnung „*purpuralis nubigena* ab. *grossmanni* Rühl (n. em.)". Diese Aberration wurde in späteren Veröffentlichungen als ab. *grossmanni* (Rühl) Burgeff, 1914, zitiert. Diese Methode wurde ursprünglich von Burgeff (loc. cit., 5: 37) eingeführt, der die Bezeichnung „n. em." (= nomen emendatum) verwendete. Dies bedeutet keine eigentliche Emendation, die entweder die Berichtigung eines falsch geschriebenen Namens oder eine falsche nachfolgende Verbesserung der Schreibweise eines Namens bedeutet. Burgeff's Methode Aberrationen zu benennen empfehlen wir nicht, aber wir haben diese auf diese Weise beschriebenen in den Katalog aufgenommen. Sie datieren von der Veröffentlichung des 2. Autors, welchem der Name zugeschrieben wird. Solche Aberrationen werden als separate Einheiten und ihre Beschreibungen als neu betrachtet. Dies darf nicht mit Falschbestimmungen verwechselt werden.

Mit Ausnahme von *Z. ephialtes* Linné ist wenig über die Genetik der verschiedenen Formen und Aberrationen der *Zygaena* Arten bekannt. Dryja (1959, Badania nad Polimorfizmem Krásnika Zmiennego) hat über *ephialtes* eine umfassende Studie veröffentlicht. Sein Werk war das Ergebnis von Zuchtexperimenten in 32 Jahren. Dryja hat gezeigt, dass jede der beiden Basisformen (peucedanoid und ephialtoid) von einem Paar unabhängig vererbter allelomorphen Genen abhängt, und dass die peucedanoiden Formen gegenüber den ephialtoiden Formen dominant sind. Die peucedanoiden und die ephialtoiden Formen können entweder rot oder gelb sein, die erstere Farbe ist gegenüber der letzteren dominant. Jede dieser Farben hängt von einem 2. Paar unabhängig vererbter Allelen ab. Diese Charaktere zeigten in F_2, F_3 und Rückkreuzungen typische Mendel'sche Trennung. Die peucedanoiden und ephialtoiden Formen können fünf- oder sechsfleckig sein. Es wurde gefunden, dass fünffleckige Formen gegenüber sechsfleckigen dominant sind und dass jede von einem 3. Paar unabhängig vererbter Allelen abhängt. Burgeff und Bovey haben gleiche Ergebnisse erhalten; die Referenzen zu ihren Publikationen sind in diesem Katalog enthalten.

Einige Arten der Gattung *Zygaena* sind nahe verwandt und die Genitalien zeigen keine sehr deutlichen Verschiedenheiten. Die Annahme einer Art wechselt je nach dem Autor. Wenn Kreuzungen zwischen nahe verwandten Arten vorgenommen werden, könnte es sich zeigen, dass die Hybriden (F_1) sich fruchtbar kreuzen und eine weitere Generation (F_2) erzeugen. Wenn Zucht-Experimente solche Resultate bringen, würden diese andeuten, dass die Trennung dieser Arten nicht gerechtfertig ist. Die nahe verwandten Arten *trifolii* Esper und *lonicerae* Scheven haben fast identische Genitalien und die geringen Genitaldifferenzen sind nicht immer konstant. Wenn diese 2 Arten aber gekreuzt werden, können die resultierenden Hybriden sich ebenfalls kreuzen und fruchtbare Nachkommenschaft erzeugen. Die Raupen der 2 Arten sind jedoch verschieden in der Färbung und in der Struktur der Beborstung und solche Verschiedenheiten sind nach unserer Meinung von beträchtlichem taxonomischem Wert. Nahe verwandte Arten, wie *trifolii* and *lonicerae*, haben eine grosse, geographische Verbreitung, welche übereinandergreift. Eine grosse Zahl gut abgegrenzter Unterarten hat sich entwickelt und jede dieser Unterarten hat die Basischaraktere entweder von *trifolii* oder *lonicerae*. Die Gruppierung dieser Arten in eine Art würde nach unserer Meinung ganz ungerechtfertig sein.

Obgleich 22 Untergattungen (Subgenera) bei der Gattung *Zygaena* errichtet wurden, sind die Arten im Katalog unter 3 Untergattungen (Subgenera) gruppiert, nämlich *Mesembrynus* Hübner, *Agrumenia* Hübner und *Zygaena* Fabricius. Diese 3 Untergattungen (Subgenera) sind in Sektionen oder Artgruppen aufgeteilt, von denen viele ehedem als Untergattungen geführt wurden. Die verbleibenden subgenerischen Namen werden jetzt in Synonymie verwendet. Die Ordnung im Katalog folgt im allgemeinen der Klassifizierung des senior Autors (Reiss, 1958, Z. wien. ent. Ges., 43: 140 und folgende), der seine Folgerungen auf die Morphologie, äussere Charaktere wie Flügelflecken und Färbung und die Biologie der Arten gründete. Die Arten im Subgenus *Mesembrynus* werden als die primitivsten angesehen. Dieser Untergattung folgt das Subgenus *Agrumenia*, dem das Subgenus *Zygaena* sich anreiht. Diese letztere Untergattung enthält die Arten, die wir als die entwicklungsgeschichtlich jüngsten ansehen. Die Färbung und die Gestaltung der Vorderflügelflecke stützen diese Theorie. Bei den Arten von *Mesembrynus*, die als die primitivsten angesehen werden, sind die Vorderflügelflecke einfärbig cremefarben, gelb, orange, hellrot oder karmin. Bei den meisten Arten von *Agrumenia* haben die Vorderflügelflecke weisslichcreme, cremefarbene oder hellgelbe Umrandung. Es wird angenommen, dass die rote Färbung der Vorderflügelflecke bei *Agrumenia* von der Mitte aus begann. Die gelbliche oder weissliche Umrandung erschien somit als der Rest der ehemaligen Färbung der Vorderflügelflecke, welche, wie schon oben erwähnt, bei den im Entstehungsalter älteren primitiveren Arten von *Mesembrynus*, gelb, orange, hellrot oder karmin sind. Die nahe verwandten Arten: *Z. banghaasi* Burgeff und *cocandica* Erschoff sind gute Beispiele. Nur bei den als jünger angesehenen Arten von *Agrumenia*, wie z. B. *exulans* Reiner & Hohenwarth und *loti* Denis & Schiffermüller fehlt die Umrandung der Vorderflügelflecke normalerweise. Wenn sie vorhanden ist, wird die Fleckenumrandung nur bei aberrierenden Stücken gefunden. Wir wissen, dass die meisten der entwicklungsgeschichtlich als jüngste anzusehenden Arten, zusammengefasst im Subgenus *Zygaena*, diese cremefarbene oder weissliche Umrandung der Vorderflügelflecke verloren haben. Es ist aber interessant, festzustellen, dass gelegentlich Stücke von *filipendulae* Linné und *transalpina* Esper mit Spuren von weisslicher Umrandung der Vorderflügelflecke gefangen wurden. Diese Formen kommen nur als sehr seltene Aberrationen in Italien vor und deuten eine mögliche Verbindung mit den primitiveren *Agrumenia* Arten an.

Bei den meisten Arten der Untergattungen *Mesembrynus* und *Agrumenia* ist ein roter Halskragen und Hinterleibsring vorhanden. Der Hinterleibsring tritt auch bei den primitiveren Arten der Untergattung *Zygaena* auf. Bei den entwicklungsgeschichtlich jüngeren Arten bei *Zygaena* indessen, kommt der Hinterleibsring nur bei aberrativen Stücken vor und wurde überhaupt noch nicht bei *ramburii* Herrich-Schäffer, *trifolii* und *lonicerae* gefunden. Diese 3 Arten werden als entwicklungsgeschichtlich jüngste der ganzen Gattung *Zygaena* angesehen.

Die Raupen der Arten von *Mesembrynus* leben an Pflanzen der Familien *Umbelliferae*, *Compositae* und *Labiatae*. Die Raupen der Arten von *Agrumenia* leben meistens an hartblätterigen *Papilionaceae*, die Raupen der Arten von *Zygaena* an weichblätterigen *Papilionaceae*. Ein Studium der Raupen, Puppen und Kokons, soweit sie bekannt sind, hat wertvolle Hinweise für die Einordnung der Arten in Artgruppen oder Sektionen ergeben.

Die Gewohnheiten der Arten bei Tag und ihre Ruheplätze bei Nacht sind ebenfalls erwähnenswert. Die als jünger im Entstehungsalter angenommenen Arten sitzen, wenn sie nicht fliegen oder fressen, an Blumen und ruhen dort einzeln, aber öfters zu mehreren versammelt, bei Regenwetter und bei Nacht. Die im Entstehungsalter als älter angenommenen Arten fliegen kurze Zeit im Sonnenschein und nach der Nahrungsaufnahme sitzen sie einzeln an Stämmen, Sträuchern und Pflanzen.

Es wird angenommen, dass die Arten der Untergattung (Subgenus) *Mesembrynus*, deren Raupen an *Umbelliferae*, *Compositae* und *Labiatae* leben, in der Tertiärzeit (möglicherweise im mittleren Miozän) entstanden sind. Die Untergattung (Subgenus) *Agrumenia* könnte ziemlich früh entstanden sein, vielleicht am Ende des Miozän, nachdem der Übergang von Zygaenenraupen auf *Papilionaceae* erfolgt war. Die im Entstehungsalter als jüngste angenommene Untergattung (Subgenus) *Zygaena* könnte sich etwa von der Mitte des Pliozäns an entwickelt haben. Es ist nicht möglich zu beweisen, ob ein zweiter Übergang von Umbelliferen- und Compositenfressern auf *Papilionaceae* erfolgt ist.

Viele im Entstehungsalter ältere Arten von *Mesembrynus*, *Agrumenia* und *Zygaena* hatten wahrscheinlich eine viel weitere Verbreitung, als sie es jetzt haben. Die gegenwärtige mehr lokalisierte und unterbrochene Verbreitung wurde wahrscheinlich durch das Eiszeitalter verursacht. Es wird angenommen, dass die im Entstehungsalter ältesten Arten in der Mittelmeerregion, und um sie herum, entstanden sind und sich von hier bis zu den westlichen und nördlichen Teilen Europas, bis zur Sahara in Nordafrika und nordöstlich von Kleinasien bis Zentralasien und Sibirien ausbreiteten. Soweit wir wissen, hat nur eine Art, *niphona* Butler, Japan erreicht. Die Verbreitung der Gattung *Zygaena* ist palaearktisch mit Ausnahme von 2 Arten: *rubricollis* Hampson und *transpamirina* Koch von Chitral.

In diesem Katalog sind die Unterarten (Subspecies) jeder Art nach ihrer geographischen Verbreitung geordnet, annährend vom Süden nach Norden und vom Osten nach Westen. Der Typenfundort jeder Unterart ist angegeben wie er ursprünglich vom Autor zitiert wurde (zur Vermeidung von Übersetzungsirrtümern), und ist in der Höhe des subspezifischen Namens in der Spalte rechts angegeben. Die Namen der Formen und Aberrationen sind unter der Unterart (Subspecies), zu der sie gehören, angeführt. Die subspecifischen Namen tragen die Bezeichnung „ssp." und die Aberrationsnamen die Bezeichnung „ab." Wir haben davon abgesehen, die Bezeichnung „f." (= forma) für die Aberrationen zu benützen, weil wir einer Form einen höheren Status als einer Aberration zubilligen, z. B. Saisonformen (f.t. = forma tempestatis). Die Bezeichnung „f." wird aber angewendet, um die ständig wiederkehrenden Formen der polymorphen Art *ephialtes* zu bezeichnen. Die Bezeichnung „f.loc." (= forma alicuius loci) wird gebraucht, um eine Rasse innerhalb eines Subspecies-Bereichs in einem einzelnen andersgearteten Biotop, in dem sich eine von der Subspecies hinreichend verschiedene lokale Population, die einen Namen verdient, vorfindet, zu bezeichnen. Diese Art der Bezeichnung basiert auf der von Rothschild & Jordan (1903, Novit. zool., 9, Supplement) angewendeten. Die Bezeichnung „var.", welche ganz unterschiedslos von vielen Autoren verwendet ist, um sowohl Unterarten (Subspecies) als auch geographische Rassen und Aberrationen zu bezeichnen, ist hier nicht verwendet worden, weil sie keinen Status in der Nomenclatur hat. Jüngere Synonyme werden in Kursivschrift unter ihren betreffenden älteren Synonymen angeführt. Jüngere Homonyme werden ebenso in Kursivschrift unter einem verfügbaren vorhandenen Namen angereiht.

Wir haben manchmal Schwierigkeiten gehabt bei der Feststellung der Daten der Veröffentlichung von gewissen Namen von Esper, Hübner und Herrich-Schäffer. Esper's "Die Schmetterlinge" wurde in verschiedenen Teilen herausgegeben. Wegen der Daten der Veröffentlichung dieser verschiedenen Teile haben wir die Arbeit von Sherborn & Woodward (1901, Ann. Mag. nat. Hist., (7) 7: 137-140) zu Rate gezogen. Die Daten der Veröffentlichung der verschiedenen Teile von Hübner's „Sammlung europäischer Schmetterlinge" und „Verzeichniss bekannter Schmettlinge [sic]" wurden Hemming's vorzüglichen Werk „Hübner" entnommen, das in 2 Bänden von der Royal Entomological Society of London im Jahre 1937 veröffentlicht wurde. Die Daten der in Herrich-Schäffer's „Systematische Bearbeitung der Schmetterlinge von Europa" veröffentlichten Namen wurden ebenfalls aus Hemming's Werk entnommen. Hinweise auf Zeitschriften wurden abgekürzt, die Abkürzungen erfolgten soweit als möglich nach der „World List of Scientific Periodicals, 1900-1950". Buchtitel und andere Veröffentlichungen werden unabgekürzt angeführt. Referenzen und die richtige Buchstabierung aller Namen sind mit der originalen Literatur verglichen worden.

Wir sind Mrs. Sylvia M. Tremewan, der Gattin des jüngeren Autor's, sehr dankbar für die wertvolle Mitarbeit in der Zusammenstellung des Inhalt-Verzeichnisses für den Katalog, welches über 3400 Namen enthält und eine ausserordentliche Aufgabe darstellte.

HUGO REISS,	W. GERALD TREMEWAN,
7 Stuttgart 1,	56, Hart Road,
Traubenstrasse 15B,	Byfleet,
Westdeutschland.	Surrey, England.

10. September, 1966.

ZYGAENA FABRICIUS

Zygaena Fabricius, 1775, Systema Entomologiae, p. 550. Type-species: **Sphinx filipendulae** Linné, 1758 (**Zygaena filipendulae** (Linné)), by subsequent designation, Latreille, 1810, Considérations Générales, p. 441. Reiss, 1958, Z. wien. ent. Ges., **43:** 140-147, 155-163, 181-183. Alberti, 1958, Mitt. zool. Mus. Berl., **34:** 245-396, figs. 1-4, pls. 1-32; 1959, ibidem, **35:** 203-242, pls. 33-64. Tremewan, 1961, Ent. Rec., **73:** 200.

Anthrocera Scopoli, 1777, Introductio ad Historiam naturalem, **10**: 414. Type-species: **Sphinx filipendulae** Linné, 1758 (**Anthrocera filipendulae** (Linné)), by subsequent designation, Westwood, 1840, Synopsis of the Genera of British Insects, p. 89.

Subg. **Mesembrynus** Hübner

Mesembrynus Hübner, [1819], Verzeichniss bekannter Schmettlinge [sic], p. 119. Type-species: **Zygaena pluto** Ochsenheimer, 1808 (= **Zygaena purpuralis** Brünnich, 1763), by subsequent designation, Tremewan, 1961, Ent. Rec., **73**: 201.

Hesychia Hübner, [1819], Verzeichniss bekannter Schmettlinge [sic], p. 116. Type-species: **Sphinx laeta** Hübner, 1790, by subsequent designation, Holik & Sheljuzhko, 1953, Mitt. münch. ent. Ges., **43**: 219.

Hyala Burgeff, 1926, in Strand, Lepid. Cat., **33**: 15 (preoccupied by **Hyala** Adams, 1852, Ann. Mag. nat. Hist., (2) **10**: 359 [Mollusca]). Type-species: **Zygaena loyselis** Oberthür, 1876, by subsequent designation, Tremewan, 1961, Ent. Rec., **73**: 201.

Santolinophaga Burgeff, 1926, in Strand, Lepid. Cat., **33**: 18. Type-species: **Zygaena corsica** Boisduval, 1828, by subsequent designation, Tremewan, 1961, Ent. Rec., **73**: 201.

Peucedanophila Burgeff, 1926, in Strand, Lepid. Cat., **33**: 19. Type-species: **Sphinx cynarae** Esper, 1789, by subsequent designation, Tremewan, 1961, Ent. Rec., **73**: 201.

Coelestis Burgeff, 1926, in Strand, Lepid. Cat., **33**: 29. Type-species: **Zygaena cuvieri** Boisduval, 1828, by subsequent designation, Tremewan, 1961, Ent. Rec., **73**: 201.

Yasumatsuia Strand, 1936, Folia Zool. Hydrobiol., Riga, **9**: 167 (nomen novum for *Hyala* Burgeff). Type-species: **Zygaena loyselis** Oberthür, 1876, by subsequent designation, Tremewan, 1961, Ent. Rec., **73**: 201.

Cirsiphaga Holik, 1953, Ent. Z., **62**: 153. Type-species: **Sphinx brizae** Esper, 1797, by original designation, Holik, 1953, loc. cit.

Mesembrynoidea Holik & Sheljuzhko, 1958, Mitt. münch. ent. Ges., **48**: 271. Type-species: **Zygaena cambysea** Lederer, 1870, by original designation, Holik & Sheljuzhko, 1958, loc. cit.

SECTION 1

rubicundus Hübner

Distribution: Central and southern Italy.

ssp. **rubicundus** Hübner, [13th March 1814]-[31st December 1817], Sammlung europäischer Schmetterlinge, **2**, pl. 30, fig. 137. Freyer, 1835, Neuere Beiträge zur Schmetterlingskunde, **3**: 13, pl. 200, fig. 3. Perlini, 1905, Forme di Lepidotteri esclusivamente Italiane, p. 49, pl. 2, fig. 10. Spuler, 1906, in Hofmann, Die Schmetterlinge Europas, **2**: 153, pl. 77, fig. 2. Oberthür, 1910, Études de Lépidoptérologie comparée, **4**: 422. Querci, 1912, in Oberthür, Études de Lépidoptérologie comparée, **6**: 143, 163. Verity, 1922, Ent. Rec., **34**: 30. Reiss, 1930, in Seitz, Die Gross-Schmetterlinge der Erde, Supplement, **2**: 7, pl. 1e. Haaf, 1952, Veröff. zool. Staatssamml. Münch., **2**: 152, 157, pl. 14. Alberti, 1958, Mitt. zool. Mus. Berl., **34**: 335. — Mittel- und Süd-italien.

ab. **polygalaeformis** Verity, 1916, Boll. Soc. ent. ital., **47**: 71.

ab. **pseudofaitensis** Stauder, 1929, Ent. Z., **43**: 6. Tremewan, 1961, Bull. Brit. Mus. (nat. Hist.) Ent., **10** (7): 243, pl. 50, fig. 3.

ab. **pallescens** Stauder, 1915, Iris, **29**: 32.

ssp. **tarentensis** Dujardin, 1965, Entomops, Nice, no. 2: 59, fig. — Gioia del Colle (B. d. Terzi), Puglie, Italie Sud, 400 m.

Biology

Burgeff, 1950, Portug. acta biol., (A) Goldschmidt: 663-728. Holik, 1937, Lambillionea, **37**: 15-24, 32-45, 80-91; 1938, Ent. Rdsch., **55**: 349-354, 382-384; 1943, Ent. Z., **57**: 41-45; 1946, Rev. franç. Lépid., **10**: 250-261, 273-280; 1953, Ent. Z., **62**: 155. Oberthür, 1911, Études de Lépidoptérologie comparée, **5** (1), pl. 85, figs. 824, 825. Querci, 1912, in Oberthür, Études de Lépidoptérologie comparée, **6**: 144, 163. Reiss, 1930, in Seitz, Die Gross-Schmetterlinge der Erde, Supplement, **2**: 7; 1958, Z. wien. ent. Ges., **43**: 155.

SECTION 2

cambysea Lederer

Distribution: Iran, Transcaspia, Armenia.

ssp. **cambysea** Lederer, 1870, Horae Soc. ent. Ross., **6**: 86, pl. 5, fig. 6. Seitz, 1908, Die Gross-Schmetterlinge der Erde, **2**: 26, pl. 6i. Reiss, 1932, Int. ent. Z., **26**: 275, 280, fig.; 1933, in Seitz, Die Gross-Schmetterlinge der Erde, Supplement, **2**: 258; 1937, Ent. Rdsch., **55**: 18, fig. a1 (ssp.?). Koch, 1937, Ent. Z., **51**: 20, 40, figs. 4-12. Holik & Sheljuzhko, 1953, Mitt. münch. ent. Ges., **43**: 208. Alberti, 1958, Mitt. zool. Mus. Berl., **34**: 343. — Astrabad, Nordiran.

ab. **pseudorosacea** Koch, 1937, Ent. Z., **51**: 20, fig. 13.

ssp. **hafis** Reiss, 1938, Ent. Rdsch., **55**: 251, figs. a1, a2. Holik & Sheljuzhko, 1953, Mitt. münch. ent. Ges., **43**: 209. — Strasse Chiraz-Kazeroun, Fort Sine-Sefid, Südiran, 2200 m.

ssp. **rosacea** Romanoff, 1884, Mémoires sur les Lépidoptères, **1**: 79. Seitz, 1908, Die Gross-Schmetterlinge der Erde, **2**: 26, pl. 6i. Reiss, 1932, Int. ent. Z., **26**: 275, 280, figs.; 1933, in Seitz, Die Gross-Schmetterlinge der Erde, Supplement, **2**: 259. Koch, 1937, Ent. Z., **51**: 36. Haaf, 1952, Veröff. zool. Staatssamml. Münch., **2**: 152, 157, pl. 14 (as *cambysea* Lederer). Holik & Sheljuzhko, 1953, Mitt. münch. ent. Ges., **43**: 206. — Istissou près Erivan, Armenia.

ab. **cingulata** Dziurzyński, 1908, Berl. ent. Z., **53**: 13, 46.

ab. **pseudocambysea** Reiss, 1932, Int. ent. Z., **26**: 276, fig.

ab. **bipuncta** Holik & Sheljuzhko, 1953, Mitt. münch. ent. Ges., **43**: 207.

ab. **totirubra** Reiss, 1941, Mitt. münch. ent. Ges., **31**: 987.

ab. **flava** Holik & Sheljuzhko, 1953, Mitt. münch. ent. Ges., **43**: 207.

Biology

Christoph, 1872, Stettin. ent. Ztg., **33**: 213. Holik, 1937, Lambillionea, **37**: 15-24, 32-45, 80-91; 1938, Ent. Rdsch., **55**: 349-354, 382-384; 1946, Rev. franç. Lépid., **10**: 250-261, 273-280; 1953, Ent. Z., **62**: 158. Reiss, 1958, Z. wien. ent. Ges., **43**: 155.

SECTION 3

seitzi Reiss

Distribution: Iran.

ssp. **seitzi** Reiss, 1938, Ent. Rdsch., **55**: 291, figs. d4, e1. Brandt, 1938, Ent. Rdsch., **55**: 673, pl. 5, fig. 25. Haaf, 1952, Veröff. — Fort Sine-Sefid, Strasse zool. Staatssamml. Münch., **2**: 151, 156, pl. 6. Holik & Sheljuzhko, 1955, Mitt. münch. ent. Ges., **44/45**: 49. Alberti, 1958, Mitt. zool. Mus. Berl., **34**: 340. — Chiraz-Kazeroun, Südiran, 2200 m.

ssp. **escaleraiana** Holik, 1958, Ent. Z., **68**: 17. Holik & Sheljuzhko, 1958, Mitt. münch. ent. Ges., **48**: 272. — ?Chindaar vallée, Haut-Kharoum, Iran.

Biology

Holik, 1938, Mitt. münch. ent. Ges., **28**: 393; 1953, Ent. Z., **62**: 190. Reiss, 1958, Z. wien. ent. Ges., **43**: 155. Wiltshire, 1952, Bull. Soc. Fouad. Ent., **36**: 176, fig. 5.

tamara Christoph

Distribution: Armenia, Iraq, Asia Minor.

ssp. **tamara** Christoph, 1889, Horae Soc. ent. Ross., **23**: 300; 1889, in Romanoff, Mémoires sur les Lépidoptères, **5**: 196, pl. 9, figs. 2a, b. Seitz, 1908, Die Gross-Schmetterlinge der Erde, **2**: 26, pl. 7b. Reiss, 1933, Int. ent. Z., **26**: 490, 505, figs.; 1933, in Seitz, Die Gross-Schmetterlinge der Erde, Supplement, **2**: 262. Koch, 1938, Ent. Z., **51**: 345. Holik, 1941, Ent. Z., **54**: 209. Haaf, 1952, Veröff. zool. Staatssamml. Münch., **2**: 151, 156, pl. 6. Holik & Sheljuzhko, 1955, Mitt. münch. ent. Ges., **44/45**: 37, 40. Alberti, 1958, Mitt. zool. Mus. Berl., **34**: 340. — Ordubat, Armenisches Bergland.

ab. **quadripuncta** Reiss, 1933, Int. ent. Z., **26**: 491, 505, fig.

ab. **confluens** Koch, 1938, Ent. Z., **51**: 345.

ab. **daemon** Christoph, 1893, Iris, **6**: 88. Seitz, 1907, Die Gross-Schmetterlinge der Erde, **2**: 26, pl. 7b. Koch, 1940, Ent. Z., **54**: 199.

ab. **rubra** Rebel, 1901, in Staudinger & Rebel, Catalog der Lepidopteren des Palaearctischen Faunengebietes, p. 386. Seitz, 1907, Die Gross-Schmetterlinge der Erde, **2**: 26, pl. 7b.

ssp. **ochtsii** Koch, 1940, Ent. Z., **54**: 200. Holik & Sheljuzhko, 1955, Mitt. münch. ent. Ges., **44/45**: 44. — Ochtsi [Ochtshi], Zangezur Gebirge, Armenisches Bergland.

ab. **quadripuncta** Koch, 1940, Ent. Z., **54**: 200.

ab. **latecingulata** Holik & Sheljuzhko, 1955, Mitt. münch. ent. Ges., **44/45**: 45.

ab. **latefasciata** Holik & Sheljuzhko, 1955, Mitt. münch. ent. Ges., **44/45**: 45.

ab. **basifractistrigata** Holik & Sheljuzhko, 1955, Mitt. münch. ent. Ges., **44/45**: 46.

ssp. **daralagezi** Holik & Sheljuzhko, 1955, Mitt. münch. ent. Ges., **44/45**: 43. — Azizbekov (Pashalu), Daralagëz Gebirge, Armenisches Bergland, 1650 m.

ab. **rubifrons** Holik & Sheljuzhko, 1955, Mitt. münch. ent. Ges., **44/45**: 44.

ab. **mediofractistrigata** Holik & Sheljuzhko, 1955, Mitt. münch. ent. Ges., **44/45**: 44.

ab. **bifractistrigata** Holik & Sheljuzhko, 1955, Mitt. münch. ent. Ges., **44/45**: 44.

ab. **confluens** Koch, 1938, Ent. Z., **51**: 346.

ab. **aurantiaca** Holik & Sheljuzhko, 1955, Mitt. münch. ent. Ges., **44/45**: 43.

ssp. **placida** Bang-Haas, 1913, Iris, **27**: 108. Burgeff, 1914, Mitt. münch. ent. Ges., **5**: 50, pl. 2, figs. 170, 171, pl. 6, figs. 70-72. Reiss, 1930, in Seitz, Die Gross-Schmetterlinge der Erde, Supplement, **2**: 21, pl. 2h; 1933, ibidem, **2**: 262. Holik, 1941, Ent. Z., **54**: 211. Holik & Sheljuzhko, 1955, Mitt. münch. ent. Ges., **44/45**: 46. — Wansee Gebiet, Kleinasien.

Biology

Holik, 1937, Lambillionea, **37**: 15-24, 32-45, 80-91; 1938, Ent. Rdsch., **55**: 320-323, 331-333; 1938, Mitt. münch. ent. Ges., **28**: 389. pl. 8, fig. 4, pl. 9, fig. 9; 1946, Rev. franç. Lépid., **10**: 250-261, 273-280; 1953, Ent. Z., **62**: 189. Reiss, 1958, Z. wien. ent. Ges., **43**: 155. Wiltshire, 1957, The Lepidoptera of Iraq, p. 100; 1957, Z. wien. ent. Ges., **42**: 154, pl. 12, figs. 8, 9.

fredi Reiss

Distribution: Southern Iran.

fredi Reiss, 1938, Ent. Rdsch., **55**: 290, fig. d3. Brandt, 1938, Ent. Rdsch., **55**: 673, pl. 5, fig. 24. Holik & Sheljuzhko, 1955, Mitt. münch. ent. Ges., **44/45**: 61. Alberti, 1958, Mitt. zool. Mus. Berl., **34**: 341. — Fort Sine-Sefid, Strasse Chiraz-Kazeroun, Südiran, 2200 m.

cacuminum Christoph

Distribution: Northern Iran.

cacuminum Christoph, 1877, Horae Soc. ent. Ross., **12**: 243, pl. 6, fig. 17. Seitz, 1907, Die Gross-Schmetterlinge der Erde, **2**: 26, pl. 6k. Reiss, 1933, Int. ent. Z., **26**: 488, 505, figs.; 1933, in Seitz, Die Gross-Schmetterlinge der Erde, Supplement, **2**: 261. Holik, 1940, Ent. Z., **54**: 203. Koch, 1941, Mitt. münch. ent. Ges., **31**: 564. Holik & Sheljuzhko, 1955, Mitt. münch. ent. Ges., **44/45**: 58. Alberti, 1958, Mitt. zool. Mus. Berl., **34**: 341. — Höchstgelegene Abhänge bei Schakuh, Nordiran, 3400 m.

ab. **nigra** Dziurzyński, 1908, Berl. ent. Z., **53**: 47.

Biology

Christoph, 1877, Horae Soc. ent. Ross., **12**: 243.

speciosa Reiss

Distribution: Northern Iran.

ssp. **speciosa** Reiss, 1937, Ent. Rdsch., **54**: 466, figs. a2, b2. Holik & Sheljuzhko, 1955, Mitt. münch. ent. Ges., **44/45**: 60. Alberti, 1958, Mitt. zool. Mus. Berl., **34**: 342. — Hecarcaltal, Elbursgebirge, Nordiran, 2800-3200 m.
ab. **quadripuncta** Reiss, 1937, Ent. Rdsch., **54**: 468.
ab. **eradiata** Reiss, 1937, Ent. Rdsch., **54**: 468, fig. c2.
ssp. **suleimanicola** Reiss, 1938, Mitt. münch. ent. Ges., **27**: 165. Holik & Sheljuzhko, 1955, Mitt. münch. ent. Ges., **44/45**: 61. Tremewan, 1961, Bull. Brit. Mus. (nat. Hist.) Ent., **10** (7): 304, pl. 57, fig. 11. — Särdabtal (Hećerćam), Elbursgebirge, Nordiran, 4200 m.
ab. **quadripuncta** Reiss, 1938, Mitt. münch. ent. Ges., **27**: 165.
ab. **paupera** Reiss, 1938, Mitt. münch. ent. Ges., **27**: 166.

Biology

Holik, 1938, Mitt. münch. ent. Ges., **28**: 391; 1938, Ent. Rdsch., **55**: 320-323, 331-333; 1953, Ent. Z., **62**: 189. Reiss, 1937, Ent. Rdsch., **54**: 468; 1938, Mitt. münch. ent. Ges., **27**: 166; 1958, Z. wien. ent. Ges., **43**: 155.

manlia Lederer
Distribution: Iran, Iraq, Transcaspia, Armenia.
ssp. **manlia** Lederer, 1870, Horae Soc. ent. Ross., **6**: 87, pl. 5, fig. 7. Seitz, 1907, Die Gross-Schmetterlinge der Erde, **2**: 26, pl. 6h. Reiss, 1933, Int. ent. Z., **26**: 487; 1933, in Seitz, Die Gross-Schmetterlinge der Erde, Supplement, **2**: 260; 1937, Ent. Rdsch., **55**: 19; 1938, ibidem, **55**: 313, figs. b4, c1. Koch, 1941, Mitt. münch. ent. Ges., **31**: 560. Haaf, 1952, Veröff. zool. Staatssamml. Münch., **2**: 151, 156, pl. 6. Holik & Sheljuzhko, 1955, Mitt. münch. ent. Ges., **44/45**: 51, 54. Alberti, 1958, Mitt. zool. Mus. Berl., **34**: 341. Reiss & Tremewan, 1960, Bull. Brit. Mus. (nat. Hist.) Ent., **9** (10): 459, pl. 22, fig. 2, pl. 24, figs. 9, 10. — Hadschyabad (Astrabad), Nordiran.
ab. **confluens** Koch, 1941, Mitt. münch. ent. Ges., **31**: 562.
ssp. **belutschistani** Koch, 1941, Mitt. münch. ent. Ges., **31**: 563. Holik & Sheljuzhko, 1955, Mitt. münch. ent. Ges., **44/45**: 55. Reiss, 1960, Ent. Z., **70**: 78, 79, figs. 3a, b. — Kouh i Taftan (Khach), Belutschistan, Iran, 2500 m.
ab. **medioseparata** Reiss, 1960, Ent. Z., **70**: 77.

ssp. **taftanica** Reiss, 1960, Ent. Z., **70**: 76, figs. 1, 2, 4a, b. — Kouh i Taftan (Khach), Belutschistan, Iran, 4000 m.

ssp. **schahrudensis** Koch, 1941, Mitt. münch. ent. Ges., **31**: 563. Holik & Sheljuzhko, 1955, Mitt. münch. ent. Ges., **44/45**: 54. — Schahrud (Ebene), Nordiran.

ssp. **turkmenica** Reiss, 1933, Int. ent. Z., **26**: 490, 505, figs.; 1933, in Seitz, Die Gross-Schmetterlinge der Erde, Supple- — Jablonowka, Achal-Tekke, Transcaspien, 2000 m.

ment, **2**: 261, pl. 16i (as *turkmena*). Koch, 1936, Iris, **50**: 43, pl. 2, figs. 9-12. Holik, 1940, Ent. Z., **54**: 204. Haaf, 1952, Veröff. zool. Staatssamml. Münch., **2**, pl. 6. Holik & Sheljuzhko, 1955, Mitt. münch. ent. Ges., **44/45**: 55.

ab. **pseudomanlia** Koch, 1936, Iris, **50**: 41, pl. 2, figs. 13, 14.

ab. **totirubra** Holik, 1940, Ent. Z., **54**: 205.

totarubra Holik & Sheljuzhko, 1955, Mitt. münch. ent. Ges., **44/45**: 56.

ssp. **araxis** Koch, 1936, Iris, **50**: 41, pl. 2, figs. 15-19; 1938, Ent. Z., **51**: 347. Holik, 1940, Ent. Z., **54**: 204. Holik & Sheljuzhko, 1955, Mitt. münch. ent. Ges., **44/45**: 52. — Nus-nus bei Ordubat, Nachitshevan, Armenisches Bergland.

ab. **hedwigi** Koch, 1936, Iris, **50**: 42, pl. 2, fig. 20.

ab. **pseudoturkmenica** Koch, 1936, Iris, **50**: 42, pl. 2, fig. 21.

ssp. **daralagezica** Holik & Sheljuzhko, 1955, Mitt. münch. ent. Ges., **44/45**: 53. — Terkesh; Azizbekov; Güartshin; Sultanbek; Daralagëz Gebirge, Armenisches Bergland

ab. **rubriscapulis** Reiss, 1941, Z. wien. EntVer., **26**: 63.

ab. **confluens** Holik & Sheljuzhko, 1955, Mitt. münch. ent. Ges., **44/45**: 53.

ab. **omniconfluens** Holik & Sheljuzhko, 1955, Mitt. münch. ent. Ges., **44/45**: 53.

ab. **totarubra** Holik & Sheljuzhko, 1955, Mitt. münch. ent. Ges., **44/45**: 53.

Biology

Holik, 1937, Lambillionea, **37**: 15-24, 32-45, 80-91; 1938, Ent. Rdsch., **55**: 320-323, 331-333; 1938, Mitt. münch. ent. Ges., **28**: 390, pl. 8, figs. 5-7, pl. 9, figs. 8, 10; 1946, Rev. franç. Lépid., **10**: 250-261, 273-280; 1953, Ent. Z., **62**: 189.

afghanica Reiss

Distribution: North-west Afghanistan.

afghanica Reiss, 1940, Ent. Z., **54**: 105, figs. Holik & Sheljuzhko, 1955, Mitt. münch. ent. Ges., **44/45**: 56. Reiss & Tremewan, 1960, Bull. Brit. Mus. (nat. Hist.) Ent., **9** (10): 459, pl. 24, fig. 11. Reiss & Schulte, 1962, Ent. Z., **72**: 49; 1964, ibidem, **74**: 154. — Alpenwiesenzone, Firuskuhi-Mont., Afghanistan, 2800-3000 m.

ab. **medioseparata** Reiss, 1940, Ent. Z., **54**: 106, fig.

ab. **ornata** Reiss & Schulte, 1962, Ent. Z., **72**: 50.

excellens Reiss

Distribution: North-west Afghanistan.

excellens Reiss, 1940, Ent. Z., **54**: 106, fig. Holik & Sheljuzhko, 1955, Mitt. münch. ent. Ges., **44/45**: 57. Reiss & Schulte, 1964, Ent. Z., **74**: 154. — Alpenwiesenzone, Firuskuhi-Mont., Afghanistan, 2800-3000 m.

Biology

Holik, 1953, Ent. Z., **62**: 190.

hindukuschi Koch

Distribution: North-east Afghanistan.

hindukuschi Koch, 1937, Ent. Z., **51**: 64, 71, figs. 19-21. Holik & Sheljuzhko, 1955, Mitt. münch. ent. Ges., **44/45**: 62. Alberti, 1958, Mitt. zool. Mus. Berl., **34**: 342. — Chodja-Mahomed, Hindukusch, Afghanistan, 3800-4000 m.

Biology

Holik, 1938, Ent. Rdsch., **55**: 320-323, 331-333; 1953, Ent. Z., **62**: 190.

superba Reiss & Schulte

Distribution: North-east Afghanistan.

superba Reiss & Schulte, 1964, Ent. Z., **74**: 154, 155, figs. 1-4. — Bala Quran, Anjuman Valley, N.E. Afghanistan, 11500 ft.

rubricollis Hampson

Distribution: Chitral.

rubricollis Hampson, 1900, J. Bombay nat. Hist. Soc., **13**: 224. Reiss, 1930, in Seitz, Die Gross-Schmetterlinge der Erde, Supplement, **2**: 21, pl. 2h. Koch, 1937, Ent. Z., **51**: 64, 72, fig. 18 (after Hampson). Holik & Sheljuzhko, 1955, Mitt. münch. ent. Ges., **44/45**: 58. Alberti, 1958, Mitt. zool. Mus. Berl., **34**: 342. Reiss & Tremewan, 1960, Bull. Brit. Mus. (nat. Hist.) Ent., **9** (10): 459, pl. 22, fig. 1, pl. 24, figs. 7, 8. Tremewan, 1961, Bull. Brit. Mus. (nat. Hist.) Ent., **10** (7): 244, pl. 50, fig. 8, pl. 64, figs. 1, 2. — Shishi Kuh Valley, Chitral, 9000-1400 ft.

Biology

Holik, 1938, Ent. Rdsch., **55**: 320-323, 331-333; 1953, Ent. Z., **62**: 190.

cuvieri Boisduval

Distribution: Iraq, Asia Minor, Armenia, Syria, Transcaspia.

ssp. **cuvieri** Boisduval, 1828, Essai sur une Monographie des Zygénides, p. 53, pl. 3, fig. 6; 1834, Icones historique des Lépidoptères nouveaux ou peu connus, **2**: 76. Staudinger, 1879, Horae Soc. ent. Ross., **14**: 323. Seitz, 1907, Die Gross-Schmetterlinge der Erde, **2**: 26, pl. 6h. Reiss, 1933, in Seitz, Die Gross-Schmetterlinge der Erde, Supplement, **2**: 260. Koch, 1937, Ent. Z., **51**: 37. Holik, 1940, Ent. Z., **54**: 203. Koch, 1941, Mitt. münch. ent. Ges., **31**: 556. Holik & Shel- — Environs d'Amaden, Perse [?Amadia, Iraq].

juzhko, 1955, Mitt. münch. ent. Ges., **44/45**: 26. Alberti, 1958, Mitt. zool. Mus. Berl., **34**: 339. Tremewan, 1961, Bull. Brit. Mus. (nat. Hist.) Ent., **10** (7): 244, pl. 50, fig. 1.

ab. **pseudolibani** Holik, 1940, Ent. Z., **54**: 203.

ab. **fractistrigata** Holik & Sheljuzhko, 1955, Mitt. münch. ent. Ges., **44/45**: 30.

ab. **confluens** Oberthür, 1896, Études d'Entomologie, **20**: 46, pl. 7, fig. 112. Tremewan, 1961, Bull. Brit. Mus. (nat. Hist.) Ent., **10** (7): 244, pl. 50, fig. 2.

ssp. **libani** Burgeff, 1914, Mitt. münch. ent. Ges., **5**: 77, pl. 2, fig. 172, pl. 6, figs. 68, 69. Reiss, 1930, in Seitz, Die Gross-Schmetterlinge der Erde, Supplement, **2**: 21, pl. 2h; 1932, Int. ent. Z., **26**: 277, 280, fig. Koch, 1937, Ent. Z., **51**: 38. Holik & Sheljuzhko, 1955, Mitt. münch. ent. Ges., **44/45**: 32. — Beirut, Libanon.

ab. **separata** Koch, 1941, Mitt. münch. ent. Ges., **31**: 558.

ab. **totarubra** Dziurzyński, 1908, Berl. ent. Z., **53**: 12, 45 (as *totisrubra*); 1909, ibidem, **53**: 251; 1909, Jber. wien. ent. Ver., **19**: 135, pl. 1, fig. 6.

ssp. **melitensis** Koch, 1941, Mitt. münch. ent. Ges., **31**: 558. Holik & Sheljuzhko, 1955, Mitt. münch. ent. Ges., **44/45**: 31. — Malatia, Taurus, Kleinasien.

ab. **confluens** Koch, 1941, Mitt. münch. ent. Ges., **31**: 558.

ab. **totarubra** Koch, 1941, Mitt. münch. ent. Ges., **31**: 558.

ssp. **okhtchaperdica** Reiss, 1941, Z. wien. EntVer., **26**: 63. Romanoff, 1884, Mémoires sur les Lépidoptères, **1**: 80, pl. 4, fig. 6. Reiss, 1932, Int. ent. Z., **26**: 280, fig. (as *cuvieri* Boisduval var.). Koch, 1941, Mitt. münch. ent. Ges., **31**: 559. Holik & Sheljuzhko, 1955, Mitt. münch. ent. Ges., **44/45**: 29. — Okhtchaperd bei Erivan, Armenien.

ab. **confluens** Reiss, 1941, Z. wien. EntVer., **26**: 63.

ssp. **achaltekkensis** Koch, 1937, Ent. Z., **51**: 38, figs. 18, 19. Holik & Sheljuzhko, 1955, Mitt. münch. ent. Ges., **44/45**: 33. — Jablonowka, Achal-Tekke, Transkaspien.

Biology

Holik, 1937, Lambillionea, **37**: 15-24, 32-45, 80-91; 1938, Ent. Rdsch., **55**: 320-323, 331-333; 1938, Mitt. münch. ent. Ges., **28**: 390; 1946, Rev. franç. Lépid., **10**: 250-261, 273-280; 1953, Ent. Z., **62**: 188; 1953, ibidem, **63**: 31. Holik & Sheljuzhko, 1955, Mitt. münch. ent. Ges., **44/45**: 27. Reiss, 1958, Z. wien. ent. Ges., **43**: 155. Wiltshire, 1935, Ent. Rec., Supplement, **47**: (1), pl. 3, fig. 1; 1957, The Lepidoptera of Iraq, p. 99.

lydia Staudinger

Distribution: Asia Minor (Taurus).

ssp. **lydia** Staudinger, 1887, Berl. ent. Z., **31**: 36. Oberthür, 1896, Études d'Entomologie, **20**: 51, pl. 8, fig. 139 (as *haberhaueri* Lederer). Seitz, 1907, Die Gross-Schmetterlinge der Erde, **2**: — Malatia, Taurus, Kleinasien.

27, pl. 7a; 1912, ibidem, **2**: 443. Reiss, 1930, in Seitz, Die Gross-Schmetterlinge der Erde, Supplement, **2**: 23; 1933, ibidem, **2**: 261, pl. 16k; 1931, Int. ent. Z., **25**: 359, figs. Koch, 1941, Mitt. münch. ent. Ges., **31**: 565. Holik & Sheljuzhko, 1955, Mitt. münch. ent. Ges., **44/45**: 63. Alberti, 1958, Mitt. zool. Mus. Berl., **34**: 337.

ssp. **hadjinensis** Reiss, 1931, Int. ent. Z., **25**: 342, 359, figs.; 1933, in Seitz, Die Gross-Schmetterlinge der Erde, Supplement, **2**: 261, pl. 16k. Koch, 1941, Mitt. münch. ent. Ges., **31**: 566. Holik & Sheljuzhko, 1955, Mitt. münch. ent. Ges., **44/45**: 64. — Hadjin, Taurus, Kleinasien.

Biology

Holik, 1938, Ent. Rdsch., **55**: 320-323, 331-333.

SECTION 4

laeta Hübner

Distribution: Asia Minor, the Balkans, Southern Russia, Ukraine, Crimea, Podolia, Bohemia, Moravia, Hungary, Lower Austria.

ssp. **danieli** Reiss, July, 1935, Int. ent. Z., **29**: 159. Holik, 1935, Sborn. ent. Odd. nár. Mus. Praze, **13**: 56, 64, fig. 37. Holik & Sheljuzhko, 1953, Mitt. münch. ent. Ges., **43**: 225. — Marasch, Taurus, Kleinasien, 600-1000 m.

laetissima Holik, December, 1935, Sborn. ent. Odd. nár. Mus. Praze, **13**: 56, 64, figs. 35, 36.

ssp. **akschehirensis** Reiss, 1929, Int. ent. Z., **23**: 151. Holik, 1935, Sborn. ent. Odd. nár. Mus. Praze, **13**: 56, 64, fig. 38. Holik & Sheljuzhko, 1953, Mitt. münch. ent. Ges., **43**: 225. — Ak-Schehir, Kleinasien, 900-1100 m.

ssp. **orientis** Burgeff, 1926, Mitt. münch. ent. Ges., **16**: 41. Reiss, 1930, in Seitz, Die Gross-Schmetterlinge der Erde, Supplement, **2**: 23. Holik, 1935, Sborn. ent. Odd. nár. Mus. Praze, **13**: 56, 63, figs. 29, 30; 1938, Mitt. münch. ent. Ges., **27**: 135. Daniel, 1964, Prirod. Muz. Skopje, no. 2: 13. — Bogdanzi, Nicolic, Dojran See, Mazedonien.

ab. **eos** Holik, 1938, Mitt. münch. ent. Ges., **27**: 135.

ssp. **occidentissima** Holik, 1935, Sborn. ent. Odd. nár. Mus. Praze, **13**: 56, 63, figs. 6-15. Povolný & Gregor, 1946, Folia ent., Brno, **9**, Supplement, 12: 79, 98. Komárek, 1951, Acta Soc. ent. Bohem. (Čsl.), **48**: 179. Forster & Wohlfahrt, 1958, Die Schmetterlinge Mitteleuropas, **3**: 94, pl. 10, figs. 21, 26 (as *laeta* Hübner). — Bei Kolin, Böhmen.

moraviensis Povolný & Gregor, 1946, Folia ent., Brno, **9**, Supplement, 12: 89, 98, figs. Holik, 1939, Sborn. ent. Odd. nár. Mus. Praze, **17**: 42. Gregor & Povolný, 1955, Sborn. ent. Odd. nár. Mus. Praze, **30**: 266, 274, pl. 5, fig. 4.

ab. **omnicingulata** Povolný & Gregor, 1946, Folia ent., Brno, **9**, Supplement, 12: 89, 98, pl. 2, fig. 5f.

ab. **pseudomannerheimi** Burgeff, 1926, Mitt. münch. ent. Ges., **16**: 41. Reiss & Tremewan, 1964, Ent. Rec., **76**: 129.

pseudoorientis Holik, 1935, Sborn. ent. Odd. nár. Mus. Praze, **13**: 57, 64, fig. 21. Reiss & Tremewan, 1964, Ent. Rec., **76**: 129.

ab. **semireversa** Holik, 1935, Sborn. ent. Odd. nár. Mus. Praze, **13**: 56, 64, figs. 16-20.

ssp. **laeta** Hübner, 1790, Beiträge zur Geschichte der Schmetterlinge, **2**: 88, pl. 2, fig. H; 1796, Sammlung europäischer Schmetterlinge, **2**, pl. 6, figs. 34, 35; 1806, ibidem, Der Ziefer, p. 85. Ochsenheimer, 1808, Die Schmetterlinge von Europa, **2**: 100. Boisduval, 1828, Essai sur une Monographie des Zygénides, p. 104, pl. 6, fig. 7; 1834, Icones historique des Lépidoptères nouveaux ou peu connus, **2**: 77, pl. 55, fig. 2. Duponchel, 1835, in Godart & Duponchel, Histoire naturelle des Lépidoptères ou Papillons de France, Supplement, **2**: 82, pl. 7, fig. 4. Freyer, 1839, Neuere Beiträge zur Schmetterlingskunde, **3**: 12, pl. 200, fig. 1; 1858, ibidem, **7**: 64, pl. 637, fig. 1. Spuler, 1906, in Hofmann, Die Schmetterlinge Europas, **2**: 163, pl. 72, fig. 10, pl. 77, fig. 28. Seitz, 1907, Die Gross-Schmetterlinge der Erde, **2**: 26, pl. 6k, 7a. Reiss, 1930, in Seitz, Die Gross-Schmetterlinge der Erde, Supplement, **2**: 23, pl. 21; 1933, ibidem, **2**: 270. Berger, 1931, Z. öst. EntVer., **16**: 83. Sterzl, 1931, Z. Ver. NatBeob., Wien, **6**: 1, pl. 14, figs. 1, 2. Holik, 1935, Sborn. ent. Odd. nár. Mus. Praze, **13**: 56, 63, figs. 1-5; 1939, Ann. Mus. zool. Polon., **12**: 32, pl. 2, fig. 61, pl. 5, figs. 168-170. Haaf, 1952, Veröff. zool. Staatssamml. Münch., **2**: 152, 157, pl. 14. Alberti, 1958, Mitt. zool. Mus. Berl., **34**: 330. — Wiener Becken, Österreich.

ab. **reversa** Burgeff, 1914, Mitt. münch. ent. Ges., **5**: 52, pl. 2, fig. 167, pl. 6, fig. 56. Sterzl, 1931, Z. Ver. NatBeob., Wien, **6**: 3, pl. 14, fig. 6.

ab. **pseudocuvieri** Silbernagel, 1937, Acta Soc. ent. Bohem. (Čsl.), **34**: 10, fig. 1.

ab. **eos** Sterzl, 1924, Verh. zool.-bot. Ges. Wien, **73**: 15; 1931, Z. Ver. NatBeob., Wien, **6**: 3, pl. 14, fig. 4.

Biology

Freyer, 1858, Neuere Beiträge zur Schmetterlingskunde, **6**: 64, pl. 637, fig. 1. Holik, 1937, Lambillionea, **37**: 15-24, 32-45, 80-91; 1938, Ent. Rdsch., **55**: 320-323, 331-333; 1938, ibidem, **55**: 349-354, 382-384; 1946, Rev. franç. Lépid., **10**: 250-261, 273-280; 1953, Ent. Z., **62**: 183. Hrubý, 1964, Prodromus Lepidopter Slovenska, p. 477. Pujman, 1952,

Acta Soc. ent. Bohem. (Čsl.), **49**: 212. Reiss, 1930, in Seitz, Die Gross-Schmetterlinge der Erde, Supplement, **2**: 23; 1958, Z. wien. ent. Ges., **43**: 155. Spuler, 1910, in Hofmann, Die Raupen der Schmetterlinge Europas, pl. 10, figs. 6a, b, c. Sterzl, 1931, Z. Ver. NatBeob., Wien, **6**: 1-4.

SECTION 5

centaureae Fischer-Waldheim

Distribution: Southern Russia, Ukraine.

ssp. **centaureae** Fischer-Waldheim, 1832, Nouv. Mém. Soc. Nat. Moscou, **2**: 358, pl. 21, fig. 4. Herrich-Schäffer, 1846, Systematische Bearbeitung der Schmetterlinge von Europa, **2**: 39, pl. 8, figs. 57, 58. Bartel, 1902, Iris, **15**: 227. Spuler, 1906, in Hofmann, Die Schmetterlinge Europas, **2**: 157, pl. 77, fig. 11b. Seitz, 1907, Die Gross-Schmetterlinge der Erde, **2**: 22, pl. 5d. Burgeff, 1914, Mitt. münch. ent. Ges., **5**: 47. Reiss, 1930, in Seitz, Die Gross-Schmetterlinge der Erde, Supplement, **2**: 16, pl. 2a; 1933, ibidem, **2**: 256; 1932, Ent. Rdsch., **49**: 165, 169, pl. 1, fig. Holik & Sheljuzhko, 1955, Mitt. münch. ent. Ges., **44/45**: 76. Alberti, 1958, Mitt. zool. Mus. Berl., **34**: 330.

Bei Sarepta, Wolga, Südrussland.

evandrus Heydenreich, 1851, Lepidopterorum Europaeorum Catalogus Methodicus, p. 22 (nomen nudum).

ab. **mannerheimi** Chardiny, 1836, Rev. Ent. (Silbermann), **4**: 194, pl. 37. Herrich-Schäffer, 1851, Systematische Bearbeitung der Schmetterlinge von Europa, **2**, pl. 15, fig. 104; 1852, ibidem, **6**: 45. Reiss, 1930, in Seitz, Die Gross-Schmetterlinge der Erde, Supplement, **2**: 16.

inversa Burgeff, 1926, Mitt. münch. ent. Ges., **16**: 29.

ssp. **ukrainica** Sheljuzhko, 1924, Mitt. münch. ent. Ges., **14**: 34. Reiss, 1930, in Seitz, Die Gross-Schmetterlinge der Erde, Supplement, **2**: 16. Holik & Reiss, 1932, in Holik, Iris, **46**: 117, pl. 1, figs. 32, 33. Reiss, 1932, Ent. Rdsch., **49**: 165, 169, pl. 1, fig.; 1933, in Seitz, Die Gross-Schmetterlinge der Erde, Supplement, **2**: 256. Haaf, 1952, Veröff. zool. Staatssamml. Münch., **2**: 152, 157, pl. 14 (as *centaureae* Fischer-Waldheim). Holik & Sheljuzhko, 1955, Mitt. münch. ent. Ges., **44/45**: 78.

Bei Kijev, Ukraine.

ab. **privata** Sheljuzhko, 1924, Mitt. münch. ent. Ges., **14**: 35.

ab. **parvimaculata** Sheljuzhko, 1924, Mitt. münch. ent. Ges., **14**: 34.

ab. **cynaraeformis** Sheljuzhko, 1924, Mitt. münch. ent. Ges., **14**: 29.

Biology

Burgeff, 1950, Portug. acta biol., (A) Goldschmidt: 663-728. Holik, 1937, Lambillionea, **37**: 15-24, 32-45, 80-91; 1953, Ent. Z., **62**: 188; 1953, ibidem, **63**: 31. Reiss, 1930, in Seitz, Die Gross-Schmetterlinge der Erde, Supplement, **2**: 16; 1958, Z. wien. ent. Ges., **43**: 155.

cynarae Esper

Distribution: Russia, Urals, Transcaucasia, Hungary, Dalmatia, Poland, Austria, Bohemia, northern Italy, southern Tyrol, southern France, Germany.

ssp. **uralensis** Herrich-Schäffer, 1846, Systematische Bearbeitung der Schmetterlinge von Europa, **2**: 34; 1847, ibidem, **2**, pl. 12, fig. 85. Reiss, 1933, in Seitz, Die Gross-Schmetterlinge der Erde, Supplement, **2**: 256. Holik & Sheljuzhko, 1955, Mitt. münch. ent. Ges., **44/45**: 72. — Ural.

ssp. **transuralica** Holik & Sheljuzhko, 1955, Mitt. münch. ent. Ges., **44/45**: 73. Reiss, 1932, Ent. Rdsch., **49**: 163, pl. 1, figs. (as *uralensis* Herrich-Schäffer); 1933, in Seitz, Die Gross-Schmetterlinge der Erde, Supplement, **2**: 256 (as *uralensis* Herrich-Schäffer). — Kalkanova, Urgunnerwald, Osthang des Urals, 830 m.

ssp. **baschkirica** Holik, 1939, Rev. franç. Lépid., **9**: 274, pl. 7, figs. 20-22. Holik & Sheljuzhko, 1955, Mitt. münch. ent. Ges., **44/45**: 71. — Bei Uzjan, Westural.

ssp. **samarensis** Holik, 1939, Rev. franç. Lépid., **9**: 275, pl. 7, figs. 23, 24. Holik & Sheljuzhko, 1955, Mitt. münch. ent. Ges., **44/45**: 73. — Stavropol, Samara, Südostrussland.

ssp. **pinskensis** Burgeff, 1914, Mitt. münch. ent. Ges., **5**: 46, pl. 5, figs. 25, 26. Reiss, 1930, in Seitz, Die Gross-Schmetterlinge der Erde, Supplement, **2**: 15. Holik, 1939, Ann. Mus. zool. Polon., **12**: 37, pl. 5, figs. 165-167. — Sumpfgebiet von Pinsk, Westrussland.

ssp. **centrorossica** Holik & Sheljuzhko, 1955, Mitt. münch. ent. Ges., **44/45**: 68. — Sosnovka-Shuberskoje, Voronezh, Zentralrussland.

ssp. **adzharensis** Holik & Sheljuzhko, 1955, Mitt. münch. ent. Ges., **44/45**: 75. Reiss, 1935, Int. ent. Z., **29**: 149, 232, fig. (as *cynarae* ssp.). Koch, 1939, Mitt. münch. ent. Ges., **29**: 408. — Adzhara Gebirge, Transkaukasien.

ssp. **sylvana** Przegendza, 1932, Ent. Z., **46**: 112, figs. 29-32. Holik & Reiss, 1932, in Holik, Iris, **46**: 117, pl. 1, figs. 27-31. Reiss, 1933, in Seitz, Die Gross-Schmetterlinge der Erde, Supplement, **2**: 255. Holik & Sheljuzhko, 1955, Mitt. münch. ent. Ges., **44/45**: 69. — Bei Kijev, Ukraine.

ab. **tricingulata** Holik & Reiss, 1932, in Holik, Iris, **46**: 116.

ab. **rubrianata** Holik & Reiss, 1932, in Holik, Iris, **46**: 116.
ab. **basiconfluens** Sheljuzhko, 1941, Acta Mus. zool. Kijev, **1**: 62, 95, 101.
ab. **analielongata** Sheljuzhko, 1941, Acta Mus. zool. Kijev, **1**: 62, 95, 101.
ab. **confluens** Holik & Reiss, 1932, in Holik, Iris, **46**: 116.
analiconfluens Sheljuzhko, 1941, Acta Mus. zool. Kijev, **1**: 62, 95, 101.
medioconfluens Holik & Sheljuzhko, 1955, Mitt. münch. ent. Ges. **44/45**: 70 (nomen nudum).
ab. **apicaliconfluens** Sheljuzhko, 1941, Acta Mus. zool. Kijev, **1**: 62, 95, 101.
ab. **omniconfluens** Holik & Reiss, 1932, in Holik, Iris, **46**: 116.
ab. **flava** Holik & Sheljuzhko, 1955, Mitt. münch. ent. Ges., **44/45**: 70.

ssp. **cynarae** Esper, 1789, Die Schmetterlinge, Supplement, **2**(2): 2, pl. 37, figs. 2-4. Hübner, [1796]-[24th December 1799], Sammlung europäischer Schmetterlinge, **2**, pl. 17, fig. 80; 1806, ibidem, Der Ziefer, p. 82. Ochsenheimer, 1808, Die Schmetterlinge von Europa, **2**: 42. Boisduval, 1828, Essai sur une Monographie des Zygénides, p. 49, pl. 3, fig. 4; 1834, Icones historique des Lépidoptères nouveaux ou peu connus, **2**: 52, pl. 54, fig. 3. Duponchel, 1835, in Godart & Duponchel, Histoire naturelle des Lépidoptères ou Papillons de France, Supplement, **2**: 60, pl. 5, fig. 6. Freyer, 1842, Neuere Beiträge zur Schmetterlingskunde, **4**: 106, pl. 350, fig. 1. Spuler, 1906, in Hofmann, Die Schmetterlinge Europas, **2**: 157, pl. 75, fig. 46, pl. 77, fig. 11. Seitz, 1907, Die Gross-Schmetterlinge der Erde, **2**: 22, pl. 5b, 5c. Oberthür, 1910, Études de Lépidoptérologie comparée, **4**: 428. Reiss, 1930, in Seitz, Die Gross-Schmetterlinge der Erde, Supplement, **2**: 15. Holik, 1932, Iris, **46**: 115, pl. 1, figs. 23-26. Schwingenschuss, 1952, Z. wien. ent. Ges., **36**: 135. — Lemberg, Galizien.

genistae Herrich-Schäffer, 1846, Systematische Bearbeitung der Schmetterlinge von Europa, **2**: 35. Reiss, 1930, in Seitz, Die Gross-Schmetterlinge der Erde, Supplement, **2**: 15 (footnote).
ab. **tricingulata** Holik, 1939, Ann. Mus. zool. Polon., **12**: 38.
ab. **semianulata** Holik, 1939, Ann. Mus. zool. Polon., **12**: 38.
ab. **deanulata** Holik, 1939, Ann. Mus. zool. Polon., **12**: 38.
ab. **analiconfluens** Holik, 1939, Ann. Mus. zool. Polon., **12**: 37.
ab. **fumata** Holik, 1939, Ann. Mus. zool. Polon., **12**: 40.

ssp. **adriatica** Burgeff, 1926, Mitt. münch. ent. Ges., **16**: 29. Reiss, 1930, in Seitz, Die Gross-Schmetterlinge der Erde, Supplement, **2**: 16. Holik, 1938, Mitt. münch. ent. Ges., **27**: 136. Meier, 1957, NachrBl. bayer. Ent., **6**: 85. — Zara, Dalmatien.

ab. **confluens** Burgeff, 1914, Mitt. münch. ent. Ges., **5**: 46.

ssp. **parvisi** Dujardin, 1965, Entomops, Nice, no. 2: 58, fig.
Farini d'Olmo, Emilie, Italie, 400-500 m.

ssp. **strongyla** Dujardin, 1965, Entomops, Nice, no. 2: 58, fig.
Arquata Scrivia (au Nord de Gênes, Piemont méridional), Apennins ligures, Italie.

ssp. **turatii** Standfuss, 1892, Mitt. schweiz. ent. Ges., **8**: 368. Seitz, 1907, Die Gross-Schmetterlinge der Erde, **2**: 22, pl. 5c. Oberthür, 1910, Études de Lépidoptérologie comparée, **4**: 431. Rocci, 1915, Atti Soc. ligust. Sci. nat. geogr., **25**: 120, pl. 1, fig. 9d. Reiss, 1930, in Seitz, Die Gross-Schmetterlinge der Erde, Supplement, **2**: 16, pl. 1l.
Genua, Ligurien, Italien.

ab. **deminiata** Rocci, 1914, Atti Soc. ligust. Sci. nat. geogr., **24**: 113.

deminiata Rocci, 1914, Soc. ent., **29**: 41.

ab. **depuncta** Rocci, 1914, Atti Soc. ligust. Sci. nat. geogr., **24**: 113; 1915, ibidem, **25**: 125, 130, pl. 1, fig. 11d.

depuncta Rocci, 1914, Soc. ent., **29**: 41.

ab. **semiconfluens** Rocci, 1914, Atti Soc. ligust. Sci. nat. geogr., **24**: 113; 1915, ibidem, **25**: 125, 130, pl. 1, fig. 10d.

semiconfluens Rocci, 1914, Soc. ent., **29**: 41.

ab. **conjuncta** Rocci, 1914, Atti Soc. ligust. Sci. nat. geogr., **24**: 113.

conjuncta Rocci, 1914, Soc. ent., **29**: 41.

ab. **unita** Rocci, 1915, Atti Soc. ligust. Sci. nat. geogr., **25**: 125.

ab. **rubra** Rocci, 1915, Atti Soc. ligust. Sci. nat. geogr., **25**: 126.

ab. **semirubra** Rocci, 1915, Atti Soc. ligust. Sci. nat. geogr., **25**: 126.

ab. **diaphana** Rocci, 1915, Atti Soc. ligust. Sci. nat. geogr., **25**: 127.

ab. **bicolor** Rocci, 1915, Atti Soc. ligust. Sci. nat. geogr., **25**: 128.

ab. **cynaroides** Rocci, 1915, Atti Soc. ligust. Sci. nat. geogr., **25**: 123.

ssp. **tusca** Verity, 1930, Mem. Soc. ent. ital., **9**: 15. Reiss, 1933, in Seitz, Die Gross-Schmetterlinge der Erde, Supplement, **2**: 256.
Pian di Mugnone, Firenze, Italia.

ab. **confluens** Cannaviello, 1903, Rev. Soc. ent. namur., p. 44.

ssp. **humilis** Rocci, 1926, Boll. Soc. ent. ital., **58**:71. Reiss, 1930, in Seitz, Die Gross-Schmetterlinge der Erde, Supplement, **2**: 16.
Val Bisagno, Liguria, Italia, 800-900 m.

ab. **achilleaeformis** Rocci, 1926, Boll. Soc. ent. ital., **58**: 70.	
ab. **ligusticaeformis** Rocci, 1926, Boll. Soc. ent. ital., **58**: 70.	
ssp. **taurinorum** Verity, 1930, Mem. Soc. ent. ital., **9**: 15. Reiss, 1933, in Seitz, Die Gross-Schmetterlinge der Erde, Supplement, **2**: 256.	Torino, Piedmont, Italia.
ssp. **tolmezzana** Meier, 1957, NachrBl. bayer. Ent., **6**: 84.	Tolmezzo, Friaul, Oberitalien.
ssp. **vallettensis** Reiss, 1958, Bull. Soc. ent. Mulhouse, p. 61. *vallettensis* Reiss, 1958, Z. wien. ent. Ges., **43**: 157 (nomen nudum).	Les Vallettes près Tourettes-sur-Loup, Alpes-Maritimes, France méridionale, 300 m.
ssp. **eurogramma** Dujardin, 1965, Entomops, Nice, no. 2: 56, fig.	? St. Barnabé (col de Vence), Alpes-Maritimes, France, 1000 m.
ssp. **curtyi** Dujardin, 1965, Entomops, Nice, no. 2: 55, fig.	St. Basile; Antibes; Valmasque; Castellaras, Mougins; Valbonne, Alpes-Maritimes, France, ca. 200 m.
ssp. **ceriana** Burgeff, 1926, Mitt. münch. ent. Ges., **16**: 28. Reiss, 1930, in Seitz, Die Gross-Schmetterlinge der Erde, Supplement, **2**: 16, pl. 2a.	San Remo (Tal von Ceriana), Oberitalien.
ssp. **florianii** Dujardin, 1965, Entomops, Nice, no. 2: 54, fig.	Mazaugues; St. Baume; Var, France.
ssp. **goberti** Le Charles, 1952, Rev. franç. Lépid., **13**: 219. Holik, 1953, NachrBl. bayer. Ent., **2**: 47.	Grenoble, Isère, France.
ssp. **waltharii** Burgeff, 1926, Mitt. münch. ent. Ges., **16**: 28. Reiss, 1930, in Seitz, Die Gross-Schmetterlinge der Erde, Supplement, **2**: 16. Meier, 1957, NachrBl. bayer. Ent., **6**: 85.	Bozen, Trient, Südtirol.
ab. **semiannulata** Rocci, 1915, Atti Soc. ligust. Sci. nat. geogr., **25**: 124.	
ab. **deannulata** Rocci, 1915, Atti Soc. ligust. Sci. nat. geogr., **25**: 125.	
ssp. **austriaca** Schwingenschuss, 1952, Z. wien. ent. Ges., **36**: 136. Meier, 1957, NachrBl. bayer. Ent., **6**: 86.	Theyerner-Höhe bei Herzogenburg, Südhänge, Nieder-Österreich, 350-400 m.

ssp. **millefolii** Borkhausen, 1789, Naturgeschichte der Europäischen Schmetterlinge, **2**: 239. Holik, 1936, Ent. Rdsch., **53**: 406. — Wien, Österreich.

ssp. **wachauensis** Leinfest, 1952, Ent. Z., **61**: 190. Meier, 1957, NachrBl. bayer. Ent., **6**: 86. — Weissenkirchen, Wachau, Nieder-Österreich.

ssp. **franconica** Holik, 1936, Ent. Rdsch., **53**: 407. Burgeff, 1926, Mitt. münch. ent. Ges., **16**: 27 (as *veronicae* Borkhausen). Reiss, 1930, in Seitz, Die Gross-Schmetterlinge der Erde, Supplement, **2**: 15, pl. 2a (as *veronicae* Borkhausen); 1939, Ent. Z., **53**: 114. Haaf, 1952, Veröff. zool. Staatssamml. Münch., **2**: 152, 157, pl. 14 (as *cynarae* Esper). Forster & Wohlfahrt, 1958, Die Schmetterlinge Mitteleuropas, **3**: 93, pl. 9, figs. 43, 44. Heuser & Jöst, 1959, Mitt. Pollichia, (3) **6**: 124. — Marburg; Schweinfurt; Ludwigshafen Hockenheim bei Mannheim; Darmstadt; Süddeutschland.

ab. **aureoviridis** Burgeff, 1926, Mitt. münch. ent. Ges., **16**: 28.

ab. **coniuncta** Spuler, 1906, in Hofmann, Die Schmetterlinge Europas, **2**: 157.

ab. **confluens** Burgeff, 1914, Mitt. münch. ent. Ges., **5**: 46.

ab. **totirubra** Reiss, 1939, Ent. Z., **53**: 114.

ssp. **pusztae** Burgeff, 1926, Mitt. münch. ent. Ges., **16**: 27. Reiss, 1930, in Seitz, Die Gross-Schmetterlinge der Erde, Supplement, **2**: 15. Forster & Wohlfahrt, 1958, Die Schmetterlinge Mitteleuropas, **3**: 93, pl. 9, fig. 45. — Peszér-Alsódahas, Budapest; Gödölo, Ungarn.

cinarae Esper, 1797, Die Schmetterlinge, Supplement, **2** (2): 26 (preoccupied by **cynarae** Esper, 1789).

ab. **rubrianata** Burgeff, 1906, Ent. Z., **20**: 154.

ab. **tricingulata** Burgeff, 1906, Ent. Z., **20**: 154.

ab. **confluens** Burgeff, 1906, Ent. Z., **20**: 154, fig. 1.

Biology

Burgeff, 1912, Z. wiss. InsektBiol., **8**: 124; 1926, Mitt. münch. ent. Ges., **16**: 28; 1950, Portug. acta biol., (A) Goldschmidt: 663-728; 1951, Biol. Zbl., **70**: 1-23. Holik, 1937, Lambillionea, **37**: 15-24, 32-45, 80-91; 1953, Ent. Z., **62**: 183; 1953, ibidem, **62**: 188 (**goberti**); 1953, ibidem, **63**: 31. Hrubý, 1964, Prodromus Lepidopter Slovenska, p. 476. Koch, 1955, Wir Bestimmen Schmetterlinge, **2**: 58, 59, pl. 1, fig. 9. Reiss, 1930, in Seitz, Die Gross-Schmetterlinge der Erde, Supplement, **2**: 15, 16; 1958, Z. wien. ent. Ges., **43**: 155. Spuler, 1910, in Hofmann, Die Raupen der Schmetterlinge Europas, Nachtrag, pl. 9, fig. 21.

corsica Boisduval

Distribution: Corsica, Sardinia.

ssp. **corsica** Boisduval, 1828, Essai sur une Monographie des Zygénides, p. 81, pl. 5, fig. 2. Rambur, 1832, Ann. Soc. ent. Fr., **1**: 267, pl. 7, figs. 5, 6. Boisduval, 1834, Icones historique des Lépidoptères nouveaux ou peu connus, **2**: 58, pl. 55, fig. 9. Duponchel, 1835, in Godart & Duponchel, Histoire naturelle des Lépidoptères ou Papillons de France, Supplement, **2**: 87, pl. 7, fig. 7. Herrich-Schäffer, 1843, Systematische Bearbeitung der Schmetterlinge von Europa, **2**, pl. 1, figs. 5, 6; 1846, ibidem, **2**: 37. Perlini, 1905, Forme di Lepidotteri esclusivamente Italiane, p. 52, pl. 1, fig. 13. Spuler, 1906, in Hofmann, Die Schmetterlinge Europas, **2**: 157, pl. 77, fig. 14. Seitz, 1907, Die Gross-Schmetterlinge der Erde, **2**: 24, pl. 6d. Oberthür, 1910, Études de Lépidoptérologie comparée, **4**: 442. Alberti, 1958, Mitt. zool. Mus. Berl., **34**: 329. Tremewan, 1961, Bull. Brit. Mus. (nat. Hist.) Ent., **10** (7): 245, pl. 50, fig. 4, pl. 58, figs. 1, 2. — Les montagnes de l'île de Corse.

ssp. **sardiniensis** Holik, 1936, Lambillionea, **36**: 226. Reiss, 1930, in Seitz, Die Gross-Schmetterlinge der Erde, Supplement, **2**: 15, pl. 1n (as *corsica* Boisduval). Haaf, 1952, Veröff. zool. Staatssamml. Münch., **2**: 152, 157, pl. 15 (as *corsica* Boisduval). Tremewan, 1961, Bull. Brit. Mus. (nat. Hist.) Ent., **10** (7): 304, pl. 57, fig. 9. — Aritzo, Sardaigne.

ab. **minor** Bytinski-Salz, 1937, Mem. Soc. ent. ital., **15**: 196.

ab. **bielongata** Bytinski-Salz, 1937, Mem. Soc. ent. ital., **15**: 196.

ab. **albogrisea** Reiss, 1941, Mitt. münch. ent. Ges., **31**: 996.

ab. **confluens** Reiss, 1920, Int. ent. Z., **14**: 116.

Biology

Boisduval, Rambur & Graslin, 1832, Collection iconographique et historique des Chenilles, pl. 3, figs. 4, 5. Burgeff, 1950, Portug. acta biol., (A) Goldschmidt: 663-728. Holik, 1937, Lambillionea, **37**: 15-24, 32-45, 80-91; 1938, Ent. Rdsch., **55**: 349-354, 382-384; 1946, Rev. franç. Lépid., **10**: 250-261, 273-280; 1953, Ent. Z., **62**: 191. Mann, 1855, Verh. zool.-bot. Ver. Wien, **5**: 538. Rambur, 1832, Ann. Soc. ent. Fr., **2**: 267. Reiss, 1958, Z. wien. ent. Ges., **43**: 156. Spuler, 1910, in Hofmann, Die Raupen der Schmetterlinge Europas, pl. 9, figs. 23a, b.

SECTION 7

aurata Blachier

Distribution: Morocco.

ssp. **aurata** Blachier, 1905, Bull. Soc. ent. Fr., p. 213; 1908, Ann. Soc. ent. Fr., **77**: 221; 1913, Bull. Soc. lépid. Genève, **2**: 255, pl. 20, fig. 10. Reiss, 1930, in Seitz, Die Gross-Schmetterlinge der Erde, Supplement, **2**: 13, pl. 11. Alberti, 1958, Mitt. zool. Mus. Berl., **34**: 332. — Tizi Gourza, Grand Atlas, Maroc, 3000-4000 m.

ssp. **opaca** Blachier, 1908, Ann. Soc. ent. Fr., **77**: 220; 1913, Bull. Soc. lépid. Genève, **2**: 255, pl. 20, fig. 9. Reiss, 1930, in Seitz, Die Gross-Schmetterlinge der Erde, Suppl., **2**: 13. — Amez-miz, au pied de l'Atlas, Maroc.

ssp. **tachdirtica** Reiss, 1943, Z. wien. ent. Ges., **28**: 314, pl. 37, figs. 56, 57. Haaf, 1952, Veröff. zool. Staatssamml. Münch., **2**: 151, 156, pl. 5 (as *aurata* Blachier). — Tachdirt, im oberen Iminenetal, Grand Atlas, Marokko, 2700 m.

ab. **apicaliconfluens** Reiss, 1943, Z. wien. ent. Ges., **28**: 314.

ab. **interrupta** Reiss, 1943, Z. wien. ent. Ges., **28**: 314.

ssp. **blachieri** Rothschild, 1931, Novit. zool., **36**: 199. Tremewan, 1961, Bull. Brit. Mus. (nat. Hist.) Ent., **10** (7): 245, pl. 50, fig. 5. — Tizi-n-Tichka Great Atlas, Morocco, 2450 m.

Biology

Holik, 1938, Ent. Rdsch., **55**: 349-354, 382-384; 1953, Ent. Z., **62**: 182. Reiss, 1958, Z. wien. ent. Ges., **43**: 155. Wiegel, 1965, Mitt. münch. ent. Ges., **55**: 123.

loyselis Oberthür

Distribution: Algeria, Morocco.

ssp. **occidentis** Burgeff, 1926, Mitt. münch. ent. Ges., **16**: 25 (nomen novum for *occidentalis* Oberthür). Reiss, 1930, in Seitz, Die Gross-Schmetterlinge der Erde, Supplement, **2**: 12. — Géryville, Algérie.

occidentalis Oberthür, 1916, Études de Lépidoptérologie comparée, **12**: 208 (preoccupied by **occidentalis** Oberthür, 1907, ssp. of **hippocrepidis** Hübner); 1890, Études d'Entomologie, **13**: 20, pl. 8, figs. 76-78. Tremewan, 1961, Bull. Brit. Mus. (nat. Hist.) Ent., **10** (7): 245, pl. 50, fig. 7.

ssp. **loyselis** Oberthür, 1876, Études d'Entomologie, **1**: 34; 1888, ibidem, **12**: 24; 1890, ibidem, **13**: 20, pl. 8, figs. 79, 80. Seitz, 1907, Die Gross-Schmetterlinge der Erde, **2**: 20, pl. 4f, g. Oberthür, 1910, Études de Lépidoptérologie comparée, **4**: 444. Rothschild, 1917, Novit. zool., **24**: 332. Oberthür, 1922, Études de Lépidoptérologie comparée, **19**: 157. Reiss, 1930, in Seitz, Die Gross-Schmetterlinge der Erde, Supplement, **2**: 12. Haaf, 1952, Veröff. zool. Staatssamml. Münch., **2**: 151, 156, pl. 5. Alberti, 1958, Mitt. zool. Mus. Berl., **34**: 333. Tremewan, 1961, Bull. Brit. Mus. (nat. Hist.) Ent., **10** (7): 245, pl. 50, fig. 6, pl. 58, figs. 3, 4. — Lambessa, Algérie.

ab. **confluens** Dziurzyński, 1902, Iris, **15**: 336; 1904, Jber. wien. ent. Ver., **14**: 47, pl. 2, fig. 4.

ssp. **fracticingulata** Rothschild, 1925, Ann. Mag. nat. Hist., (9) **15**: 679. Reiss, 1930, in Seitz, Die Gross-Schmetterlinge der Erde, Supplement, **2**: 13, pl. 1l; 1943, Z. wien. ent. Ges., **28**: 354. Tremewan, 1961, Bull. Brit. Mus. (nat. Hist.) Ent., **10** (7): 246, pl. 50, fig. 9. — Hte. Réraya, Great Atlas, Morocco.

ssp. **olivacea** Rothschild, 1925, Ann. Mag. nat. Hist., (9) **15**: 680. Reiss, 1930, in Seitz, Die Gross-Schmetterlinge der Erde, Supplement, **2**: 13, pl. 1l; 1943, Z. wien. ent. Ges., **28**: 354. Tremewan, 1961, Bull. Brit. Mus. (nat. Hist.) Ent., **10** (7): 246, pl. 50, fig. 10. — Taza, north-east of Fez, Middle Atlas, Morocco.

ssp. **montana** Rothschild, 1925, Bull. Soc. Sci. nat. Maroc., **5**: 140. Reiss, 1930, in Seitz, Die Gross-Schmetterlinge der Erde, Supplement, **2**: 13, pl. 1k; 1943, Z. wien. ent. Ges., **28**: 353. Tremewan, 1961, Bull. Brit. Mus. (nat. Hist.) Ent., **10** (7): 246, pl. 50, fig. 11. — Near Azrou, Middle Atlas, Morocco, 1300 m.

ssp. **xauensis** Reiss, 1943, Z. wien. ent. Ges., **28**: 353, pl. 37, figs. 58, 59. Reisser, 1933, Eos, Madr., **9**: 279 (as *ungemachi* Le Cerf). — Izilan; A'fas-ka, Rif, Marokko, 1350 m.

Biology

Holik, 1938, Ent. Rdsch., **55**: 349-354, 382-384; 1953, Ent. Z., **62**: 182. Reiss, 1930, in Seitz, Die Gross-Schmetterlinge der Erde, Supplement, **2**: 13; 1958, Z. wien. ent. Ges., **43**: 155.

ungemachi Le Cerf

Distribution: Morocco.

ungemachi Le Cerf, 1923, Bull. Soc. ent. Fr., p. 200. Reiss, 1930, in Seitz, Die Gross-Schmetterlinge der Erde, Supplement, **2**: 12, pl. 1k; 1943, Z. wien. ent. Ges., **28**: 352, pl. 35, fig. 13. Alberti, 1958, Mitt. zool. Mus. Berl., **34**: 333. — Oulmès, Maroc.

thevestis Staudinger

Distribution: Algeria.

ssp. **thevestis** Staudinger, 1887, Berl. ent. Z., **31**: 33. Oberthür, 1888, Études d'Entomologie, **12**: 26; 1890, ibidem, **13**: 20, pl. 8, fig. 75. Seitz, 1907, Die Gross-Schmetterlinge der Erde, **2**: 20, pl. 4f. Rothschild, 1917, Novit. zool., **24**: 337. Reiss, 1930, in Seitz, Die Gross-Schmetterlinge der Erde, Supplement, **2**: 13; 1933, ibidem, **2**: 254. Alberti, 1958, Mitt. zool. Mus. Berl., **34**: 332. — Lambessa, Algerien.

ssp. **centrialgeria** Reiss, 1933, in Seitz, Die Gross-Schmetterlinge der Erde, Supplement, **2**: 254. — Géryville; Guelt es Stel, Mittel-algerien.

Biology

Burgeff, 1913, Ent. Z., **27**: 181. Reiss, 1930, in Seitz, Die Gross-Schmetterlinge der Erde, Supplement, **2**: 14.

cadillaci Oberthür

Distribution: Morocco.

cadillaci Oberthür, 1921, Études de Lépidoptérologie comparée, **18** (1): 62, pl. T, fig. 2; 1922, ibidem, **19**: 158, pl. 535, figs. 4450-4452. Alberti, 1958, Mitt. zool. Mus. Berl., **34**: 332. Tremewan, 1961, Bull. Brit. Mus. (nat. Hist.) Ent., **10** (7): 247, pl. 50, fig. 15. — Forêt d'Azrou, Moyen Atlas, Maroc.

confluens Reiss, 1943, Z. wien. ent. Ges., **28**: 310, pl. 35, fig. 9.

ab. **pseudofavonia** Reiss, 1964, Coridon, (A) **6**: 2; 1930, in Seitz, Die Gross-Schmetterlinge der Erde, Supplement, **2**: 13, pl. 1m (as *cadillaci* Oberthür); 1933, ibidem, **2**: 254.

staudingeri Austaut

Distribution: Algeria.

staudingeri Austaut, 1878, Petites Nouv. Ent., **2**: 243. Oberthür, 1881, Études d'Entomologie, **6**: 70. Seitz, 1907, Die Gross-Schmetterlinge der Erde, **2**: 20, pl. 4g. Reiss, 1933, in Seitz, Die Gross-Schmetterlinge der Erde, Supplement, **2**: 254. Haaf, 1952, Veröff. zool. Staatssamml. Münch., **2**: 151, 156, pl. 5. Alberti, 1958, Mitt. zool. Mus. Berl., **34**: 331. — Nemours, Oran, Algérie.

favonia Freyer

Distribution: Morocco, Algeria.

ssp. **ahmarensis** Reiss, 1943, Z. wien. ent. Ges., **28**: 311, pl. 35, figs. 5, 6, 10, 11. — Tizi s'Tkrine, DjebelAhmar, Mittel Atlas, Marokko, 1750 m.

ab. **tricingulata** Reiss, 1943, Z. wien. ent. Ges., **28**: 310, pl. 35, figs. 4, 7.

ssp. **maroccensis** Reiss, 1930, in Seitz, Die Gross-Schmetterlinge der Erde, Supplement, **2**: 13, pl. 1l (nomen novum for *intermedia* Rothschild). — Tizi Gourza, Great Atlas, Morocco.

intermedia Rothschild, 1917, Novit. zool., **24**: 336 (preoccupied by **intermedia** Rocci, 1914, ssp. of **transalpina** Esper). Tremewan, 1961, Bull. Brit. Mus. (nat. Hist.) Ent., **10** (7): 247, pl. 50, fig. 17.

ssp. **borreyi** Oberthür, 1922, Études de Lépidoptérologie comparée, **19**: 157, pl. 535, figs. 4453, 4454. Reiss, 1930, in Seitz, Die Gross-Schmetterlinge der Erde, Supplement, **2**: 13, pl. 1m. Tremewan, 1961, Bull. Brit. Mus. (nat. Hist.) Ent., **10** (7): 247, pl. 50, fig. 14. — Chabat-el-Hamma, région des Zemmours, Maroc.

ssp. **kabylica** Reiss, 1941, Z. wien. EntVer., **26**: 289, pl. 31, figs. 2, 3. Reisser, 1933, Eos, Madr., **9**: 280 (as *littoralis* Rothschild). — Xauen; A'faska; Izilan, Rif, Marokko.

ab. **latemarginata** Reiss, 1941, Z. wien. EntVer., **26**: 289, pl. 31, fig. 1.

ab. **pseudoborreyi** Reiss, 1941, Z. wien. EntVer., **26**: 289.

ssp. **littoralis** Rothschild, 1917, Novit. zool., **24**: 336. Reiss, 1930, in Seitz, Die Gross-Schmetterlinge der Erde, Supplement, **2**: 13, pl. 11; 1941, Z. wien. EntVer., **26**: 288, pl. 31. fig. 4. Tremewan, 1961, Bull. Brit. Mus. (nat. Hist.) Ent., **10** (7): 247, pl. 50, fig. 16. — Mogador, Morocco.

cingulata Reiss, 1943, Z. wien. ent. Ges., **28**: 353. Tremewan, 1961, Bull. Brit. Mus. (nat. Hist.) Ent., **10** (7): 247, pl. 57, fig. 8.

ssp. **vitrina** Staudinger, 1887, Berl. ent. Z., **31**: 32. Seitz, 1907, Die Gross-Schmetterlinge der Erde, **2**: 20, pl. 4g. Reiss, 1930, in Seitz, Die Gross-Schmetterlinge der Erde, Supplement, **2**: 13. Haaf, 1952, Veröff. zool. Staatssamml. Münch., **2**: 151, 156, pl. 5 (as *favonia* Freyer). Alberti, 1958, Mitt. zool. Mus. Berl., **34**: 332. — Bei Constantine, Prov. Constantine, Algerien.

ab. **valentini** Bruand, 1846, Ann. Soc. ent. Fr., (2) **4**: 201, pl. 8, fig. 1. Dziurzyński, 1907, Jber. wien. ent. Ver., **17**: 83, pl. 2, fig. 2 (after Bruand).

ab. **powelli** Oberthür, 1909, Études de Lépidoptérologie comparée, **3**, pl. 29, fig. 175; 1910, ibidem, **4**: 446. Tremewan, 1961, Bull. Brit. Mus. (nat. Hist.) Ent., **10** (7): 246, pl. 50 fig. 13.

ssp. **sebdouensis** Przegendza, 1932, Ent. Z., **46**: 117, figs. 13-16. Reiss, 1933, in Seitz, Die Gross-Schmetterlinge der Erde, Supplement, **2**: 254. — Sebdou, Oran, Algerien.

ssp. **favonia** Freyer, 1845, Neuere Beiträge zur Schmetterlingskunde, **5**: 76, pl. 428, fig. 1. Oberthür, 1888, Études d'Entomologie, **12**: 25; 1890, ibidem, **13**: 20, pl. 8, figs. 74, 82-88. Seitz, 1907, Die Gross-Schmetterlinge der Erde, **2**: 20, pl. 4g; 1912, ibidem, **2**: 441. Oberthur, 1909, Études de Lépidoptérologie comparée, **3**, pl. 29, fig. 174; 1910, ibidem, **4**: 445. Holl, 1912, Bull. Soc. Hist. nat. Afr. N., **4**: 116. Reiss, 1930, in Seitz, Die Gross-Schmetterlinge der Erde, Supplement, **2**: 13; 1943, Z. wien. ent. Ges., **28**: 308. Burgeff, 1950, Portug. acta biol., (A) Goldschmidt: 670, figs. c1; 1951, Biol. Zbl., **70**: 4, figs. 2c. Alberti, 1958, Mitt. zool. Mus. Berl., **34**: 332. — Algier, Algerien [Türkei irrtümlich]

cedri Bruand, 1846, Ann. Soc. ent. Fr., (2) **4**: 202, pl. 8, fig. 2.

mediterranea Herrich-Schäffer, 1846, Systematische Bearbeitung der Schmetterlinge von Europa, **2**: 38; 1852, ibidem, **6**: 45.

ab. **pseudostaudingeri** Burgeff, 1926, in Strand, Lepid. Cat., **33**: 16.

ab. **flava** Rothschild, 1917, Novit. Zool., **24**: 336. Tremewan, 1961, Bull. Brit. Mus. (nat. Hist.) Ent., **10** (7): 246, pl. 50, fig. 12.

Biology

Burgeff, 1913, Ent. Z., **27**: 180, pl. 4; 1950, Portug. acta biol., (A) Goldschmidt: 663-728; 1951, Biol. Zbl., **70**: 1-23. Holik, 1937, Lambillionea, **37**: 15-24, 32-45, 80-91; 1938, Ent. Rdsch., **55**: 349-354, 382-384; 1953, Ent. Z., **62**: 182. Oberthür, 1912, Études de Lépidoptérologie comparée, **6**, pl. 135, fig. 182. Reiss, 1930, in Seitz, Die Gross-Schmetterlinge der Erde, Supplement, **2**: 13; 1958, Z. wien. ent. Ges., **43**: 155.

sarpedon Hübner

Distribution: Balearic Islands, Portugal, Spain, France.

ssp. **confluenta** Reiss, 1927, Int. ent. Z., **21**: 290; 1930, in Seitz, Die Gross-Schmetterlinge der Erde, Supplement, **2**: 14, pl. 1m; 1933, ibidem, **2**: 254; 1936, Ent. Rdsch., **54**: 29, 92, pl. 2, figs. Koch, 1945, Eos, Madr., **20**: 342. Agenjo, 1952, Fáunula Lepidopterológica Almeriense, p. 158. Reiss, 1966, Ent. Rec., **78**: 137. — Totana, Sierra de Espuña, Murcia, Spanien.

ampla Marten, 1957, Ent. Z., **67**: 274. Reiss, 1966, Ent. Rec., **78**: 137.

ab. **pseudotrimaculata** Burgeff, 1914, Mitt. münch. ent. Ges., **5**: 46, pl. 2, fig. 153, pl. 5, figs. 23, 24.

ab. **pseudohispanica** Reiss, 1927, Int. ent. Z., **21**: 290.

ab. **confluens** Reiss, 1941, Mitt. münch. ent. Ges., **31**: 996.

f. loc. **kampfi** Marten, 1957, Ent. Z., **67**: 273. — Hügel bei Altea, Alicante, Spanien.

ssp. **algecirensis** Reiss, 1927, Int. ent. Z., **21**: 290; 1930, in Seitz, Die Gross-Schmetterlinge der Erde, Supplement, **2**: 14, pl. 1m, n; 1936, Ent. Rdsch., **54**: 29, 92, pl. 2, figs. Koch, 1945, Eos, Madr., **20**: 341. — Algeciras, Südspanien.

ab. **confluens** Dziurzyński, 1902, Iris, **15**: 336.

ssp. **zapateri** Reiss, 1936, Ent. Rdsch., **54**: 57, pl. 2, figs. Koch, 1945, Eos, Madr., **20**: 344. — Albarracin, Teruel, Spanien.

ab. **quinquepuncta** Reiss, 1936, Ent. Rdsch., **54**: 58.

ab. **pseudohispanica** Reiss, 1936, Ent. Rdsch., **54**: 58.

ab. **pseudovariabilis** Reiss, 1936, Ent. Rdsch., **54**: 58.

ab. **rubrior** Reiss, 1936, Ent. Rdsch., **54**: 58.

ab. **totirubra** Burgeff, 1926, Mitt. münch. ent. Ges., **16**: 26.

ssp. **bethunei** Romei, July/August, 1927, Ent. Rec., **39**:107. Reiss, 1930, in Seitz, Die Gross-Schmetterlinge der Erde, Supplement, **2**: 14, pl. 1m. Koch, 1945, Eos, Madr., **20**: 344. — Guadix, Sierra Nevada, Spain, 3500 ft.

bethunei Reiss, November, 1927, Int. ent. Z., **21**: 290.

ab. **quinquepuncta** Reiss, 1927, Int. ent. Z., **21**: 289.

ssp. **balearica** Boisduval, 1828, Essai sur une Monographie des — Les îles Baléares.

Zygénides, p. 39, pl. 2, fig. 5. Tremewan, 1961, Bull. Brit. Mus. (nat. Hist.) Ent., **10** (7): 249.

ssp. **hispanica** Rambur, 1866, Catalogue systématique des Lépidoptères de l'Andalousie, p. 167. Oberthür, 1910, Études de Lépidoptérologie comparée, **4**: 454. Reiss, 1930, in Seitz, Die Gross-Schmetterlinge der Erde, Supplement, **2**: 14, pl. 1m; 1936, Ent. Rdsch., **54**: 29. Koch, 1945, Eos, Madr., **20**: 341. Agenjo, 1945, Eos, Madr., **20**: 347. Marten, 1957, Ent. Z., **67**: 279.
Granada, Andalousie, Espagne méridionale.

granadina Marten, 1957, Ent. Z., **67**: 275. Schmidt-Koehl, 1965, Ent. Z., **75**: 281.

ab. **quinquepuncta** Reiss, 1927, Int. ent. Z., **21**: 289.

ab. **rubrior** Reiss, 1927, Int. ent. Z., **21**: 289.

ssp. **lusitanica** Reiss, 1936, Ent. Rdsch., **54**: 30, pl. 2, figs.
Guarda, Portugal, 1000 m.

ssp. **variabilis** Burgeff, 1926, Mitt. münch. ent. Ges., **16**: 25. Reiss, 1930, in Seitz, Die Gross-Schmetterlinge der Erde, Supplement, **2**: 14. Koch, 1945, Eos, Madr., **20**: 344.
Barcelona, Spanien.

ab. **azona** Reiss, 1927, Int. ent. Z., **21**: 289.

ab. **quinquepuncta** Reiss, 1927, Int. ent. Z., **21**: 289.

ab. **puncta** Reiss, 1927, Int. ent. Z., **21**: 289; 1930, in Seitz, Die Gross-Schmetterlinge der Erde, Supplement, **2**: 14, pl. 4n.

ab. **nigrata** Reiss, 1920, Int. ent. Z., **14**: 116.

f. loc. **cristallina** Marten, 1957, Ent. Z., **67**: 263.
Tarragona, Spanien.

f. loc. **pobleti** Marten, 1957, Ent. Z., **67**: 263.
Prades am Montsant, Spanien, 900-1000 m.

ssp. **anadipsia** Marten, 1957, Ent. Z., **67**: 240.
Gebirge (litoral) südlich von Tortosa, Spanien, 1000 m.

ssp. **tipula** Marten, 1957, Ent. Z., **67**: 262. Reiss, 1966, Ent. Rec., **78**: 140, pl. 5, figs. 9, 10.
Tortosa, Katalonien, Spanien, 100-200 m.

ssp. **escorialica** Reiss, 1936, Ent. Rdsch., **54**: 30, pl. 2, figs. Koch, 1945, Eos, Madr., **20**: 342.
El Escorial, Madrid, Spanien.

ssp. **musza** Marten, 1957, Ent. Z., **67**: 280.
Almadén de la Plata, Sevilla, Spanien, 500 m.

ssp. **irene** Marten, 1957, Ent. Z., **67**: 239. — Ripoll-Campdevánol, Spanien.

ssp. **cimbali** Marten, 1957, Ent. Z., **67**: 277. — Irúrzun, Navarra, Spanien, 450-500 m.

f. loc. **plumula** Marten, 1957, Ent. Z., **67**: 277. — Umgebung des Ortes Penches, La Bureba, Spanien, 600 m.

ssp. **subalmanzorica** Koch, 1945, Eos, Madr., **20**: 344; 1948, ibidem, **24**: 328. — San Rafael, 1200 m.; Cercedilla, 1460 m.; Sierra de Gredos, España.

ssp. **rianoica** Tremewan, 1961, Ent. Rec., **73**: 1; 1963, ibidem, **75**: 1, pl. 1, figs. 1, 2. — Riano, Leon, Spain, 3650 ft.

ssp. **carmencita** Oberthür, 1910, Études de Lépidoptérologie comparée, **4**: 457; 1923, ibidem, **20**: 60. Bernardi & Viette, 1959, Entomologiste, **15**: 4. Tremewan, 1960, Ent. Rec., **72**: 207; 1961, Bull. Brit. Mus. (nat. Hist.) Ent., **10** (7): 248, pl. 50, fig. 18. — Vernet-les-Bains, Pyrénées-Orientales, France.

ab. **vernetensis** Oberthür, 1884, Études d'Entomologie, **8**: 28. Graslin, 1863, Ann. Soc. ent. Fr., (4) **3**: 336, pl. 8, fig. 1. Oberthür, 1896, Études d'Entomologie, **20**: 47, pl. 8, fig. 143 (as *trimaculata* Esper). Seitz, 1907, Die Gross-Schmetterlinge der Erde, **2**: 20, pl. 4f. Reiss, 1930, in Seitz, Die Gross-Schmetterlinge der Erde, Supplement, **2**: 14. Bernardi & Viette, 1959, Entomologiste, **15**: 5. Tremewan, 1960, Ent. Rec., **72**: 207; 1961, Bull. Brit. Mus. (nat. Hist.) Ent., **10** (7): 248, pl. 50, fig. 19.

ab. **azona** Reiss, 1927, Int. ent. Z., **21**: 289.

ab. **flava** Oberthür, 1896, Études d'Entomologie, **20**: 43, pl. 8, fig. 142. Tremewan, 1961, Bull. Brit. Mus. (nat. Hist.) Ent., **10** (7): 248, pl. 50, fig. 20.

ssp. **crozesi** Marten, 1957, Ent. Z., **67**: 238. — Parisot, Tarn, Frankreich.

ssp. **sarpedon** Hübner, 1790, Beiträge zur Geschichte der Schmetterlinge, **2**: 85, pl. 1, fig. C; 1796, Sammlung europäischer Schmetterlinge, **2**: 13, pl. 2, fig. 9; 1806, ibidem, Der Ziefer, p. 83. Burgeff, 1926, Mitt. münch. ent. Ges., **16**: 25. Reiss, 1930, in Seitz, Die Gross-Schmetterlinge der — Montpellier, Provence, Frankreich.

Erde, Supplement, **2**: 14. Burgeff, 1950, Portug. acta biol., (A) Goldschmidt: 670, figs. c2; 1951, Biol. Zbl., **70**: 4, figs. 2c. Reiss, 1958, Bull. Soc. ent. Mulhouse, p. 46. Alberti, 1958, Mitt. zool. Mus. Berl., **34**: 334. Tremewan, 1960, Ent. Rec., **72**: 206.

ab. **micingulata** Reiss, 1927, Int. ent. Z., **21**: 289.

ab. **azona** Reiss, 1927, Int. ent. Z., **21**: 289.

ab. **trimaculata** Esper, 1793, Die Schmetterlinge, Supplement, **2** (2): 16, pl. 40, figs. 7, 8. Seitz, 1907, Die Gross-Schmetterlinge der Erde, **2**: 20, pl. 4f. Reiss, 1930, in Seitz, Die Gross-Schmetterlinge der Erde, Supplement, **2**: 14. Tremewan, 1960, Ent. Rec., **72**: 207.

ssp. **xerophila** Dujardin, 1956, Bull. mens. Soc. linn. Lyon, **25**: 260. Dufay, 1966, Bull. mens. Soc. linn. Lyon, **35**: 73.

St. Auban, environs de Sisteron, Basses-Alpes, France, 450 m.

ssp. **leuzensis** Dujardin, 1956, Bull. mens. Soc. linn. Lyon, **25**: 260. Loritz, 1961, Bull. Soc. ent. Mulhouse, p. 83-102, fig.; 1962, ibidem, p. 17-26.

Mt. Pacanaglia (Mt. Leuze), environs de Nice, Alpes-Maritimes, France, 500 m.

ab. **nigroabdomine** Le Charles, 1930/35, in Lhomme, Catalogue des Lépidoptères de France et de Belgique, **1**: 674.

ab. **azona** Reiss, 1958, Bull. Soc. ent. Mulhouse, p. 47.

ab. **quinquepuncta** Reiss, 1958, Bull. Soc. ent. Mulhouse, p. 47.

ab. **rubrior** Reiss, 1958, Bull. Soc. ent. Mulhouse, p. 47.

ssp. **pictonorum** Bernardi & Viette, 1959, Entomologiste, **15**: 4. Reiss, 1930, in Seitz, Die Gross-Schmetterlinge der Erde, Supplement, **2**: 14, pl. 4m (as *carmencita* Oberthür).

Longeville, Vendée, France.

ab. **azona** Reiss, 1927, Int. ent. Z., **21**: 289.

Biology

Abeille, 1909, Mém. Soc. linn. Provence, **1**: 8-9. Abicot, 1849, Ann. Soc. ent. Fr., (2) **7**: 175, pl. 6, fig. 3. Boisduval, Rambur & Graslin, 1832, Collection iconographique et historique des Chenilles, pl. 3, figs. 2, 3. Burgeff, 1950, Portug. acta biol., (A) Goldschmidt: 663-728; 1951, Biol. Zbl., **70**: 1-23. Foulquier, 1918, in Oberthür, Études de Lépidoptérologie comparée, **16**: 262. Holik, 1937, Lambillionea, **37**: 15-24, 32-45, 80-91; 1938, Ent. Rdsch., **55**: 349-354, 382-384; 1946, Rev. franç. Lépid., **10**: 250-261, 273-280; 1953, Ent. Z., **62**: 158. Rambur, 1866, Catalogue systématique des Lépidoptères de l'Andalousie, p. 167. Reiss, 1930, in Seitz, Die Gross-Schmetterlinge der Erde, Supplement, **2**: 14; 1958, Z. wien. ent. Ges., **43**: 155. Roüast, 1883, Catalogue des Chenilles européennes connues, p. 22. Spuler, 1910, in Hofmann, Die Raupen der Schmetterlinge Europas, pl. 9, fig. 20.

contaminei Boisduval

Distribution: Sierra de Gredos, Picos de Europa, Hautes-Pyrénées, Basses-Pyrénées.

ssp. **almanzorica** Reiss, 1936, Ent. Rdsch., **54**: 59, pl. 2, figs. Koch, 1945, Eos, Madr., **20**: 343; 1948, ibidem, **24**: 327.

Almanzor Gebiet, Sierra de Gredos, Spanien, 1900-2000 m.

ssp. **penalabrica** Fernández, 1929, Mem. Soc. esp. Hist. nat., **15**: 599, figs. 8, 9. Tremewan, 1961, Bull. Brit. Mus. (nat. Hist.) Ent., **10** (7): 249, pl. 50, fig. 22.

ab. **cingulata** Fernández, 1929, Mem. Soc. esp. Hist. nat., **15**: 599.

ab. **semiconfluens** Fernández, 1929, Mem. Soc. esp. Hist. nat., **15**: 599, fig. 10.

Peñalabra; Pico de las Tres Aguas; Picos de Europa, España, 1300 m.

ssp. **asturica** Reiss, 1936, Ent. Rdsch., **54**: 59, pl. 2, fig. Koch, 1948, Eos, Madr., **24**: 328. Tremewan, 1961, Ent. Rec., **73**: 2.

La Liebana; Treviso, 1200 m. (Picos de Europa), Nordspanien.

ssp. **contaminei** Boisduval, 1834, Icones historique des Lépidoptères nouveaux ou peu connus, **2**: 48, pl. 53, figs. 4, 5. Duponchel, 1835, in Godart & Duponchel, Histoire naturelle des Lépidoptères ou Papillons de France, Supplement, **2**: 51, pl. 5, fig. 2. Herrich-Schäffer, 1843, Systematische Bearbeitung der Schmetterlinge von Europa, **2**, pl. 1, fig. 1; 1846, ibidem, **2**: 33. Oberthür, 1884, Études d'Entomologie, **8**: 29. Spuler, 1906, in Hofmann, Die Schmetterlinge Europas, **2**: 155, pl. 75, fig. 43, pl. 77, fig. 7. Seitz, 1907, Die Gross-Schmetterlinge der Erde, **2**: 20, pl. 4e. Oberthür, 1910, Études de Lépidoptérologie comparée, **4**: 458. Reiss, 1930, in Seitz, Die Gross-Schmetterlinge der Erde, Supplement, **2**: 15, pl. 1n; 1933, ibidem, **2**: 255. Koch, 1948, Eos, Madr., **24**: 330. Alberti, 1958, Mitt. zool. Mus. Berl., **34**: 334. Tremewan, 1961, Bull. Brit. Mus. (nat. Hist.) Ent., **10** (7): 249, pl. 50, fig. 21, pl. 58, figs. 5, 6. Tremewan & Manley, 1965, Ent. Rec., **77**: 4.

Environs de Barèges, Hautes-Pyrénées, France.

heegeri Heydenreich, 1851, Lepidopterorum Europaeorum Catalogus Methodicus, p. 21 (nomen nudum).

pennina Rambur, 1866, Catalogue systématique des Lépidoptères de l'Andalousie, p. 169. Bernardi & Viette, 1961, Bull. mens. Soc. linn. Lyon, **30**: 140. Tremewan, 1963, Ent. Rec., **75**: 167, figs. 1, 2.

ab. **cingulata** Fernández, 1929, Mem. Soc. esp. Hist. nat., **15**: 599, fig. 11.

Biology

Holik, 1937, Lambillionea, **37**: 15-24, 32-45, 80-91; 1938, Ent. Rdsch., **55**: 349-354, 382-384; 1953, Ent. Z., **62**: 158. Oberthür, 1910, Études de Lépidoptérologie comparée, **4**: 458. Reiss, 1930, in Seitz, Die Gross-Schmetterlinge der Erde, Supplement, **2**: 15; 1958, Z. wien. ent. Ges., **43**: 155. Ribbe, 1909/12, Iris, **23**: 356. Roüast, 1883, Catalogue des Chenilles européennes connues, p. 22.

punctum Ochsenheimer

Distribution: Asia Minor, Crete, Rhodes, Balkan peninsula, southern Russia, Caucasus, Transcaucasia, Hungary, Lower Austria, Moravia, Podolia, Italy, Sicily.

ssp. **anatoliensis** Reiss, 1929, Int. ent. Z., **23**: 148; 1930, ibidem, **24**: 249; 1930, in Seitz, Die Gross-Schmetterlinge der Erde, Supplement, **2**: 15, pl. 1n; 1935, Int. ent. Z., **29**: 142. Holik, 1938, Mitt. münch. ent. Ges., **27**: 133. Holik & Sheljuzhko, 1953, Mitt. münch. ent. Ges., **43**: 217.	Ak-Schehir, Kleinasien, 1000-1500 m.
ab. **cingulata** Reiss, 1930, in Seitz, Die Gross-Schmetterlinge der Erde, Supplement, **2**: 15.	
ab. **dystreptoides** Reiss, 1929, Int. ent. Z., **23**: 148.	
ssp. **malatina** Dziurzyński, 1902, Iris, **15**: 337. Seitz, 1907, Die Gross-Schmetterlinge der Erde, **2**: 21, pl. 4h. Reiss, 1930, Int. ent. Z., **24**: 250; 1930, in Seitz, Die Gross-Schmetterlinge der Erde, Supplement, **2**: 15. Holik & Sheljuzhko, 1953, Mitt. münch. ent. Ges., **43**: 216.	Malatia, Taurus, Kleinasien.
ssp. **kefersteinii** Herrich-Schäffer, 1846, Systematische Bearbeitung der Schmetterlinge von Europa, **2**: 31, pl. 11, fig. 77. Holik, 1938, Mitt. münch. ent. Ges., **27**: 131-132.	Bei Canea, Insel Kreta.
ssp. **kalavrytica** Reiss, 1962, Ent. Z., **72**: 217, 219. ab. **apicaliseparata** Reiss, 1962, Ent. Z., **72**: 218. ab. **pseudoscupensis** Reiss, 1962, Ent. Z., **72**: 219. ab. **rubrior** Reiss, 1962, Ent. Z., **72**: 219.	Umgebung von Kalavryta, Peloponnes (Morea), Griechenland, 750 m.
ssp. **athenae** Reiss, 1962, Ent. Z., **72**: 219, 220. ab. **pseudoscupensis** Reiss, 1962, Ent. Z., **72**: 219.	Daphni, Umgebung von Athen, Griechenland.
ssp. **rhodosica** Reiss, 1962, Ent. Z., **72**: 220, 222.	Am Fuss des Fileremo; Maritsa, Insel Rhodos.
ssp. **scupensis** Koch, 1942, Iris, **56**: 94. Daniel, 1958, Fragmenta Balcanica, **2**: 39; 1964, Prirod. Muz. Skopje, no. 2: 12.	Ueskiib, Mazedonien.
ssp. **dalmatina** Boisduval, 1834, Icones historique des Lépidoptères nouveaux ou peu connus, **2**: 45, pl. 54, fig. 2. Holik, 1935, Ent. Rdsch., **53**: 56; 1936, Lambillionea, **36**: 50; 1938,	Raguse, Dalmatie.

Mitt. münch. ent. Ges., **27**: 127. Bernardi & Viette, 1960, Bull. mens. Soc. linn. Lyon, **29**: 244. Tremewan, 1961, Bull. Brit. Mus. (nat. Hist.) Ent., **10** (7): 283, pl. 54, fig. 18, pl. 63, figs. 3, 4 (in error). Holik, 1961, Bull. Soc. ent. Mulhouse, p. 52. Reiss & Tremewan, 1962, Bull. Soc. ent. Mulhouse, p. 39-43, figs. 2, 3. Tremewan & Reiss, 1964, Ent. Rec., **76**: 1.

ssp. **kolbi** Reiss, 1933, in Seitz, Die Gross-Schmetterlinge der Erde, Supplement, **2**: 255. Holik, 1938, Mitt. münch. ent. Ges., **27**: 127. — Süsak Trsat, Fiume, Dalmatien.

ssp. **dystrepta** Fischer-Waldheim, 1832, Nouv. Mém. Soc. Nat. Moscou, (2) **2**: 359, pl. 21, fig. 3. Romanoff, 1879, Horae Soc. ent. Ross., **14**: 489. Spuler, 1906, in Hofmann, Die Schmetterlinge Europas, **2**: 156, pl. 77, fig. 8a. Seitz, 1907, Die Gross-Schmetterlinge der Erde, **2**: 21, pl. 4h. Reiss, 1930, Int. ent. Z., **24**: 249; 1930, in Seitz, Die Gross-Schmetterlinge der Erde, Supplement, **2**: 15, pl. 1n. Holik & Sheljuzhko, 1953, Mitt. münch. ent. Ges., **43**: 213. Alberti & Soffner, 1962, Mitt. münch. ent. Ges., **52**: 174. — Sarepta, Südrussland.

eversmanni Heydenreich, 1851, Lepidopterorum Europaeorum Catalogus Methodicus, p. 21 (nomen nudum).

ab. **cingulata** Burgeff, 1914, Mitt. münch. ent. Ges., **5**: 46, pl. 5, fig. 27.

ssp. **chersonesica** Reiss, 1941, Z. wien. EntVer., **26**: 59. Holik & Sheljuzhko, 1953, Mitt. münch. ent. Ges., **43**: 213 (as *chersonensis*). — Simferopol, Krim.

ssp. **kremkyi** Holik, 1939, Ann. Mus. zool. Polon., **12**: 32, pl. 5, figs. 141-146. — Krzywe, Podolien.

ab. **pseudodystrepta** Holik, 1939, Ann. Mus. zool. Polon., **12**: 31.

ssp. **punctum** Ochsenheimer, 1808, Die Schmetterlinge von Europa, **2**: 36. Hübner, [1808]-[20th June 1813], Sammlung europäischer Schmetterlinge, **2**, pl. 26, fig. 119. Boisduval, 1828, Essai sur une Monographie des Zygénides, p. 33, pl. 2, fig. 2; 1834, Icones historique des Lépidoptères nouveaux ou peu connus, **2**: 46, pl. 53, fig. 3. Duponchel, 1835, in Godart & Duponchel, Histoire naturelle des Lépidoptères ou Papillons de France, Supplement, **2**, pl. 5, fig. 1. Spuler, 1906, in Hofmann, Die Schmetterlinge Europas, **2**: 155, pl. 77, fig. 8. Seitz, 1907, Die Gross-Schmetterlinge der Erde, **2**: 20, pl. 4g. Oberthür, 1910, Études de Lépidoptérologie comparée, **4**: 459. Reiss, 1930, in Seitz, Die Gross-Schmetterlinge der Erde, Supplement, **2**: 14; 1933, ibidem, **2**: 254. Holik, 1939, Sborn. ent. Odd. nár. Mus. Praze, **17**: 41. Povolný & Gregor, 1946, Folia ent., Brno, **9**, Supplement, 12: 26, 94. Gregor & Povolný, 1955, Sborn. ent. Odd. nár. Mus. Praze, **30**: 259, 271, pl. 3, fig. 5. Alberti, 1958, Mitt. zool. Mus. Berl., **34**: 334. — Ungarn; Nieder-Österreich.

Forster & Wohlfahrt, 1958, Die Schmetterlinge Mitteleuropas, **3**: 90, pl. 9, figs. 35, 40.

ab. **pseudocontamineoides** Burgeff, 1926, in Strand, Lepid. Cat., **33**: 18.

ab. **pseudodystrepta** Reiss, 1931, Int. ent. Z., **25**: 97. Popescu-Gorj, 1964, Catalogue de la Collection de Lépidoptères „Prof. A. Ostrogovich", p. 66.

ssp. **isaszeghensis** Reiss, 1929, Int. ent. Z., **22**: 357. — Isaszeg, Pesth, Ungarn.

ssp. **zangherii** Dujardin, 1965, Entomops, Nice, no. 2: 59, fig. — Tiriolo, 600 m.; Cerenzia, 600 m.; Capo Rizzuto, 50-100 m., Calabria, Italie Sud.

ssp. **faitensis** Stauder, 1929, Ent. Z., **43**: 30. Reiss, 1930, in Seitz, Die Gross-Schmetterlinge der Erde, Supplement, **2**: 14. Tremewan, 1961, Bull. Brit. Mus. (nat. Hist.) Ent., **10** (7): 249, pl. 50, fig. 23. — Mt. Faito, Sorrento, Italien.

ab. **pseudorubicundus** Stauder, 1929, Ent. Z., **43**: 30. Tremewan, 1961, Bull. Brit. Mus. (nat. Hist.) Ent., **10** (7): 249, pl. 50, fig. 24.

ab. **flava** Reiss, 1941, Mitt. münch. ent. Ges., **31**: 995.

f. loc. **superdystrepta** Verity, 1930, Mem. Soc. ent. ital., **9**: 13. Reiss, 1933, in Seitz, Die Gross-Schmetterlinge der Erde, Supplement, **2**: 255. — Esperia, Monti Aurunci (Caserta), Italia.

f. loc. **microdystrepta** Verity, 1930, Mem. Soc. ent. ital., **9**: 13. Reiss, 1933, in Seitz, Die Gross-Schmetterlinge der Erde, Supplement, **2**: 255. — Monti Aurunci, Valle del Petrella (Caserta), Italia, 1200 m.

ssp. **excelsior** Verity, 1930, Mem. Soc. ent. ital., **9**: 12. Reiss, 1933, in Seitz, Die Gross-Schmetterlinge der Erde, Supplement, **2**: 255. — Pizzo Tre Vescovi, Monti Sibillini (Piceno), Italia, 1700 m.

ab. **rubicundiformis** Verity, 1930, Mem. Soc. ent. ital., **9**: 13.

ssp. **italaparva** Verity, 1930, Mem. Soc. ent. ital., **9**: 14. Reiss, 1933, in Seitz, Die Gross-Schmetterlinge der Erde, Supplement, **2**: 255. — Pian di Mugnone, Firenze, Italia.

ssp. **itala** Burgeff, 1926, Mitt. münch. ent. Ges., **16**: 27 (nomen novum for *italica* Rebel). Reiss, 1930, in Seitz, Die Gross-Schmetterlinge der Erde, Supplement, **2**: 14; 1933, ibidem, **2**: 255. — Livorno, Montenero, Italien, 200 m.

italica Rebel, 1901, in Staudinger & Rebel, Catalog der Lepidopteren des Palaearctischen Faunengebietes, p. 381 (preoccupied by **italica** Caradja, 1895, ssp. of **viciae** Denis & Schiffermüller). Perlini, 1905, Forme di Lepidotteri esclusivamente Italiane, p. 52, pl. 4, fig. 16. Seitz, 1907, Die Gross-Schmetterlinge der Erde, **2**: 21, pl. 4h.

ab. **pseudodystrepta** Burgeff, 1926, in Strand, Lepid. Cat., **33**: 18.

ssp. **ledereri** Rambur, 1866, Catalogue systématique des Lépidoptères de l'Andalousie, p. 169, pl. 1, fig. 9. Tremewan, 1963, Ent. Rec., **75**: 166. — Sizilien [Sierra de Ronda irrtümlich].

contamineoides Staudinger, 1871, in Staudinger & Wocke, Catalog der Lepidopteren des Europaeischen Faunengebiets, p. 46. Freyer, 1852, Neuere Beiträge zur Schmetterlingskunde, **6**: 39, pl. 506, fig. 1 (as *contaminei* Boisduval). Oberthür, 1884, Études d'Entomologie, **8**: 28. Perlini, 1905, Forme di Lepidotteri esclusivamente Italiane, p. 52, pl. 3, fig. 13. Reiss, 1930, in Seitz, Die Gross-Schmetterlinge der Erde, Supplement, **2**: 14, pl. 1n. Tremewan, 1963, Ent. Rec., **75**: 166.

Biology

Burgeff, 1950, Portug. acta biol., (A) Goldschmidt: 663-728; 1951, Biol. Zbl., **70**: 1-23. Holik, 1936, Ent. Rdsch., **54**: 39-40; 1937, Lambillionea, **37**: 15-24, 32-45, 80-91; 1938, Ent. Rdsch., **55**: 349-354, 382-384; 1946, Rev. franç. Lépid., **10**: 250-261, 273-280; 1953, Ent. Z., **62**: 158. Hrubý, 1964, Prodromus Lepidopter Slovenska, p. 476. Oberthür, 1910, Études de Lépidoptérologie comparée, **4**: 459. Reiss, 1958, Z. wien. ent. Ges., **43**: 155. Spuler, 1910, in Hofmann, Die Raupen der Schmetterlinge Europas, pl. 48, fig. 12.

zuleima Pierret

Distribution: Tunisia, Algeria, Morocco.

zuleima Pierret, 1837, Ann. Soc. ent. Fr., **6**: 22, pl. 1, fig. 8. Oberthür, 1876, Études d'Entomologie, **1**: 33; 1888, ibidem, **12**: 24; 1890, ibidem, **13**: 19, pl. 8, fig. 81. Seitz, 1907, Die Gross-Schmetterlinge der Erde, **2**: 19, pl. 4d. Oberthür, 1910, Études de Lépidoptérologie comparée, **4**: 441. Holl, 1912, Bull. Soc. Hist. nat. Afr. N., **4**: 114. Reiss, 1930, in Seitz, Die Gross-Schmetterlinge der Erde, Supplement, **2**: 12, pl. 1k; 1943, Z. wien. ent. Ges., **28**: 307. Haaf, 1952, Veröff. zool. Staatssamml. Münch., **2**: 152, 157, pl. 15. Alberti, 1958, Mitt. zool. Mus. Berl., **34**: 331. Barragué, 1961, Alexanor, **2**: 130, fig. — Environs de Bône, Constantine, Algérie.

ludicra Lucas, 1849, Exploration Scientifique de l'Algérie, **3**: 373. pl. 3, fig. 1.

ab. **trimaculata** Barragué, 1961, Alexanor, **2**: 136.

ab. **nigromarginata** Barragué, 1961, Alexanor, **2**: 136.

ab. **decolorata** Barragué, 1961, Alexanor, **2**: 136.

ab. **confluens** Dziurzyński, 1908, Berl. ent. Z., **53**: 17.
confluens Dziurzyński, 1906, Ent. Z., **19**: 185 (nomen nudum).
ab. **flavescens** Rothschild, 1917, Novit. zool., **24**: 334. Tremewan, 1961, Bull. Brit. Mus. (nat. Hist.) Ent., **10** (7): 250, pl. 50, fig. 25.
f. t. **aestiva** Burgeff, 1914, Mitt. münch. ent. Ges., **5**: 45, pl. 5, figs. 19-22. Djebel-Afrane, Tunesien.

Biology

Barragué, 1961, Alexanor, **2**: 130-134 (including genetics of ab. **flavescens** Rothschild). Burgeff, 1913, Ent. Z., **27**: 189, fig. 3. Holik, 1937, Lambillionea, **37**: 15-24, 32-45, 80-91; 1938, ibidem, **38**: 51-58, 79-88, 95-102; 1953, Ent. Z., **62**: 183. Reiss, 1930, in Seitz, Die Gross-Schmetterlinge der Erde, Supplement, **2**: 12; 1958, Z. wien. ent. Ges., **43**: 156.

SECTION 8

araratensis Reiss
Distribution: Asia Minor.
ssp. **araratensis** Reiss, 1935, Int. ent. Z., **29**: 139. Romanoff, 1884, Mémoires sur les Lépidoptères, **1**: 78 (as *brizae* Esper). Holik, 1939, Ent. Rdsch., **56**: 70. Koch, 1939, Mitt. münch. ent. Ges., **29**: 404. Haaf, 1952, Veröff. zool. Staatssamml. Münch., **2**: 151, 156, pl. 5. Holik & Sheljuzhko, 1955, Mitt. münch. ent. Ges., **44/45**: 90. Alberti, 1958, Mitt. zool. Mus. Berl., **34**: 336. Reiss, 1961, Ent. Z., **71**: 205, figs. 1, 2a, 2b, 3. Kasikoparan, Westarmenien, Kleinasien.
ab. **confluens** Reiss, 1935, Int. ent. Z., **29**: 140.
ssp. **lycaonica** Reiss, 1935, Int. ent. Z., **29**: 141, 232, figs. Holik & Sheljuzhko, 1955, Mitt. münch. ent. Ges., **44/45**: 91. Bulghar-Maden, nordwestlich vom Adana, Kleinasien.

Biology

Holik, 1938, Ent. Rdsch., **55**: 349-354, 382-384.

adsharica Reiss
Distribution: Caucasus, Transcaucasia.
ssp. **adsharica** Reiss, 1935, Int. ent. Z., **29**: 140. Burgeff, 1914, Mitt. münch. ent. Ges., **5**: 44, pl. 5, figs. 12-15 (as *erebus* Staudinger). Reiss, 1930, in Seitz, Die Gross-Schmetterlinge der Erde, Supplement, **2**: 9, pl. 1h (as *erebaea* Burgeff); 1933, ibidem, **2**: 251 (as *erebaea* Burgeff). Koch, 1939, Mitt. münch. ent. Ges., **29**: 405. Holik, 1939, Ann. Mus. zool. Polon., **13**: 252, pl. 23, fig. 25 (as *araratensis* Reiss). Haaf, 1952, Veröff. zool. Staatssamml. Münch., **2**: 151, 156, pl. 5 (as *erebaea* Achalzich (Chambobel) und Adshara Gebirge, Transkaukasien.

Burgeff). Reiss, 1953, Z. wien. ent. Ges., **38**: 138, 141, figs. 7-9. Holik & Sheljuzhko, 1955, Mitt. münch. ent. Ges., **44/45**: 92. Alberti, 1958, Mitt. zool. Mus. Berl., **34**: 336; 1964, Mitt. münch. ent. Ges., **54**: 262.

ab. **interrupta** Reiss, 1935, Int. ent. Z., **29**: 140, 231, fig.

ab. **omniconfluens** Koch, 1939, Mitt. münch. ent. Ges., **29**: 405.

ssp. **shemachensis** Holik & Sheljuzhko, 1955, Mitt. münch. ent. Ges., **44/45**: 95. — Demeretshi bei Shemacha, Gouv. Baku, Transkaukasien.

Biology

Holik, 1938, Ent. Rdsch., **55**: 349-354, 382-384.

corycia Staudinger

Distribution: Asia Minor, Syria, Palestine.

ssp. **corycia** Staudinger, 1878, Horae Soc. ent. Ross., **14**: 318. Seitz, 1907, Die Gross-Schmetterlinge der Erde, **2**: 19, pl. 4d. Reiss, 1932, Int. ent. Z., **26**: 269, figs. Holik & Sheljuzhko, 1955, Mitt. münch. ent. Ges., **44/45**: 85. Alberti, 1958, Mitt. zool. Mus. Berl., **34**: 336. — Dschichmam, Deressi, Magnesia, Kleinasien.

ssp. **adanensis** Reiss, 1929, Int. ent. Z., **22**: 357; 1930, in Seitz, Die Gross-Schmetterlinge der Erde, Supplement, **2**: 9, pl. 1h; 1933, ibidem, **2**: 251; 1932, Int. ent. Z., **26**: 271, 279, fig. Haaf, 1952, Veröff. zool. Staatssamml. Münch., **2**: 151, 156, pl. 5 (as *corycia* Staudinger). Holik & Sheljuzhko, 1955, Mitt. münch. ent. Ges., **44/45**: 86. — Hadschin, Vil. Adana, Kleinasien.

ssp. **brussensis** Reiss, 1929, Int. ent. Z., **22**: 357; 1930, in Seitz, Die Gross-Schmetterlinge der Erde, Supplement, **2**: 9. Holik & Sheljuzhko, 1955, Mitt. münch. ent. Ges., **44/45**: 89. — Brussa, Kleinasien.

ssp. **staudingeriana** Reiss, 1932, Int. ent. Z., **26**: 270, figs; 1933, in Seitz, Die Gross-Schmetterlinge der Erde, Supplement, **2**: 251; 1953, Z. wien. ent. Ges., **38**: 138, 141, figs. 10-12. Holik & Sheljuzhko, 1955, Mitt. münch. ent. Ges., **44/45**: 87. Tremewan, 1961, Bull. Brit. Mus. (nat. Hist.) Ent., **10** (7): 304, pl. 57, fig. 10. — Bscharre, Libanon, 1300-1850 m.

ssp. **wiltshirei** Bytinski-Salz, 1936, Ent. Rec., Supplement, **48**: 1. Holik & Sheljuzhko, 1955, Mitt. münch. ent. Ges., **44/45**: 87. Ellison & Wiltshire, 1939, Trans. R. ent. Soc. Lond., **88**: 25, pl. 1, fig. 25. Le Charles, 1939, Rev. franç. Lépid., **9**: 262, figs. 1, 6, pl. 5, figs. 1-5, pl. 6, figs. 2, 5 (as *corycia* Staudinger). — Kineseh, Lebanon.

ssp. **amseli** Bytinski-Salz, 1936, Ent. Rec., Supplement, **48**: 2. Holik & Sheljuzhko, 1955, Mitt. münch. ent. Ges., **44/45**: 88. — Ain Karem near Jerusalem, Palestine.

Biology

Holik, 1938, Ent. Rdsch., **55**: 349-354, 382-384; 1953, Ent. Z., **62**: 190.

brizae Esper

Distribution: Hungary, Lower Austria, Moravia, southern Poland, Macedonia, Bosnia.

ssp. **brizae** Esper, 1797, Die Schmetterlinge, Supplement, **2** (2): 27, pl. 43, figs. 3, 4. Hübner, [July 1803]-[15th November 1806], Sammlung europäischer Schmetterlinge, **2**, pl. 18, fig. 85; 1806, ibidem, Der Ziefer, p. 78. Boisduval, 1828, Essai sur une Monographie des Zygénides, p. 35, pl. 2, fig. 3; 1834, Icones historique des Lépidoptères nouveaux ou peu connus, **2**: 42, pl. 52, fig. 6. Duponchel, 1835, in Godart & Duponchel, Histoire naturelle des Lépidoptères ou Papillons de France, Supplement, **2**: 55, pl. 5, fig. 4. Spuler, 1906, in Hofmann, Die Schmetterlinge Europas, **2**: 154, pl. 77, fig. 4. Seitz, 1907, Die Gross-Schmetterlinge der Erde, **2**: 19, pl. 4c. Reiss, 1930, in Seitz, Die Gross-Schmetterlinge der Erde, Supplement, **2**: 9, pl. 1g. Le Charles, 1939, Rev. franç. Lépid., **9**: 262, figs. 2, 4, pl. 5, figs. 11-15, pl. 6, figs. 1, 4. Holik, 1939, Ann. Mus. zool. Polon., **12**: 21, pl. 1, fig. 34. Haaf, 1952, Veröff. zool. Staatssamml. Münch., **2**: 151, 156, pl. 5. Reiss, 1953, Z. wien. ent. Ges., **38**: 141, pl. 10, figs. 15, 16. Holik & Sheljuzhko, 1955, Mitt. münch. ent. Ges., **44/45**: 82. Alberti, 1958, Mitt. zool. Mus. Berl., **34**: 336; 1964, Mitt. münch. ent. Ges., **54**: 262, fig. a. — Ofen, Ungarn.

flava Nickerl, 1897, Verzeichnis der Insekten Böhmen's, **5**: 7 (nomen nudum). Holik, 1929, Int. ent. Z., **23**: 4.

ab. **rubrianata** Burgeff, 1906, Ent. Z., **20**: 154.

ab. **cingulata** Burgeff, 1906, Ent. Z., **20**: 154.

cingulata Dziurzyński, 1906, Ent. Z., **19**: 185 (nomen nudum).

cingulata Seitz, 1907, Die Gross-Schmetterlinge der Erde, **2**: 19. Dziurzyński, 1908, Berl. ent. Z., **53**: 17.

cingulata Dziurzyński, 1908, Verh. zool.-bot. Ges. Wien, **58**: 73.

ab. **interrupta** Hirschke, 1905, Jber. wien. ent. Ver., **16**: 93.

ab. **confluens** Spuler, 1906, in Hofmann, Die Schmetterlinge Europas, **2**: 154.

confluens Dziurzyński, 1906, Ent. Z., **19**: 185 (nomen nudum).

confluens Dziurzyński, 1908, Berl. ent. Z., **53**: 17.

ssp. **ochrida** Holik, 1937, Mitt. münch. ent. Ges., **27**: 5. Daniel, 1964, Prirod. Muz. Skopje, no. 2: 13. — Ochrid, Petrin Planina, Mazedonien.

ssp. **alamuntis** Koch, 1942, Iris, **56**: 95. Holik, 1939, Sborn. ent. Odd. nár. Mus. Praze, **17**: 41. Povolný, 1945, Folia ent., Brno, **8**: 77. Povolný & Gregor, 1946, Folia ent., Brno, **9**, Supplement, 12: 17, 93. Reiss, 1953, Z. wien. ent. Ges., **38**: — Bielkowitz (Belkovice), Olmütz, Mähren.

138, 141, pl. 10, figs. 17, 18. Slabý, 1954, Acta Soc. ent. Bohem. (Čsl.), **50**: 67. Gregor & Povolný, 1955, Sborn. ent. Odd. nár. Mus. Praze, **30**: 259, 271, pl. 3, fig. 4. Reiprich, 1960, Motýle Slovenska, p. 300, pl. 36, figs. A3, B1.

ab. **longicosta** Slabý, 1954, Acta Soc. ent. Bohem. (Čsl.), **50**: 74.

Biology

Burgeff, 1950, Portug. acta biol., (A) Goldschmidt: 663-728. Holik, 1937, Lambillionea, **37**: 15-24, 32-45, 80-91; 1938, Ent. Rdsch., **55**: 349-354, 382-384; 1953, Ent. Z., **62**: 190. Hrubý, 1964, Prodromus Lepidopter Slovenska, p. 476. Reiss, 1958, Z. wien. ent. Ges., **43**: 155. Rogenhofer, 1884, Verh. zool.-bot. Ges. Wien, **34**: 154.

vesubiana Le Charles

Distribution: Drôme, Hautes-Alpes, Basses-Alpes, Alpes-Maritimes, France.

ssp. **vesubiana** Le Charles, 1933, Bull. Soc. ent. Fr., **38**: 253; 1939, Rev. franç. Lépid., **9**: 262, figs. 3, 5, pl. 5, figs. 6-10, pl. 6, figs. 3, 6. Praviel, 1944, Rev. franç. Lépid., **10**: 146. Reiss, 1953, Ent. Z., **63**: 104; 1953, Z. wien. ent. Ges., **38**: 137, 141, pl. 10, figs. 1-6. Alberti, 1958, Mitt. zool. Mus. Berl., **34**: 335. Leinfest, 1963, Bull. Soc. ent. Fr., **68**: 60. Dufay, 1965, Alexanor, **4**: 105; 1966, Bull. mens. Soc. linn. Lyon, **35**: 73. — St. Martin-de-Vésubie, Alpes-Maritimes, France.

ab. **amplarubra** Dujardin, 1956, Bull. mens. Soc. linn. Lyon, **25**: 254.

ssp. **droitica** Dujardin, 1956, Bull. mens. Soc. linn. Lyon, **25**: 253. Dufay, 1955, Rev. franç. Lépid., **15**: 12; 1960, Alexanor, **1**: 237. Leinfest, 1963, Bull. Soc. ent. Fr., **68**: 61. Dufay, 1965, Alexanor, **4**: 105, fig. 1 (distribution map); 1966, Bull. mens. Soc. linn. Lyon, **35**: 73. — Céüze (environs de Gap), Hautes-Alpes, France, 1500 m.

ab. **salmonea** Dujardin, 1956, Bull. mens. Soc. linn. Lyon, **25**: 254.

Biology

Dufay, 1960, Alexanor, **1**: 237; 1965, ibidem, **4**: 107. Dujardin, 1956, Bull. mens. Soc. linn. Lyon, **25**: 254. Holik, 1938, Ent. Rdsch., **55**: 349-354, 382-384. Reiss, 1953, Z. wien. ent. Ges., **38**: 137; 1953, Ent. Z., **63**: 104; 1958, Z. wien. ent. Ges., **43**: 155; 1958, Bull. Soc. ent. Mulhouse, p. 59.

SECTION 9

erythrus Hübner

Distribution: Sicily, Italy, south-east France.

ssp. **saportae** Boisduval, 1829, Monographie des Zygénides, — Sicilie.

Errata et Addenda, p. 1; 1828, Essai sur une Monographie des Zygénides, p. 29 (partim), pl. 1, fig. 7 (as *minos* Denis & Schiffermüller); 1829, Europaeorum Lepidopterorum Index Methodicus, Errata et Addenda, p. 2; 1834, Icones historique des Lépidoptères nouveaux ou peu connus, **2**: 38, pl. 52, figs. 2, 3. Tremewan, 1962, Ent. Rec., **74**: 125; 1966, ibidem, **78**: 29.

albipes Verity, 1916, Bull. Soc. ent. Fr., p. 289; 1922, Ent. Rec., **34**: 31. Reiss, 1930, in Seitz, Die Gross-Schmetterlinge der Erde, Supplement, **2**: 7, pl. 1e. Tremewan, 1961, Bull. Brit. Mus. (nat. Hist.) Ent., **10** (7): 250, pl. 50, fig. 26; 1966, Ent. Rec., **78**: 29.

ab. **erythraeformis** Verity, 1916, Bull. Soc. ent. Fr., p. 289. Tremewan, 1961, Bull. Brit. Mus. (nat. Hist.) Ent., **10** (7): 250, pl. 50, fig. 27.

ab. **cingulata** Reiss, 1920, Int. ent. Z., **14**: 115.

ssp. **irpinoides** Burgeff, 1926, Mitt. münch. ent. Ges.. **16**: 13. Reiss, 1930, in Seitz, Die Gross-Schmetterlinge der Erde, Supplement, **2**: 7, pl. 1f; 1933, ibidem, **2**: 249. — Mte. Sirente, Abruzzen, Italien.

ab. **magna** Seitz, 1907, Die Gross-Schmetterlinge der Erde, **2**: 18, pl. 4a. Burgeff, 1914, Mitt. münch. ent. Ges., **5**: 42.

ab. **hirpina** Zickert, 1905, Ent. Z., **19**: 117.

ssp. **erythrus** Hübner, [July 1803]-[15th November 1806], Sammlung europäischer Schmetterlinge, **2**, pl. 18, fig. 87; 1806, ibidem, Der Ziefer, p. 77. Herrich-Schäffer, 1844, Systematische Bearbeitung der Schmetterlinge von Europa, **2**, pl. 6, fig. 44; 1846, ibidem, **2**: 30. Spuler, 1906, in Hofmann, Die Schmetterlinge Europas, **2**: 153, pl. 75, fig. 39. Oberthür, 1910, Études de Lépidoptérologie comparée, **4**: 423. Reiss, 1930, in Seitz, Die Gross-Schmetterlinge der Erde, Supplement, **2**: 7. Haaf, 1952, Veröff. zool. Staatssamml. Münch., **2**: 151, 156, pl. 4. Alberti, 1958, Mitt. zool. Mus. Berl., **34**: 343. Tremewan, 1966, Ent. Rec.. **78**: 29. — Die Gegend des Vesuvs, Italien.

ab. **verityi** Stefanelli, 1909, Boll. Soc. ent. ital., **40**: 256.

ssp. **miserrima** Verity, 1922, Ent. Rec., **34**: 31. Reiss, 1930, in Seitz, Die Gross-Schmetterlinge der Erde, Supplement, **2**: 7. Verity, 1946, Redia, **31**: 56. — Mte. Musiné near Torino, Italy.

ssp. **janua** Storace, 1956, Boll. Soc. ent. ital., **86**: 139; 1963, ibidem, **92**: 37, figs. 3-6. — Chiappeto (Genoa), Liguria, Italia.

ab. **minor** Rocci, 1926, Boll. Soc. ent. ital., **58**: 65.

ssp. **pedemontana** Rocci, 1926, Boll. Soc. ent. ital., **58**: 65. Reiss, 1930, in Seitz, Die Gross-Schmetterlinge der Erde, Supplement, **2**: 7. Storace, 1963, Boll. Soc. ent. ital., **92**: 37, fig. 7. — Alpi di Piedmont, Italia.

ssp. **actae** Burgeff, 1926, Mitt. münch. ent. Ges., **16**: 12. Reiss, — Mentone (Menton), Riviera, Frankreich.

1930, in Seitz, Die Gross-Schmetterlinge der Erde, Supplement, **2**: 7, pl. 1e. Storace, 1963, Boll. Soc. ent. ital., **92**: 37, figs. 1, 2.

ssp. **azurica** Reiss, 1958, Bull. Soc. ent. Mulhouse, p. 47. Loritz, 1961, Bull. Soc. ent. Mulhouse, p. 83-102, fig.; 1962, ibidem, p. 17-26.

Mt. Pacanaglia (Mt. Leuze), Alpes-Maritimes, France, 450 m.

azurica Reiss, 1958, Z. wien. ent. Ges., **43**: 156 (nomen nudum).

ab. **pseudoerythrus** Reiss, 1958, Bull. Soc. ent. Mulhouse, p. 48.

ab. **latemarginata** Reiss, 1958, Bull. Soc. ent. Mulhouse, p. 48.

ssp. **dujardini** Leinfest, 1965, Entomops, Nice, no. 3: 74, figs. Dufay, 1966, Bull. mens. Soc. linn. Lyon, **35**: 73.

Env. de Digne (Camp de Bès; Eaux Chaudes, etc.), Basses-Alpes, France.

ab. **interrupta** Leinfest, 1965, Entomops, Nice, no. 3: 75.

Biology

Abeille, 1909, Mém. Soc. linn. Provence, **1**: 6-7. Boisduval, Rambur & Graslin, 1832, Collection iconographique et historique des Chenilles, pl. 3, fig. 1. Burgeff, 1950, Portug. acta biol., (A) Goldschmidt: 663-728; 1951, Biol. Zbl., **70**: 1-23. Foulquier, 1918, in Oberthür, Études de Lépidoptérologie comparée, **16**: 262. Holik, 1937, Lambillionea, **37**: 15-24, 32-45, 80-91; 1938, Ent. Rdsch., **55**: 349-354, 382-384; 1943, Ent. Z., **57**: 41-45; 1953, ibidem, **62**: 156. Leinfest, 1965, Entomops, Nice, no. 3: 75. Mann, 1859, Wien. ent. Monatschr., **3**: 92. Millière, 1869/74, Iconographie et Description de Chenilles et Lépidoptères, **3**: 65, pl. 107, figs. 9-11. Reiss, 1930, in Seitz, Die Gross-Schmetterlinge der Erde, Supplement, **2**: 7; 1958, Bull. Soc. ent. Mulhouse, p. 47; 1958, Z. wien. ent. Ges., **43**: 155. Roüast, 1883, Catalogue des Chenilles européennes connues, p. 22. Spuler, 1910, in Hofmann, Die Raupen der Schmetterlinge Europas, pl. 9, figs. 18a, b. Zeller, 1847, Isis von oken, p. 296.

alpherakyi Sheljuzhko

Distribution: Caucasus.

ssp. **alpherakyi** Sheljuzhko, 1936, Folia Zool. Hydrobiol., Riga, **9**: 17. Holik, 1940, Ent. Z., **54**: 201. Holik & Sheljuzhko, 1953, Mitt. münch. ent. Ges., **43**: 199. Alberti, 1958, Mitt. zool. Mus. Berl., **34**: 343.

Berg Shachdag bei Kurush an der südlichen Dagestan-Grenze, Kaukasus, 10000-12000 ft.

ab. **purpuraliformis** Holik, 1940, Ent. Z., **54**: 202.

ab. **mediointerrupta** Holik, 1940, Ent. Z., **54**: 202.

ab. **latomarginata** Holik, 1940, Ent. Z., **54**: 202.

ab. **plutonia** Holik, 1940, Ent. Z., **54**: 202.

ssp. **ossetica** Holik, 1939, Ann. Mus. zool. Polon., **13**: 248, pl. 23, figs. 1, 2, 9, 10. Holik & Sheljuzhko, 1953, Mitt.

Kara-ugom, Nord-Ossetien, Kauka-

münch. ent. Ges., **43**: 201. Tremewan, 1961, Bull. Brit. Mus. (nat. Hist.) Ent., **10** (7): 305, pl. 57, fig. 12. — sus, 1800-2500 m.

ab. **purpuraliformis** Holik, 1939, Ann. Mus. zool. Polon., **13**: 251, pl. 23, figs. 3, 4.

ab. **mediointerrupta** Holik, 1939, Ann. Mus. zool. Polon., **13**: 251, pl. 23, figs. 6, 7.

ab. **latomarginata** Holik, 1939, Ann. Mus. zool. Polon., **13**: 252, pl. 23, fig. 8.

ab. **plutonia** Holik, 1939, Ann. Mus. zool. Polon., **13**: 252, pl. 23, fig. 5.

Biology

Holik, 1938, Ent. Rdsch., **55**: 349-354, 382-384; 1943, Ent. Z., **57**: 41-45; 1953, ibidem, **62**: 157.

diaphana Staudinger

Distribution: Southern Russia, Asia Minor, Syria, Peloponnesia, Austria, east and north France, Germany, Bohemia, Moravia, Poland, Denmark, southern Scandinavia.

ssp. **sareptensis** Rebel, 1901, in Staudinger & Rebel, Catalog der Lepidopteren des Palaearctischen Faunengebietes, p. 380. Seitz, 1907, Die Gross-Schmetterlinge der Erde, **2**: 19, pl. 4b. Holik, 1939, Rev. franç. Lépid., **9**: 271, pl. 7, figs. 1, 2. Reiss, 1941, Mitt. münch. ent. Ges., **31**: 987. Holik & Sheljuzhko, 1953, Mitt. münch. ent. Ges., **43**: 167. Alberti, 1958, Mitt. zool. Mus. Berl., **34**: 348. Tremewan, 1958, Ent. Gaz., **9**: 183; 1961, Bull. Brit. Mus. (nat. Hist.) Ent., **10** (7): 305, pl. 57, fig. 14. — Sarepta, Südrussland.

sareptensis Krulikowsky, 1897, Soc. ent., **12**: 1 (nomen nudum).

ab. **redlichi** Krulikowsky, 1893, Bull. Soc. Nat. Moscou, **6**: 24.

cingulata Holik, 1939, Rev. franç. Lépid., **9**: 271.

ab. **citrina** Oberthür, 1910, Études de Lépidoptérologie comparée, **4**: 424. Tremewan, 1961, Bull. Brit. Mus. (nat. Hist.) Ent., **10** (7): 251, pl. 50, fig. 28.

ssp. **clavigera** Burgeff, 1914, Mitt. münch. ent. Ges., **5**: 44, pl. 2, figs. 152, 160, pl. 5, figs. 8-11. Reiss, 1930, in Seitz, Die Gross-Schmetterlinge der Erde, Supplement. **2**: 9, pl. 1g. Holik, 1941, Mitt. münch. ent. Ges., **31**: 774. Holik & Sheljuzhko, 1953, Mitt. münch. ent. Ges., **43**: 189. Reiss & Tremewan, 1960, Bull. Brit. Mus. (nat. Hist.) Ent., **9** (10): 460. — Akbès (Eibes), Syrien.

ssp. **chamurli** Koch, 1934, Iris, **48**: 192. Reiss, 1935, Int. ent. Z., **29**: 122, 231, fig. Koch, 1936, Ent. Z., **50**: 398. Holik, 1941, Mitt. münch. ent. Ges., **31**: 777. Holik & Sheljuzhko, 1953, Mitt. münch. ent. Ges., **43**: 182. Reiss & Tremewan, 1960, Bull. Brit. Mus. (nat. Hist.) Ent., **9** (10): 461. — Chamurlu-Dagh, West-armenien, 2900 m.

ssp. **martirosensis** Koch, 1942, Iris, **56**: 95. Holik & Sheljuzhko, — Martiros, Daralagëz Gebirge, Armenisches

1953, Mitt. münch. ent. Ges., **43**: 180. Reiss & Tremewan, 1960, Bull. Brit. Mus. (nat. Hist.) Ent., **9** (10): 461.

Bergland, 1800 m.

ssp. **diaphana** Staudinger, 1887, Berl. ent. Z., **31**: 31. Seitz, 1907, Die Gross-Schmetterlinge der Erde, **2**: 19, pl. 4c. Holik, 1941, Mitt. münch. ent. Ges., **31**: 773. Holik & Sheljuzhko, 1953, Mitt. münch. ent. Ges., **43**: 186. Alberti, 1958, Mitt. zool. Mus. Berl., **34**: 348. Tremewan, 1958, Ent. Gaz., **9**: 183; 1961, Bull. Brit. Mus. (nat. Hist.) Ent., **10** (7): 305, pl. 57, fig. 13.

Hadjin, Taurus, Kleinasien, 2000 m.

ssp. **peloponnesica** Holik, 1937, Mitt. münch. ent. Ges., **27**: 4; 1941, ibidem, **31**: 770. Daniel, 1958, Fragmenta Balcanica, **2**: 39. Reiss, 1962, Ent. Z., **72**: 222.

Mt. Chelmos, Peloponnes, Griechenland, 1700-1800 m.

ab. **rubrotecta** Holik, 1941, Mitt. münch. ent. Ges., **31**: 770.

ssp. **agnoetica** Dujardin, 1965, Entomops, Nice, no. 2: 60, fig.

Les Achards (vallée de la Durance), Hautes-Alpes, France Sud-Est, 900-1000 m.

ssp. **incognita** Reiss, 1940, Stettin. ent. Ztg., **101**: 12, pl. 2, figs. e3, e4, e5. Holik, 1941, Mitt. münch. ent. Ges., **31**: 751. Dufay, 1966, Bull. mens. Soc. linn. Lyon, **35**: 73.

Digne, Basses-Alpes, Frankreich.

ssp. **normanna** Verity, 1922, Ent. Rec., **34**: 34. Dupont, 1900, Bull. Soc. Sci. nat. Elbeuf, **18**: 58 (as *minos* Denis & Schiffermüller); 1925, ibidem, **43**: 129; 1927, ibidem, **45**: 72. Reiss, 1930, in Seitz, Die Gross-Schmetterlinge der Erde, Supplement, **2**: 8, pl. 1f. Derenne, 1934, Amat. Papillons, **7**: 147.

Pont de l'Arche, Eure, northern France.

ab. **incisa** Verity, 1922, Ent. Rec., **34**: 34.

ssp. **renneri** Reiss, 1940, Stettin. ent. Ztg., **101**: 10, pl. 2, figs. b6, c1, d4; 1940, Jh. Ver. vaterl. Naturk. Württemb., **96**: 95, figs. 2a, c; 1941, Mitt. münch. ent. Ges., **31**: 989, pl. 34, figs. B2, C2; 1949, Entomon, **1**: 169. Haaf, 1952, Veröff. zool. Staatssamml. Münch., **2**: 151, 156, pl. 4 (as *sareptensis* Krulikowsky). Forster & Wohlfahrt, 1958, Die Schmetterlinge Mitteleuropas, **3**: 89, pl. 9, figs. 25, 30.

Gailenkirchen bei Schwäb. Hall; Weikersheim, Tauber; Württemberg, Süddeutschland.

ab. **plutonia** Reiss, 1941, Mitt. münch. ent. Ges., **31**: 991.

ab. **apicefusca** Reiss, 1940, Stettin. ent. Ztg., **101**: 11; 1940, Jh. Ver. vaterl. Naturk. Württemb., **96**: 95, fig. 2b.

ssp. **varior** Reiss, 1940, Stettin. ent. Ztg., **101**: 9, pl. 2, fig. e1.

Denzerheide und Montabaur, Westerwald, Westdeutschland.

ab. **analiinterrupta** Vorbrodt, 1913, in Vorbrodt & Müller-Rutz, Die Schmetterlinge der Schweiz, **2**: 252, fig. 36.

ab. **interrupta** Reiss, 1940, Stettin. ent. Ztg., **101**: 10.

ab. **divisa** Vorbrodt, 1913, in Vorbrodt & Müller-Rutz, Die Schmetterlinge der Schweiz, **2**: 252.

ab. **quinquemaculata** Reiss, 1940, Stettin. ent. Ztg., **101**: 10.

ab. **apicefusca** Reiss, 1940, Stettin. ent. Ztg., **101**: 10.

ab. **grisescens** Reiss, 1940, Stettin. ent. Ztg., **101**: 10, pl. 2, fig. e2.
ab. **sexmaculata** Reiss, 1940, Stettin. ent. Ztg., **101**: 10; 1930, in Seitz, Die Gross-Schmetterlinge der Erde, Supplement, **2**: 7, pl. 4m (as *sexmaculata* Burgeff).
ab. **paupera** Reiss, 1930, in Seitz, Die Gross-Schmetterlinge der Erde, Supplement, **2**: 7, pl. 4m.
ab. **rubrior** Reiss, 1940, Stettin. ent. Ztg., **101**: 10.

ssp. **allgavica** Reiss, 1941, Mitt. münch. ent. Ges., **31**: 991, pl. 34, figs. A3, B3. — Ummendorf, Warthausen bei Biberach (Riss), Württembergisches Oberland, Süddeutschland.
ab. **interrupta** Reiss, 1941, Mitt. münch. ent. Ges., **31**: 991.
ab. **plutonia** Reiss, 1941, Mitt. münch. ent. Ges., **31**: 991.
ab. **grisescens** Reiss, 1941, Mitt. münch. ent. Ges., **31**: 992.
ab. **apicefusca** Reiss, 1941, Mitt. münch. ent. Ges., **31**: 992.

ssp. **vindobonensis** Reiss, 1940, Stettin. ent. Ztg., **101**: 12, pl. 2, figs. c2, c3. — Wiener-Neustadt, Österreich.
ab. **apicefusca** Reiss, 1940, Stettin. ent. Ztg., **101**: 12.

ssp. **moraviensis** Reiss, 1940, Stettin. ent. Ztg., **101**: 11, pl. 2, figs. d1, d2, d3. Povolný, 1945, Folia ent., Brno, **8**: 76. Povolný & Gregor, 1946, Folia ent., Brno, **9**, Supplement, 12: 7. Gregor & Povolný, 1955, Sborn. ent. Odd. nár. Mus. Praze, **30**: 258, 271, pl. 3, fig. 2. — Kletten, Mähren, 361 m.
ab. **interrupta** Reiss, 1941, Mitt. münch. ent. Ges., **31**: 994.
ab. **sexmacula** Reiss, 1941, Mitt. münch. ent Ges., **31**: 994.
ab. **grisescens** Reiss, 1941, Mitt. münch. ent. Ges., **31**: 994.
ab. **apicefusca** Reiss, 1941, Mitt. münch. ent. Ges., **31**: 994.
ab. **apicalirubrior** Reiss, 1941, Mitt. münch. ent. Ges., **31**: 994.
ab. **rubrior** Reiss, 1941, Mitt. münch. ent. Ges., **31**: 994.

ssp. **hellmanni** Reiss, 1940, Stettin. ent. Ztg., **101**: 9, pl. 2, figs. a1, a2, a3, a4. — Rüdzanny, Masuren.
ab. **interrupta** Reiss, 1964, Coridon, (A) **6**: 4.
ab. **sexmacula** Reiss, 1964, Coridon, (A) **6**: 4, fig. 1.

ssp. **scholzi** Reiss, 1941, Mitt. münch. ent. Ges., **31**: 989, pl. 34, figs. C1, A2; 1940, Stettin. ent. Ztg., **101**: 9, pl. 2, fig. b5. — Bei Guben, Ostdeutschland.
ab. **interrupta** Reiss, 1941, Mitt. münch. ent. Ges., **31**: 990.
ab. **sexmacula** Reiss, 1941, Mitt. münch. ent. Ges., **31**: 990.
ab. **apicefusca** Reiss, 1941, Mitt. münch. ent. Ges., **31**: 990.
ab. **flava** Reiss, 1941, Mitt. münch. ent. Ges., **31**: 990.

ssp. **pimpinellae** Reiss, 1940, Stettin. ent. Ztg., **101**: 4, figs. 3, 4, C2, pl. 1, figs. A2, B2, D2, pl. 2, figs. a5, a6, b1, b2, b3, b4; 1941, Märk. Tierw., **4**: 282, 286, figs. 3, 4; 1941, Mitt. münch. ent. Ges., **31**: 988, pl. 34, figs. A1, B1. Kuserau, 1942, Ent. Z., **56**: 55. Reiss, 1943, Ent. Z.. **57**: 134. Tremewan, 1960, Ent. Rec., **72**: 207. Reiss, 1961, Ent. Z., **71**: 144. — Rüdersdorf bei Berlin, Norddeutschland.

pimpinellae Guhn, 1932, Ent. Jb., **41**: 89 (infrasubspecific). Reiss, 1933, in Seitz, Die Gross-Schmetterlinge der Erde, Supplement, **2**: 250. Koch, 1942, Z. wien. EntVer., **27**: 40. Povolný, 1951, Acta Acad. sci. nat. Morav., **23**: 387, pls. 1-4, figs.; 1956, Z. wien. ent. Ges., **41**: 225. Alberti, 1957, NachrBl. bayer. Ent., **6**: 49; 1958, Ent. Z., **68**: 4; 1958, Mitt. zool. Mus. Berl., **34**: 348. Forster & Wohlfahrt, 1958, Die Schmetterlinge Mitteleuropas. **3**: 89, pl. 9, figs. 24, 29.

ab. **analiinterrupta** Reiss, 1964, Coridon, (A) **6**: 4.

ab. **interrupta** Staudinger, 1871, in Staudinger & Wocke, Catalog der Lepidopteren des europaeischen Faunengebietes, p. 45. Seitz, 1907, Die Gross-Schmetterlinge der Erde, **2**: 19, pl. 4b. Tremewan, 1958, Ent. Gaz., **9**: 184.

ab. **quinquemaculata** Reiss, 1940, Stettin. ent. Ztg., **101**: 8.

ab. **semipaupera** Reiss, 1940, Stettin. ent. Ztg., **101**: 8.

ab. **apicefusca** Reiss, 1940, Stettin. ent. Ztg., **101**: 8.

ab. **rubrior** Reiss, 1940, Stettin. ent. Ztg., **101**: 8.

ssp. **hoffmeyeri** Reiss, 1961, Ent. Z., **71**: 136, 144. Hoffmeyer & Knudsen, 1941, Flora og Fauna, **47**: 7, figs. 1, 3, 6, 8, 11 (as *pimpinellae* Guhn). Hoffmeyer, 1948, De Danske Spindere, p. 153, pl. 16, fig. 6, pl. 23, fig. 4 (as *sareptensis* Krulikowsky); 1958, Ent. Gaz., **9**: 197, pl. 10, figs. (as *pimpinellae* Guhn). Reiss, 1960, in Hoffmeyer, De Danske Spindere, ed. 2, p. 208, pl. 16, fig. 6, pl. 24, fig. 4. — Nord Seeland (Klintsö), Dänemark.

ab. **rubrianata** Reiss, 1961, Ent. Z., **71**: 147.

ab. **grisescens** Reiss, 1961, Ent. Z., **71**: 147.

ab. **apicefusca** Reiss, 1961, Ent. Z., **71**: 146, fig. 1.

Biology

Alberti, 1957, Dtsch. ent. Z. (N.F.), **4**: 1-7; 1957, NachrBl. bayer. Ent., **6**: 49-54 (**pimpinellae**); 1958, Ent. Z., **68**: 4-8. Burgeff, 1950, Portug. acta biol., (A) Goldschmidt: 663-728. Gregor & Povolný, 1955, Sborn. ent. Odd. nár. Mus. Praze, **30**: 271. Holik, 1943, Ent. Z., **57**: 41-45 (**sareptensis** & **pimpinellae**); 1946, Rev. franç. Lépid., **10**: 250-261, 273-280; 1953, Ent. Z., **62**: 157 (**sareptensis**). Koch, 1955, Wir Bestimmen Schmetterlinge, **2**: 58. 59. Reiss, 1933, in Seitz, Die Gross-Schmetterlinge der Erde, Supplement, **2**: 250; 1940, Stettin. ent. Ztg., **101**: 1-13 (**pimpinellae**); 1943, Ent. Z., **57**: 135 (**pimpinellae**); 1955, Z. wien. ent. Ges., **40**: 284-286, pls. 28, 30; 1958, ibidem, **43**: 155 (**sareptensis** & **pimpinellae**).

smirnovi Christoph

Distribution: Northern Iran, Transcaspia.

ssp. **smirnovi** Christoph, 1884, in Romanoff, Mémoires sur les Lépidoptères, **1**: 108, pl. 6, figs. 6a, b. Seitz, 1907, Die Gross- — Nuchur (Achal-Tekke), Transkaspien.

Schmetterlinge der Erde, **2**: 19, pl. 4c. Burgeff, 1914, Mitt. münch. ent. Ges., **5**: 44. Reiss, 1930, in Seitz, Die Gross-Schmetterlinge der Erde, Supplement, **2**: 9, pl. 1g; 1933, Int. ent. Z., **26**: 475, 505, fig.; 1941, Mitt. münch. ent. Ges., **31**: 994. Holik & Sheljuzhko, 1953, Mitt. münch. ent. Ges., **43**: 202. Alberti, 1958, Mitt. zool. Mus. Berl., **34**: 348. Reiss & Tremewan, 1960, Bull. Brit. Mus. (nat. Hist). Ent., **9** (10): 460, pl. 22, figs. 3. 4, pl. 23, figs. 11-14.

ssp. **persica** Burgeff, 1926, Mitt. münch. ent. Ges., **16**: 15. Reiss, 1930, in Seitz, Die Gross-Schmetterlinge der Erde, Supplement, **2**: 9, pl. 1g; 1933, Int. ent. Z., **26**: 476, fig.; 1937, Ent. Rdsch., **54**: 453; 1938, Mitt. münch. ent. Ges., **27**: 164; 1941, Mitt. münch. ent. Ges., **31**: 994. Haaf, 1952, Veröff. zool. Staatssamml. Münch., **2**: 148, 151, 156, pl. 4 (as *smirnovi* Christoph). Holik & Sheljuzhko, 1953, Mitt. münch. ent. Ges., **43**: 204. — Nordiran (?Aschabad).

ab. **pseudosmirnovi** Reiss, 1941, Mitt. münch. ent. Ges., **31**: 995.

Biology

Christoph, 1884, in Romanoff, Mémoires sur les Lépidoptères, **1**: 109. Holik, 1938, Ent. Rdsch., **55**: 349-354, 382-384; 1953, Ent. Z., **62**: 157. Reiss, 1958, Z. wien. ent. Ges., **43**: 155.

purpuralis Brünnich

Distribution: Central Asia, central and southern Russia, Caucasus, Transcaucasia, Asia Minor, Europe (excluding Spain), Denmark, British Isles.

ssp. **tianschanica** Burgeff, 1926, Mitt. münch. ent. Ges., **16**: 14. Reiss, 1930, in Seitz, Die Gross-Schmetterlinge der Erde, Supplement, **2**: 9, pl. 4n. Holik, 1941, Mitt. münch. ent. Ges., **31**: 780. Holik & Sheljuzhko, 1953, Mitt. münch. ent. Ges., **43**: 192. — Aksu; Kouldzha, Tian-shan, Zentralasien.

ab. **flava** Dziurzyński, 1908, Berl. ent. Z., **53**: 12, 16; 1909, Jber. wien. ent. Ver., **19**: 135, pl. 1, fig. 3.

ssp. **naryna** Burgeff, 1926, Mitt. münch. ent. Ges., **16**: 14. Reiss, 1930, in Seitz, Die Gross-Schmetterlinge der Erde, Supplement, **2**: 9. Holik, 1941, Mitt. münch. ent. Ges., **31**: 780. Holik & Sheljuzhko, 1953, Mitt. münch. ent. Ges., **43**: 193. — Naryn Gebiet, Zentralasien.

ssp. **talassica** Holik & Sheljuzhko, 1953, Mitt. münch. ent. Ges., **43**: 192. — West Talasskij Ala-Tau, Syr Darja, Zentralasien, 2500 m.

ssp. **kasakstana** Holik, 1939, Rev. franç. Lépid., **9**: 273, pl. 7, figs. 5, 6; 1941, Mitt. münch. ent. Ges., **31**: 780. Holik & Sheljuzhko, 1953, Mitt. münch. ent. Ges., **43**: 193. — Targaisk, Kandyk Tau, Kasakstan, Zentralasien, 1400 m.

ssp. **barthai** Reiss, 1929, Int. ent. Z., **23**: 148; 1930, in Seitz, Die Gross-Schmetterlinge der Erde, Supplement, **2**: 9, pl. 1g; 1935, Int. ent. Z., **29**: 121, 231, figs. Holik, 1941, Mitt. münch. ent. Ges., **31**: 772. Holik & Sheljuzhko, 1953, Mitt. münch. ent. Ges., **43**: 187. Reiss & Tremewan, 1960, Bull. Brit. Mus. (nat. Hist.) Ent., **9** (10): 460.	Sultan Dagh bei Akschehir, Kleinasien, 2000 m.
ssp. **pseudodiaphana** Tremewan, 1958, Ent. Gaz., **9**: 184; 1961, Bull. Brit. Mus. (nat. Hist.) Ent., **10** (7): 252, pl. 51, fig. 2.	Karacabey near Brussa, Asia Minor.
ssp. **rosea** Burgeff, 1914, Mitt. münch. ent. Ges., **5**: 44. Staudinger, 1887, Berl. ent. Z., **31**: 32 (as *polygalae* Esper). Reiss & Tremewan, 1964, Ent. Rec., **76**: 130.	Malatia, Taurus, Kleinasien.
rosalis Burgeff, 1926, Mitt. münch. ent. Ges., **16**: 14. Reiss, 1930, in Seitz, Die Gross-Schmetterlinge der Erde, Supplement, **2**: 9. Holik & Sheljuzhko, 1953, Mitt. münch. ent. Ges., **43**: 184. Reiss & Tremewan, 1964, Ent. Rec., **76**: 130.	
ssp. **villosa** Burgeff, 1914, Mitt. münch. ent. Ges., **5**: 43, pl. 2, figs. 151, 159, pl. 5, figs. 4-7. Reiss, 1930, in Seitz, Die Gross-Schmetterlinge der Erde, Supplement, **2**: 8, pl. 4n. Holik, 1941, Mitt. münch. ent. Ges., **31**: 779. Holik & Sheljuzhko, 1953, Mitt. münch. ent. Ges., **43**: 177.	Achalzich (Chambobel), Transkaukasien.
ssp. **sultanbeki** Holik, 1941, Mitt. münch. ent. Ges., **31**: 779. Holik & Sheljuzhko, 1953, Mitt. münch. ent. Ges., **43**: 180.	Sultanbek, Daralagëz Gebirge, Armenisches Bergland.
ssp. **agridaghi** Holik, 1941, Mitt. münch. ent. Ges., **31**: 778. Holik & Sheljuzhko, 1953, Mitt. münch. ent. Ges., **43**: 183.	Agri-Dagh, West-Armenien, 2500-3000 m.
ssp. **tirabzona** Sheljuzhko, 1936, Folia Zool. Hydrobiol., Riga, **9**: 17. Holik & Sheljuzhko, 1953, Mitt. münch. ent. Ges., **43**: 184.	Villajet Trapezunt, West Armenien.
ssp. **zangezuri** Holik & Sheljuzhko, 1953, Mitt. münch. ent. Ges., **43**: 181.	Dorf Ochtshi bei Kafan, Zangezur Gebirge, Ost-Armenien.
ssp. **alagirica** Holik & Sheljuzhko, 1953, Mitt. münch. ent. Ges., **43**: 173.	Uruch-Tal, Ach-Sau, Kaukasus, 2000 m.
ssp. **dagestana** Sheljuzhko, 1936, Folia Zool. Hydrobiol., Riga, **9**: 16. Holik, 1941, Mitt. münch. ent. Ges., **31**: 774. Holik & Sheljuzhko, 1953, Mitt. münch. ent. Ges., **43**: 174.	Derbent in der Dagestan-Ebene, Nord-Kaukasus (An der Westküste des Kaspischen Meeres).

ssp. **strandiana** Sheljuzhko, 1936, Folia Zool. Hydrobiol., Riga, **9**: 15. Holik, 1941, Mitt. münch. ent. Ges., **31**: 774. Holik & Sheljuzhko, 1953, Mitt. münch. ent. Ges., **43**: 171. — Teberda-Täler; Dzhemagat-Täler; Nord-Kaukasus.

ssp. **chatiparae** Sheljuzhko, 1936, Folia Zool. Hydrobiol., Riga, **9**: 16. Holik & Sheljuzhko, 1953, Mitt. münch. ent. Ges., **43**: 171. — Chatipara-Berg im Teberda-Gebiet, Nord-Kaukasus, 2300-2700 m.

ssp. **kislovodskana** Sheljuzhko, 1936, Folia Zool. Hydrobiol., Riga, **9**: 14. Holik, 1941, Mitt. münch. ent. Ges., **31**: 774. Holik & Sheljuzhko, 1953, Mitt. münch. ent. Ges., **43**: 169. — Kislovodsk, Nord-Kaukasus.

f. loc. **sanguinalis** Sheljuzhko, 1936, Folia Zool. Hydrobiol., Riga, **9**: 15. Holik & Sheljuzhko, 1953, Mitt. münch. ent. Ges., **43**: 170. — Pjatigorsk, Nord-Kaukasus.

ssp. **ingens** Burgeff, 1926, Mitt. münch. ent. Ges., **16**: 14. Reiss, 1930, in Seitz, Die Gross-Schmetterlinge der Erde, Supplement, **2**: 8, pl. 1g. Koch, 1939, Mitt. münch. ent. Ges., **29**: 398. Holik, 1941, Mitt. münch. ent. Ges., **31**: 779. Holik & Sheljuzhko, 1953, Mitt. münch. ent. Ges., **43**: 176. — Tiflis, Georgien, Transkaukasien.

ssp. **alagezi** Holik & Sheljuzhko, 1953, Mitt. münch. ent. Ges., **43**: 179. — Berg Alagëz, ca. 40 km. nordwestlich von Erivan, Armenisches Bergland.

ssp. **simferopolica** Reiss, 1939, Ent. Z., **53**: 113. Holik & Sheljuzhko, 1953, Mitt. münch. ent. Ges., **43**: 162. — Simferopol, Krim.

ab. **apicefusca** Reiss, 1939, Ent. Z., **53**: 113.

ssp. **thracica** Holik, 1937, Mitt. münch. ent. Ges., **26**: 174; 1941, ibidem, **31**: 771. — Bansko, Pirin Planina, Bulgarien.

ssp. **rebeli** Drenowski, 1928, Spis. blg. Akad., **37**: 211. Reiss & Tremewan, 1964, Ent. Rec., **76**: 130. — Čepelare und Schirokaláka, Rhodope Gebirge, Bulgarien, 1400-1700 m.

drenowskii Holik, 1937, Mitt. münch. ent. Ges., **27**: 1; 1941, ibidem, **31**: 771. Reiss & Tremewan, 1964, Ent. Rec., **76**: 130.

ssp. **slivnenska** Koch, 1942, Iris, **56**: 91. — Sliven, Mittel-Bulgarien.

ssp. **hellena** Burgeff, 1926, Mitt. münch. ent. Ges., **16**: 14 (nomen novum for *graeca* Tutt). Reiss, 1930, in Seitz, Die Gross-Schmetterlinge der Erde, Supplement, **2**: 8. Holik, 1937, Mitt. münch. ent. Ges., **27**: 3. — Mt. Parnassos, Greece.

graeca Tutt, 1895, Ent. Rec., **6**: 273 (preoccupied by **graeca** Staudinger, 1871, ssp. of **carniolica** Scopoli); 1899, A Natural

History of the British Lepidoptera, **1**: 434. Tremewan, 1961, Bull. Brit. Mus. (nat. Hist.) Ent., **10** (7): 252, pl. 51, fig. 1.

ssp. **dojranica** Burgeff, 1926, Mitt. münch. ent. Ges., **16**: 13. Reiss, 1930, in Seitz, Die Gross-Schmetterlinge der Erde, Supplement, **2**: 8. Holik, 1937, Mitt. münch. ent. Ges., **26**: 173. Daniel, 1964, Prirod. Muz. Skopje, no. 2: 11. — Nicolic, Dojran See; Plaguscha Planina (ca. 1000 m.), Mazedonien.

ab. **rubrianata** Burgeff, 1926, Mitt. münch. ent. Ges., **16**: 13.

ssp. **bukuwkyi** Holik, 1937, Mitt. münch. ent. Ges., **26**: 173; 1941, ibidem, **31**: 771. Daniel, 1958, Fragmenta Balcanica, **2**: 39; 1964, Prirod. Muz. Skopje, no. 2: 11. — Ochrid, Petrin Planina, Serbisch Mazedonien.

ssp. **slavonica** Holik & Koch, 1937, in Holik, Mitt. münch. ent. Ges., **26**: 169. — Beočin, Fruška-Gora, Ost-Slavonien (Syrmien), ca. 500 m.

ssp. **lathyri** Boisduval, 1828, Essai sur une Monographie des Zygénides, p. 32, pl. 2, fig. 1. Holik, 1935, Ent. Rdsch., **53**: 56; 1936, Lambillionea, **36**: 50; 1937, Mitt. münch. ent. Ges., **27**: 128. Bernardi & Viette, 1960, Bull. mens. Soc. linn. Lyon, **29**: 244. Tremewan, 1961, Bull. Brit. Mus. (nat. Hist.) Ent., **10** (7): 252. Holik, 1961, Bull. Soc. ent. Mulhouse, p. 51. Reiss & Tremewan, 1962, Bull. Soc. ent. Mulhouse, p. 42, fig. 1. — Raguse, Dalmatie.

ssp. **bosniaca** Burgeff, 1914, Mitt. münch. ent. Ges., **5**: 43, pl. 5, figs. 1-3. Reiss, 1930, in Seitz, Die Gross-Schmetterlinge der Erde, Supplement, **2**: 8, pl. 4n. Holik, 1937, Mitt. münch. ent. Ges., **26**: 171. — Vlasic Gebirge, Bosnien, 1800 m.

ssp. **mirabilis** Verity, 1922, Ent. Rec., **34**: 32. Reiss, 1930, in Seitz, Die Gross-Schmetterlinge der Erde, Supplement, **2**: 8, pl. 1f. Holik, 1941, Mitt. münch. ent. Ges., **31**: 759. — San Fili di Cosenza, Calabria, Italy, 900 m.

ab. **rubrotecta** Verity, 1922, Ent. Rec., **34**: 31.

ssp. **austronubigena** Verity, 1946, Redia, **31**: 58. — Gran Sasso, Abruzzi, Italia, 1300-1500 m.

ssp. **querciana** Holik, 1941, Mitt. münch. ent. Ges., **31**: 758. — Mte. Sibillini, Bolognola, Italien.

ab. **rubroanata** Holik, 1941, Mitt. münch. ent. Ges., **31**: 758.

ab. **plutonia** Holik, 1941, Mitt. münch. ent. Ges., **31**: 758.

ab. **rubrotecta** Holik, 1941, Mitt. münch. ent. Ges., **31**: 758.

ssp. **fiorii** Costantini, 1916, Atti Soc. Nat. Mat. Modena, (5) **3**: 17. Reiss, 1930, in Seitz, Die Gross-Schmetterlinge der Erde, Supplement, **2**: 8. Holik, 1941, Mitt. münch. ent. Ges., **31**: 756. — Cimone; Fiumalbo, Tagliole, Appennino emilia, Italia.

ab. **rubrivalga** Rostagno, 1911, Boll. Soc. Zool. ital., (2) **12**: 105.

ssp. **rocciana** Reiss, 1930, in Seitz, Die Gross-Schmetterlinge der Erde, Supplement, **2**: 8, pl. 4m. ab. **erythrusoides** Rocci, 1918, Atti Soc. ligust. Sci. nat. geogr., **28**: 143, pl. 3, fig. 1c. ab. **apicefusca** Rocci, 1918, Atti Soc. ligust. Sci. nat. geogr., **28**: 142. ab. **viridescens** Rocci, 1918, Atti Soc. ligust. Sci. nat. geogr., **28**: 142.	Genua, Ligurien, Nord-Italien.
ssp. **lombarda** Holik, 1941, Mitt. münch. ent. Ges., **31**: 744.	Bernate; Turbigo, Galliaté, Cameri am Oberlauf des Ticino, Lombardei, Italien.
ssp. **taurinensis** Rocci, 1926, Boll. Soc. ent. ital., **58**: 67. Reiss, 1930, in Seitz, Die Gross-Schmetterlinge der Erde, Supplement, **2**: 8. Rocci, 1941, Boll. Ist. Ent. Univ. Bologna, **13**: 111, pl. 2, figs. 1, 2.	Collina di Torino, Italia, 650 m.
ssp. **levannica** Rocci, 1941, Boll. Ist. Ent. Univ. Bologna, **13**: 112, pl. 2, figs. 5-7.	Levanna, Forno Alpi Graie, Piedmont, Italia, 1300 m.
ssp. **intrepida** Rocci, 1941, Boll. Ist. Ent. Univ. Bologna, **13**: 112, pl. 2, figs. 3-4.	Sestrière et Clavière, Piedmont, Italia, 1500-2000 m.
ssp. **altalpica** Dujardin, 1965, Entomops, Nice, no. 2: 61, fig.	Champsaure-le-Roy (Molines, La Severaissette), massif du Champsaure, Hautes-Alpes, France Sud-Est, 1300-1600 m.
ssp. **hyporea** Dujardin, 1965, Entomops, Nice, no. 2: 61, fig. Dufay, 1966, Bull. mens. Soc. linn. Lyon, **35**: 74.	Montagne de Lure (près refuge), 1400 m.; Col de Jurs; Archail, Basses-Alpes; Col de Reychassat; Col de Perty; Col St. Jean; Col de Négron, Drôme, France.

ssp. **anglardi** Dujardin, 1965, Entomops, Nice, no. 2: 63, fig.	Beaune-le-Froid, Puy-de-Dôme (Massif Central), France centrale.
ssp. **magnalpina** Verity, 1922, Ent. Rec., **34**: 33. Oberthür, 1896, Études d'Entomologie, **20**: 54, pl. 7, fig. 124 (as *minos* Denis & Schiffermüller). Reiss, 1930, in Seitz, Die Gross-Schmetterlinge der Erde, Supplement, **2**: 8. Holik, 1941, Mitt. münch. ent. Ges., **31**: 753. Koch, 1948, Eos, Madr., **24**: 320. Tremewan & Manley, 1965, Ent. Rec., **77**: 4.	Gèdre, Hautes-Pyrénées, France, 1000 m.
hirsuta Holik, 1941, Mitt. münch. ent. Ges., **31**: 755. Koch, 1948, Eos, Madr., **24**: 320.	
ssp. **margitae** Koch, 1942, Iris, **56**: 91; 1948, Eos, Madr., **24**: 322.	Vernet-les-Bains, Ostpyrenäen, Frankreich.
ssp. **parvalpina** Verity, 1922, Ent. Rec., **34**: 33. Reiss, 1930, in Seitz, die Gross-Schmetterlinge der Erde, Supplement, **2**: 8, pl. 1f. Holik, 1941, Mitt. münch. ent. Ges., **31**: 749. Loritz, 1961, Bull. Soc. ent. Mulhouse, p. 83-102, figs.; 1962, ibidem, p. 17-26.	Valdieri, Piedmontese Maritime Alps, Italy, 1375 m.
ssp. **colmianica** Dujardin, 1965, Entomops, Nice, no. 2: 62, fig.	La Colmiane, 1500 m., territoire de la commune de Valdeblore, au-dessus de St. Martin-Vésubie, Alpes-Maritimes, France Sud-Est.
ab. **boursini** Le Charles, 1927, Encycl. ent., (B) 3, Lepidoptera, **2**: 151, pl. 9, fig. 7.	
ssp. **pozziae** Dujardin, 1965, Entomops, Nice, no. 2: 64, fig.	Fusio (dans le Tessin), Suisse méridionale, 1300 m.
ssp. **erythroides** Przegendza, 1932, Ent. Z., **46**: 112, figs. 1-4. Reiss, 1933, in Seitz, Die Gross-Schmetterlinge der Erde, Supplement, **2**: 250. Holik, 1941, Mitt. münch. ent. Ges., **31**: 743. Reiss, 1950, Jber. naturf. Ges. Graubünden, **82**: 99, fig. 2.	Menaggio, Comer See, Italien, 200 m.
ab. **purachilleae** Vorbrodt & Müller-Rutz, 1917, Mitt. schweiz. ent. Ges., **12**: 496.	
ssp. **isarca** Verity, 1922, Ent. Rec., **34**: 32. Dannehl, 1929, Ent. Z., **43**: 14. Reiss, 1930, in Seitz, Die Gross-Schmetterlinge	Isarco Valley, South Tyrol.

der Erde, Supplement, **2**: 8. Holik, 1941, Mitt. münch. ent. Ges., **31**: 737.

ssp. **carsica** Rocci, 1926, Boll. Soc. ent. ital., **58**: 67. Reiss, 1930, in Seitz, Die Gross-Schmetterlinge der Erde, Supplement, **2**: 8. Holik, 1937, Mitt. münch. ent. Ges., **26**: 169.

Opčina, regione carsica, Istria.

ssp. **nubigena** Lederer, 1853, Verh. zool.-bot. Ver. Wien, **2**: 70, 93. Spuler, 1906, in Hofmann, Die Schmetterlinge Europas, **2**: 154, pl. 77, fig. 3a. Seitz, 1907, Die Gross-Schmetterlinge der Erde, **2**: 19, pl. 4c. Daniel, 1932, in Osthelder, Die Schmetterlinge Südbayerns, **1**: 565. Holik, 1941, Mitt. münch. ent. Ges., **31**: 729. Reiss, 1950, Jber. naturf. Ges. Graubünden, **82**: 101, fig. 1.

Oberhalb des Pasterz-Gletschers, Grossglockner, Österreich.

ab. **parvimaculata** Vorbrodt, 1913, in Vorbrodt & Müller-Rutz, Die Schmetterlinge der Schweiz, **2**: 251, fig. 33.

ab. **plutonia** Holik, 1941, Mitt. münch. ent. Ges., **31**: 730.

ab. **mediointerrupta** Vorbrodt, 1913, in Vorbrodt & Müller-Rutz, Die Schmetterlinge der Schweiz, **2**: 252, fig. 34.

ab. **immaculata** Hellweger, 1914, Die Gross-Schmetterlinge Nordtirols, p. 310. Holik, 1941, Mitt. münch. ent. Ges., **31**: 732.

ab. **costalielongata** Vorbrodt, 1914, in Vorbrodt & Müller-Rutz, Die Schmetterlinge der Schweiz, Nachtrag, **2**: 647.

ab. **omniconfluens** Vorbrodt, 1913, in Vorbrodt & Müller-Rutz, Die Schmetterlinge der Schweiz, **2**: 251.

ab. **flava** Vorbrodt, 1913, in Vorbrodt & Müller-Rutz, Die Schmetterlinge der Schweiz, **2**: 251.

grossmanni Burgeff, 1914, Mitt. münch. ent. Ges., **5**: 43.

citrina Holik, 1941, Mitt. münch. ent. Ges., **31**: 729.

ssp. **carnica** Verity, 1930, Mem. Soc. ent. ital., **9**: 12. Reiss, 1933, in Seitz, Die Gross-Schmetterlinge der Erde, Supplement, **2**: 250. Holik, 1941, Mitt. münch. ent. Ges., **31**: 741.

Sappada, Alpi Carniche, Italia, 1300 m.

ssp. **nubigenella** Koch, 1942, Iris, **56**: 92.

Pinzolo; Val Genova; Val Nambino, Adamello Alpen.

ssp. **purpurella** Reiss, 1953, Z. wien. ent. Ges., **38**: 263, pl. 18, figs. A1, B1, C1.

Col Pralongia, Ladinia, Dolomiten, 1900-2100 m.

ab. **latemarginata** Reiss, 1953, Z. wien. ent. Ges., **38**: 264, pl. 18, fig. C2.

ab. **plutonia** Reiss, 1953, Z. wien. ent. Ges., **38**: 264, pl. 18, figs. A2, B2.

ab. **rubrotecta** Reiss, 1953, Z. wien. ent. Ges., **38**: 264, pl. 18, fig. A3.

ssp. **rhaetomontana** Holik, 1941, Mitt. münch. ent. Ges., **31**: 733.

Bei Vent, Ötztaler Alpen, 2000 m.

ssp. **labacensis** Holik, 1941, Mitt. münch. ent. Ges., **31**: 731.
ab. **plutonia** Holik, 1941, Mitt. münch. ent. Ges., **31**: 732.
ab. **rubrotecta** Holik, 1941, Mitt. münch. ent. Ges., **31**: 732.

Bei Laibach, Krain, Österreich.

ssp. **jurae** Verity, 1922, Ent. Rec., **34**: 34. Reiss, 1930, in Seitz, Die Gross-Schmetterlinge der Erde, Supplement, **2**: 8.
ab. **rubrofimbriata** Verity, 1922, Ent. Rec., **34**: 35.
ab. **plutoides** Reiss, 1922, Int. ent. Z., **16**: 67.
ab. **grisescens** Burgeff, 1906, Ent. Z., **20**: 154.

Dombresson, Swiss Jura, 1000 m.

ssp. **subalpicola** Reiss, 1940, Stettin. ent. Ztg., **101**: 20, pl. 3, fig. 6a. Holik, 1941, Mitt. münch. ent. Ges., **31**: 734.
ab. **latemarginata** Reiss, 1940, Stettin. ent. Ztg., **101**: 20.

Kochel, Bayerische Alpen, 600-800 m.

ssp. **bezauensis** Reiss, 1940, Stettin. ent. Ztg., **101**: 20, pl. 3, figs. 6b-6e; 1950, Jber. naturf. Ges. Graubünden, **82**: 101.
ab. **latemarginata** Reiss, 1940, Stettin. ent. Ztg., **101**: 21.

Klausberg bei Bezau, Bregenzerwald, Österreich, 600-1200 m.

ssp. **zermattensis** Holik, 1941, Mitt. münch. ent. Ges., **31**: 745.
ab. **rubrianata** Holik, 1941, Mitt. münch. ent. Ges., **31**: 746.

Zermatt, Schweiz.

ssp. **lautareti** Holik, 1941, Mitt. münch. ent. Ges., **31**: 752.

Col de Lautaret, Dauphiné Alpen, Frankreich, 2000 m.

ssp. **scabiosae** Scheven, 1777, Der Naturforscher, Halle, **10**: 97. Schäffer, 1766, Icones Insectorum circa Ratisbonam indigenorum, **1**, pl. 16, figs. 4, 5. Reiss, 1933, in Seitz, Die Gross-Schmetterlinge der Erde, Supplement, **2**: 250. Holik, 1935, Int. ent. Z., **29**: 61. Reiss, 1935, Int. ent. Z., **29**: 169. Holik, 1935, Int. ent. Z., **29**: 195. Reiss, 1940, Stettin. ent. Ztg., **101**: 19, pl. 3, figs. 4e, 4f, 5a. Kuserau, 1942, Ent. Z., **56**: 54. Koch, 1942, Z. wien. EntVer., **27**: 40. Haaf, 1952, Veröff. zool. Staatssamml. Münch., **2**: 151, 156, pl. 4 (as *purpuralis* Brünnich). Reiss, 1955, Z. wien. ent. Ges., **40**: 291, pl. 29. Tremewan & Reiss, 1964, Ent. Rec., **76**: 1.

Bei Regensburg, Bayern (Fränkischer Jura; Schwäbische Alb).

serpylli Borkhausen, 1789, Naturgeschichte der Europäischen Schmetterlinge nach systematischer Ordnung, **2**: 163; 1793, Rheinisches Magazin, **1**: 640.
reissoides Koch, 1942, Z. wien. EntVer., **27**: 44.
ab. **cingulata** Reiss, 1964, Coridon, (A) **6**: 3.
ab. **rubrianata** Reiss, 1964, Coridon, (A) **6**: 3.
ab. **apicefusca** Reiss, 1940, Stettin. ent. Ztg., **101**: 20.
ab. **latemarginata** Reiss, 1949, Entomon, **1**: 170.
ab. **plutoides** Reiss, 1964, Coridon, (A) **6**: 3.
ab. **interrupta** Reiss, 1964, Coridon, (A) **6**: 3.
ab. **sexmaculata** Reiss, 1964, Coridon, (A) **6**: 3.

ab. **quinquemaculata** Reiss, 1964, Coridon, (A) **6**: 3.
ab. **paupera** Reiss, 1964, Coridon, (A) **6**: 3.
ab. **confluens** Burgeff, 1906, Ent. Z., **20**: 153.
ab. **omniconfluens** Reiss, 1964, Coridon, (A) **6**: 3.
ab. **flava** Reiss, 1964, Coridon, (A) **6**: 2.
ab. **flavolatemarginata** Reiss, 1949, Entomon, **1**: 170; 1930, in Seitz, Die Gross-Schmetterlinge der Erde, Supplement, **2**: 7, pl. 1f (as *grossmanni* Rühl).
ab. **flavointerrupta** Reiss, 1949, Entomon, **1**: 170.
ab. **flavoapicefusca** Reiss, 1964, Coridon, (A) **6**: 2.
ab. **flavoplutoides** Reiss, 1964, Coridon, (A) **6**: 2.
ab. **flavofimbriata** Reiss, 1964, Coridon, (A) **6**: 2.
ab. **brunnea** Reiss, 1964, Coridon, (A) **6**: 2.
ab. **nigra** Reiss, 1929, Int. ent. Z., **22**: 356.

ssp. **pythia** Fabricius, 1777, Genera Insectorum, p. 275. Reiss, 1937, in Schneider, Jh. Ver. vaterl. Naturk. Württemb., **93**: 124; 1940, Stettin. ent. Ztg., **101**: 18; 1949, Entomon, **1**: 170. Heuser & Jöst, 1959, Mitt. Pollichia, (3) **6**: 123. Zimsen, 1964, The Type Material of I. C. Fabricius, p. 531. — Mitteldeutschland [Willrodaer Forst bei Erfurt, Thüringen].

pilosellae Esper, 1780, Die Schmetterlinge, **2**, pl. 24, figs. 2a, b; 1781, ibidem, **2**: 186.

pasiphae Meigen, 1830, Systematische Beschreibung der Europäischen Schmetterlinge, **2**: 78, pl. 57, fig. 11.

ab. **semicincta** Lambillion, 1909, Rev. Soc. ent. namur., **9**: 67.
ab. **dilatata** Burgeff, 1906, Ent. Z., **20**: 153.
ab. **apicefusca** Reiss, 1940, Stettin. ent. Ztg., **101**: 20.
ab. **latemarginata** Reiss, 1940, Stettin. ent. Ztg., **101**: 19.
ab. **semiinterrupta** Lambillion, 1909, Rev. Soc. ent. namur., **9**: 67.
ab. **segregata** Spuler, 1906, in Hofmann, Die Schmetterlinge Europas, **2**: 154.
ab. **marginata** Burgeff, 1906, Ent. Z., **20**: 153.

ssp. **minos** Denis & Schiffermüller, 1775, Ankündigung eines systematischen Werkes von den Schmetterlingen der Wienergegend, p. 45; 1776, Systematischen Verzeichniss der Schmetterlinge der Wienergegend, p. 45. Fabricius, 1781, Species Insectorum, **2**: 158. Schrank, 1785, in Fuessly, Neues Magazin für die Liebhaber der Entomologie, **2**: 208. Hübner, 1790, Beiträge zur Geschichte der Schmetterlinge, **2**: 20, pl. 3, fig. O. Denis & Schiffermüller, 1801, Systematisches Verzeichniss von den Schmetterlingen der Wiener Gegend, **1**: 35. Ochsenheimer, 1808, Die Schmetterlinge von Europa, **2**: 25. Werneburg, 1864, Beiträge zur Schmetterlingskunde, **1**: 398, 399, 500. Dujardin, 1953, Bull. mens. Soc. linn. Lyon, **22**: 246. Bernardi & Viette, 1960, Bull. mens. Soc. linn. Lyon, **29**: 245. — Wien, Österreich.

ab. **plutonia** Verity, 1922, Ent. Rec., **34**: 34.

ab. **brunnea** Dziurzyński, 1918, Z. öst. EntVer., **3**: 20.
ab. **alba** Sterzl, 1921, Z. öst. EntVer., **6**: 59.
ssp. **pluto** Ochsenheimer, 1808, Die Schmetterlinge von Europa, **2**: 26. Boisduval, 1828, Essai sur une Monographie des Zygénides, p. 31, pl. 2, fig. 4. Holik, 1937, Mitt. münch. ent. Ges., **26**: 170. Forster & Wohlfahrt, 1958, Die Schmetterlinge Mitteleuropas, **3**: 88, pl. 9, figs. 33, 38. Reiprich, 1960, Motýle Slovenska, p. 299, pl. 56, fig. A2.
Budapest, Ungarn.
pythia Hübner, [July 1803]-[15th November 1806], Sammlung europäischer Schmetterlinge, **2**, pl. 18, fig. 88; 1806, ibidem, Der Ziefer, p. 78; 1819, Verzeichniss bekannter Schmettlinge [sic], p. 119 (preoccupied by **pythia** Fabricius, 1777, ssp. of **purpuralis** Brünnich).
ssp. **fatrensis** Reiss, 1940, Stettin. ent. Ztg., **101**: 19, pl. 3, figs. 5c-5e. Holik, 1941, Mitt. münch. ent. Ges., **31**: 764; 1942, Ent. Z., **55**: 237-238.
Arvaer Magura, Fatra, Mähren.
ab. **cingulata** Burgeff, 1906, Ent. Z., **20**: 153.
ab. **rubrianata** Burgeff, 1906, Ent. Z., **20**: 153.
ab. **latemarginata** Reiss, 1940, Stettin. ent. Ztg., **101**: 19, pl. 3, fig. 5f.
latomarginata Holik, 1941, Mitt. münch. ent. Ges., **31**: 765.
ab. **mediointerrupta** Holik, 1941, Mitt. münch. ent. Ges., **31**: 766.
ab. **quinquemaculata** Burgeff, 1906, Ent. Z., **20**: 153.
quinquemaculata Burgeff, 1914, Mitt. münch. ent. Ges., **5**: 42.
ab. **sexmaculata** Burgeff, 1906, Ent. Z., **20**: 153.
ab. **grossmanni** Rühl, 1898, Ent. Z., **12**: 117.
ssp. **masovica** Holik, 1939, Ann. Mus. zool. Polon., **12**: 15, pl. 1, figs. 3, 4.
Tannenberg, Masuren, Ostpreussen.
ab. **rubrianata** Holik, 1939, Ann. Mus. zool. Polon., **12**: 15.
ssp. **cracoviensis** Holik, 1932, Iris, **46**: 111, pl. 1, figs. 1-4. Reiss, 1933, in Seitz, Die Gross-Schmetterlinge der Erde, Supplement, **2**: 250. Holik, 1939, Ann. Mus. zool. Polon., **12**: 17, pl. 1, figs. 17-22.
Kraków, Westgalizien, Polen.
ab. **reducta** Dąbrowski, 1965, Acta zool. cracov., **10**: 151, fig. 136i; pl. 9, fig. 5 (transitional) (p. 93 as *reducens*).
ab. **bicolor** Dąbrowski, 1965, Acta zool. cracov., **10**: 153, pl. 11, fig. 23.
ab. **carnifera** Ziegler, 1911, Int. ent. Z., **5**: 139.
ab. **hyperpunctata** Schnaider, 1950, Bull. ent. Pologne, **19**: 243.
ssp. **tomaszowiensis** Holik, 1939, Ann. Mus. zool. Polon., **12**: 20, pl. 1, figs. 12-16.
Tomaszów, Lublin, Polen.

ssp. **subcarpathica** Holik, 1941, Mitt. münch. ent. Ges., **31**: 767. Reiss, 1943, Ent. Z., **57**: 136.
Luchatal (Chlucha dolina), Galgoczer Gebirge (Veterné holé).

ab. **mediointerrupta** Holik, 1941, Mitt. münch. ent. Ges., **31**: 767.
ab. **plutonia** Holik, 1941, Mitt. münch. ent. Ges., **31**: 767.
ab. **grisescens** Holik, 1941, Mitt. münch. ent. Ges., **31**: 767.
ssp. **strecsnoensis** Holik, 1941, Mitt. münch. ent. Ges., **31**: 763. Strecsno
ab. **pauperata** Holik, 1941, Mitt. münch. ent. Ges., **31**: 764. Ovár, Karpathen (Kleine Fatra).

ssp. **kijevana** Przegendza, 1932, Ent. Z., **46**: 112, figs. 5-8. Holik & Reiss, 1932, in Holik, Iris, **46**: 113, pl. 1, figs. 7-11. Reiss, 1933, in Seitz, Die Gross-Schmetterlinge der Erde, Supplement, **2**: 250. Holik & Sheljuzhko, 1953, Mitt. münch. ent. Ges., **43**: 159. Tschary, Teterew, Kijev, Ukraine.
ab. **cingulata** Holik & Reiss, 1932, in Holik, Iris, **46**: 113.
semicingulata Sheljuzhko, 1941, Acta Mus. zool. Kijev, **1**: 58, 95, 101. Holik & Sheljuzhko, 1953, Mitt. münch. ent. Ges., **43**: 160.
ab. **rubrotecta** Holik & Reiss, 1932, in Holik, Iris, **46**: 113.
ssp. **reissi** Burgeff, 1926, Mitt. münch. ent. Ges., **16**: 13. Reiss, 1921, Int. ent. Z., **15**: 118 (as *heringi* Zeller). Reiss & Tremewan, 1964, Ent. Rec., **76**: 130. Osterode, Ostpreussen.
reissiana Burgeff, 1926, in Strand, Lepid. Cat., **33**: 8. Reiss, 1930, in Seitz, Die Gross-Schmetterlinge der Erde, Supplement, **2**: 8, pl. 1f. Holik, 1939, Ann. Mus. zool. Polon., **12**: 14, pl. 1, figs. 1, 2. Reiss, 1940, Stettin. ent. Ztg., **101**: 14, pl. 3, figs. 1a-1f. Reiss & Tremewan, 1964, Ent. Rec., **76**: 130.
ssp. **guhni** Reiss, 1940, Stettin. ent. Ztg., **101**: 15 (3, 13), figs 1, 2, C1, pl. 1, figs. A1, B1, D1, pl. 3, figs. 3a, 3c, 3d; 1941, Märk. Tierw., **4**: 283, 284, figs. 1, 2. Bei Spandau (Berlin), Norddeutschland.
ab. **mediointerrupta** Reiss, 1940, Stettin. ent. Ztg., **101**: 15.
ab. **sexmaculata** Reiss, 1941, Mitt. münch. ent. Ges., **31**: 987.
ab. **plutonia** Reiss, 1940, Stettin. ent. Ztg., **101**: 15.
ab. **apicefusca** Reiss, 1940, Stettin. ent. Ztg., **101**: 15, pl. 3, fig. 3b.
ab. **omniconfluens** Reiss, 1940, Stettin. ent. Ztg., **101**: 15.
ab. **aurantiaca** Guhn, 1932, Ent. Jb., **41**: 90.
ab. **brunescens** Guhn, 1932, Ent. Jb., **41**: 90.
ssp. **neumanni** Reiss, 1940, Stettin. ent. Ztg., **101**: 16, pl. 3, figs. 3e, 3f, 4a, 4b. Strausberg bei Berlin, Norddeutschland.
ab. **plutonia** Reiss, 1940, Stettin. ent. Ztg., **101**: 16.
ab. **omniconfluens** Reiss, 1940, Stettin. ent. Ztg., **101**: 16.
ssp. **heringi** Zeller, 1844, Stettin. ent. Ztg., **5**: 42. Reiss, 1940, Stettin. ent. Ztg., **101**: 16, pl. 3, figs. 2a-2f. Tremewan, 1958, Ent. Gaz., **9**: 184; 1961, Bull. Brit. Mus. (nat. Hist.) Ent., **10** (7): 252, pl. 50, fig. 32. Stettin, Pommern, Norddeutschland.
ssp. **purpuralis** Brünnich, 1763, in Pontoppidan, Den Danske Atlas, **1**: 686, pl. 30; 1761, Prodromus Insectologiae siaelland- Insel Seeland, Dänemark [Adserbo].

icae, p. 29. Reiss, 1933, Int. ent. Z., **26**: 505, pl. 9, fig.; 1933, in Seitz, Die Gross-Schmetterlinge der Erde, Supplement, **2**: 249. Hoffmeyer & Knudsen, 1941, Flora og Fauna, **47**: 7, figs. 2, 5, 7, 10, 11. Koch, 1942, Z. wien. EntVer., **27**: 40. Urbahn, 1942, Ent. Z., **56**: 112. Reiss, 1943, Ent. Z., **57**: 134. Hoffmeyer, 1948, De Danske Spindere, p. 151, pl. 16, fig. 5, pl. 23, fig. 3, Povolný, 1951, Acta Acad. sci. nat. Morav., **23**: 387, pls. 1-4, figs.; 1956, Z. wien. ent. Ges. **41**: 225-231. Svensson, 1957, Opusc. Ent., **22**: 156, figs. 14, 17, 18, pl. 2, fig. 16. Reiss, 1958, Ent. Z., **68**: 144. Alberti, 1958, Mitt. zool. Mus. Berl., **34**: 348. Hoffmeyer, 1958, Ent. Gaz., **9**: 197, pl. 10, figs. Forster & Wohlfahrt, 1958, Die Schmetterlinge Mitteleuropas, **3**: 88, pl. 9, fig. 31. Hering, 1959, Ent. Z., **69**: 10. Hoffmeyer, 1960, De Danske Spindere, ed. 2, p. 205, pl. 16, fig. 5, pl. 24, fig. 3.

ab. **latemarginata** Reiss, 1933, in Seitz, Die Gross-Schmetterlinge der Erde, Supplement, **2**: 249.

ssp. **segontii** Tremewan, 1958, Ent. Gaz., **9**: 188. South, 1908, The Moths of the British Isles, **2**: 334, pl. 146, figs. 1, 2. Tremewan, 1960, Ent. Gaz., **11**: 187; 1961, Bull. Brit. Mus. (nat. Hist.) Ent., **10** (7): 251, pl. 50, fig. 29; 1961, Coridon, (A) **1**: 2, pl. C1, fig. 1. — Abersoch, Caernarvonshire, North Wales.

ab. **separata** Tutt, 1899, A Natural History of the British Lepidoptera, **1**: 434. Tremewan, 1961, Bull. Brit. Mus. (nat. Hist.) Ent., **10** (7): 251.

ab. **obscura** Tutt, 1899, A Natural History of the British Lepidoptera, **1**: 434. Alberti, 1955, Ent. Z., **65**: 89-91. Tremewan, 1961, Bull. Brit. Mus. (nat. Hist.) Ent., **10** (7): 251, pl. 50, fig. 30; 1961, Coridon, (A) **1**: 2, pl. C1, fig. 2.

ssp. **caledonensis** Reiss, 1931, Int. ent. Z., **25**: 341; 1931, ibidem, **25**: 359, figs.; 1933, in Seitz, Die Gross-Schmetterlinge der Erde, Supplement, **2**: 249. Holik, 1941, Mitt. münch. ent. Ges., **31**: 760. Tremewan, 1958, Ent. Gaz., **9**: 187; 1960, ibidem, **11**: 187; 1961, Coridon, (A) **1**: 2, pl. C1, fig. 5. Tremewan & Manley, 1964, Ent. Rec., **76**: 149-153. Richardson, 1965, Ent. Rec., **77**: 17. Tremewan, 1965, Ent. Gaz., **16**: 119-124. — Oban, Argyllshire, West Schottland.

ssp. **hibernica** Reiss, 1933, in Seitz, Die Gross-Schmetterlinge der Erde, Supplement, **2**: 249. Birchall, 1866, Ent. mon. Mag., **3**: 33 (as *nubigena* Lederer [partim]). Holik, 1941, Mitt. münch. ent. Ges., **31**: 760. Tremewan, 1958, Ent. Gaz., **9**: 187; 1960, ibidem, **11**: 186; 1961, Coridon, (A) **1**: 2, pl. C1, fig. 3. South, 1961, The Moths of the British Isles, **2**: 328, pl. 129, figs. 1, 2. Baynes, 1964, A Revised Catalogue of Irish Macrolepidoptera (Butterflies and Moths), p. 88. — Ardrahan, Co. Galway, Irland.

ab. **lutescens** Tutt, 1899, A Natural History of the British

Lepidoptera, **1**: 434. Tremewan, 1961, Bull. Brit. Mus. (nat. Hist.) Ent., **10** (7): 251.

f. loc. **sabulosa** Tremewan, 1960, Ent. Gaz., **11**: 186. Newman, 1861, Zoologist, **19**: 7677 (as *nubigena* Lederer). Birchall, 1866, Ent. mon. Mag., **3**: 33 (as *nubigena* Lederer [partim]). Tremewan, 1961, Bull. Brit. Mus. (nat. Hist.) Ent., **10** (7): 252, pl. 50, fig. 31; 1961, Coridon, (A) **1**: 2, pl. C1, fig. 4. Baynes, 1964, A Revised Catalogue of Irish Macrolepidoptera (Butterflies and Moths), p. 88. — Ballyvaughan, Co. Clare, Ireland.

Biology

Alberti, 1957, Dtsch. ent. Z., (N.F.), **4**: 1-7; 1957, NachrBl. bayer. Ent., **6**: 49-54; 1958, Ent. Z., **68**: 4-8. Barrett, 1895, The Lepidoptera of the British Islands, **2**: 117, pl. 58, fig. 4b. Boisduval, Rambur & Graslin, 1832, Collection iconographique et historique des Chenilles, pl. 3, figs. 6, 7. Buckler, 1886, The Larvae of the British Butterflies and Moths, **2**: 9, pl. 18, figs. 4, 4a. Burgeff, 1912, Z. wiss. InsektBiol., **8**: 123; 1950, Portug. acta biol., (A) Goldschmidt: 663-728; 1951, Biol. Zbl., **70**: 1-23. Dorfmeister, 1854, Verh. zool.-bot. Ver. Wien, **4**: 477. Döring, 1955, Zur Morphologie der Schmetterlingseier, p. 120, pl. 17, fig. 252. Esper, 1793, Die Schmetterlinge, Supplement, **2** (2): 14, pl. 40, figs. 3-6. Freyer, 1833, Neuere Beiträge zur Schmetterlingskunde, **1**: 156, pl. 86, fig. 1. Gregor & Povolný, 1955, Sborn. ent. Odd. nár. Mus. Praze, **30**: 271. Holik, 1937, Lambillionea, **37**: 15-24, 32-45, 80-91; 1938, Ent. Rdsch., **55**: 320-323, 331-333; 1938, ibidem, **55**: 349-354, 382-384; 1943, Ent. Z., **57**: 41-45; 1946, Rev. franç. Lépid., **10**: 250-261, 273-280; 1953, Ent. Z., **62**: 156. Hrubý, 1964, Prodromus Lepidopter Slovenska, p. 475. Hyde, 1964, Animals, **3** (6): 162-164, fig. 5. Kiefer, 1933, Int. ent. Z., **27**: 252-256; 1934, ibidem, **27**: 521-524. Koch, 1955, Wir Bestimmen Schmetterlinge, **2**: 58, 59, pl. 1, fig. 6, pl. 15, fig. 6. Millière, 1869/74, Iconographie et Description de Chenilles et Lépidoptères, **3**: 66, pl. 107, fig. 12. Ochsenheimer, 1808, Die Schmetterlinge von Europa, **2**: 22. Reiss, 1930, in Seitz, Die Gross-Schmetterlinge der Erde, Supplement, **2**: 9; 1940, Stettin. ent. Ztg., **101**: 1-3, 14-21; 1943, Ent. Z., **57**: 135; 1955, Z. wien. ent. Ges., **40**: 285, 291, pl. 29; 1958, ibidem, **43**: 155. Roüast, 1883, Catalogue des Chenilles européennes connues, p. 22. Sarlet, 1964, Mém. Soc. r. ent. Belg., **29**: 6. South, 1961, The Moths of the British Isles, **2**: 328, pl. 132, fig. 1. Spuler, 1910, in Hofmann, Die Raupen der Schmetterlinge Europas, pl. 9, fig. 19. Tremewan & Manley, 1964, Ent. Rec., **76**: 149-153. Tutt, 1899, A Natural History of the British Lepidoptera, **1**: 437.

Subg. **Agrumenia** Hübner

Agrumenia Hübner, [1819], Verzeichniss bekannter Schmettlinge [sic], p. 116. Type-species: **Sphinx onobrychis** Denis & Schiffermüller, 1775 (= **Zygaena carniolica** Scopoli, 1763), by subsequent designation, Tremewan, 1961, Ent. Rec., **73**: 202.

Lycastes Hübner, [1819], Verzeichniss bekannter Schmettlinge [sic], p. 118. Type-species: **Sphinx exulans** Reiner & Hohenwarth, 1792, by subsequent designation, Tremewan, 1961, Ent. Rec., **73**: 202.

Lictoria Burgeff, 1926, in Strand, Lepid. Cat., **33**: 20. Type-species: **Sphinx achilleae** Esper, 1781 (=**Zygaena loti** Denis & Schiffermüller, 1775), by subsequent designation, Tremewan, 1961, Ent. Rec., **73**: 202.

Agrumenoidea Holik, 1937, Ent. Z., **51**: 132. Type-species: **Zygaena johannae** Le Cerf, 1923, by original designation, Holik, 1937, loc. cit.

Coelestina Holik, 1953, Ent. Z., **63**: 15. Type-species: **Sphinx sedi** Fabricius, 1787, by original designation, Holik, 1953, loc. cit.

SECTION 1

johannae Le Cerf

Distribution: Morocco.

ssp. **johannae** Le Cerf, 1923, Bull. Soc. ent. Fr., p. 224. Reiss, 1930, in Seitz, Die Gross-Schmetterlinge der Erde, Supplement, **2**: 44, pl. 4m. Zerny, 1935, Mém. Soc. Sci. nat. Maroc, **42**: 102. Holik, 1937, Ent. Z., **51**: 131. Reiss, 1944, Z. wien. ent. Ges., **29**: 54, pl. 38, figs. 88, 89. Haaf, 1952, Veröff. zool. Staatssamml. Münch., **2**: 151, 154, 156, pl. 7. Alberti, 1958, Mitt. zool. Mus. Berl., **34**: 302. — Vallée de l'Imminen, Grand Atlas, Maroc, 2700-2870 m.

ab. **latestrigata** Schwingenschuss, 1935, in Zerny, Mém. Soc. Sci. nat. Maroc, **42**: 102.

ab. **interrupta** Schwingenschuss, 1935, in Zerny, Mém. Soc. Sci. nat. Maroc, **42**: 103.

ab. **albescens** Schwingenschuss, 1935, in Zerny, Mém. Soc. Sci. nat. Maroc, **42**: 103.

ab. **rubribasalis** Schwingenschuss, 1935, in Zerny, Mém. Soc. Sci. nat. Maroc, **42**: 103.

ab. **aurantiaca** Schwingenschuss, 1935, in Zerny, Mém. Soc. Sci. nat. Maroc, **42**: 103.

ab. **flava** Schwingenschuss, 1935, in Zerny, Mém. Soc. Sci. nat. Maroc, **42**: 103.

ssp. **turbeti** Le Cerf, 1929, Bull. Soc. ent. Fr., p. 263. Reiss, 1930, in Seitz, Die Gross-Schmetterlinge der Erde, Supplement, **2**: — Crête de l'ich Bou Naçeur

44, pl. 4m; 1944, Z. wien. ent. Ges., **29**: 55, pl. 38, figs. 90, 91. Tremewan, 1961, Bull. Brit. Mus. (nat. Hist.) Ent., **10** (7): 305, pl. 57, fig. 15. (Gabberal ou Guelb er Rahal), Moyen Atlas, Maroc, 3300-3400 m.

Biology

Holik, 1953, Ent. Z., **63**: 21. Reiss, 1958, Z. wien. ent. Ges., **43**: 158. Wiegel, 1965, Mitt. münch. ent. Ges., **55**: 126, 132-134, pl. 5, figs. 29-31. Zerny, 1935, Mém. Soc. Sci. nat. Maroc, **42**: 102.

SECTION 2

felix Oberthür

Distribution: Algeria, Tripoli, Tunisia.

ssp. **felix** Oberthür, 1876, Études d'Entomologie, **1**: 36; 1878, ibidem, **3**: 41, pl. 5, fig. 4; 1888, ibidem, **12**: 26; 1890, ibidem, **13**: 22, pl. 7, fig. 68; 1910, Études de Lépidoptérologie comparée, **4**: 611. Rothschild, 1917, Novit. zool., **24**: 340. Reiss, 1933, in Seitz, Die Gross-Schmetterlinge der Erde, Supplement, **2**: 272 (as *mauretanica* Staudinger); 1944, Z. wien. ent. Ges., **29**: 15, pl. 47, fig. XV (as *mauretanica* Staudinger). Alberti, 1958, Mitt. zool. Mus. Berl., **34**: 304 (as *mauretanica* Staudinger). Tremewan, 1961, Bull. Brit. Mus. (nat. Hist.) Ent., **10** (7): 255, pl. 51, fig. 10, pl. 59, figs. 3, 4 (Neotype designation invalid); 1962, Ent. Rec., **74**: 125, pl. 2, figs. 1, 16-18 (Lectotype ♂). Lambessa, Algérie.

eudaemon Mabille, 1885, Bull. Soc. philom. Paris, (7) **9**: 57. Bernardi & Viette, 1961, Bull. mens. Soc. linn. Lyon, **30**: 142.

andalusiae Burgeff, 1914, Mitt. münch. ent. Ges., **5**: 53. Spuler, 1906, in Hofmann, Die Schmetterlinge Europas, **2**: 163. Reiss, 1930, in Seitz, Die Gross-Schmetterlinge der Erde, Supplement, **2**: 26. Reiss & Tremewan, 1964, Ent. Rec., **76**: 130.

pseudofelix Reiss, 1933, in Seitz, Die Gross-Schmetterlinge der Erde, Supplement, **2**: 272.

ab. **mauretanica** Staudinger, 1887, Berl. ent. Z., **31**: 37, 38. Allard, 1867, Ann. Soc. ent. Fr., (4) **7**: 316 (as *faustina* Ochsenheimer). Oberthür, 1888, Études d'Entomologie, **12**: 26. Seitz, 1907, Die Gross-Schmetterlinge der Erde, **2**: 28, pl. 8a.

ab. **pseudofaustula** Reiss, 1933, in Seitz, Die Gross-Schmetterlinge der Erde, Supplement, **2**: 272.

ssp. **barraguei** Dujardin, 1964, Entomops, Nice, no. 1: 18, fig. Tikjda, Djurdjura, 1300 m.; Col de Tirourda, Djurdjura, 1700 m., Algérie.

ssp. **constantinensis** Reiss & Tremewan, 1964, Ent. Rec., **76**: 131 (nomen novum for *faustula* Reiss). — Provinz Constantine, Algerien.

faustula Reiss, 1933, in Seitz, Die Gross-Schmetterlinge der Erde, Supplement, **2**: 272 (preoccupied by *faustula* Rambur, 1866, = **genevensis** Millière, 1861, ssp. of **fausta** Linné). Reiss & Tremewan, 1964, Ent. Rec., **76**: 131.

faustula Staudinger, 1887, Berl. ent. Z., **31**: 37 (infrasubspecific). Seitz, 1907, Die Gross-Schmetterlinge der Erde, **2**: 28, pl. 8a. Rothschild, 1917, Novit. zool., **24**: 340. Reiss & Tremewan, 1964, Ent. Rec., **76**: 131.

ssp. **quercina** Burgeff, 1926, Mitt. münch. ent. Ges., **16**: 45. Reiss, 1930, in Seitz, Die Gross-Schmetterlinge der Erde, Supplement, **2**: 26, pl. 2n. Chneour, 1950, Bull. Soc. Hist. nat. Afr. N., **41**: 44. Haaf, 1952, Veröff. zool. Staatssamml. Münch., **2**: 151, 154, 156, pl. 7 (as *felix* Oberthür). — Tagiura bei Sidi Messri, Tripolitanien.

ab. **cingulata** Burgeff, 1926, Mitt. münch. ent. Ges., **16**: 45. Reiss, 1930, in Seitz, Die Gross-Schmetterlinge der Erde, Supplement, **2**: 26.

ab. **ornata** Burgeff, 1926, Mitt. münch. ent. Ges., **16**: 45. Reiss, 1930, in Seitz, Die Gross-Schmetterlinge der Erde, Supplement, **2**: 26.

ssp. **silvestrii** Romei, 1927, Boll. Lab. Zool. Portici, **20**: 280. Reiss, 1930, in Seitz, Die Gross-Schmetterlinge der Erde, Supplement, **2**: 26, pl. 2o. — Sidi Messri, Tripolitania.

ab. **confluens** Reiss, 1929, Int. ent. Z., **22**: 358; 1930, in Seitz, Die Gross-Schmetterlinge der Erde, Supplement, **2**: 26.

Biology

Burgeff, 1913, Ent. Z., **27**: 175, fig. 1, pls. 2, 5. Holik, 1953, Ent. Z., **63**: 22. Reiss, 1930, in Seitz, Die Gross-Schmetterlinge der Erde, Supplement, **2**: 26; 1958, Z. wien. ent. Ges., **43**: 158.

beatrix Przegendza

Distribution: West Algeria, Morocco.

ssp. **beatrix** Przegendza, 1932, Ent. Z., **46**: 113, figs. 9-12. Seitz, 1907, Die Gross-Schmetterlinge der Erde, **2**: 28, pl. 7k (as *felix* Oberthür). Reiss, 1933, in Seitz, Die Gross-Schmetterlinge der Erde, Supplement, **2**: 272 (as *felix* Oberthür); 1944, Z. wien. ent. Ges., **29**: 16, pl. 36, figs. 40, 41 (as *felix* Oberthür). Alberti, 1958, Mitt. zool. Mus. Berl., **34**: 304 (as *felix* Oberthür). Tremewan, 1961, Bull. Brit. Mus. (nat. Hist.) Ent., **10** (7): 255, pl. 51, fig. 10, pl. 59, figs. 3, 4 (as *felix* Oberthür); 1962, Ent. Rec., **74**: 126. — Sebdou, Provinz Oran, Algerien.

ab. **pseudomauretanica** Reiss, 1933, in Seitz, Die Gross-Schmetterlinge der Erde, Supplement, **2**: 272.

ssp. **felicina** Reiss, 1944, Z. wien. ent. Ges., **29**: 16, pl. 36, figs. 42, 45-47, pl. 47, fig. XIII.

Tizi s'Tkrine, Mittelatlas, Marokko.

ab. **flavipalpis** Reiss, 1944, Z. wien. ent. Ges., **29**: 19.
ab. **reducta** Reiss, 1944, Z. wien. ent. Ges., **29**: 19.
ab. **segregata** Reiss, 1944, Z. wien. ent. Ges., **29**: 19.
ab. **sextaseparata** Reiss, 1944, Z. wien. ent. Ges., **29**: 20.
ab. **dealbata** Reiss, 1944, Z. wien. ent. Ges., **29**: 20.
ab. **albiornata** Reiss, 1944, Z. wien. ent. Ges., **29**: 19, pl. 36, fig. 44.
ab. **laticincta** Reiss, 1944, Z. wien. ent. Ges., **29**: 19.
ab. **confluens** Reiss, 1944, Z. wien. ent. Ges., **29**: 19, pl. 36, fig. 43.

ssp. **pudiga** Reiss, 1944, Z. wien. ent. Ges., **29**: 20, pl. 38, fig. 86.

Bab Joana (B. Selman), Rif, Marokko, 2000 m.

ab. **reducta** Reiss, 1944, Z. wien. ent. Ges., **29**: 21, pl. 38, fig. 87.
ab. **sextaseparata** Reiss, 1944, Z. wien. ent. Ges., **29**: 21.
ab. **laticincta** Reiss, 1944, Z. wien. ent. Ges., **29**: 21.

SECTION 3

banghaasi Burgeff

Distribution: Buchara.

banghaasi Burgeff, 1927, in Bang-Haas, Horae Macrolepidopterologicae regionis palaearcticae, **1**: 56, pl. 9, figs. 18, 19. Reiss, 1930, in Seitz, Die Gross-Schmetterlinge der Erde, Supplement, **2**: 23, pl. 2k. Haaf, 1952, Veröff. zool. Staatssamml. Münch., **2**: 151, 153, 156, pl. 7. Holik & Sheljuzhko, 1956, Mitt. münch. ent. Ges., **46**: 181. Alberti, 1958, Mitt. zool. Mus. Berl., **34**: 296. Tremewan, 1961, Bull. Brit. Mus. (nat. Hist.) Ent., **10** (7): 306, pl. 57, fig. 16.

Dorf Dombratschi, südöstliche Karategin Berge, Buchara, 2000 m.

ab. **pseudococandica** Reiss, 1930, in Seitz, Die Gross-Schmetterlinge der Erde, Supplement, **2**: 23.

cocandica Erschoff

Distribution: Central Asia.

ssp. **cocandica** Erschoff, 1874, in Fedtshenko, Reise in Turkestan, **2**: 28, pl. 2, fig. 22. Grum-Grshimailo, 1890, in Romanoff, Mémoires sur les Lépidoptères, **4**: 525. Seitz, 1907, Die Gross-Schmetterlinge der Erde, **2**: 31, pl. 7g. Dziurzyński, 1908, Berl. ent. Z., **53**: 59, pl. 2, fig. 22. Burgeff, 1927, in Bang-Haas, Horae Macrolepidopterologicae regionis palaearcticae, **1**: 56. Reiss, 1930, in Seitz, Die Gross-Schmetterlinge der Erde, Supplement, **2**: 23, pl. 2k; 1933, ibidem, **2**: 270. Holik & Sheljuzhko, 1956, Mitt. münch. ent. Ges., **46**: 174. Alberti, 1958, Mitt. zool. Mus. Berl., **34**: 296.

Westlicher Alai am Flusse Kisil-su, Chanschaft Kokand, Fergana, 8000 ft.

ab. **latecingulata** Holik & Sheljuzhko, 1956, Mitt. münch. ent. Ges., **46**: 179.

ab. **nigra** Dziurzyński, 1908, Berl. ent. Z., **53**: 13, 59. Reiss, 1930, in Seitz, Die Gross-Schmetterlinge der Erde, Supplement, **2**: 23.

ab. **pseudoconserta** Holik & Sheljuzhko, 1956, Mitt. münch. ent. Ges., **46**: 179.

ab. **extrema** Dziurzyński, 1919, Z. öst. EntVer., **4**: 78.

ab. **rosea** Holik & Sheljuzhko, 1956, Mitt. münch. ent. Ges., **46**: 179.

ab. **fumosa** Reiss & Tremewan, 1964, Ent. Rec., **76**: 131. Burgeff, 1906, Ent. Z., **20**: 161, fig. 2; 1914, Mitt. münch. ent. Ges., **5**: 52, pl. 6, fig. 54.

ssp. **minor** Erschoff, 1874, in Fedtshenko, Reise in Turkestan, **2**: 29, pl. 2, fig. 23. Reiss, 1930, in Seitz, Die Gross-Schmetterlinge der Erde, Supplement, **2**: 23. Holik & Sheljuzhko, 1956, Mitt. münch. ent. Ges., **46**: 175. — Bei Djiptik (Dzhiptik), Kokand, Fergana.

ssp. **karategina** Grum-Grshimailo, 1890, in Romanoff, Mémoires sur les Lépidoptères, **4**: 525. Reiss, 1933, in Seitz, Die Gross-Schmetterlinge der Erde, Supplement, **2**: 270. Holik & Sheljuzhko, 1956, Mitt. münch. ent. Ges., **46**: 177. — Environs du Fort Obi-Garm, au nord de Baldzhuan, Karategin Mts., 1000-1400 m.

ssp. **pamira** Sheljuzhko, 1919, N. Beitr. syst. Insektenk., **1**: 130. Reiss, 1930, in Seitz, Die Gross-Schmetterlinge der Erde, Supplement, **2**: 23, pl. 2k. Holik & Sheljuzhko, 1956, Mitt. münch. ent. Ges., **46**: 180. — Ak-tash am Flusse Murgab; Alitshur; Kojtezek-Pass, Pamir.

avinoffi Hampson, 1920, Trans. ent. Soc. Lond., p. 433. Tremewan, 1961, Bull. Brit. Mus. (nat. Hist.) Ent., **10** (7): 253, pl. 51, fig. 3.

f. loc. **conserta** Grum-Grshimailo, 1890, in Romanoff, Mémoires sur les Lépidoptères, **4**: 525. Seitz, 1907, Die Gross-Schmetterlinge der Erde, **2**: 31. Reiss, 1930, in Seitz, Die Gross-Schmetterlinge der Erde, Supplement, **2**: 23; 1933, ibidem, **2**: 270. Holik & Sheljuzhko, 1956, Mitt. münch. ent. Ges., **46**: 178. — Steppes de Stipa, Darvaz, Pamir.

Biology

Holik, 1937, Lambillionea, **37**: 15-24, 32-45, 80-91; 1953, Ent. Z., **63**: 21.

kavrigini Grum-Grshimailo

Distribution: Buchara.

kavrigini Grum-Grshimailo, 1887, in Romanoff, Mémoires sur les Lépidoptères, **3**: 402; 1890, ibidem, **4**: 522, pl. 18, fig. 9. Seitz, 1907, Die Gross-Schmetterlinge der Erde, **2**: 28, pl. 7f, g. Reiss, 1933, in Seitz, Die Gross-Schmetterlinge der Erde, Supplement, **2**: 269. Holik & Sheljuzhko, 1956, Mitt. münch. — Im Dzhilian-Tau (Zeravshan) und bei Baldzhuan (Karategin), Buchara, 800-1200 m.

ent. Ges., **46**: 170. Alberti, 1958, Mitt. zool. Mus. Berl., **34**: 296.
rhodogastra Staudinger, 1889, Stettin. ent. Ztg., **50**: 24.

Biology

Holik, 1953, Ent. Z., **63**: 21.

shivacola Reiss & Schulte
Distribution: North Afghanistan.
shivacola Reiss & Schulte, 1962, Ent. Z., **72**: 50, 52, figs. 1-4; 1964, ibidem, **74**: 159. — Shiva Mts., Badakshan, Afghanistan, 6000-8000 ft.
ab. **latecingulata** Reiss & Schulte, 1962, Ent. Z., **72**: 50, 51.
ab. **rubrianata** Reiss & Schulte, 1962, Ent. Z., **72**: 51.
ab. **rubroabdominalis** Reiss & Schulte, 1962, Ent. Z., **72**: 52.
ab. **flavianata** Reiss & Schulte, 1962, Ent. Z., **72**: 50.
ab. **azona** Reiss & Schulte, 1962, Ent. Z., **72**: 50.
ab. **flavisignata** Reiss & Schulte, 1962, Ent. Z., **72**: 51, 52.
ab. **medioseparata** Reiss & Schulte, 1962, Ent. Z., **72**: 51.
ab. **sextadealbata** Reiss & Schulte, 1962, Ent. Z., **72**: 51.
ab. **sextaseparata** Reiss & Schulte, 1962, Ent. Z., **72**: 51.
ab. **apicaliconfluens** Reiss & Schulte, 1962, Ent. Z., **72**: 51.

ferganae Sheljuzhko
Distribution: Fergana.
ferganae Sheljuzhko, 1941, Z. wien. EntVer., **26**: 6, figs. 1, 2. Holik & Sheljuzhko, 1956, Mitt. münch. ent. Ges., **46**: 143. Alberti, 1958, Mitt. zool. Mus. Berl., **34**: 294. — Pag. Taulj, Besh-Aryk, Katta-otuz-ashbar, Distrikt Kokand, Fergana.
ab. (?hybr.) **pseudotruchmena** Sheljuzhko, 1941, Z. wien. EntVer., **26**: 8, fig. 3.

Biology

Holik, 1953, Ent. Z., **63**: 21.

magnifica Reiss
Distribution: Central Asia (eastern Buchara).
magnifica Reiss, 1935, Int. ent. Z., **29**: 41; 1941, Mitt. münch. ent. Ges., **31**, pl. 34, figs. B8, C8. Holik & Sheljuzhko, 1956, Mitt. münch. ent. Ges., **46**: 140. Alberti, 1958, Mitt. zool. Mus. Berl., **34**: 293. — Kurgan Tjube, Buchara (nahe der afghanischen Grenze).
ab. **rubrimacula** Reiss, 1935, Int. ent. Z., **29**: 42; 1941, Z. wien. EntVer., **26**: 63.
ab. **trimacula** Reiss, 1941, Mitt. münch. ent. Ges., **31**: 1000, pl. 34, fig. A9.

truchmena Eversmann
Distribution: Transcaspia, Central Asia.
ssp. **truchmena** Eversmann, 1854, Bull. Soc. Nat. Moscou, **27** (3): 184. Herrich-Schäffer, 1861, Neuere Schmetterlinge aus — Syr-Darja Gebiet,

Europa und den angrenzenden Ländern, p. 2, figs. 7, 8. Erschoff, 1874, in Fedtshenko, Reise in Turkestan, **2**: 28. Seitz, 1907, Die Gross-Schmetterlinge der Erde, **2**: 28, pl. 7g. Dziurzyński, 1908, Berl. ent. Z., **53**: 51, pl. 2, fig. 18. Reiss, 1930, in Seitz, Die Gross-Schmetterlinge der Erde, Supplement, **2**: 22; 1933, ibidem, **2**: 266. Holik & Sheljuzhko, 1956, Mitt. münch. ent. Ges., **46**: 138. Alberti, 1958, Mitt. zool. Mus. Berl., **34**: 293. — Transkaspien (südliche Kirgisensteppen irrtümlich).

carbuncula Burgeff, 1926, Mitt. münch. ent. Ges., **16**: 41. Reiss, 1930, in Seitz, Die Gross-Schmetterlinge der Erde, Supplement, **2**: 22. Holik & Sheljuzhko, 1956, Mitt. münch. ent. Ges., **46**: 139.

ssp. **ferganica** Holik & Sheljuzhko, 1956, Mitt. münch. ent. Ges., **46**: 142. Reiss, 1930, in Seitz, Die Gross-Schmetterlinge der Erde, Supplement, **2**: 22, pl. 2i (as *truchmena* Eversmann). Haaf, 1952, Veröff. zool. Staatssamml. Münch., **2**: 151, 153, 156, pl. 7, pl. 16 (as *truchmena* Eversmann). — Bei Margelan, Kokand, Ferganabecken.

ab. **rubriventer** Sheljuzhko, 1941, Z. wien. EntVer., **26**: 9.

ab. **omniconfluens** Sheljuzhko, 1941, Z. wien. EntVer., **26**: 9, fig. 4.

Biology

Holik, 1937, Lambillionea, **37**: 15-24, 32-45, 80-91; 1953, Ent. Z., **63**: 21.

erschoffi Staudinger

Distribution: Central Asia.

ssp. **erschoffi** Staudinger, 1887, Stettin. ent. Ztg., **48**: 76. Grum-Grshimailo, 1890, in Romanoff, Mémoires sur les Lépidoptères, **4**: 523. Reiss, 1930, in Seitz, Die Gross-Schmetterlinge der Erde, Supplement, **2**: 22; 1933, ibidem, **2**: 265, pl. 16l, m; 1933, Int. ent. Z., **26**: 505, figs.; 1933, Ent. Rdsch., **50**: 194, pl. 1, figs., pl. 2, fig. Holik & Sheljuzhko, 1956, Mitt. münch. ent. Ges., **46**: 164. Alberti, 1958, Mitt. zool. Mus. Berl., **34**: 295. — Margelan; Osh; Usgent; Fergana.

ssp. **sovinskiji** Holik & Sheljuzhko, 1956, Mitt. münch. ent. Ges., **46**: 166. — Tal des Padsha-ata (Tuzte), Bezirk Namangan, Fergana.

ssp. **kohistana** Grum-Grshimailo, 1893, Horae Soc. ent. Ross., **27**: 385; 1894, ibidem, **28**: 94. Seitz, 1907, Die Gross-Schmetterlinge der Erde, **2**: 28, pl. 7f (as *erschoffi* Staudinger). Reiss, 1933, Ent. Rdsch., **50**: 154, pl. 2, figs.; 1933, in Seitz, Die Gross-Schmetterlinge der Erde, Supplement, **2**: 265 (as *sogdiana* Erschoff var.). Holik & Sheljuzhko, 1956, Mitt. münch. ent. Ges., **46**: 163. — Shach-sara am Flusse Jagnob, nördliche Ausläufer des Hissargebirges, Turkestan.

ssp. **tashkentensis** Reiss, 1932, Int. ent. Z., **26**: 125; 1932, ibidem, **26**: 230, fig. (after Erschoff). Erschoff, 1874, in Fedtshenko, Reise in Turkestan, **2**: 28, pl. 2, fig. 21 (as *olivieri* Boisduval var.). Reiss, 1933, in Seitz, Die Gross-Schmetterlinge der Erde, Supplement, **2**: 266. Holik & Sheljuzhko, 1956, Mitt. münch. ent. Ges., **46**: 167. — Bei Tashkent und im Zaravshantal zwischen Jori und Dashty-Kazy, Turkestan. [Karzhan-tau, nordöstlich von Tashkent].

Biology

Holik, 1953, Ent. Z., **63**: 21.

merzbacheri Reiss

Distribution: Naryn region, Hindukush.

ssp. **merzbacheri** Reiss, 1933, Ent. Rdsch., **50**: 162, pl. 1, figs., pl. 2, figs.; 1933, in Seitz, Die Gross-Schmetterlinge der Erde, Supplement, **2**: 266, pl. 16m. Haaf, 1952, Veröff. zool. Staatssamml. Münch., **2**: 151, 153, 156, pl. 6 (as *fraxini* Ménétriés), pl. 16. Holik & Sheljuzhko, 1956, Mitt. münch. ent. Ges., **46**: 168. Alberti, 1958, Mitt. zool. Mus. Berl., **34**: 296. — Naryn Gebiet, Tian-Shan, 2200 m. (Semiretshje).

ab. **dealbata** Reiss, 1933, Ent. Rdsch., **50**: 170; 1933, in Seitz, Die Gross-Schmetterlinge der Erde, Supplement, **2**: 266.

ab. **confluens** Sheljuzhko, 1910, Rev. russe Ent., **9**: 385. Reiss, 1930, in Seitz, Die Gross-Schmetterlinge der Erde, Supplement, **2**: 22 (as *fraxini* Ménétriés ab.); 1933, ibidem, **2**: 266; 1933, Ent. Rdsch., **50**: 171. Holik & Sheljuzhko, 1956, Mitt. münch. ent. Ges., **46**: 170.

rubescens Burgeff, 1914, Mitt. münch. ent. Ges., **5**: 51, pl. 6, fig. 47. Reiss, 1930, in Seitz, Die Gross-Schmetterlinge der Erde, Supplement, **2**: 22 (as *scovitzii* Ménétriés ab.); 1933, Ent. Rdsch., **50**: 171. Holik & Sheljuzhko, 1956, Mitt. münch. ent. Ges., **46**: 170.

ssp. **scheibei** Kardakoff, 1937, Arb. morph. taxon. Ent. Berl., **4**: 191. Holik & Sheljuzhko, 1956, Mitt. münch. ent. Ges., **46**: 169. Alberti, 1960, Ent. Z., **70**: 174. — Pushki, Parun-Tal, Hindukush.

Biology

Holik, 1953, Ent. Z., **63**: 21.

afghana Moore

Distribution: Afghanistan.

afghana Moore, 1858/59, in Horsfield & Moore, A Catalogue of the Lepidopterous Insects in the Museum of Natural History at the East-India House, **2**: 286, pl. 7a, fig. 1. Hampson, [1892] 1893, Fauna of British India, Moths, **1**: 231. Seitz, 1907, Die Gross-Schmetterlinge der Erde, **2**: 31, pl. 8f. — Afghanistan [Kabul]. [Quetta, British Baluchistan].

Holik & Sheljuzhko, 1956, Mitt. münch. ent. Ges., **46**: 192. Alberti, 1958, Mitt. zool. Mus. Berl., **34**: 297. Reiss & Tremewan, 1960, Bull. Brit. Mus. (nat. Hist.) Ent., **9** (10): 461, pl. 22, figs. 7, 8, pl. 23, figs. 1, 2, pl. 24, figs. 12-14. Tremewan, 1961, Bull. Brit. Mus. (nat. Hist.) Ent., **10** (7): 253, pl. 51, fig. 4, pl. 64, figs. 5, 6.

mangeri Burgeff

Distribution: Afghanistan.

ssp. **mangeri** Burgeff, 1927, in Bang-Haas, Horae Macrolepidopterologicae regionis palaearcticae, **1**: 55, pl. 9, fig. 17. Reiss, 1930, in Seitz, Die Gross-Schmetterlinge der Erde, Supplement, **2**: 22, pl. 2i. Holik & Sheljuzhko, 1956, Mitt. münch. ent. Ges., **46**: 191. Alberti, 1958, Mitt. zool. Mus. Berl., **34**: 297. Reiss & Tremewan, 1960, Bull. Brit. Mus. (nat. Hist.) Ent., **9** (10): 461, pl. 22, figs. 5, 6, pl. 23, figs. 5, 6, pl. 24, figs. 15, 16. Reiss & Schulte, 1964, Ent. Z., **74**: 157. — Westlich von Kabul, Paghman mont., Afghanistan.

ab. **laticincta** Burgeff, 1927, in Bang-Haas, Horae Macrolepidopterologicae regionis palaearcticae, **1**: 55. Reiss, 1930, in Seitz, Die Gross-Schmetterlinge der Erde, Supplement, **2**: 22. Tremewan, 1961, Bull. Brit. Mus. (nat. Hist.) Ent., **10** (7): 306, pl. 57, fig. 17 (as *mangeri* Burgeff).

ssp. **panjaoica** Reiss & Schulte, 1964, Ent. Z., **74**: 158, 161, figs. 9, 10. — Ko-i-Baba Mts., Panjao, Afghanistan, 2700-3000 m.

Biology

Holik, 1953, Ent. Z., **63**: 21.

rothschildi Reiss

Distribution: Hissar Mountains.

rothschildi Reiss, 1930, in Seitz, Die Gross-Schmetterlinge der Erde, Supplement, **2**: 22, pl. 2i. Holik & Sheljuzhko, 1956, Mitt. münch. ent. Ges., **46**: 190. Reiss & Tremewan, 1960, Bull. Brit. Mus. (nat. Hist.) Ent., **9** (10): 462, pl. 22, fig. 9, pl. 23, figs. 3, 4. Tremewan, 1961, Bull. Brit. Mus. (nat. Hist.) Ent., **10** (7): 253, pl. 51, fig. 5, pl. 59, figs. 1, 2. — Fluss Jagnob, Ulaxs Capa, Hissar Gebirge.

ab. **latecincta** Reiss & Tremewan, 1960, Bull. Brit. Mus. (nat. Hist.) Ent., **9** (10): 462, pl. 22, fig. 10. Tremewan, 1961, Bull. Brit. Mus. (nat. Hist.) Ent., **10** (7): 253, pl. 51, fig. 6.

Biology

Holik, 1953, Ent. Z., **63**: 21.

nuksanensis Koch

Distribution: Border between Afghanistan and Chitral (Hindukush).

ssp. **nuksanensis** Koch, 1937, Ent. Z., **51**: 61, 64, figs. 8-17. Haaf, 1952, Veröff. zool. Staatssamml. Münch., **2**: 151, 154, 156, pl. 7. Holik & Sheljuzhko, 1956, Mitt. münch. ent. Ges., **46**: 186. Alberti, 1958, Mitt. münch. ent. Ges., **34**: 298.
Nuksan-Pass, Nordseite, Nordost-Hindukush, 3500-4000 m.

ab. **cingulata** Koch, 1937, Ent. Z., **51**: 61.

ssp. **andarabensis** Koch, 1938, Ent. Z., **51**: 399. Holik & Sheljuzhko, 1956, Mitt. münch. ent. Ges., **46**: 188.
Andarab, West-Hindukush, 4000-4500 m.

ssp. **omotoi** Reiss & Schulte, 1964, Ent. Z., **74**: 159, 160, figs. 5-8.
Bala Quran, Anjuman Valley, N. E. Afghanistan, 10500 ft.

Biology

Holik, 1953, Ent. Z., **63**: 21.

transpamirina Koch
Distribution: Chitral.

transpamirina Koch, 1936, Iris, **50**: 40, pl. 2, figs. 1-4. Holik & Sheljuzhko, 1956, Mitt. münch. ent. Ges., **46**: 190. Alberti, 1958, Mitt. zool. Mus. Berl., **34**: 298; 1960, Ent. Z., **70**: 173.
Jasin, Ost-Chitral, 2400 m.

glasunovi Grum-Grshimailo
Distribution: Hissar Mountains.

glasunovi Grum-Grshimailo, 1893, Horae Soc. ent. Ross., **27**: 386; 1894, ibidem, **28**: 94. Holik & Sheljuzhko, 1956, Mitt. münch. ent. Ges., **46**: 189. Alberti, 1958, Mitt. zool. Mus. Berl., **34**: 299.
Am Fluss Dshidshi-grut-darja (Simarch), Hissar Gebirge.

magiana Staudinger
Distribution: Hissar Mountains.

magiana Staudinger, 1889, Stettin. ent. Ztg., **50**: 23. Seitz, 1907, Die Gross-Schmetterlinge der Erde, **2**: 24, pl. 7b (as *hissariensis* Grum-Grshimailo). Reiss, 1933, Int. ent. Z., **26**: 501, figs.; 1933, in Seitz, Die Gross-Schmetterlinge der Erde, Supplement, **2**: 265. Haaf, 1952, Veröff. zool. Staatssamml. Münch., **2**: 151, 153, 156, pl. 6. Holik & Sheljuzhko, 1956, Mitt. münch. ent. Ges., **46**: 183. Alberti, 1958, Mitt. zool. Mus. Berl., **34**: 299.
Bei Magian (Maguian), Hissar Gebirge.

ab. **cingulata** Holik & Sheljuzhko, 1956, Mitt. münch. ent. Ges., **46**: 186.

ab. **hissariensis** Grum-Grshimailo, 1890, in Romanoff, Mémoires sur les Lépidoptères, **4**: 520, pl. 19, fig. 1. Seitz, 1907, Die Gross-Schmetterlinge der Erde, **2**: 24, pl. 7b (as *magiana* Staudinger). Reiss, 1933, Int. ent. Z., **26**: 502, fig.; 1933, in Seitz, Die Gross-Schmetterlinge der Erde, Supplement, **2**:

265. Holik & Sheljuzhko, 1956, Mitt. münch. ent. Ges., **46**: 185.

Biology

Holik, 1953, Ent. Z., **63**: 21.

alaica Holik & Sheljuzhko
Distribution: Alai Mountains.
alaica Holik & Sheljuzhko, 1956, Mitt. münch. ent. Ges., **46**: 186. Fluss Borsunsaj; Fluss Aksaj, Nördliche Alai-Gebirge.
ab. **cingulata** Holik & Sheljuzhko, 1956, Mitt. münch. ent. Ges., **46**: 186.

saadii Reiss
Distribution: Southern Iran.
saadii Reiss, 1938, Ent. Rdsch., **55**: 310, figs. dl, d2. Brandt, 1938, Ent. Rdsch., **55**: 673, pl. 5, figs. 22, 23. Haaf, 1952, Veröff. zool. Staatssamml. Münch., **2**: 151, 153 (as *brandti* Reiss), 156, pl. 6 (as *brandti saadii* Reiss). Holik & Sheljuzhko, 1956, Mitt. münch. ent. Ges., **46**: 135. Alberti, 1958, Mitt. zool. Mus. Berl., **34**: 293. Fort Sine-Sefid, Strasse Chiraz-Kazeroun, Südiran, 2200 m.
ab. **immaculata** Reiss, 1938, Ent. Rdsch., **55**: 311.

sengana Holik & Sheljuzhko
Distribution: Baluchistan, Iran.
sengana Holik & Sheljuzhko, 1956, Mitt. münch. ent. Ges., **46**: 136. Alberti, 1958, Mitt. zool. Mus. Berl., **34**: 298. Fort Sengan, Strasse Khach-Zahedan, Belutshistan, Iran, 1800 m.

escalerai Poujade
Distribution: Iran.
escalerai Poujade, 1900, Bull. Mus. Hist. nat. Paris, **6** (2): 68. Oberthür, 1909, Études de Lépidoptérologie comparée, **3**, pl. 28, fig. 172. Reiss, 1930, in Seitz, Die Gross-Schmetterlinge der Erde, Supplement, **2**: 22, pl. 2i (after Oberthür); 1933, ibidem, **2**: 262. Holik & Sheljuzhko, 1956, Mitt. münch. ent. Ges., **46**: 144. Reiss, 1958, Z. wien. ent. Ges., **43**: 158 (footnote 9). Alberti, 1958, Mitt. zool. Mus. Berl., **34**: 293. Tremewan, 1961, Bull. Brit. Mus. (nat. Hist.) Ent., **10** (7): 306, pl. 57, fig. 18. Chindáar (vallée), Haut-Kharoum, Perse (Iran).

formosa Herrich-Schäffer
Distribution: Asia Minor, Armenia.
ssp. **malatiana** Rebel, 1901, in Staudinger & Rebel, Catalog der Lepidopteren des Palaearctischen Faunengebietes, p. 387. Seitz, 1907, Die Gross-Schmetterlinge der Erde, **2**: 28, pl. 7h. Reiss, 1930, in Seitz, Die Gross-Schmetterlinge der Erde, Malatia, Taurus, Kleinasien.

Supplement, **2**: 22; 1933, ibidem, **2**: 269. Holik & Sheljuzhko, 1956, Mitt. münch. ent. Ges., **46**: 131.

ssp. **formosa** Herrich-Schäffer, 1852, Systematische Bearbeitung der Schmetterlinge von Europa, **6**: 45; 1851, ibidem, **2**, pl. 14, fig. 99 (non-binominal). Staudinger, 1879, Horae Soc. ent. Ross., **14**: 324. Seitz, 1907, Die Gross-Schmetterlinge der Erde, **2**: 28, pl. 7i. Reiss, 1930, in Seitz, Die Gross-Schmetterlinge der Erde, Supplement, **2**: 22; 1933, ibidem, **2**: 269. Haaf, 1952, Veröff. zool. Staatssamml. Münch., **2**: 149, 151, 153, 156, pl. 6. Holik & Sheljuzhko, 1956, Mitt. münch. ent. Ges., **46**: 129. Alberti, 1958, Mitt. zool. Mus. Berl., **34**: 297. — Amasia, Pontus, Kleinasien.

ab. **amoenoides** Holik & Sheljuzhko, 1956, Mitt. münch. ent. Ges., **46**: 130.

ssp. **hadjinica** Holik & Sheljuzhko, 1958, Mitt. münch. ent. Ges., **48**: 273 (nomen novum for *hadjinensis* Holik & Sheljuzhko). — Hadjin, Taurus, Kleinasien.

hadjinensis Holik & Sheljuzhko, 1956, Mitt. münch. ent. Ges., **46**: 131 (preoccupied by **hadjinensis** Reiss, 1931, ssp. of **lydia** Staudinger).

ssp. **kotzschi** Reiss, 1935, Int. ent. Z., **28**: 489. Holik, 1935, Ent. Z., **49**: 29. Holik & Sheljuzhko, 1956, Mitt. münch. ent. Ges., **46**: 131. Tremewan, 1961, Bull. Brit. Mus. (nat. Hist.) Ent., **10** (7): 307, pl. 57, fig. 21. — Khashkhash-Dagh, Aghri-Dagh, West-Armenien, 3200 m.

Biology

Holik, 1953, Ent. Z., **63**: 20. Reiss, 1933, in Seitz, Die Gross-Schmetterlinge der Erde, Supplement, **2**: 269; 1958, Z. wien. ent. Ges., **43**: 158. Staudinger, 1879, Horae Soc. ent. Ross., **14**: 324.

rosinae Korb

Distribution: Armenia, Transcaucasia.

ssp. **rosinae** Korb, 1902, Iris, **15**: 326. Dziurzyński, 1904, Jber. wien. ent. Ver., **14**: 52, pl. 2, fig. 14. Seitz, 1907, Die Gross-Schmetterlinge der Erde, **2**: 28, pl. 7f. Reiss, 1930, in Seitz, Die Gross-Schmetterlinge der Erde, Supplement, **2**: 22; 1933, ibidem, **2**: 269. Holik, 1935, Ent. Z., **49**: 29. Koch, 1936, Ent. Z., **50**: 398. Haaf, 1952, Veröff. zool. Staatssamml. Münch., **2**: 149, 151, 153, 156, pl. 7, pl. 16. Holik & Sheljuzhko, 1956, Mitt. münch. ent. Ges., **46**: 132. Alberti, 1958, Mitt. zool. Mus. Berl., **34**: 291. — Bei Kulp, West-Armenien, Kleinasien.

ssp. **concinna** Holik & Sheljuzhko, 1956, Mitt. münch. ent. Ges., **46**: 134. — Germatshatach, Daralagëz Gebirge (Nachitshevan), Transkaukasien, 6000-6500 ft.

Biology

Holik, 1937, Lambillionea, **37**: 15-24, 32-45, 80-91; 1938, Mitt. münch. ent. Ges., **28**: 394, pl. 8, fig. 2; 1953, Ent. Z., **63**: 21. Korb, 1902, Iris, **15**: 327. Reiss, 1933, in Seitz, Die Gross-Schmetterlinge der Erde, Supplement, **2**: 269; 1958, Z. wien. ent. Ges., **43**: 158.

brandti Reiss

Distribution: Northern Iran.

ssp. **brandti** Reiss, 1937, Ent. Rdsch., **55**: 30, fig. a3; 1938, ibidem, **55**: 313, figs. c2, c3. Brandt, 1938, Ent. Rdsch., **55**: 673, pl. 5, fig. 20. Holik & Sheljuzhko, 1956, Mitt. münch. ent. Ges., **46**: 134. Alberti, 1958, Mitt. zool. Mus. Berl., **34**: 292. — Keredj, Elburs-Gebirge, Nordiran, 1800 m.

ssp. **nissana** Reiss, 1937, Ent. Rdsch., **55**: 32; 1938, ibidem, **55**: 313, fig. c4. Brandt, 1938, Ent. Rdsch., **55**: 673, pl. 5, fig. 21. Holik & Sheljuzhko, 1956, Mitt. münch. ent. Ges., **46**: 135. — Nissa, Elburs-Gebirge, Nordiran, 2700 m.

sogdiana Erschoff

Distribution: Central Asia.

ssp. **sogdiana** Erschoff, 1874, in Fedtshenko, Reise in Turkestan, **2**: 27, pl. 2, fig. 20. Reiss, 1933, Ent. Rdsch., **50**: 150, pl. 1, figs.; 1933, in Seitz, Die Gross-Schmetterlinge der Erde, Supplement, **2**: 264, pl. 16l. Holik & Sheljuzhko, 1956, Mitt. münch. ent. Ges., **46**: 158. Alberti, 1958, Mitt. zool. Mus. Berl., **34**: 295. — Bei Tashkent, Syr-Darja, Turkestan.

ab. **latocingulata** Holik & Sheljuzhko, 1956, Mitt. münch. ent. Ges., **46**: 159.

ssp. **padshaatensis** Holik & Sheljuzhko, 1956, Mitt. münch. ent. Ges., **46**: 158. — Tal des Padsha-ata, Distrikt Namangan, Fergana.

ssp. **altissima** Burgeff, 1914, Mitt. münch. ent. Ges., **5**: 51, pl. 2, fig. 165, pl. 6, figs. 48-52. Reiss, 1930, in Seitz, Die Gross-Schmetterlinge der Erde, Supplement, **2**: 22, pl. 2i; 1933, ibidem, **2**: 265; 1933, Ent. Rdsch., **50**: 155, pl. 2, figs. Holik & Sheljuzhko, 1956, Mitt. münch. ent. Ges., **46**: 154. — Ak-Bassegha (Ak-Bossaga), Trans-Alai Kette, über 2000 m.

ab. **alba** Reiss, 1933, Ent. Rdsch., **50**: 155, pl. 1, fig.

ssp. **karatauensis** Holik & Sheljuzhko, 1956, Mitt. münch. ent. Ges., **46**: 159. Reiss, 1933, Ent. Rdsch., **50**: 155, pl. 2, figs. — Dorf Vyssokoje, Karatau, Syr-Darja.

ssp. **talassinensis** Holik & Sheljuzhko, 1958, Mitt. münch. ent. Ges., **48**: 273 (nomen novum for *talassica* Holik & Sheljuzhko). — Dzhebagly, Talasskij Ala-tau, Syr-Darja.

talassica Holik & Sheljuzhko, 1956, Mitt. münch. ent. Ges., **46**:

160 (preoccupied by **talassica** Holik & Sheljuzhko, 1953, ssp. of **purpuralis** Brünnich).

ab. **rubroabdominalis** Holik & Sheljuzhko, 1956, Mitt. münch. ent. Ges., **46**: 160.

ssp. **separata** Staudinger, 1887, Stettin. ent. Ztg., **48**: 74. Grum-Grshimailo, 1890, in Romanoff, Mémoires sur les Lépidoptères, **4**: 521. Reiss, 1930, in Seitz, Die Gross-Schmetterlinge der Erde, Supplement, **2**: 22; 1933, ibidem, **2**: 264; 1933, Ent. Rdsch., **50**: 151, pl. 1, figs. Holik & Sheljuzhko, 1956, Mitt. münch. ent. Ges., **46**: 156. — Usgent, an der Kara-darja, einem Nebenfluss des Naryn, Fergana (Alexandergebirge).

ab. **nigra** Dziurzyński, 1909, Berl. ent. Z., **53**: 250. Reiss, 1930, in Seitz, Die Gross-Schmetterlinge der Erde, Supplement, **2**: 22; 1933, ibidem, **2**: 265.

ab. **ornata** Burgeff, 1914, Mitt. münch. ent. Ges., **5**: 50, pl. 2, fig. 157, pl. 6, figs. 44, 45; 1926, ibidem, **16**: 41. Reiss, 1930, in Seitz, Die Gross-Schmetterlinge der Erde, Supplement, **2**: 22; 1933, ibidem, **2**: 264, pl. 16l; 1933, Ent. Rdsch., **50**: 148, pl. 1, fig.

ab. **confluens** Dziurzyński, 1908, Berl. ent. Z., **53**: 13, 51. Seitz, 1912, Die Gross-Schmetterlinge der Erde, **2**: 443. Reiss, 1930, in Seitz, Die Gross-Schmetterlinge der Erde, Supplement, **2**: 22; 1933, ibidem, **2**: 265.

ab. **alba** Dziurzyński, 1908, Berl. ent. Z., **53**: 13, 51; 1909, Jber. wien. ent. Ver., **19**: 136, pl. 1, fig. 8. Seitz, 1912, Die Gross-Schmetterlinge der Erde, **2**: 443. Reiss, 1930, in Seitz, Die Gross-Schmetterlinge der Erde, Supplement, **2**: 22; 1933, ibidem, **2**: 264.

ssp. **margelanensis** Reiss, 1933, Ent. Rdsch., **50**: 151, pl. 1, fig., pl. 2, fig. Staudinger, 1887, Stettin. ent. Ztg., **48**: 74. Seitz, 1907, Die Gross-Schmetterlinge der Erde, **2**: 28, pl. 7e (as *scovitzii* Ménétriés). Reiss, 1933, in Seitz, Die Gross-Schmetterlinge der Erde, Supplement, **2**: 264. Holik & Sheljuzhko, 1956, Mitt. münch. ent. Ges., **46**: 155. — Margelan, Fergana (Namangan, Osh).

ssp. **tshimganica** Holik, 1935, Ent. Rdsch., **53**: 5. Holik & Sheljuzhko, 1956, Mitt. münch. ent. Ges., **46**: 160. Tremewan, 1961, Bull. Brit. Mus. (nat. Hist.) Ent., **10** (7): 307, pl. 57, fig. 20. — Tshimgan, West-Tianshan, 90 km. nordöstlich von Tashkent, 1500-1600 m.

ab. **latecingulata** Sheljuzhko, 1935, Ent. Rdsch., **53**: 8.

ab. **pseudoseparata** Sheljuzhko, 1935, Ent. Rdsch., **53**: 7.

Biology

Grum-Grshimailo, 1890, in Romanoff, Mémoires sur les Lépidoptères, **4**: 521, pl. 19, fig. 2. Holik, 1946, Rev. franç. Lépid., **10**: 250-261, 273-280; 1953, Ent. Z., **63**: 21. Korb, 1916, Mitt. münch. ent. Ges., **7**: 22. Reiss, 1958, Z. wien. ent. Ges., **43**: 158.

fraxini Ménétriés

Distribution: Transcaucasia, southern Caucasus.

ssp. **fraxini** Ménétriés, 1832, Catalogue raisonné des Objets de Zoologie, p. 260. Boisduval, 1834, Icones historique des Lépidoptères nouveaux ou peu connus, **2**: 75. Reiss, 1933, Ent. Rdsch., **50**: 130; 1933, in Seitz, Die Gross-Schmetterlinge der Erde, Supplement, **2**: 263. Holik & Sheljuzhko, 1956, Mitt. münch. ent. Ges., **46**: 148. Alberti, 1958, Mitt. zool. Mus. Berl., **34**: 291. Reiss, 1965, Ent. Rec., **77**: 165. — Talysh (Lenkoran) am Kaspischen Meer.

ab. **scovitzii** Ménétriés, 1832, Catalogue raisonné des Objets de Zoologie, p. 260. Boisduval, 1834, Icones historique des Lépidoptères nouveaux ou peu connus, **2**: 76. Lederer, 1869/70, Ann. Soc. ent. Belg., **13**: 29. Reiss, 1933, Ent. Rdsch., **50**: 131; 1933, in Seitz, Die Gross-Schmetterlinge der Erde, Supplement, **2**: 263. Holik & Sheljuzhko, 1956, Mitt. münch. ent. Ges., **46**: 148.

cingulata Sheljuzhko, 1908, Rev. russe Ent., **7**: 234. Reiss, 1930, in Seitz, Die Gross-Schmetterlinge der Erde, Supplement, **2**: 22; 1933, ibidem, **2**: 263; 1933, Ent. Rdsch., **50**: 135. Holik & Sheljuzhko, 1956, Mitt. münch. ent. Ges., **46**: 148.

ssp. **perdita** Staudinger, 1887, Stettin. ent. Ztg., **48**: 76. Seitz, 1907, Die Gross-Schmetterlinge der Erde, **2**: 28. Dziurzyński, 1909, Jber. wien. ent. Ver., **19**: 136, pl. 1, fig. 7. Reiss, 1930, in Seitz, Die Gross-Schmetterlinge der Erde, Supplement, **2**: 22; 1933, ibidem, **2**: 263, pl. 161; 1933, Ent. Rdsch., **50**: 135, pl. 1, figs. Holik & Sheljuzhko, 1956, Mitt. münch. ent. Ges., **46**: 150. Reiss, 1965, Ent. Rec., **77**: 166. — Bei Nucha, südlicher Kaukasus.

ssp. **oribasus** Herrich-Schäffer, 1846, Systematische Bearbeitung der Schmetterlinge von Europa, **2**: 46; 1844, ibidem, **2**, pl. 4, figs. 31-34 (non-binominal). Freyer, 1842, Neuere Beiträge zur Schmetterlingskunde, **4**: 107, pl. 350, fig. 2 (as *carneolica* [sic] Scopoli). Seitz, 1907, Die Gross-Schmetterlinge der Erde, **2**: 27, pl. 7e (as *fraxini* Ménétriés). Reiss, 1933, in Seitz, Die Gross-Schmetterlinge der Erde, Supplement, **2**: 263; 1933, Ent. Rdsch., **50**: 134, pl. 1, figs. Holik & Sheljuzhko, 1956, Mitt. münch. ent. Ges., **46**: 150. Reiss, 1965, Ent. Rec., **77**: 165. — Helenendorf; Elisabethpol, Transkaukasien.

rognada Boisduval, 1848, Bull. Soc. ent. Fr., (2) **6**: xxx. Tremewan, 1961, Bull. Brit. Mus. (nat. Hist.) Ent., **10** (7): 254, pl. 51, fig. 9.

oribasus Freyer, 1852, Neuere Beiträge zur Schmetterlingskunde, **6**: 135, pl. 568, fig. 1.

ssp. **slabyiana** Reiss, 1965, Ent. Rec., **77**: 166, pl. 2, figs. 1, 2. — Tbilisi (Tiflis), Gruzia, Transcaucasia.

Biology

Holik, 1946, Rev. franç. Lépid., **10**: 250-261, 273-280; 1953, Ent. Z., **63**: 21.

chirazica Reiss

Distribution: Southern Iran.

chirazica Reiss, 1938, Ent. Rdsch., **55**: 311, figs. e2, e3. Holik & Sheljuzhko, 1956, Mitt. münch. ent. Ges., **46**: 142. Alberti, 1958, Mitt. zool. Mus. Berl., **34**: 300. — Fort Sine-Sefid, Strasse Chiraz-Kazeroun, Südiran, 2200 m. [Fort Serra-Tschenk, 2700 m].

ab. **ornata** Reiss, 1938, Ent. Rdsch., **55**: 312, fig. e4.

olivieri Boisduval

Distribution: Syria.

ssp. **olivieri** Boisduval, 1828, Essai sur une Monographie des Zygénides, p. 98, pl. 6, fig. 4. Herrich-Schäffer, 1846, Systematische Bearbeitung der Schmetterlinge von Europa, **2**: 46; 1851, ibidem, **2**, pl. 15, fig. 103 (after Boisduval). Reiss, 1933, Ent. Rdsch., **50**: 195, pl. 1, fig., pl. 2, fig.; 1933, in Seitz, Die Gross-Schmetterlinge der Erde, Supplement, **2**: 266, pl. 16m. Holik & Sheljuzhko, 1956, Mitt. münch. ent. Ges., **46**: 109. Alberti, 1958, Mitt. zool. Mus. Berl., **34**: 301. Tremewan, 1961, Bull. Brit. Mus. (nat. Hist.) Ent., **10** (7): 254, pl. 51, fig. 7, pl. 64, figs. 3, 4. — Beyrut, Syrie.

cremonae Seitz, 1907, Die Gross-Schmetterlinge der Erde, **2**: 27, pl. 7a.

ssp. **libanicola** Burgeff, 1927, in Bang-Haas, Horae Macrolepidopterologicae regionis palaearcticae, **1**: 55. Reiss, 1930, in Seitz, Die Gross-Schmetterlinge der Erde, Supplement, **2**: 23, pl. 2k; 1933, Ent. Rdsch., **50**: 203, pl. 1, fig., pl. 2, figs. Holik & Sheljuzhko, 1956, Mitt. münch. ent. Ges., **46**: 113. Tremewan, 1961, Bull. Brit. Mus. (nat. Hist.) Ent., **10** (7): 306, pl. 57, fig. 19. — Umgebung der Stadt Zahlé, östlicher Libanon, Syrien.

Biology

Holik, 1936, Ent. Rdsch., **53**: 506; 1938, ibidem, **55**: 349-354, 382-384; 1938, Mitt. münch. ent. Ges., **28**: 394, pl. 8, fig. 1; 1953, Ent. Z., **63**: 16. Reiss, 1958, Z. wien. ent. Ges., **43**: 158.

laetifica Herrich-Schäffer

Distribution: Asia Minor.

laetifica Herrich-Schäffer, 1846, Systematische Bearbeitung der Schmetterlinge von Europa, **2**: 44; 1847, ibidem, **2**, pl. 12, fig. 88. Seitz, 1907, Die Gross-Schmetterlinge der Erde, **2**: 28, pl. 7g. Reiss, 1933, Ent. Rdsch., **50**: 204, pl. 1, fig., pl. 2, fig.; — Kleinasien, Mesopotamien [?Mardin].

1933, in Seitz, Die Gross-Schmetterlinge der Erde, Supplement, **2**: 267, pl. 16m. Holik & Sheljuzhko, 1956, Mitt. münch. ent. Ges., **46**: 115. Alberti, 1958, Mitt. zool. Mus. Berl., **34**: 301.

freyeriana Reiss

Distribution: Asia Minor (Amasia, Tokat, Kara-Hissar, Ankara).

freyeriana Reiss, 1933, Ent. Rdsch., **50**: 221, pl. 1, figs., pl. 2, figs. (nomen novum for *ganimedes* Freyer); 1933, in Seitz, Die Gross-Schmetterlinge der Erde, Supplement, **2**: 268. Haaf, 1952, Veröff. zool. Staatssamml. Münch., **2**: 151, 154, 156, pl. 8, pl. 16. Holik & Sheljuzhko, 1956, Mitt. münch. ent. Ges., **46**: 126. Alberti, 1958, Mitt. zool. Mus. Berl., **34**: 301. — Bei Amasia im Karasdere und auf dem Caraman, Kleinasien.

ganimedes Freyer, 1852, Neuere Beiträge zur Schmetterlingskunde, **6**: 136, pl. 568, fig. 3 (preoccupied by **ganymedes** Herrich-Schäffer, 1851). Staudinger, 1879, Horae Soc. ent. Ross., **14**: 324 (as *ganymedes* Herrich-Schäffer).

ab. **tricolor** Holik & Sheljuzhko, 1956, Mitt. münch. ent. Ges., **46**: 128.

ab. **algarvensis** Dziurzyński, 1908, Berl. ent. Z., **53**: 7, 52. Seitz, 1912, Die Gross-Schmetterlinge der Erde, **2**: 443. Reiss, 1933, in Seitz, Die Gross-Schmetterlinge der Erde, Supplement, **2**: 268. Holik & Sheljuzhko, 1956, Mitt. münch. ent. Ges., **46**: 127.

algarvensis Dziurzyński, 1906, Ent. Z., **19**: 186 (nomen nudum).

ab. **confluens** Dziurzyński, 1908, Berl. ent. Z., **53**: 52. Reiss, 1933, in Seitz, Die Gross-Schmetterlinge der Erde, Supplement, **2**: 268.

ab. **totarubra** Holik & Sheljuzhko, 1956, Mitt. münch. ent. Ges., **46**: 128.

Biology

Holik, 1938, Ent. Rdsch., **55**: 349-354, 382-384; 1953. Ent. Z., **63**: 20. Reiss, 1933, in Seitz, Die Gross-Schmetterlinge der Erde, Supplement, **2**: 268; 1958, Z. wien. ent. Ges., **43**: 158. Staudinger, 1879, Horae Soc. ent. Ross., **14**: 324.

ganymedes Herrich-Schäffer

Distribution: Asia Minor (Taurus, Kurdistan), Armenia.

ganymedes Herrich-Schäffer, 1852, Systematische Bearbeitung der Schmetterlinge von Europa, **6**: 45; 1851, ibidem, **2**, pl. 14, figs. 100, 101 (non-binominal). Seitz, 1907, Die Gross-Schmetterlinge der Erde, **2**: 28, pl. 7h; 1912, ibidem, **2**: 443. Reiss, 1933, Ent. Rdsch., **50**: 205, pl. 1, figs., pl. 2, figs.; 1933, in Seitz, Die Gross-Schmetterlinge der Erde, Supplement, **2**: — Taurus, Kleinasien [?Zeitun].

267. Holik & Sheljuzhko, 1956, Mitt. münch. ent. Ges., **46**: 121. Alberti, 1958, Mitt. zool. Mus. Berl., **34**: 301. Tremewan, 1966, Ent. Rec., **78**: 31, figs. 1, 2, pl. 1, fig. 1.

hebe Seitz, 1907, Die Gross-Schmetterlinge der Erde, **2**: 28, pl. 7h. Holik & Sheljuzhko, 1956, Mitt. münch. ent. Ges., **46**: 119.

ab. **rubroabdominalis** Holik & Sheljuzhko, 1956, Mitt. münch. ent. Ges., **46**: 124.

ab. **confluens** Dziurzyński, 1908, Berl. ent. Z., **53**: 52. Seitz, 1912, Die Gross-Schmetterlinge der Erde, **2**: 443.

ab. **rubroabdominalisconfluens** Holik & Sheljuzhko, 1956, Mitt. münch. ent. Ges., **46**: 123.

Biology

Burgeff, 1950, Portug. acta biol., (A) Goldschmidt: 663-728. Holik, 1953, Ent. Z., **63**: 21.

dsidsilia Freyer

Distribution: Transcaucasia.

dsidsilia Freyer, 1852, Neuere Beiträge zur Schmetterlingskunde, **6**: 136, pl. 568, fig. 4. Herrich-Schäffer, 1846, Systematische Bearbeitung der Schmetterlinge von Europa, **2**: 44 (as *olivieri* Boisduval [partim]); 1852, ibidem, **6**: 46. Seitz, 1907, Die Gross-Schmetterlinge der Erde, **2**: 28, pl. 7h (as *olivieri* Boisduval); 1912, ibidem, **2**: 443. Reiss, 1933, Ent. Rdsch., **50**: 207, pl. 1, figs., pl. 2, figs.; 1933, in Seitz, Die Gross-Schmetterlinge der Erde, Supplement, **2**: 267. Holik & Sheljuzhko, 1956, Mitt. münch. ent. Ges., **46**: 124. Alberti, 1958, Mitt. zool. Mus. Berl., **34**: 301.

Helenendorf, Transkaukasien.

ab. **tricolor** Reiss, 1933, Ent. Rdsch., **50**: 208, pl. 1, fig.; 1933, in Seitz, Die Gross-Schmetterlinge der Erde, Supplement, **2**: 268, pl. 16m.

Biology

Holik, 1953, Ent. Z., **63**: 21.

haberhaueri Lederer

Distribution: Transcaucasia, ?Asia Minor (Taurus).

haberhaueri Lederer, 1870, Ann. Soc. ent. Belg., **13**: 29, 45. Seitz, 1907, Die Gross-Schmetterlinge der Erde, **2**: 26, pl. 6i. Reiss, 1933, Ent. Rdsch., **50**: 241, pl. 1, figs., pl. 2, figs.; 1933, in Seitz, Die Gross-Schmetterlinge der Erde, Supplement, **2**: 268. Holik & Sheljuzhko, 1956, Mitt. münch. ent. Ges., **46**: 107. Alberti, 1958, Mitt. zool. Mus. Berl., **34**: 301.

Montagnes de Hankynda, Transcaucasie.

ab. **cingulata** Holik & Sheljuzhko, 1956, Mitt. münch. ent. Ges., **46**: 108.

Biology

Holik, 1938, Ent. Rdsch., **55**: 349-354, 382-384; 1953, Ent. Z., **63**: 21.

optima Reiss

Distribution: Caucasus (Terek, Dagestan, Kutais).

optima Reiss, 1939, Ent. Z., **53**: 118 (nomen novum for *nobilis* Reiss). Haaf, 1952, Veröff. zool. Staatssamml. Münch., **2**: 151, 154, 156, pl. 8, pl. 16. Holik & Sheljuzhko, 1956, Mitt. münch. ent. Ges., **46**: 105. Alberti, 1958, Mitt. zool. Mus. Berl., **34**: 301. Missura; Oni; Provinz Kutais; Ossetenstrasse, Kaukasus.

nobilis Reiss, 1933, Ent. Rdsch., **50**: 144, pl. 1, figs., pl. 2, figs. (preoccupied by **nobilis** Navás, 1924, ssp. of **lonicerae** Scheven); 1933, in Seitz, Die Gross-Schmetterlinge der Erde, Supplement, **2**: 263, pl. 16l.

ab. **cingulata** Reiss, 1935, Int. ent. Z., **28**: 543.

ab. **dissoluta** Reiss, 1935, Int. ent. Z., **28**: 543.

ab. **tricolor** Reiss, 1935, Int. ent. Z., **28**: 542. Tremewan, 1961, Bull. Brit. Mus. (nat. Hist.) Ent., **10** (7): 254, pl. 51, fig. 8.

Biology

Holik, 1937, Lambillionea, **37**: 15-24, 32-45, 80-91; 1938, Ent. Rdsch., **55**: 349-354, 382-384; 1953, Ent. Z., **63**: 21. Reiss, 1958, Z. wien. ent. Ges., **43**: 158.

sedi Fabricius

Distribution: Southern Russia, (?Crimea), Bulgaria.

ssp. **sedi** Fabricius, 1787, Mantissa Insectorum, **2**: 101: 1793, Entomologia Systematica, **3**: 388. Ochsenheimer, 1808, Die Schmetterlinge von Europa, **2**: 93; 1816, ibidem, **4**: 166. Hübner, [1808]-[20th June 1813], Sammlung europäischer Schmetterlinge, **2**, pl. 28, fig. 132. Boisduval, 1828, Essai sur une Monographie des Zygénides, p. 94, 96. Freyer, 1842, Neuere Beiträge zur Schmetterlingskunde, **4**: 107, pl. 350, figs. 3, 4. Herrich-Schäffer, 1844, Systematische Bearbeitung der Schmetterlinge von Europa, **2**, pl. 6, figs. 46, 47; 1846, ibidem, **2**: 43. Staudinger, 1879, Horae Soc. ent. Ross., **14**: 325. Spuler, 1906, in Hofmann, Die Schmetterlinge Europas, **2**: 163, pl. 72, fig. 6, pl. 77, fig. 27. Seitz, 1907, Die Gross-Schmetterlinge der Erde, **2**: 26, pl. 6k. Burgeff, 1914, Mitt. münch. ent. Ges., **5**: 52. Reiss, 1933, Ent. Rdsch., **50**: 146, pl. 2, figs.; 1933, in Seitz, Die Gross-Schmetterlinge der Erde, Supplement, **2**: 262. Holik, 1939, Mitt. münch. ent. Ges., **29**: 55. Holik & Sheljuzhko, 1956, Mitt. münch. ent. Ges., **46**: 100. Alberti, 1958, Mitt. zool. Mus. Berl., **34**: 302. Zimsen, 1964, The Type Material of I. C. Fabricius, p. 546. Rossia meridionali [Sarepta, Südrussland].

ab. **cingulata** Holik, 1939, Mitt. münch. ent. Ges., **29**: 57.
ab. **dissoluta** Burgeff, 1926, Mitt. münch. ent. Ges., **16**: 41. Reiss, 1930, in Seitz, Die Gross-Schmetterlinge der Erde, Supplement, **2**: 23.
ssp. **sliwenensis** Reiss, 1933, Ent. Rdsch., **50**: 147, pl. 2, figs. Frivaldszky, 1834, Mag. Tud. Tars. Evk., **2**: 271, pl. 7, fig. 4. Lederer, 1863, Wien. ent. Monatschr., **7**: 22. Reiss, 1933, in Seitz, Die Gross-Schmetterlinge der Erde, Supplement, **2**: 262. Holik, 1939, Mitt. münch. ent. Ges., **29**: 57. — Sliwen (Slivno), Bulgarien.

Biology

Becker, 1888, Bull. Soc. Nat. Moscou, **2**: 375. Burgeff, 1950, Portug. acta biol., (A) Goldschmidt: 663-728. Holik, 1938, Ent. Rdsch., **55**: 349-354, 382-384; 1953, Ent. Z., **63**: 15. Reiss, 1958, Z. wien. ent. Ges., **43**: 158.

SECTION 4

algira Boisduval
Distribution: Tunisia, Algeria, Morocco.
ssp. **algira** Boisduval, 1834, Icones historique des Lépidoptères nouveaux ou peu connus, **2**: 75. Herrich-Schäffer, 1846, Systematische Bearbeitung der Schmetterlinge von Europa, **2**: 45, 1851, ibidem, **2**, pl. 15, fig. 106; 1852, ibidem, **6**: 46. Tremewan, 1964, Ent. Rec., **76**: 35, pl. 1, figs. 1-3. — Alger, Algérie,
algira Duponchel, 1835, in Godart & Duponchel, Histoire naturelle des Lépidoptères ou Papillons de France, Supplement, **2**: 86 (pl. 7, fig. 6 is erroneous). Oberthür, 1890, Études d'Entomologie, **13**: 22, pl. 7, figs. 59-62. Seitz, 1907, Die Gross-Schmetterlinge der Erde, **2**: 29, pl. 8a. Oberthür, 1910, Études de Lépidoptérologie comparée, **4**: 604. Holl, 1912, Bull. Soc. Hist. nat. Afr. N., **4** (6): 117. Rothschild, 1917, Novit. zool., **24**: 338. Reiss, 1944, Z. wien. ent. Ges., **29**: 46, pl. 36, figs. 48, 49, pl. 46, fig. XI. Haaf, 1952, Veröff. zool. Staatssamml. Münch., **2**: 152, 154, 156, pl. 10. Alberti, 1958, Mitt. zool. Mus. Berl., **34**: 310. Barragué, 1961, Alexanor, **2**: 130, 134, fig. Tremewan, 1964, Ent. Rec., **76**: 35.
bachagha Oberthür, 1916, Études de Lépidoptérologie comparée, **12**: 226. Rothschild, 1917, Novit. zool., **24**: 338. Reiss, 1930, in Seitz, Die Gross-Schmetterlinge der Erde, Supplement, **2**: 25. Burgeff, 1950, Portug. acta biol., (A) Goldschmidt: 670, figs. a1; 1951, Biol. Zbl., **70**: 4, figs. 2b. Tremewan, 1961, Bull. Brit. Mus. (nat. Hist.) Ent., **10** (7): 257, pl. 51, fig. 22, pl. 60, figs. 5, 6.
ab. **aurantiaca** Holl, 1912, Bull. Soc. Hist. nat. Afr. N., **4** (6):

119. Tremewan, 1961, Bull. Brit. Mus. (nat. Hist.) Ent., **10** (7): 258, pl. 51, fig. 24.

ab. **bicolor** Holl, 1912, Bull. Soc. Hist. nat. Afr. N., **4** (6): 119, fig. Tremewan, 1961, Bull. Brit. Mus. (nat. Hist.) Ent., **10** (7): 258, pl. 51, fig. 23.

ab. **concolor** Oberthür, 1881, Études d'Entomologie, **6**: 68, pl. 2, fig. 4. Seitz, 1907, Die Gross-Schmetterlinge der Erde, **2**: 29, pl. 8b. Tremewan, 1961, Bull. Brit. Mus. (nat. Hist.) Ent., **10** (7): 258.

ab. **barraguei** Reiss & Tremewan, 1964, Ent. Rec., **76**: 131 (nomen novum for *aurantiaca* Barragué).

aurantiaca Barragué, 1961, Alexanor, **2**: 135, 136 (preoccupied by **aurantiaca** Holl, 1912). Reiss & Tremewan, 1964, Ent. Rec., **76**: 131.

ab. **flava** Barragué, 1961, Alexanor, **2**: 135, 136.

ab. **brunnea** Barragué, 1961, Alexanor, **2**: 135, 136.

f. t. **postalgira** Dujardin, 1964, Entomops, Nice, no. 1: 19, fig. — Dunes de Maison Carrée, Pins Maritimes; Le Tarf; Algérie (forma aestivalis).

ssp. **exigua** Rothschild, 1917, Novit. zool., **24**: 340. Reiss & Tremewan, 1964, Ent. Rec., **76**: 131. — Batna; Lambessa; Khenchela; Algeria.

exigua Seitz, 1907, Die Gross-Schmetterlinge der Erde, **2**: 29, pl. 8a (infrasubspecific). Reiss, 1930, in Seitz, Die Gross-Schmetterlinge der Erde, Supplement, **2**: 25. Reiss & Tremewan, 1964, Ent. Rec., **76**: 131.

ssp. **kebirica** Reiss, 1944, Z. wien. ent. Ges., **29**: 49, pl. 36, figs. 50-52, pl. 47, fig. XII. — Ksar el Kebir, Hochtal des Oued Soufouloud, Mittel Atlas, Marokko, 2000 m.

Biology

Barragué, 1961, Alexanor, **2**: 134-136 (genetics of ab. **flava** Barragué, p. 135). Burgeff, 1913, Ent. Z., **27**: 170, pl. 1; 1950, Portug. acta biol., (A) Goldschmidt: 663-728; 1951, Biol. Zbl., **70**: 1-23. Holik, 1937, Lambillionea, **37**: 15-24, 32-45, 80-91; 1938, ibidem, **38**: 51-58, 79-88, 95-102; 1953, Ent. Z., **63**: 22. Holl, 1912, Bull. Soc. Hist. nat. Afr. N., **4** (6): 120. Reiss, 1930, in Seitz, Die Gross-Schmetterlinge der Erde, Supplement, **2**: 25; 1958, Z. wien. ent. Ges., **43**: 158.

elodia Powell

Distribution: Morocco.

ssp. **elodia** Powell, 1934, Bull. Soc. ent. Fr., **39**: 12. Reiss, 1944, Z. wien. ent. Ges., **29**: 52, pl. 36, fig. 55. Alberti, 1958, Mitt. zool. Mus. Berl., **34**: 310. — Vallée d'Ifrane, Moyen Atlas, Maroc, 1500 m.

ab. **latissimecincta** Reiss, 1944, Z. wien. ent. Ges., **29**: 53.

ssp. **kalypso** Marten, 1944, Z. wien. ent. Ges., **29**: 197. Rungs, 1950, Bull. Soc. Sci. nat. Maroc, **28**: 164. — Ob. Tiguisas, Rif, Marokko, 1400 m.

f. t. **jourdani** Rungs, 1950, Bull. Soc. Sci. nat. Maroc, **28**: 164. — Region d'Ouezzane, Rif occidental (forma vernalis).

Biology

Holik, 1953, Ent. Z., **63**: 25.

faustina Ochsenheimer

Distribution: Southern Portugal, south-west Spain.

ssp. **faustina** Ochsenheimer, 1808, Die Schmetterlinge von Europa, **2**: 99. Reiss, 1932, Int. ent. Z., **26**: 173; 1932, ibidem, **26**: 230, fig.; 1933, in Seitz, Die Gross-Schmetterlinge der Erde, Supplement, **2**: 271. Alberti, 1958, Mitt. zool. Mus. Berl., **34**: 310. Burgeff, 1963, Nachr. Akad. Wiss. Göttingen, **2**, mat.-phys. Kl., no. 22: 326, pl. 2, fig. 1. — Algarve, südliches Portugal. [Chiclana (Andalusien)].

ssp. **baetica** Rambur, 1839, Faune entomologique de l'Andalousie, **2**, pl. 12, fig. 9. Herrich-Schäffer, 1846, Systematische Bearbeitung der Schmetterlinge von Europa, **2**: 45, pl. 11, figs. 79, 80. Rambur, 1866, Catalogue systématique des Lépidoptères de l'Andalousie, p. 170. Seitz, 1907, Die Gross-Schmetterlinge der Erde, **2**: 29, pl. 8b. Oberthür, 1910, Études de Lépidoptérologie comparée, **4**: 624. Reiss, 1930, in Seitz, Die Gross-Schmetterlinge der Erde, Supplement, **2**: 25; 1933, ibidem, **2**: 271; 1932, Int. ent. Z., **26**: 230, figs. Tremewan, 1961, Ent. Rec., **73**: 224, pl. 7, figs. 6, 10; 1962, ibidem, **74**: 126, pl. 2, figs. 2, 19-21. Burgeff, 1963, Nachr. Akad. Wiss. Göttingen, **2**, mat.-phys. Kl., no. 22: 325, pl. 1, figs. b1-b3, b5, b6, pl. 2, figs. 4, 5. — Environs de Malaga, Espagne méridionale.

ab. **faustisimilis** Burgeff, 1963, Nachr. Akad. Wiss. Göttingen, **2**, mat.-phys. Kl., no. 22: 326, pl. 1, figs. b4, b7.

ssp. **almerica** Burgeff, 1963, Nachr. Akad. Wiss. Göttingen, **2**, mat.-phys. Kl., no. 22: 237, pl. 1, fig. c, pl. 2, figs. 6-11. — In einem felsigen Tal bei Almeria, Südspanien.

Biology

Burgeff, 1950, Portug. acta biol., (A) Goldschmidt: 663-728; 1963, Nachr. Akad. Wiss. Göttingen, **2**, mat.-phys. Kl., no. 22: 325-331. Holik, 1937, Lambillionea, **37**: 15-24, 32-45, 80-91; 1938, ibidem, **38**: 51-58, 79-88, 95-102 (**baetica**); 1953, Ent. Z., **63**: 22. Rambur, 1866, Catalogue systématique des Lépidoptères de l'Andalousie, p. 170 (**baetica**). Reiss, 1958, Z. wien. ent. Ges., **43**: 158 (**baetica**). Roüast, 1883, Catalogue des Chenilles européennes connues, p. 23 (**baetica**).

gibraltarica Tremewan

Distribution: Gibraltar.

gibraltarica Tremewan, 1961, Ent. Rec., **73**: 223, pl. 7, figs. 1, 7. Walker, 1890, Trans. ent. Soc. Lond., p. 380 (as *baetica* Rambur). Sheldon, 1908, Entomologist, **41**: 216 (as *baetica* Rambur). Ribbe, 1909/12, Iris, **23**: 358 (as *baetica* Rambur [partim]). Jacobs, 1913, Ent. mon. Mag., **49**: 234 (as *baetica* Rambur). Haaf, 1952, Veröff. zool. Staatssamml. Münch., **2**: 139, 152, 156, pls. 9, 10 (as *baetica* Rambur). Tremewan, 1962, Ent. Rec., **74**: 126, pl. 2, figs. 3, 4, 22, 23. Burgeff, 1963, Nachr. Akad. Wiss. Göttingen, **2**, mat.-phys. Kl., no. 22: 325-331, pl. 1, figs. a2-a6, pl. 2, figs. 2, 3. — Alameda Gardens, Gibraltar.

ab. **faustisimilis** Burgeff, 1963, Nachr. Akad. Wiss. Göttingen, **2**, mat.-phys. Kl., no. 22: 326, pl. 1, figs. a1, a7.

Biology

Burgeff, 1963, Nachr. Akad. Wiss. Göttingen, **2**, mat.-phys. Kl., no. 22: 325-331. Ribbe, 1909/12, Iris, **23**: 358 (as *baetica* [partim]). Sheldon, 1908, Entomologist, **41**: 216 (as *baetica*). Tremewan, 1961, Ent. Rec., **73**: 224. Walker, 1890, Trans. ent. Soc. Lond., p. 380 (as *baetica*).

murciensis Reiss

Distribution: Spain (Murcia, Alicante).

murciensis Reiss, 1922, Int. ent. Z., **15**: 179; 1930, in Seitz, Die Gross-Schmetterlinge der Erde, Supplement, **2**: 25, pl. 2m, n; 1932, Int. ent. Z., **26**: 177; 1932, ibidem, **26**: 230, figs. Haaf, 1952, Veröff. zool. Staatssamml. Münch., **2**: 139, 152, 156, pls. 9, 10. Tremewan, 1961, Ent. Rec., **73**: 224, pl. 7, figs. 2, 8. Burgeff, 1963, Nachr. Akad. Wiss. Göttingen, **2**, mat.-phys. Kl., no. 22: 327, pl. 2, figs. 12, 13. — Totana, Sierra de España, Murcia, Südspanien.

ab. **sextaseparata** Reiss, 1964, Coridon, (A) **6**: 4.

Biology

Burgeff, 1963, Nachr. Akad. Wiss. Göttingen, **2**, mat.-phys. Kl., no. 22: 325-331.

fausta Linné

Distribution: Spain, France, West Alps, central and southern Germany.

ssp. **preciosa** Reiss, 1920, Int. ent. Z., **14**: 117; 1930, in Seitz, Die Gross-Schmetterlinge der Erde, Supplement, **2**: 25, pl. 2m; 1932, Int. ent. Z., **26**: 230, figs. — Sierren um Albarracin, Spanien.

ssp. **oranoides** de Sagarra, 1925, Butll. Inst. catal. Hist. nat., (2) **5**: 274. Reiss, 1930, in Seitz, Die Gross-Schmetterlinge der Erde, Supplement, **2**: 25, pl. 2m. — Ribes i de Manllen; Montseny; Catalonia, España.

laeta Navás, 1923, Arx. Inst. Cienc. Barcelona, **8**: 26 (preoccupied by **laeta** Hübner, 1790).

oranoides Burgeff, 1926, Mitt. münch. ent. Ges., **16**: 45.

ab. **lilliputana** de Sagarra, 1925, Butll. Inst. catal. Hist. nat., (2) **5**: 274. Reiss, 1930, in Seitz, Die Gross-Schmetterlinge der Erde, Supplement, **2**: 25.

ab. **disjuncta** de Sagarra, 1925, Butll. Inst. catal. Hist. nat., (2) **5**: 274. Reiss & Tremewan, 1964, Ent. Rec., **76**: 131.

sagarrai Reiss, 1930, in Seitz, Die Gross-Schmetterlinge der Erde, Supplement, **2**: 25. Reiss & Tremewan, 1964, Ent. Rec., **76**: 132.

ab. **fractimacula** de Sagarra, 1925, Butll. Inst. catal. Hist. nat., (2) **5**: 274. Reiss, 1930, in Seitz, Die Gross-Schmetterlinge der Erde, Supplement, **2**: 25.

ab. **florella** de Sagarra, 1925, Butll. Inst. catal. Hist. nat., (2) **5**: 274.

ab. **confluens** de Sagarra, 1925, Butll. Inst. catal. Hist. nat., (2) **5**: 274.

ab. **dupuyi** Reiss, 1941, Mitt. münch. ent. Ges., **31**: 1000.

f. t. **macraria** de Sagarra, 1925, Butll. Inst. catal. Hist. nat., (2) **5**: 273. Reiss, 1930, in Seitz, Die Gross-Schmetterlinge der Erde, Supplement, **2**: 25. — Entre Llinas i St. Antoni de Vilamajor, Catalonia, España.

f. t. **microsaria** de Sagarra, 1925, Butll. Inst. catal. Hist. nat., (2) **5**: 274. Reiss, 1930, in Seitz, Die Gross-Schmetterlinge der Erde, Supplement, **2**: 25. — Entre Llinas i St. Antoni de Vilamajor, Catalonia, España.

ssp. **junceae** Oberthür, 1884, Études d'Entomologie, **8**: 32; 1888, ibidem, **12**: 24, pl. 7, figs. 49, 49a. Seitz, 1907, Die Gross-Schmetterlinge der Erde, **2**: 29, pl. 8d. Oberthür, 1910, Études de Lépidoptérologie comparée, **4**: 624. Tremewan, 1961, Bull. Brit. Mus. (nat. Hist.) Ent., **10** (7): 259, pl. 51, fig. 27. — Bois d'El Pinats près de Vernet-les-Bains, Pyrénées-Orientales, France.

ssp. **fassnidgei** Tremewan & Manley, 1965, Ent. Rec., **77**: 4. — Jaca, 2700 ft.; La Pena, 2400 ft.; Huesca, Spain.

ssp. **fernan** Agenjo, 1948, Eos, Madr., **24**: 394. Tremewan, 1963, Ent. Rec., **75**: 251. Tremewan & Manley, 1965, Ent. Rec., **77**: 4. — Mte. Santiuste, Pampliega, Burgos, España, 894 m.

ssp. **margheritae** Tremewan, 1961, Ent. Rec., **73**: 3; 1963, ibidem, **75**: 9, pl. 1, figs. 3, 4. — Riano, Leon, Spain, 3650 ft.

ssp. **fausta** Linné, 1767, Systema Naturae, ed. XII, **1** (2): 807. Esper, 1779, Die Schmetterlinge, **2**, pl. 18, figs. 1a, b; 1781, ibidem, **2**: 156. Hübner, 1796, Sammlung europäischer Schmetterlinge, **2**: 14, pl. 5, fig. 27; 1806, ibidem, Der Ziefer, p. 84; [1808]-[20th June, 1813], ibidem, **2**, pl. 26, fig. 122. Ochsenheimer, 1808, Die Schmetterlinge von Europa, **2**: 96. Boisduval, 1828, Essai sur une Monographie des Zygénides, p. 101, pl. 6, fig. 6. Herrich-Schäffer, 1846, Systematische Bearbeitung der Schmetterlinge von Europa, **2**: 45. Freyer, 1852, Neuere Beiträge zur Schmetterlingskunde, **6**: 154, pl. 578, figs. Perlini, 1905, Forme di Lepidotteri esclusivamente Italiane, p. 56, pl. 4, fig. 19. Spuler, 1906, in Hofmann, Die Schmetterlinge Europas, **2**: 163, pl. 72, fig. 13, pl. 77, fig. 32. Seitz, 1907, Die Gross-Schmetterlinge der Erde, **2**: 29, pl. 8b, c. Oberthür, 1910, Études de Lépidoptérologie comparée, **4**: 613. Holik, 1930, Int. ent. Z., **24**: 99. Reiss, 1932, Int. ent. Z., **26**: 221-230, figs.; 1933, in Seitz, Die Gross-Schmetterlinge der Erde, Supplement, **2**: 270. Burgeff, 1950, Portug. acta biol., (A) Goldschmidt: 670, figs. a2; 1951, Biol. Zbl., **70**: 4, figs. 2b. Alberti, 1958, Mitt. zool. Mus. Berl., **34**: 310. Loritz, 1961, Bull. Soc. ent. Mulhouse, p. 83-102, fig. Burgeff, 1963, Nachr. Akad. Wiss. Göttingen, **2**, mat.-phys. Kl., no. 22: 327, 328, pl. 3, figs. 1, 3-6. — Südfrankreich [Küstenregion bei Nizza, im September].

nicaeae Staudinger, 1871, in Staudinger & Wocke, Catalog der Lepidopteren des Europaeischen Faunengebiets, p. 49. Reiss, 1930, in Seitz, Die Gross-Schmetterlinge der Erde, Supplement, **2**: 25.

cerialis Burgeff, 1963, Nachr. Akad. Wiss. Göttingen, **2**, mat.-phys. Kl., no. 22: 328, pl. 3, figs. 7, 8 (nomen nudum).

ab. **segregata** Reiss, 1932, Int. ent. Z., **26**: 224; 1933, in Seitz, Die Gross-Schmetterlinge der Erde, Supplement, **2**: 270.

ab. **pygmaeoides** Burgeff, 1926, Mitt. münch. ent. Ges., **16**: 43.

ssp. **apocrypha** Le Charles, 1930/35, in Lhomme, Catalogue des Lépidoptères de France et de Belgique, **1**: 684. Burgeff, 1963, Nachr. Akad. Wiss. Göttingen, **2**, mat.-phys. Kl., no. 22: 328, pl. 3, fig. 2. Dufay, 1966, Bull. mens. Soc. linn. Lyon, **35**: 69. — Région entre Colle et Tourette, Var; Digne, Basses-Alpes, France.

dresnayi Bernardi & Viette, 1959, Entomologiste, **15**: 6 (nomen nudum).

ab. **tricolor** Oberthür, 1904, Études de Lépidoptérologie

comparée, **1**: 52, pl. 3, figs. 28, 29; 1910, ibidem, **4**: 623. Reiss, 1930, in Seitz, Die Gross-Schmetterlinge der Erde, Supplement, **2**: 24. Tremewan, 1961, Bull. Brit. Mus. (nat. Hist.) Ent., **10** (7): 258, pl. 51, fig. 25.

f. t. **parvafausta** Dujardin, 1956, Bull. mens. Soc. linn. Lyon., **25**: 261. — Thorenc, Alpes-Maritimes, France, 1250 m. (Août).

f. t. **magnafausta** Dujardin, 1956, Bull. mens. Soc. linn. Lyon, **25**: 260. — Col de Braus, Alpes-Maritimes, France, 1000 m. (Juillet).

ssp. **alpiummicans** Verity, 1926, Ent. Rec., **38**: 102. Reiss, 1930, in Seitz, Die Gross-Schmetterlinge der Erde, Supplement, **2**: 25; 1932, Int. ent. Z., **26**: 225, 229, fig. — Oulx, Val Susa, Italy, 1100 m.

ssp. **jucunda** Meissner, 1818, Allg. schweiz. Ges. nat. Anz., **2**: 35. Oberthür, 1910, Études de Lépidoptérologie comparée, **4**: 626. Reiss, 1932, Int. ent. Z., **26**: 229, figs.; 1933, in Seitz, Die Gross-Schmetterlinge der Erde, Supplement, **2**: 270. Lacreuze, 1955, Mitt. ent. Ges. Basel (N.F.), **5**: 107, figs. Daniel, 1957, Bull. Soc. ent. Mulhouse, April, 1957. — Alp Anceindaz, am Fuss der Diablerets (Alpes vaudoises), Schweiz.

ab. **cingulata** Reiss, 1922, Int. ent. Z., **16**: 84; 1930, in Seitz, Die Gross-Schmetterlinge der Erde, Supplement, **2**: 24.

ab. **bicolor** Reverdin, 1922, Bull. Soc. lépid. Genève, **5**: 46, pl. 1, fig. 5.

ab. **lutescens** Muschamp, 1905, Bull. Soc. lépid. Genève, **1**: 70.

flava Vorbrodt, 1913, in Vorbrodt & Müller-Rutz, Die Schmetterlinge der Schweiz, **2**: 277.

ssp. **fina** Burgeff, 1956, Nova Acta Leop. Carol., **18** (127): 40, pl. 6, figs. 17f-p. — Gegend von Siders (Sièrre) im Rhônetal, Schweiz.

ssp. **genevensis** Millière, 1861, Iconographie et Description de Chenilles et Lépidoptères, **1**: 237, pl. 3, figs. 1, 2. Guenée, 1865, Ann. Soc. ent. Fr., (4) **5**: 91. Millière, 1887, Ann. Soc. ent. Fr., (6) **7**: 216, pl. 5, figs. 4-6. Seitz, 1907, Die Gross-Schmetterlinge der Erde, **2**: 29, pl. 8c (as *jucunda* Meissner). Reiss, 1930, in Seitz, Die Gross-Schmetterlinge der Erde, Supplement, **2**: 24. Daniel, 1957, Bull. Soc. ent. Mulhouse, April, 1957. — Mont Salève près Genève, Suisse.

faustula Rambur, 1866, Catalogue systématique des Lépidoptères de l'Andalousie, p. 171. Bernardi & Viette, 1961, Bull. mens. Soc. linn. Lyon, **30**: 144.

mauritanica Mabille, 1885, Bull. Soc. philom. Paris, (7) **9**: 57. Bernardi & Viette, 1961, Bull. mens. Soc. linn. Lyon, **30**: 144.

mabillei Kirby, 1892, A Synonymic Catalogue of Lepidoptera Heterocera (Moths), p. 77. Bernardi & Viette, 1961, Bull. mens. Soc. linn. Lyon, **30**: 144.

ab. **pygmaeoides** Blachier, 1906, Ann. Soc. ent. Fr., **75**: 22; 1914, Bull. Soc. lépid. Genève, **3**: 84, pl. 2, fig. 11. Reiss, 1930, in Seitz, Die Gross-Schmetterlinge der Erde, Supplement, **2**: 24.

ab. **segregata** Blachier, 1905, Bull. Soc. ent. Fr., p. 52; 1906, Ann. Soc. ent. Fr., **75**: 22. Seitz, 1907, Die Gross-Schmetterlinge der Erde, **2**: 29. Blachier, 1914, Bull. Soc. lépid. Genève, **3**: 84, pl. 2, fig. 3.

parvimaculata Vorbrodt, 1913, in Vorbrodt & Müller-Rutz, Die Schmetterlinge der Schweiz, **2**: 278.

ab. **simplex** Muschamp, 1909, Bull. Soc. lépid. Genève, **1**: 368.

ssp. **lacrymans** Burgeff, 1914, Mitt. münch. ent. Ges., **5**: 53, pl. 2, figs. 168, 169, pl. 6, figs. 61-67. Zeller, 1877, Stettin. ent. Ztg., **38**: 321. Reiss, 1930, in Seitz, Die Gross-Schmetterlinge der Erde, Supplement, **2**: 24, pl. 21; 1933, ibidem, **2**: 270; 1932, Int. ent. Z., **26**: 229, figs.; 1950, Jber. naturf. Ges. Graubünden, **82**: 111, fig. 19. Daniel, 1957, Bull. Soc. ent. Mulhouse, April, 1957. Forster & Wohlfahrt, 1958, Die Schmetterlinge Mitteleuropas, **3**: 95, pl. 10, figs. 15, 20. — Filisur, Graubünden, Schweiz.

ssp. **monacensis** Daniel, 1932, in Osthelder, Die Schmetterlinge Südbayerns, **1**: 577, pl. 21, figs. 25, 26. Reiss, 1933, in Seitz, Die Gross-Schmetterlinge der Erde, Supplement, **2**: 270. Bergmann, 1953, Die Grossschmetterlinge Mitteldeutschlands, **3**, pl. 66, figs. D6, D7. Daniel, 1957, Bull. Soc. ent. Mulhouse, April, 1957. — Bei Wolfratshausen, Südbayern, Süddeutschland.

ssp. **fortunata** Rambur, 1866, Catalogue systématique des Lépidoptères de l'Andalousie, p. 172. Oberthür, 1910, Études de Lépidoptérologie comparée, **4**: 615. Reiss, 1930, in Seitz, Die Gross-Schmetterlinge der Erde, Supplement, **2**: 24, pl. 2m; 1933, ibidem, **2**: 270; 1932, Int. ent. Z., **26**: 229, figs. — Angoulème, Charente; Fontainebleau, Seine et Marne; France.

ab. **major** Pionneau, 1939, Échange, **55**: 27.

ab. **bernieri** Gouin, 1923, Act. Soc. linn. Bordeaux, **74** (1): 169.

ab. **dupuyi** Oberthür, 1907, Ann. Soc. ent. Fr., **76**: 45; 1909, Études de Lépidoptérologie comparée, **3**, pl. 22, fig. 109; 1910, ibidem, **4**: 616. Seitz, 1912, Die Gross-Schmetterlinge der Erde, **2**: 444. Reiss, 1930, in Seitz, Die Gross-Schmetterlinge der Erde, Supplement, **2**: 24. Tremewan, 1961, Bull. Brit. Mus. (nat. Hist.) Ent., **10** (7): 259, pl. 51, fig. 28.

ab. **melusina** Oberthür, 1909, Études de Lépidoptérologie comparée, **3**, pl. 22, fig. 108; 1910, ibidem, **4**: 616. Reiss, 1930, in Seitz, Die Gross-Schmetterlinge der Erde, Supplement, **2**: 24. Tremewan, 1961, Bull. Brit. Mus. (nat. Hist.) Ent., **10** (7): 259, pl. 51, fig. 29.

f. t. **autumnalis** Burgeff, 1914, Mitt. münch. ent. Ges., **5**: 53, pl. 6, figs. 57-60. Reiss, 1930, in Seitz, Die Gross-Schmetterlinge der Erde, Supplement, **2**: 25, pl. 2m. — Auzay, Vendée, Frankreich [September und Oktober].

ssp. **perornata** Le Charles, 1960, Bull. Soc. ent. Fr., **65**: 102 (as *pernornafa*, corrected to **perornata** in index, p. 316). Dupont, 1900, Bull. Soc. Sci. nat. Elbeuf, **18**: 75; 1925, ibidem, **43**: 137. — Lardy, Seine-et-Oise; Orgemont, Saclas; France.

ssp. **rhodana** Dujardin, 1964, Entomops, Nice, no. 1: 20, fig. — Mt. Thou (environs de Lyon), Rhône, France centrale.

ab. **lugdunensis** Millière, 1859, Iconographie et Description de Chenilles et Lépidoptères, **1**: 85, pl. 4, fig. 4. Seitz, 1907, Die Gross-Schmetterlinge der Erde, **2**: 29.

ssp. **suevica** Reiss, 1920, Int. ent. Z., **14**: 117; 1926, Die Zygaenen Deutschlands, p. 33, pl. 1, fig.; 1930, in Seitz, Die Gross-Schmetterlinge der Erde, Supplement, **2**: 24, pl. 2m; 1933, ibidem, **2**: 271; 1932, Int. ent. Z., **26**: 230, figs. Daniel, 1932, in Osthelder, Die Schmetterlinge Südbayerns, pl. 21, figs. 27-30. Reiss, 1937, in Schneider, Jh. Ver. vaterl. Naturk. Württemb., **93**: 127; 1949, Entomon, **1**: 171. — Umgebung des Hohenneuffen, Schwäbische Alb, Süddeutschland, 550 m.

ab. **rubrothoraxalis** Reiss, 1949, Entomon, **1**: 171.

ab. **pygmaeoides** Burgeff, 1926, Mitt. münch. ent. Ges., **16**: 43. Reiss, 1930, in Seitz, Die Gross-Schmetterlinge der Erde, Supplement, **2**: 24.

ab. **segregata** Reiss, 1941, Mitt. münch. ent. Ges., **31**: 1000.

ab. **dupuyi** Reiss, 1964, Coridon, (A) **6**: 4.

ab. **inversa** Reiss, 1922, Int. ent. Z., **16**: 84; 1930, in Seitz, Die Gross-Schmetterlinge der Erde, Supplement, **2**: 24; 1941, Mitt. münch. ent. Ges., **31**: 1000.

ab. **fereflava** Reiss, 1941, Mitt. münch. ent. Ges., **31**: 1000.

ab. **fereimmaculata** Reiss, 1964, Coridon, (A) **6**: 4.

ab. **flava** Reiss, 1926, Int. ent. Z., **20**: 215; 1926, Die Zygaenen Deutschlands, p. 34, pl. 1, fig.; 1930, in Seitz, Die Gross-Schmetterlinge der Erde, Supplement, **2**: 24, pl. 21 (as *lugdunensis* Millière); 1933, ibidem, **2**: 271; 1941, Z. wien. EntVer., **26**: 63.

ab. **brunnea** Reiss, 1941, Mitt. münch. ent. Ges., **31**: 999.

ssp. **agilis** Reiss, 1932, Int. ent. Z., **26**: 227, 231, figs.; 1926, Die Zygaenen Deutschlands, p. 33, pl. 1, fig. (as *fausta* Linné). Holik, 1930, Int. ent. Z., **24**: 99. Daniel, 1932, in Osthelder, Die Schmetterlinge Südbayerns, pl. 21, fig. 31. Reiss, 1933, in Seitz, Die Gross-Schmetterlinge der Erde, Supplement, **2**: 271. Mergard, 1950, Ent. Z., **60**: 144. Haaf, 1951, Ent. Z., **61**: 89-92, fig. (distribution map); 1952, Veröff. zool. Staatssamml. Münch., **2**: 151, 154, 156 (as *fausta* Linné), pl. 9. — Kunitzburg bei Jena; Apolda; Arnstadt; Thüringen, Mitteldeutschland.

Bergmann, 1953, Die Grossschmetterlinge Mitteldeutschlands, **3**: 28, pl. 65, figs. A3, B1-B4, C1-C4, D1-D4, E1-E3, pl. 66, figs. D4, D5. Forster & Wohlfahrt, 1958, Die Schmetterlinge Mitteleuropas, **3**: 95, pl. 10, figs. 14, 19.

ab. **brunnea** Oberthür, 1909, Études de Lépidoptérologie comparée, **3**, pl. 29, fig. 178; 1910, ibidem, **4**: 621. Reiss, 1930, in Seitz, Die Gross-Schmetterlinge der Erde, Supplement, **2**: 24. Tremewan, 1961, Bull. Brit. Mus. (nat. Hist.) Ent.. **10** (7): 259, pl. 51, fig. 26.

Biology

Abeille, 1909, Mém. Soc. linn. Provence, **1**: 21-24. Burgeff, 1912, Z. wiss. InsektBiol., **8**: 186; 1950, Portug. acta biol., (A) Goldschmidt: 663-728; 1951, Biol. Zbl., **70**: 1-23; 1963, Nachr. Akad. Wiss. Göttingen, **2**, mat.-phys. Kl., no. 22: 325-331. Daniel, 1957, Bull. Soc. ent. Mulhouse, April, 1957. Döring, 1955, Zur Morphologie der Schmetterlingseier, p. 120, pl. 18, fig. 255. Freyer, 1845, Stettin. ent. Ztg., **6**: 24; 1852, Neuere Beiträge zur Schmetterlingskunde, **6**: 154, pl. 578. Holik, 1937, Lambillionea, **37**: 15-24, 32-45, 80-91; 1938, ibidem, **38**: 51-58, 79-88, 95-102; 1953, Ent. Z., **63**: 21. Koch, 1955, Wir Bestimmen Schmetterlinge, **2**: 60, 61, pl. 1, fig. 11, pl. 15, fig. 11. Millière, 1869/74, Iconographie et Description de Chenilles et Lépidoptères, **3**: 63, pl. 107, figs. 7, 8; 1887, Ann. Soc. ent. Fr., (6) **7**: 216, figs. 4-6. Oberthür, 1907, Ann. Soc. ent. Fr., **76**: 44. Ochsenheimer, 1808, Die Schmetterlinge von Europa, **2**: 96. Rambur, 1866, Catalogue systématique des Lépidoptères de l'Andalousie, p. 171. Reiss, 1926, Die Zygaenen Deutschlands, p. 7; 1930, in Seitz, Die Gross-Schmetterlinge der Erde, Supplement, **2**: 25; 1958, Z. wien. ent. Ges., **43**: 158. Roüast, 1883, Catalogue des Chenilles européennes connues, p. 23. Spuler, 1910, in Hofmann, Die Raupen der Schmetterlinge Europas, pl. 10, figs. 7a, b. Vorbrodt, 1913, in Vorbrodt & Müller-Rutz, Die Schmetterlinge der Schweiz, **2**: 277.

excelsa Rothschild

Distribution: Algeria.

excelsa Rothschild, 1917, Novit. zool., **24**: 340; 1920, ibidem, **26**: 356, pl. 1, figs. 12, 13; 1925, Ann. Mag. nat. Hist., (9) **15**: 678; 1925, ibidem, (9) **16**: 269. Reiss, 1930, in Seitz, Die Gross-Schmetterlinge der Erde, Supplement, **2**: 25, pl. 2n; 1944, Z. wien. ent. Ges., **29**: 49, pl. 37, fig. 72. Reiss & Tremewan, 1960, Bull. Brit. Mus. (nat. Hist.) Ent., **9** (10): 463, pl. 22, fig. 22, pl. 24, figs. 1, 2, pl. 25, figs. 1, 2. Tremewan,

Djebel Mekter near Aïn Sefra, West Algeria, 1600-1900 m.

1961, Bull. Brit. Mus. (nat. Hist.) Ent., **10** (7): 255, pl. 51, fig. 11, pl. 59, figs. 5, 6.

alluaudi Oberthür

Distribution: Morocco.

ssp. **alluaudi** Oberthür, 1922, Études de Lépidoptérologie comparée, **19**: 159, pl. 545, figs. 4583, 4584. Rothschild, 1925, Ann. Mag. nat. Hist., (9) **16**: 269. Reiss, 1944, Z. wien. ent. Ges., **29**: 49, pl. 35, figs. 17-20, pl. 47, fig. XIV. Alberti, 1958, Mitt. zool. Mus. Berl., **34**: 209. Reiss & Tremewan, 1960, Bull. Brit. Mus. (nat. Hist.) Ent., **9** (10): 463, pl. 22, fig. 23, pl. 24, figs. 3, 4, pl. 25, figs. 3, 4. Tremewan, 1961, Bull. Brit. Mus. (nat. Hist.) Ent., **10** (7): 255, pl. 51, fig. 13, pl. 60, figs. 3, 4. — Auprès du poste de Bou Angher, Moyen Atlas, Maroc, 2000 m.

ab. **felicinoides** Reiss, 1944, Z. wien. ent. Ges., **29**: 51, pl. 37, fig. 68.

ab. **exarcuata** Reiss, 1944, Z. wien. ent. Ges., **29**: 51, pl. 35, fig. 21.

ab. **quintaseparata** Reiss, 1944, Z. wien. ent. Ges., **29**: 51.

ab. **semiconfluens** Reiss, 1944, Z. wien. ent. Ges., **29**: 51.

ab. **confluens** Reiss, 1944, Z. wien. ent. Ges., **29**: 50, pl. 35, fig. 22; 1930, in Seitz, Die Gross-Schmetterlinge der Erde, Supplement, **2**: 26, pl. 2n (as *alluaudi* Oberthür).

ssp. **inula** Reiss, 1944, Z. wien. ent. Ges., **29**: 52, pl. 37, figs. 66, 67. Haaf, 1952, Veröff. zool. Staatssamml. Münch., **2**: 152, 157, pl. 15 (as *alluaudi* Oberthür). — Berge nordöstlich Bab Tazza, Rif, Marokko, 1700 m.

ab. **pseudoalluaudi** Reiss, 1944, Z. wien. ent. Ges., **29**: 52.

Biology

Holik, 1953, Ent. Z., **63**: 25.

hilaris Ochsenheimer

Distribution: Southern Portugal, Spain, southern France, west Alps to Piedmont and Liguria.

ssp. **hilaris** Ochsenheimer, 1808, Die Schmetterlinge von Europa, **2**: 101. Hübner, [1808]-[20th June 1813], Sammlung europäischer Schmetterlinge, **2**, pl. 26, fig. 123. Boisduval, 1828, Essai sur une Monographie des Zygénides, p. 99, pl. 6, fig. 5; 1834, Icones historique des Lépidoptères nouveaux ou peu connus, **2**: 74, pl. 55, fig. 1. Duponchel, 1835, in Godart & Duponchel, Histoire naturelle des Lépidoptères ou Papillons de France, Supplement, **2**: 84, pl. 7, fig. 5. Herrich-Schäffer, 1846, Systematische Bearbeitung der Schmetterlinge von Europa, **2**: 45. Spuler, 1906, in Hofmann, Die Schmetterlinge Europas, **2**: 163, pl. 72, fig. 11, pl. 77, fig. 30. Seitz, 1907, Die Gross-Schmetterlinge der Erde, **2**: 28, pl. 7i. Oberthür, 1910, Études de Lépidoptérologie comparée, **4**: 599. Reiss, — Südliches Portugal [Algarve, Faro].

1930, in Seitz, Die Gross-Schmetterlinge der Erde, Supplement, **2**: 24; 1933, ibidem, **2**: 270; 1932, Int. ent. Z., **26**: 197-199; 1932, ibidem, **26**: 230, figs. Alberti, 1958, Mitt. zool. Mus. Berl., **34**: 309.

ssp. **aphrodisia** Burgeff, 1926, Mitt. münch. ent. Ges., **16**: 43. Rambur, 1839, Faune entomologique de l'Andalousie, **2**, pl. 12, fig. 6 (as *hilaris* Ochsenheimer). Reiss, 1930, in Seitz, Die Gross-Schmetterlinge der Erde, Supplement, **2**: 24, pl. 21. Schmidt-Koehl, 1965, Ent. Z., **75**: 282. — Alg. de la Lluvia, Granada, Südspanien.

ab. **exarcuata** Burgeff, 1926, Mitt. münch. ent. Ges., **16**: 43. Reiss, 1930, in Seitz, Die Gross-Schmetterlinge der Erde, Supplement, **2**: 24.

ab. **inversa** Burgeff, 1926, Mitt. münch. ent. Ges., **16**: 43. Reiss, 1930, in Seitz, Die Gross-Schmetterlinge der Erde, Supplement, **2**: 24.

ssp. **lucifera** Reiss, 1936, Ent. Rdsch., **54**: 72, pl. 2, figs. — Sierren um Albarracin, Spanien, 1500-1750 m.

ab. **exarcuata** Reiss, 1936, Ent. Rdsch., **54**: 73, pl. 2, fig.

ab. **gelpkei** Reiss, 1936, Ent. Rdsch., **54**: 73.

ab. **pseudocatalonica** Reiss, 1936, Ent. Rdsch., **54**: 73.

ab. **inversa** Reiss, 1936, Ent. Rdsch., **54**: 73.

ssp. **leonica** Tremewan, 1961, Ent. Rec., **73**: 3; 1963, ibidem, **75**: 2, pl. 1, figs. 5, 6. — Riano, Leon, Spain, 3650 ft.

ssp. **escorialensis** Oberthür, 1884, Études d'Entomologie, **8**: 33; 1888, ibidem, **12**: 23, pl. 7, fig. 48. Seitz, 1907, Die Gross-Schmetterlinge der Erde, **2**: 28, pl. 7k. Oberthür, 1910, Études de Lépidoptérologie comparée, **4**: 601. Reiss, 1932, Int. ent. Z., **26**: 230, fig. Tremewan, 1961, Bull. Brit. Mus. (nat. Hist.) Ent., **10** (7): 257, pl. 51, fig. 21. — Escorial, Espagne.

ab. **falleri** Reiss, 1922, Int. ent. Z., **15**: 180; 1930, in Seitz, Die Gross-Schmetterlinge der Erde, Supplement, **2**: 24.

ssp. **catalonica** de Sagarra, 1924, Trab. Mus. Cienc. nat. Barcelona, **11** (1): 18. Reiss, 1930, in Seitz, Die Gross-Schmetterlinge der Erde, Supplement, **2**: 24. Haaf, 1952, Veröff. zool. Staatssamml. Münch., **2**: 152, 157, pl. 15 (as *hilaris* Ochsenheimer). — St. Pere de Vilamajor, Barcelona, España.

cataloniana Burgeff, 1926, Mitt. münch. ent. Ges., **16**: 42.

ab. **erubescens** de Sagarra, 1924, Trab. Mus. Cienc. nat. Barcelona, **11** (1): 17, pl. 4, fig. 1.

ssp. **piemontica** Reiss, 1941, Z. wien. EntVer., **26**: 63. — Col di Sestriere, Piemont, Ober-Italien, 1600-1900 m.

ssp. **milae** Storace, 1961, Boll. Soc. ent. ital., **91**: 143; 1956, ibidem, **86**: 140, figs. B4-B8; 1962, ibidem, **92**: 102, figs. 4-8. — Genova-Camandoli; Genova-

ab. **erythraea** Storace, 1956, Boll. Soc. ent. ital., **86**: 141, fig. B11; 1962, ibidem, **92**: 105, fig. 11. — Chiapetto; S. Margherita Ligure; Liguria centrale, Italia.

ssp. **chrysophaea** Le Charles, 1930/35, in Lhomme, Catalogue des Lépidoptères de France et de Belgique, **1**: 683. Reiss, 1930, in Seitz, Die Gross-Schmetterlinge der Erde, Supplement, **2**, pl. 21 (as *galliae* Oberthür); 1958, Bull. Soc. ent. Mulhouse, p. 49. Forster & Wohlfahrt, 1958, Die Schmetterlinge Mitteleuropas, **3**: 94, pl. 10, figs. 13, 18 (as *ga'liae* Oberthür). De Bros, 1964, Mitt. ent. Ges. Basel (N.F.), **14**: 159. Dufay, 1966, Bull. mens. Soc. linn. Lyon, **35**: 69. — Fontgaillard, 1000 m.; Digne; Basses-Alpes, France.

ab. **pallida** Seitz, 1907, Die Gross-Schmetterlinge der Erde, **2**: 28, pl. 7i. Tremewan, 1962, Ent. Rec., **74**: 127, pl. 2, fig. 5.

ssp. **ononidis** Millière, 1878, Petites Nouv. Ent., **2**: 249; 1879, Mem. Soc. Sci. nat. Cannes, **8**: 112, pl. 5, figs. 6-10. Oberthür, 1910, Études de Lépidoptérologie comparée, **4**: 603. Burgeff, 1926, Mitt. münch. ent. Ges., **16**: 43. Reiss, 1930, in Seitz, Die Gross-Schmetterlinge der Erde, Supplement, **2**: 24, pl. 21; 1958, Bull. Soc. ent. Mulhouse, p. 49. — Sur une colline de la vallée du Cannet, Cannes, Alpes-Maritimes, France.

ab. **tricolor** Oberthür, 1909, Études de Lépidoptérologie comparée, **3**, pl. 29, fig. 180; 1910, ibidem, **4**: 603. Reiss, 1930, in Seitz, Die Gross-Schmetterlinge der Erde, Supplement, **2**: 24. Tremewan, 1961, Bull. Brit. Mus. (nat. Hist.) Ent., **10** (7): 256, pl. 51, fig. 18.

ab. **aurantiaca** Oberthür, 1910, Études de Lépidoptérologie comparée, **4**: 603. Tremewan, 1961, Bull. Brit. Mus. (nat. Hist.) Ent., **10** (7): 257, pl. 51, fig. 20.

ab. **foulquieri** Oberthür, 1909, Études de Lépidoptérologie comparée, **3**, pl. 29, fig. 179; 1910, ibidem, **4**: 603. Seitz, 1912, Die Gross-Schmetterlinge der Erde, **2**: 443. Tremewan, 1961, Bull. Brit. Mus. (nat. Hist.) Ent., **10** (7): 257, pl. 51, fig. 19.

ssp. **superononidis** Dujardin, 1964, Entomops, Nice, no. 1: 19, fig. — Haut Gréolières, Alpes-Maritimes, France, 850 m.

ssp. **nicaeica** Reiss, 1958, Bull. Soc. ent. Mulhouse, p. 48.
nicaeica Reiss, 1958, Z. wien. ent. Ges., **43**: 159 (nomen nudum).
ab. **sextaseparata** Reiss, 1958, Bull. Soc. ent. Mulhouse, p. 49.
ab. **medioseparata** Reiss, 1958, Bull. Soc. ent. Mulhouse, p. 49.
ab. **pseudoononidis** Reiss, 1958, Bull. Soc. ent. Mulhouse, p. 49.
— Mt. Leuze (Mt. Pacanaglia) près de Nice, Alpes-Maritimes, France.

ssp. **galliae** Oberthür, 1910, Études de Lépidoptérologie comparée, **4**: 602; 1888, Études d'Entomologie, **12**: 23, pl. 7, figs. 48a, b, c; 1896, ibidem, **20**: 51. Reiss, 1958, Bull. Soc. — Vernet-les-Bains, Pyrénées-Orientales, France.

ent. Mulhouse, p. 48, 49. Tremewan, 1961, Bull. Brit. Mus. (nat. Hist.) Ent., **10** (7): 256, pl. 51, fig. 14.

ab. **confluens** Oberthür, 1896, Études d'Entomologie, **20**: 51, pl. 7, fig. 129. Seitz, 1907, Die Gross-Schmetterlinge der Erde, **2**: 28, pl. 7i. Tremewan, 1961, Bull. Brit. Mus. (nat. Hist.) Ent., **10** (7): 256, pl. 51, fig. 17.

confluens Spuler, 1906, in Hofmann, Die Schmetterlinge Europas, **2**: 163.

coniuncta Spuler, 1910, in Hofmann, Die Schmetterlinge Europas, **2**: 494.

ab. **bicolor** Oberthür, 1896, Études d'Entomologie, **20**: 51, pl. 7, fig. 130. Seitz, 1907, Die Gross-Schmetterlinge der Erde, **2**: 28, pl. 7k. Oberthür, 1910, Études de Lépidoptérologie comparée, **4**: 603. Tremewan, 1961, Bull. Brit. Mus. (nat. Hist.) Ent., **10** (7): 256, pl. 51, fig. 15.

ab. **unicolor** Oberthür, 1896, Études d'Entomologie, **20**: 51, pl. 7, fig. 131. Seitz, 1907, Die Gross-Schmetterlinge der Erde, **2**: 28, pl. 7k. Oberthür, 1910, Études de Lépidoptérologie comparée, **4**: 603. Tremewan, 1961, Bull. Brit. Mus. (nat. Hist.) Ent., **10** (7): 256, pl. 51, fig. 16.

Biology

Burgeff, 1950, Portug. acta biol., (A) Goldschmidt: 663-728; 1951, Biol. Zbl., **70**: 1-23. Holik, 1937, Lambillionea, **37**: 15-24, 32-45, 80-91; 1953, Ent. Z., **63**: 21. Millière, 1878, Petites Nouv. Ent., **2**: 249; 1878/79, Mém. Soc. Sci. nat. Cannes, **8**: 112, pl. 5, figs. 6-10. Reiss, 1958, Bull. Soc. ent. Mulhouse, p. 48; 1958, Z. wien. Ent. Ges., **43**: 158. Roüast, 1883, Catalogue des Chenilles européennes connues, p. 23.

marcuna Oberthür

Distribution: Algeria, Morocco.

ssp. **marcuna** Oberthür, 1888, Études d'Entomologie, **12**: 27; 1890, ibidem, **13**: 22, pl. 7, fig. 58. Seitz, 1907, Die Gross-Schmetterlinge der Erde, **2**: 29, pl. 8b. Oberthür, 1910, Études de Lépidoptérologie comparée, **4**: 613. Reiss, 1930, in Seitz, Die Gross-Schmetterlinge der Erde, Supplement, **2**: 25; 1933, ibidem, **2**: 271; 1944, Z. wien. ent. Ges., **29**: 10, pl. 37, figs. 69, 70, pl. 46, fig. X. Alberti, 1958, Mitt. zool. Mus. Berl., **34**: 307. Reiss & Tremewan, 1960, Bull. Brit. Mus. (nat. Hist.) Ent., **9** (10): 463, pl. 22, fig. 20, pl. 24, figs. 5, 6, pl. 25, figs. 5, 6. Tremewan, 1961, Bull. Brit. Mus. (nat. Hist.) Ent., **10** (7): 255, pl. 51, fig. 12, pl. 60, figs. 1, 2.

Marcouna près Lambessa, Constantine, Algérie.

ssp. **ahmarica** Reiss, 1944, Z. wien. ent. Ges., **29**: 11, pl. 37, fig. 71.

Militär Posten 1565, Djebel Ahmar, Mittel-Atlas, Marokko, 1700-1765 m.

Biology

Holik, 1953, Ent. Z., **63**: 25.

tingitana Reiss
Distribution: Morocco (Rif).

ssp. **tingitana** Reiss, 1937, Ent. Rdsch., **54**: 469, fig. d4. Reisser, 1933, Eos, Madr., **9**: 281, pl. 6, figs. 51, 52, 55 (as *maroccana* Rothschild). Reiss, 1941, Z. wien. EntVer., **26**: 290, pl. 31, fig. 7; 1944, Z. wien. ent. Ges., **29**: 12, pl. 37, figs. 73, 74, pl. 46, fig. VIII. Haaf, 1952, Veröff. zool. Staatssamml. Münch., **2**: 152, 154, 157, pl. 11 (as *maroccana* Rothschild). Alberti, 1958, Mitt. zool. Mus. Berl., **34**: 307. — Izilan und A'Faska, Rif, Marokko.

ab. **latemarginata** Reiss, 1944, Z. wien. ent. Ges., **29**: 14.

ab. **laticincta** Reiss, 1944, Z. wien. ent. Ges., **29**: 13, pl. 37, fig. 75.

ab. **medioconfluens** Reiss, 1944, Z. wien. ent. Ges., **29**: 13, pl. 37, fig. 76.

ab. **sexmacula** Reiss, 1941, Z. wien. EntVer., **26**: 290, pl. 31, figs. 8, 9. Reisser, 1933, Eos, Madr., **9**: 281, pl. 6, figs. 53, 54, 56 (as *maroccana* Rothschild). Reiss, 1944, Z. wien. ent. Ges., **29**: 13, pl. 37, fig. 77.

ab. **pseudomarcuna** Reiss, 1944, Z. wien. ent. Ges., **29**: 13, pl. 37, figs. 78, 79.

ssp. **oreodoxa** Marten, 1944, Z. wien. ent. Ges., **29**: 195. Alberti, 1958, Mitt. zool. Mus. Berl., **34**: 308. — Magómassiv, Rif, Marokko, 1400-1800 m.

ab. **tigrina** Marten, 1944, Z. wien. Ent. Ges., **29**: 195.

ab. **meleagris** Marten, 1944, Z. wien. ent. Ges., **29**: 196.

Biology

Holik, 1953, Ent. Z., **63**: 25.

zoraida Reiss
Distribution: Morocco (Rif).

zoraida Reiss, 1943, Z. wien. ent. Ges., **28**: 358, pl. 38, fig. 85. — Djebel Laxchab, West Rif, Marokko, 2000 m.

orana Duponchel
Distribution: Tunisia, Algeria, Sardinia, Morocco.

ssp. **orana** Duponchel, 1835, in Godart & Duponchel, Histoire naturelle des Lépidoptères ou Papillons de France, Supplement, **2**: 145, pl. 12, fig. 8. Oberthür, 1888, Études d'Entomologie, **12**: 27; 1890, ibidem, **13**: 24, pl. 7, fig. 52. Seitz, 1907, Die Gross-Schmetterlinge der Erde, **2**: 30, pl. 8h; 1912, ibidem, **2**: 444. Oberthür, 1910, Études de Lépidoptérologie comparée, **4**: 637. Rothschild, 1917, Novit. zool., **24**: 341. — Oran, Algérie.

Reiss, 1930, in Seitz, Die Gross-Schmetterlinge der Erde, Supplement, **2**: 26. Alberti, 1958, Mitt. zool. Mus. Berl., **34**: 305.

ab. **barbara** Herrich-Schäffer, 1846, Systematische Bearbeitung der Schmetterlinge von Europa, **2**: 47; 1844, ibidem, **2**, pl. 4, figs. 29, 30 (non-binominal). Seitz, 1907, Die Gross-Schmetterlinge der Erde, **2**: 30, pl. 8i.

nedroma Oberthür, 1881, Études d'Entomologie, **6**: 68, pl. 3, fig. 3. Tremewan, 1961, Bull. Brit. Mus. (nat. Hist.) Ent., **10** (7): 260, pl. 51, fig. 31.

minor Seitz, 1907, Die Gross-Schmetterlinge der Erde, **2**: 30.

minima Seitz, 1907, Die Gross-Schmetterlinge der Erde, **2**, pl. 8i.

ab. **oberthueri** Bethune-Baker, 1888, Trans. ent. Soc. Lond., p. 118 (as *oberthüri*). Tremewan, 1961, Bull. Brit. Mus. (nat. Hist.) Ent., **10** (7): 260, pl. 51, fig. 30.

ssp. **lahayei** Oberthür, 1890, Études d'Entomologie, **13**: 24, pl. 7, fig. 53; 1911, Études de Lépidoptérologie comparée, **5** (1): 320, pl. 63, figs. 593, 594. Seitz, 1912, Die Gross-Schmetterlinge der Erde, **2**: 444. Reiss, 1930, in Seitz, Die Gross-Schmetterlinge der Erde, Supplement, **2**: 26, pl. 2o (after Oberthür). Tremewan, 1961, Bull. Brit. Mus. (nat. Hist.) Ent., **10** (7): 260, pl. 51, fig. 33. — Géryville, Algérie.

ab. **powelli** Oberthür, 1911, Études de Lépidoptérologie comparée, **5** (1), pl. 63, fig. 592. Seitz, 1912, Die Gross-Schmetterlinge der Erde, **2**: 444. Reiss, 1930, in Seitz, Die Gross-Schmetterlinge der Erde, Supplement, **2**: 26. Tremewan, 1961, Bull. Brit. Mus. (nat. Hist.) Ent., **10** (7): 261, pl. 51, fig. 34.

ssp. **allardi** Oberthür, 1878, Études d'Entomologie, **3**: 41, pl. 5, fig. 5; 1890, ibidem, **13**: 24, pl. 7, figs. 54-56. Seitz, 1907, Die Gross-Schmetterlinge der Erde, **2**: 30, pl. 8i. Rothschild, 1917, Novit. zool., **24**: 341. Reiss, 1930, in Seitz, Die Gross-Schmetterlinge der Erde, Supplement, **2**: 26; 1943, Z. wien. ent. Ges., **28**: 358, pl. 38, figs. 83, 84. Haaf, 1952, Veröff. zool. Staatssamml. Münch., **2**: 152, 154, 157, pl. 11. Tremewan, 1961, Bull. Brit. Mus. (nat. Hist.) Ent., **10** (7): 261, pl. 51, fig. 35. — Marcouna près Lambessa, Algérie.

ssp. **limitans** Rothschild, 1917, Novit. zool., **24**: 341. Reiss, 1930, in Seitz, Die Gross-Schmetterlinge der Erde, Supplement, **2**: 26, pl. 2o. Tremewan, 1961, Bull. Brit. Mus. (nat. Hist.) Ent., **10** (7): 260, pl. 51, fig. 32. — Bône, Constantine, Algeria.

f.t. **autumnalis** Burgeff, 1926, Mitt. münch. ent. Ges., **16**: 46. Reiss, 1930, in Seitz, Die Gross-Schmetterlinge der Erde, Supplement, **2**: 26. — Sidi bou Said und Hamman el Lif bei Tunis, Tunesien.

ssp. **sardoa** Mabille, 1892, Bull. Soc. ent. Fr., **61**: cl. Seitz, 1907, Die Gross-Schmetterlinge der Erde, **2**: 30. Turati, 1919, Nat. sicil., **23**: 241. Wagner, 1919, Ent. Mitt., **8**: 183. Reiss, 1930, in Seitz, Die Gross-Schmetterlinge der Erde, Supplement, **2**: 26, pl. 2o. Haaf, 1952, Veröff. zool. Staatssamml. Münch., **2**: 152, 154, 157, pl. 11 (as *orana* Duponchel). — Sardinien [Gran Torre].

ab. **cingulata** Turati, 1919, Nat. sicil., **23**: 242. Reiss, 1930, in Seitz, Die Gross-Schmetterlinge der Erde, Supplement, **2**: 26.

ab. **colligata** Turati, 1919, Nat. sicil., **23**: 242. Reiss, 1930, in Seitz, Die Gross-Schmetterlinge der Erde, Supplement, **2**: 26.

ab. **pulchra** Krausse, 1915, Arch. Naturgesch., **81** (A) (2): 123. Reiss, 1930, in Seitz, Die Gross-Schmetterlinge der Erde, Supplement, **2**: 26.

suffusa Turati, 1919, Nat. sicil., **23**: 342.

ssp. **contristans** Oberthür, 1922, Études de Lépidoptérologie comparée, **19**: 158, pl. 535, figs. 4455, 4456. Reiss, 1930, in Seitz, Die Gross-Schmetterlinge der Erde, Supplement, **2**: 26, pl. 3a (after Oberthür); 1943, Z. wien. ent. Ges., **28**: 355, pl. 36, fig. 27. Tremewan, 1961, Bull. Brit. Mus. (nat. Hist.) Ent., **10** (7): 261, pl. 51, fig. 36. — Mrassine, Maroc.

ssp. **hajebensis** Reiss & Tremewan, 1960, Bull. Brit. Mus. (nat. Hist.) Ent., **9** (10): 464, pl. 22, figs. 11, 12. Reiss, 1930, in Seitz, Die Gross-Schmetterlinge der Erde, Supplement, **2**: 26 (as *rothschildiana* Reiss [partim]); 1943, Z. wien. ent. Ges., **28**: 356, pl. 38, figs. 80, 81, pl. 45, fig. II, pl. 46, fig. VII (as *rothschildiana* Reiss [partim]). Tremewan, 1961, Bull. Brit. Mus. (nat. Hist.) Ent., **10** (7): 261, pl. 51, fig. 37. — El Hajeb, west slopes of Middle Atlas, Morocco, 900-1000 m.

ssp. **tatla** Reiss, 1943, Z. wien. ent. Ges., **28**: 357, pl. 36, figs. 28, 29. — Umgebung von Tetuan, Rif, Marokko.

ab. **rubricosta** Reiss, 1943, Z. wien. ent. Ges., **28**: 357.

ab. **morena** Reiss, 1943, Z. wien. ent. Ges., **28**: 357, pl. 38, fig. 96.

Biology

Burgeff, 1913, Ent. Z., **27**: 177, fig. 2, pl. 3 (**allardi**); 1950, Portug. acta biol., (A) Goldschmidt: 663-728. Holik, 1953, Ent. Z., **63**: 23. Oberthür, 1912, Études de Lépidoptérologie comparée, **6**, pl. 135, figs. 1183-1185. Reiss, 1930, in Seitz, Die Gross-Schmetterlinge der Erde, Supplement, **2**: 27; 1958, Z. wien. ent. Ges., **43**: 158.

marteni Reiss

Distribution: Morocco (Rif).

marteni Reiss, 1943, Z. wien. ent. Ges., **28**: 365, pl. 36, figs. 32, 34, pl. 38, figs. 92, 93; 1943, ibidem, **28**: 367, pl. 45, fig. VI. Haaf, 1952, Veröff. zool. Staatssamml. Münch., **2**: 152, 154, 157, pl. 11. Alberti, 1958, Mitt. zool. Mus. Berl., **34**: 309. — Bab Joana (B. Selman), Rif, Marokko, 1900-2000 m.

ab. **benica** Reiss, 1943, Z. wien. ent. Ges., **28**: 367, pl. 36, fig. 33.
ab. **reducta** Reiss, 1943, Z. wien. ent. Ges., **28**: 367, pl. 38, fig. 95.
ab. **henna** Reiss, 1943, Z. wien. ent. Ges., **28**: 367, pl. 38, fig. 94.

Biology

Holik, 1953, Ent. Z., **63**: 25.

youngi Rothschild
Distribution: Morocco.

youngi Rothschild, 1925, Bull. Soc. Sci. nat. Maroc, **5**: 338. Reiss, 1930, in Seitz, Die Gross-Schmetterlinge der Erde, Supplement, **2**: 27, pl. 3a; 1943, Z. wien. ent. Ges., **28**: 362, pl. 36, figs. 36, 38, 39, pl. 45, figs. IV, V, pl. 46, fig. IX. Alberti, 1958, Mitt. zool. Mus. Berl., **34**: 308. Reiss & Tremewan, 1960, Bull. Brit. Mus. (nat. Hist.) Ent., **9** (10): 464, pl. 22, fig. 13, pl. 23, figs. 7, 8, pl. 25, figs. 7, 8. Tremewan, 1961, Bull. Brit. Mus. (nat. Hist.) Ent., **10** (7): 262, pl. 51, fig. 38, pl. 61, figs. 1, 2. Wiegel, 1965, Mitt. münch. ent. Ges., **55**: 136.
Above Azrou, Middle Atlas, Morocco, 1800 m.

media Rothschild, 1925, Bull. Soc. Sci. nat. Maroc, **5**: 338. Reiss & Tremewan, 1960, Bull. Brit. Mus. (nat. Hist.) Ent., **9** (10): 464. Tremewan, 1961, Bull. Brit. Mus. (nat. Hist.) Ent., **10** (7): 262, pl. 51, fig. 39.

rothschildiana Reiss, 1930, in Seitz, Die Gross-Schmetterlinge der Erde, Supplement, **2**: 26. Reiss & Tremewan, 1960, Bull. Brit. Mus. (nat. Hist.) Ent., **9** (10): 464.

ab. **dealbata** Reiss, 1943, Z. wien. ent. Ges., **28**: 364, pl. 36, fig. 37.
ab. **rubricosta** Reiss, 1943, Z. wien. ent. Ges., **28**: 364, pl. 36, fig. 35.
ab. **confluens** Reiss, 1943, Z. wien. ent. Ges., **28**: 364.
f. loc. **bekretica** Reiss, 1943, Z. wien. ent. Ges., **28**: 364.
Bekrit, südlich von Fez, Mittel Atlas, Marokko, 1400 m.

Biology

Holik, 1953, Ent. Z., **63**: 25.

harterti Rothschild
Distribution: Morocco.

harterti Rothschild, 1925, Bull. Soc. Sci. nat. Maroc, **5**: 338. Reiss, 1930, in Seitz, Die Gross-Schmetterlinge der Erde, Supplement, **2**: 27, pl. 3a; 1941, Z. wien. EntVer., **26**: 289, pl. 31, figs. 10-12; 1943, Z. wien. ent. Ges., **28**: 360, pl. 36, figs. 53, 54, pl. 45, fig. III. Le Charles, 1946, Rev. franç. Lépid., **10**: 342, pl. 14, fig. 1. Alberti, 1958, Mitt. zool. Mus.
Azrou, Middle Atlas, Morocco, 1300 m.

Berl., **34**: 308. Reiss & Tremewan, 1960, Bull. Brit. Mus. (nat. Hist.) Ent., **9** (10): 464, pl. 22, fig. 14, pl. 23, figs. 9, 10, pl. 25, figs. 9, 10. Tremewan, 1961, Bull. Brit. Mus. (nat. Hist.) Ent., **10** (7): 262, pl. 52, fig. 1, pl. 61, figs. 3, 4. Wiegel, 1965, Mitt. münch. ent. Ges., **55**: 136.

ab. **rubricosta** Reiss, 1943, Z. wien. ent. Ges., **28**: 362.

Biology

Holik, 1953, Ent. Z., **63**: 25.

maroccana Rothschild

Distribution: Morocco.

ssp. **maroccana** Rothschild, 1917, Novit. zool., **24**: 342. Reiss, 1930, in Seitz, Die Gross-Schmetterlinge der Erde, Supplement, **2**: 27, pl. 2o. Le Charles, 1946, Rev. franç. Lépid., **10**: 342, pl. 14, fig. 2. Reiss & Tremewan, 1960, Bull. Brit. Mus. (nat. Hist.) Ent., **9** (10): 465, pl. 22, fig. 15, pl. 25, fig. 11. Tremewan, 1960, Ent. Rec., **72**: 208, figs. 1, 2; 1961, Bull. Brit. Mus. (nat. Hist.) Ent., **10** (7): 262, pl. 52, fig. 2, pl. 64, fig. 7. Wiegel, 1965, Mitt. münch. ent. Ges., **55**: 135.

Great Atlas, Morocco (Mogador is probably erroneous)

Biology

Burgeff, 1950, Portug. acta biol., (A) Goldschmidt: 663-728.
Holik, 1953, Ent. Z., **63**: 25.

lucasi Le Charles

Distribution: Morocco.

lucasi Le Charles, 1946, Rev. franç. Lépid., **10**: 343, pl. 14, figs. 3-9. Reiss, 1943, Z. wien. ent. Ges., **28**: 359, pl. 38, fig. 82 (as *maroccana* Rothschild [partim]). Wiegel, 1965, Mitt. münch. ent. Ges., **55**: 139.

Dodes du Todeah, Haut-Atlas, Maroc.

gundafica Reiss & Tremewan

Distribution: Morocco.

gundafica Reiss & Tremewan, 1960, Bull. Brit. Mus. (nat. Hist.) Ent., **9** (10): 465, pl. 22, fig. 16, pl. 25, fig. 16. Reiss, 1943, Z. wien. ent. Ges., **28**: 359-360 (as *maroccana* Rothschild [partim]). Wiegel, 1965, Mitt. münch. ent. Ges., **55**: 144.

Dard Goundafi (Kazba Gundafa,) on the edge of Oued n'Fis, Great Atlas, Morocco.

freudei Daniel

Distribution: South-east Spain.

freudei Daniel, 1960, Opusc. Zool., no. 46: 1, figs. 1a-1d, 2a, 2b. Tremewan, 1963, Ent. Rec., **75**: 3. Reiss, 1963, Stuttgart. Beitr. Naturk., no. 122: 1, 2, figs. 3, 4.

ab. **confluens** Reiss, 1963, Stuttgart. Beitr. Naturk., no. 122: 2, figs. 1, 2.

Ödland bei Alicante, Prov. Alicante, Spanien,

Biology

Tremewan, 1963, Ent. Rec., **75**: 3.

occitanica de Villers

Distribution: Spain, southern France to Liguria (Savona).

ssp. **vandalitia** Burgeff, 1926, Mitt. münch. ent. Ges., **16**: 62. Reiss, 1930, in Seitz, Die Gross-Schmetterlinge der Erde, Supplement, **2**: 31. — Umgebung von Granada, Südspanien.

ab. **pseudooccitanica** Reiss, 1922, Int. ent. Z., **15**: 180.

ab. **pseudodisjuncta** Reiss, 1921, Int. ent. Z., **15**: 40; 1930, in Seitz, Die Gross-Schmetterlinge der Erde, Supplement, **2**: 31.

ab. **nigra** Dziurzyński, 1914, Jber. wien. ent. Ver., **24**: X; 1908, Berl. ent. Z., **53**: 58, pl. 2, fig. 20.

ab. **miniosa** Wagner, 1919, Int. ent. Z., **13**: 160.

ab. **albicans** Staudinger, 1861, in Staudinger & Wocke, Catalog der Lepidopteren Europa's und der angrenzenden Länder, p. 22. Rambur, 1839, Faune entomologique de l'Andalousie, pl. 12, fig. 10. Millière, 1879, Mém. Soc. Sci. nat. Cannes, **8**: 116, pl. 5, fig. 13. Spuler, 1906, in Hofmann, Die Schmetterlinge Europas, **2**: 165, pl. 77, fig. 34a. Seitz, 1907, Die Gross-Schmetterlinge der Erde, **2**: 30, pl. 8k. Dziurzyński, 1908, Berl. ent. Z., **53**: 59, pl. 2, fig. 21. Burgeff, 1926, Mitt. münch. ent. Ges., **16**: 62.

ab. **extrema** Reiss, 1922, Int. ent. Z., **15**: 180; 1930, in Seitz, Die Gross-Schmetterlinge der Erde, Supplement, **2**: 31, pl. 3f.

ssp. **eulalia** Burgeff, 1926, Mitt. münch. ent. Ges., **16**: 63. Reiss, 1930, in Seitz, Die Gross-Schmetterlinge der Erde, Supplement, **2**: 31, pl. 3f. Agenjo, 1952, Fáunula Lepidopterólogica Almeriense, p. 160. — Sta. Eulalia, Prov. Murcia, Spanien.

ab. **albicans** Reiss, 1965, Ent. Rec., **77**: 86.

f. t. **inexpectata** Le Charles, 1946, Bull. Soc. ent. Fr., **51**: 83, pl. 2, cols. 6-8, figs. — Murcie (Septembre- Octobre forme).

ssp. **valenciaca** Reiss, 1965, Ent. Rec., **77**: 85, pl. 1, figs. 1, 2. — Torres, Sagunto, Valencia, Spain, 250 m.

ssp. **arragonica** Holik & Sheljuzhko, 1956, Mitt. münch. ent. Ges., **46**: 210 (nomen novum for *iberica* Staudinger). — Umgebung von Barcelona, Katalonien, Spanien.

iberica Staudinger, 1871, in Staudinger & Wocke, Catalog der Lepidopteren des Europaeischen Faunengebiets, p. 50 (preoccupied by **iberica** Kolenati, 1846, ssp. of **carniolica** Scopoli). Spuler, 1906, in Hofmann, Die Schmetterlinge Europas, **2**: 165, pl. 77, fig. 34c. Seitz, 1907, Die Gross-Schmetterlinge der Erde, **2**: 30, pl. 8i. Haaf, 1952, Veröff. zool. Staatssamml. Münch., **2**: 152, 154, 157, pl. 11 (as *occitanica* de Villers).

ab. **azona** Spuler, 1906, in Hofmann, Die Schmetterlinge Europas, **2**: 165. Seitz, 1907, Die Gross-Schmetterlinge der Erde, **2**: 30. Reiss, 1930, in Seitz, Die Gross-Schmetterlinge der Erde, Supplement, **2**: 31.
ab. **octonotata** Reiss, 1930, in Seitz, Die Gross-Schmetterlinge der Erde, Supplement, **2**: 31.
ab. **ornata** de Sagarra, 1924, Trab. Mus. Cienc. nat. Barcelona, **11** (1): 19, pl. 4, fig. 3. Reiss, 1930, in Seitz, Die Gross-Schmetterlinge der Erde, Supplement, **2**: 31.
ornata Burgeff, 1926, Mitt. münch. ent. Ges., **16**: 63.
ab. **nigra** Reiss, 1921, Int. ent. Z., **15**: 39; 1922, ibidem, **15**: 180; 1930, in Seitz, Die Gross-Schmetterlinge der Erde, Supplement, **2**: 31, pl. 3f.
ab. **albipunctata** Vartian, 1960, Z. Arb. Gem. öst. Ent., **12**: 132, fig. 11.
ab. **cataloniae** Reiss, 1921, Int. ent. Z., **15**: 39; 1930, in Seitz, Die Gross-Schmetterlinge der Erde, Supplement, **2**: 31, pl. 3f.
ab. **tarragonensis** Dziurzyński, 1914, Jber. wien. ent. Ver., **24**: XI.

ssp. **burgosensis** Tremewan, 1963, Ent. Rec., **75**: 251. Tremewan & Manley, 1965, Ent. Rec., **77**: 5. — Oña, Burgos, Spain, 2000 ft.
f. **lutea** Tremewan & Manley, 1965, Ent. Rec., **77**: 5, 6.
ab. **elisae** Tremewan & Manley, 1965, Ent. Rec., **77**: 6.

ssp. **huescacola** Tremewan & Manley, 1965, Ent. Rec., **77**: 6. — Sierra de la Pena, 3600 ft; Jaca, 2700 ft. La Pena, 2400 ft.; Huesca, Spain.

ssp. **disiuncta** Spuler, 1906, in Hofmann, Die Schmetterlinge Europas, **2**: 165, pl. 77, fig. 34b. Reiss, 1930, in Seitz, Die Gross-Schmetterlinge der Erde, Supplement, **2**: 31, pl. 3f (as *disjuncta*); 1937, Ent. Rdsch., **54**: 455, fig. c4 (as *disjuncta*). — Algecares, Ostpyrenäen.
ab. **nigra** Dziurzyński, 1910, Int. ent. Z., **4**: 200.
ab. **flaveola** Dziurzyński, 1914, Jber. wien. ent. Ver., **24**: X.

ssp. **occitanica** de Villers, 1789, Caroli Linnaei Entomologia, **2**: 114, pl. 4, fig. 21. Ochsenheimer, 1808, Die Schmetterlinge von Europa, **2**: 95. Boisduval, 1828, Essai sur une Monographie des Zygénides, p. 97, pl. 6, fig. 3. Spuler, 1906, in Hofmann, Die Schmetterlinge Europas, **2**: 165, pl. 72, fig. 16, pl. 77, fig. 34. Seitz, 1907, Die Gross-Schmetterlinge der Erde, **2**: 30, pl. 8i. Oberthür, 1910, Études de Lépidoptérologie comparée, **4**: 633. Reiss, 1953, Ent. Z., **63**: 112. Alberti, 1958, Mitt. zool. Mus. Berl., **34**: 306. — France meridionale [Peyreleau, Aveyron].

phacae Hübner, [July 1803]-[21st December 1806], Sammlung

europäischer Schmetterlinge, **2**, pl. 22, figs. 106, 107; 1806, ibidem, Der Ziefer, p. 84.

ab. **pseudoiberica** Burgeff, 1926, in Strand, Lepid. Cat., **33**: 47. Reiss, 1930, in Seitz, Die Gross-Schmetterlinge der Erde, Supplement, **2**: 31.

f. t. **praematura** Przegendza, 1932, Ent. Z., **46**: 116, figs. 21-24. Reiss, 1933, in Seitz, Die Gross-Schmetterlinge der Erde, Supplement, **2**: 274; 1958, Bull. Soc. ent. Mulhouse, p. 50.

Bei Ventimiglia, Ligurien, Italien.

ssp. **merulae** Storace, 1962, Boll. Soc. ent. ital., **91**: 153.

Marnia di Andera, nella valetta del Rio Croso, Liguria, Italia, 25-80 m.

ssp. **azurensis** Reiss, 1958, Bull. Soc. ent. Mulhouse, p. 50. Loritz, 1961, Bull. Soc. ent. Mulhouse, p. 83-102.

azurensis Reiss, 1958, Z. wien. ent. Ges., **43**: 159 (nomen nudum).

Mt. Leuze (Mt. Pacanaglia) près Nice, Alpes-Maritimes, France, 400-577 m.

ssp. **tourrettica** Reiss, 1953, Ent. Z., **63**: 111.

ab. **pseudooccitanica** Reiss, 1953, Ent. Z., **63**: 112.

ab. **quinquemaculata** Reiss, 1953, Ent. Z., **63**: 112.

Les Vallettes bei Tourettes-sur-Loup, Alpes-Maritimes, Frankreich, 250-300 m.

ssp. **arida** Dujardin, 1956, Bull. mens. Soc. linn. Lyon, **25**: 261. Dufay, 1966, Bull. mens. Soc. linn. Lyon, **35**: 69.

ab. **bicolor** Oberthür, 1909, Études de Lépidoptérologie comparée, **3**, pl. 22, fig. 114. Tremewan, 1961, Bull. Brit. Mus. (nat. Hist.) Ent., **10** (7): 263, pl. 52, fig. 3.

St. Auban près de Sisteron, Basses-Alpes, France, 450 m.

Biology

Abeille, 1909, Mém. Soc. linn. Provence, **1**: 20-21. Boisduval, Rambur & Graslin, 1832, Collection iconographique et historique des Chenilles, pl. 4, figs. 4-6. Burgeff, 1950, Portug. acta biol., (A) Goldschmidt: 663-728; 1951, Biol. Zbl., **70**: 1-23. Holik, 1937, Lambillionea, **37**: 15-24, 32-45, 80-91; 1938, Ent. Rdsch., **55**: 320-323, 331-333; 1953, Ent. Z., **63**: 24. Millière, 1878/79, Mém. Soc. Sci. nat. Cannes, **8**: 116, pl. 5, figs. 11-14. Rambur, 1829, Annales des Sciences d'Observation, **2**: (268), pl. 5, figs. 5-8. Reiss, 1930, in Seitz, Die Gross-Schmetterlinge der Erde, Supplement, **2**: 31; 1953, Ent. Z., **63**: 111; 1958, Bull. Soc. ent. Mulhouse, p. 50. Ribbe, 1909/12, Iris, **23**: 358. Roüast, 1883, Catalogue des Chenilles européennes connues, p. 23. Spuler, 1910, in

Hofmann, Die Raupen der Schmetterlinge Europas, pl. 10, fig. 9.

carniolica Scopoli

Distribution: Central Asia, Siberia, northern Iran, Syria, Asia Minor, Europe (excluding British Isles, Scandinavia).

ssp. **demavendi** Holik, 1936, Ent. Rdsch., **54**: 7; 1937, Lambillionea, **37**: 214, pl. 12, figs. 1-4. Reiss, 1937, Ent. Rdsch., **54**: 468, figs. a4, d3. Schwingenschuss, 1939, Ent. Z., **53**: 95. Holik & Sheljuzhko, 1956, Mitt. münch. ent. Ges., **46**: 230.
Osthang des Demavend, Elburs Gebirge, Nordiran, 3200-4000 m.

ab. **azona** Holik, 1936, Ent. Rdsch., **54**: 9; 1937, Lambillionea, **37**: 214.

ab. **stoechadoides** Holik, 1936, Ent. Rdsch., **54**: 9; 1937, Lambillionea, **37**: 214, pl. 12, fig. 7.

ab. **dupuyi** Holik, 1936, Ent. Rdsch., **54**: 8; 1937, Lambillionea, **37**: 214, pl. 12, fig. 5. Reiss, 1937, Ent. Rdsch., **54**: 455, fig. b4.

ab. **albomaculata** Schwingenschuss, 1939, Ent. Z., **53**: 96.

ab. **octonotata** Holik, 1937, Lambillionea, **37**: 214, pl. 12, fig. 6.

ssp. **vandarbanensis** Reiss, 1938, Mitt. münch. ent. Ges., **27**: 168. Holik & Sheljuzhko, 1956, Mitt. münch. ent. Ges., **46**: 231.
Särdabtal (Vandarban), Elburs-Gebirge, Tacht i Suleiman-Gruppe, Nordiran, 2500-2700 m.

ssp. **transiens** Staudinger, 1887, Berl. ent. Z., **31**: 40. Seitz, 1907, Die Gross-Schmetterlinge der Erde, **2**: 30. Burgeff, 1914, Mitt. münch. ent. Ges., **5**: 59. Reiss, 1930, in Seitz, Die Gross-Schmetterlinge der Erde, Supplement, **2**: 30, pl. 3e; 1933, ibidem, **2**: 273; 1933, Int. ent. Z., **26**: 492, fig.; 1937, Ent. Rdsch., **55**: 40, figs. a4, b1. Holik, 1937, Lambillionea, **37**: 213, pl. 12, figs. 11, 12. Holik & Sheljuzhko, 1956, Mitt. münch. ent. Ges., **46**: 228.
Schahkuh (Shakuh), Nordiran.

ssp. **jenissejensis** Holik & Sheljuzhko, 1956, Mitt. münch. ent. Ges., **46**: 235.
Minussinsk, Gouv. Jenissej, Zentral Sibirien.

ssp. **rueckbeili** Sheljuzhko, 1919, N. Beitr. syst. Insektenk., **1**: 130. Reiss, 1930, in Seitz, Die Gross-Schmetterlinge der Erde, Supplement, **2**: 31, pl. 3f; 1933, ibidem, **2**: 274. Holik & Sheljuzhko, 1956, Mitt. münch. ent. Ges., **46**: 233.
Semiretshje (nördlicher Teil), Bezirk Dzharkent (Tyshkan; Kok-Tass; Dorf Karatjube), Zentralasien.

ssp. **nuratanya** Burgeff, 1926, Mitt. münch. ent. Ges., **16**: 62. Reiss, 1930, in Seitz, Die Gross-Schmetterlinge der Erde, Supplement, **2**: 30, pl. 3e. Holik & Sheljuzhko, 1956, Mitt. münch. ent. Ges., **46**: 232. — Dorf Jani-Kurgan, Nuratanyntau, Samarkand, Zentralasien, 2500 m.

ssp. **praestans** Oberthür, 1910, Études de Lépidoptérologie comparée, **4**: 637; 1896, Études d'Entomologie, **20**: 51, pl. 7, figs. 115, 116. Holik, 1937, Lambillionea, **37**: 212. Koch, 1938, Ent. Z., **52**: 87. Holik & Sheljuzhko, 1956, Mitt. münch. ent. Ges., **46**: 221. Tremewan, 1961, Bull. Brit. Mus. (nat. Hist.) Ent., **10** (7): 265, pl. 52, fig. 12. — Akbès (Eibes), Syrie.

ssp. **eibesiana** Koch, 1938, Ent. Z., **52**: 87. Holik, 1937, Lambillionea, **37**, pl. 12, fig. 20. Holik & Sheljuzhko, 1956, Mitt. münch. ent. Ges., **46**: 222. — Akbès (Eibes), Syrien.

ab. **quinquemaculata** Koch, 1938, Ent. Z., **52**: 87.

ssp. **incompta** Koch, 1938, Ent. Z., **52**: 88, 89. Reiss, 1930, in Seitz, Die Gross-Schmetterlinge der Erde, Supplement, **2**: 30, pl. 3e (as *praestans* Oberthür). Holik & Sheljuzhko, 1956, Mitt. münch. ent. Ges., **46**: 225. — Umgebung von Beirut, Syrien.

ssp. **illiterata** Koch, 1938, Ent. Z., **52**: 88, 89. Holik & Sheljuzhko, 1956, Mitt. münch. ent. Ges., **46**: 225. — Libanon, Syrien.

ab. **octonotata** Koch, 1938, Ent. Z., **52**: 89.

ssp. **siehei** Holik & Sheljuzhko, 1956, Mitt. münch. ent. Ges., **46**: 224. — Mersina, Taurus, Kleinasien.

ssp. **taurica** Staudinger, 1878, Horae Soc. ent. Ross., **14**: 326. Seitz, 1907, Die Gross-Schmetterlinge der Erde, **2**: 30, pl. 8g. Reiss, 1930, in Seitz, Die Gross-Schmetterlinge der Erde, Supplement, **2**: 30. Holik, 1937, Lambillionea, **37**: 210. Koch, 1938, Ent. Z., **52**: 67. Holik & Sheljuzhko, 1956, Mitt. münch. ent. Ges., **46**: 224. — Bei Gülek, Taurus, Kleinasien.

ab. **erythrosoma** Röber, 1897, Ent. Nachr., **23**: 272. Reiss, 1930, in Seitz, Die Gross-Schmetterlinge der Erde, Supplement, **2**: 30.

rubroabdominalis Koch, 1938, Ent. Z., **52**: 68. Holik & Sheljuzhko, 1956, Mitt. münch. ent. Ges., **46**: 224.

ssp. **antitaurica** Holik, 1942, Mitt. naturw. Inst. Sofia, **15**: 255. Holik & Sheljuzhko, 1956, Mitt. münch. ent. Ges., **46**: 220. — Antitaurus, Kleinasien, 3300 m.

ssp. **wiedemannii** Ménétriés, 1839, Mém. Acad. Sci. St-Petersb. (Sci. math., phys., nat.), (6) **3**: 50, pl. 2, fig. 10. Staudinger, 1879, Horae Soc. ent. Ross., **14**: 325. Oberthür, 1896, Études d'Entomologie, **20**: 52. Seitz, 1907, Die Gross-Schmetterlinge der Erde, **2**: 30, pl. 8h. Holik, 1937, Lambillionea, **37**: 209; — Entre Constantinople et le Balkan.

1939, Mitt. münch. ent. Ges., **29**: 192. Holik & Sheljuzhko, 1956, Mitt. münch. ent. Ges., **46**: 219.

ssp. **amasina** Staudinger, 1879, Horae Soc. ent. Ross., **14**: 326; 1887, Berl. ent. Z., **31**: 40. Seitz, 1907, Die Gross-Schmetterlinge der Erde, **2**: 30, pl. 8h. Reiss, 1930, in Seitz, Die Gross-Schmetterlinge der Erde, Supplement, **2**: 30. Holik, 1937, Lambillionea, **37**: 210, pl. 12, figs. 16-19. Koch, 1938, Ent. Z., **52**: 61. Holik & Sheljuzhko, 1956, Mitt. münch. ent. Ges., **46**: 217.
Amasia, Pontus, Kleinasien.

ab. **pseudowiedemanni** Burgeff, 1914, Mitt. münch. ent. Ges., **5**: 59. Reiss, 1930, in Seitz, Die Gross-Schmetterlinge der Erde, Supplement, **2**: 30.

ab. **pseudosuavis** Koch, 1938, Ent. Z., **52**: 62.

ab. **dupuyi** Burgeff, 1914, Mitt. münch. ent. Ges., **5**: 59. Reiss, 1930, in Seitz, Die Gross-Schmetterlinge der Erde, Supplement, **2**: 30.

ab. **venusta** Schultz, 1906, Soc. ent., **20**: 170. Seitz, 1912, Die Gross-Schmetterlinge der Erde, **2**: 444.

ssp. **suavis** Burgeff, 1926, Mitt. münch. ent. Ges., **16**: 62. Reiss, 1930, in Seitz, Die Gross-Schmetterlinge der Erde, Supplement, **2**: 30, pl. 3d. Holik, 1937, Lambillionea, **37**: 211. Koch, 1938, Ent. Z., **52**: 59. Holik & Sheljuzhko, 1956, Mitt. münch. ent. Ges., **46**: 219.
Marasch; Hadjin; Zeitun; Taurus, Kleinasien.

ab. **unicingulata** Koch, 1938, Ent. Z., **52**: 60.

ab. **rubroabdominalis** Koch, 1938, Ent. Z., **52**: 60.

ab. **pseudoamasina** Koch, 1938, Ent. Z., **52**: 60.

ssp. **euxina** Holik & Sheljuzhko, 1956, Mitt. münch. ent. Ges., **46**: 206.
Novorossijsk, Westküste, Kaukasusgebiet.

ab. **tenuicingulata** Holik & Sheljuzhko, 1956, Mitt. münch. ent. Ges., **46**: 207.

ssp. **zhicharevi** Holik & Sheljuzhko, 1956, Mitt. münch. ent. Ges., **46**: 207.
Kislovodsk, Terek-Gebiet, Vorberge des Kaukasus.

ab. **tenuicingulata** Holik & Sheljuzhko, 1956, Mitt. münch. ent. Ges., **46**: 207.

ssp. **lesgina** Holik & Sheljuzhko, 1956, Mitt. münch. ent. Ges., **46**: 208.
Dorf Ussuchtshaj bei Achty, Dagestan, Kaukasus, 1200 m.

ssp. **tuapsica** Reiss, 1941, Z. wien. EntVer., **26**: 62. Holik & Sheljuzhko, 1956, Mitt. münch. ent. Ges., **46**: 208.
Bei Tuapse, Küstengebiet des Schwarzen Meeres.

ab. **pseudotaurica** Reiss, 1941, Z. wien. EntVer., **26**: 62.

ssp. **amabilis** Reiss, 1921, Int. ent. Z., **15**: 21; 1930, in Seitz, Die Gross-Schmetterlinge der Erde, Supplement, **2**: 30, pl. 3e; 1933, ibidem, **2**: 273. Holik, 1937, Lambillionea, **37**: 213.
Kasikoparan. West Armenien.

Reiss, 1941, Z. wien. EntVer., **26**: 62. Holik & Sheljuzhko, 1956, Mitt. münch. ent. Ges., **46**: 217.

ssp. **ordubadina** Koch, 1936, Ent. Z., **50**: 398; 1936, Iris, **50**: 43, pl. 2, figs. 5-8. Holik, 1936, Lambillionea, **36**, pl. 9, figs. 18-20; 1937, ibidem, **37**: 212. Koch, 1937, Ent. Z., **51**: 46, 64, figs. 5-7. Holik & Sheljuzhko, 1956, Mitt. münch. ent. Ges., **46**: 213. — Beim Dorf Inaclü, Berge Alagëz, Armenien (Daralagëz irrtümlich).

ab. **rubroabdominalis** Holik & Sheljuzhko, 1956, Mitt. münch. ent. Ges., **46**: 215.

ab. **medioconfluens** Holik & Sheljuzhko, 1956, Mitt. münch. ent. Ges., **46**: 215.

ssp. **tkatshukovi** Holik & Sheljuzhko, 1956, Mitt. münch. ent. Ges., **46**: 215. — Dorf Ochtshi, Zangezur Gebirge, Armenisches Bergland, 2100-2250 m.

ssp. **achalzichensis** Reiss, 1935, Int. ent. Z., **29**: 161. Koch, 1939, Mitt. münch. ent. Ges., **29**: 411. Holik & Sheljuzhko, 1956, Mitt. münch. ent. Ges., **46**: 211. — Chambobel, Adshara Gebirge, Transkaukasien.

ssp. **iberica** Kolenati, 1846, Meletemata entomologica, **5**: 94. Holik & Sheljuzhko, 1956, Mitt. münch. ent. Ges., **46**: 209. — Iberia, Landschaft am oberen Kyros (=Kura), Georgien.

ssp. **alta** Reiss, 1921, Int. ent. Z., **15**: 21; 1930, in Seitz, Die Gross-Schmetterlinge der Erde, Supplement, **2**: 30, pl. 3e. Holik, 1937, Lambillionea, **37**: 213. Koch, 1939, Mitt. münch. ent. Ges., **29**: 411. Reiss, 1941, Z. wien. EntVer., **26**: 62. Koch, 1942, Z. wien. EntVer., **27**: 44. Holik & Sheljuzhko, 1956, Mitt. münch. ent. Ges., **46**: 210. — Missura, Oni, Ossetenstrasse, Kutais, Kaukasus (Waldzone der Grusienberge irrtümlich).

ssp. **crymaea** Stauder, 1925, Ent. Anz., **5**: 86. Reiss, 1930, in Seitz, Die Gross-Schmetterlinge der Erde, Supplement, **2**: 30. Holik & Sheljuzhko, 1956, Mitt. münch. ent. Ges., **46**: 203. Tremewan, 1966, Ent. Rec., **78**: 32, pl. 1, fig. 2. — Feodossia, Krim (Krym).

ssp. **europaea** Burgeff, 1926, Mitt. münch. ent. Ges., **16**: 61. Reiss, 1930, in Seitz, Die Gross-Schmetterlinge der Erde, Supplement, **2**: 30. Koch, 1938, Ent. Z., **52**: 67. Holik, 1939, Mitt. münch. ent. Ges., **29**: 202. Holik & Sheljuzhko, 1956, Mitt. münch. ent. Ges., **46**: 227. — Therapia bei Konstantinopel, Bosporus, Türkei.

ab. **pseudoleonhardi** Holik, 1939, Mitt. münch. ent. Ges., **29**: 202.

ab. **dealbata** Holik, 1939, Mitt. münch. ent. Ges., **29**: 202.

ssp. **misoriensis** Koch, 1937, Ent. Z., **51**: 39, 63, figs. 21-26. — Misoria am

Holik, 1939, Mitt. münch. ent. Ges., **29**: 202. Holik & Sheljuzhko, 1956, Mitt. münch. ent. Ges., **46**: 227. — Schwarzen Meer, 60 km. südlich von Varna.

ssp. **uralia** Burgeff, 1926, Mitt. münch. ent. Ges., **16**: 62 (nomen novum for *uralensis* Krulikowsky). Reiss, 1930, in Seitz, Die Gross-Schmetterlinge der Erde, Supplement, **2**: 30; 1933, ibidem, **2**: 274; 1932, Ent. Rdsch., **49**: 165, pl. 1, figs. Holik & Sheljuzhko, 1956, Mitt. münch. ent. Ges., **46**: 204, 205. — Abhänge des mittleren Urals, Russland.

uralensis Krulikowsky, 1897, Soc. ent., **12**: 1 (preoccupied by **uralensis** Herrich-Schäffer, 1846, ssp. of **cynarae** Esper). Holik & Sheljuzhko, 1956, Mitt. münch. ent. Ges., **46**: 204.

ssp. **cruenta** Pallas, 1773, Reise durch verschiedene Provinzen des Russischen Reichs, **2**: 732. Holik & Sheljuzhko, 1956, Mitt. münch. ent. Ges., **46**: 205. — In australibus ad Volgam et Irtin [Saratov; Sarepta; Südrussland].

ssp. **caliacrae** Reiss, 1931, Int. ent. Z., **25**: 98; 1933, in Seitz, Die Gross-Schmetterlinge der Erde, Supplement, **2**: 273. Holik, 1939, Mitt. münch. ent. Ges., **29**: 201. Popescu-Gorj, 1964, Catalogue de la Collection de Lépidoptères „Prof. A. Ostrogovich", p. 67, pl. 5, figs. 19, 20. — Bei Balcic, südliche Dobrudscha, Rumänische Silberküste.

ab. **azona** Reiss, 1931, Int. ent. Z., **25**: 99; 1933, in Seitz, Die Gross-Schmetterlinge der Erde, Supplement, **2**: 273. Popescu-Gorj, 1964, Catalogue de la Collection de Lépidoptères „Prof. A. Ostrogovich", p. 68.

ab. **laticlavia** Reiss, 1931, Int. ent. Z., **25**: 99; 1933, in Seitz, Die Gross-Schmetterlinge der Erde, Supplement, **2**: 273. Popescu-Gorj, 1964, Catalogue de la Collection de Lépidoptères „Prof. A. Ostrogovich", p. 68.

laticlava Holik, 1939, Mitt. münch. ent. Ges., **29**: 201.

ab. **rubroabdominalis** Reiss, 1931, Int. ent. Z., **25**: 99; 1933, in Seitz, Die Gross-Schmetterlinge der Erde, Supplement, **2**: 273. Popescu-Gorj, 1964, Catalogue de la Collection de Lépidoptères „Prof. A. Ostrogovich", p. 68.

ab. **securigera** Reiss, 1931, Int. ent. Z., **25**: 99; 1933, in Seitz, Die Gross-Schmetterlinge der Erde, Supplement, **2**: 273. Popescu-Gorj, 1964, Catalogue de la Collection de Lépidoptères „Prof. A. Ostrogovich", p. 68.

ssp. **graeca** Staudinger, 1870, Horae Soc. ent. Ross., **7**: 105. Seitz, 1907, Die Gross-Schmetterlinge der Erde, **2**: 30. Wagner, 1919, Ent. Mitt., **8**: 179. Reiss, 1921, Int. ent. Z., **15**: 39. Koch, 1938, Ent. Z., **51**: 400. Holik, 1939, Mitt. münch. ent. Ges., **29**: 189. Daniel, 1958, Fragmenta Balcanica, **2**: 40. — Parnass, Griechenland.

ssp. **eurythaenica** Holik, 1939, Mitt. münch. ent. Ges., **29**: 191. — Veluchi Gebirge, Griechenland.

ssp. **tiranica** Holik, 1939, Mitt. münch. ent. Ges., **29**: 189.	Kruja, Tirana, Zentral Albanien.
ssp. **alibotensis** Holik, 1939, Mitt. münch. ent. Ges., **29**: 196.	Alibotus Gebirge an der griechisch-bulgarischen Grenze, 80 km östlich vom Dojran See.
ssp. **rumelica** Holik, 1939, Mitt. münch. ent. Ges., **29**: 195.	Stanimaka südlich von Philippopel, Ostrumelien.
ssp. **subonobrychis** Holik, 1939, Mitt. münch. ent. Ges., **29**: 200. ab. **acingulata** Holik, 1939, Mitt. münch. ent. Ges., **29**: 200. ab. **securigera** Holik, 1939, Mitt. münch. ent. Ges., **29**: 200.	Varna, an der Küste des Schwarzen Meeres (Mitte bis Ende Juni).
f. t. **hyperonobrychis** Holik, 1939, Mitt. münch. ent. Ges., **29**: 200. ab. **rubroabdominalis** Holik, 1939, Mitt. münch. ent. Ges., **29**: 200. ab. **amoena** Holik, 1939, Mitt. münch. ent. Ges., **29**: 200.	Varna, an der Küste des Schwarzen Meeres (Ende Juli bis Anfang August).
ssp. **tonzanica** Holik, 1939, Mitt. münch. ent. Ges., **29**: 194. ab. **acingulata** Holik, 1939, Mitt. münch. ent. Ges., **29**: 194. ab. **securigera** Holik, 1939, Mitt. münch. ent. Ges., **29**: 194.	Slivno, Tundza Tal, Bulgarien.
ssp. **paeoniae** Burgeff, 1926, Mitt. münch. ent. Ges., **16**: 61. Reiss, 1930, in Seitz, Die Gross-Schmetterlinge der Erde, Supplement, **2**: 30, pl. 3d. Koch, 1937, Ent. Z., **51**: 39, 64, figs. 1-4. Holik, 1939, Mitt. münch. ent. Ges., **29**: 185. Daniel, 1964, Prirod. Muz. Scopje, no. 2: 15.	Bei Nicolic und Volovec, Dojran-Seegebiet, Süd-Mazedonien.
ssp. **scopjina** Burgeff, 1926, Mitt. münch. ent. Ges., **16**: 61. Reiss, 1930, in Seitz, Die Gross-Schmetterlinge der Erde, Supplement, **2**: 30. Holik, 1939, Mitt. münch. ent. Ges., **29**: 185. Daniel, 1964, Prirod. Muz. Scopje, no. 2: 15.	Umgebung von Ueskūb, Mazedonien.
ssp. **djakovensis** Holik, 1939, Mitt. münch. ent. Ges., **29**: 187.	Djakova, Jugoslav.-Albanische Grenze.
ssp. **jadrana** Holik, 1939, Mitt. münch. ent. Ges., **29**: 180. ab. **octonotata** Holik, 1939, Mitt. münch. ent. Ges., **29**: 180.	Zara, Dalmatien.
ssp. **herzegovinea** Burgeff, 1926, Mitt. münch. ent. Ges., **16**: 50. Reiss, 1930, in Seitz, Die Gross-Schmetterlinge der Erde, Supplement, **2**: 27. Holik, 1939, Mitt. münch. ent. Ges., **29**: 68.	Vucija bara bei Gacko, Herzegowina, 1300 m.

ssp. **leonhardiana** Holik, 1939, Mitt. münch. ent. Ges., **29**: 67. — Korična, Bosnien.

ab. **philamoena** Reiss, 1941, Mitt. münch. ent. Ges., **31**: 997.

ssp. **onobrychoidea** Burgeff, 1926, Mitt. münch. ent. Ges., **16**: 50. Reiss, 1930, in Seitz, Die Gross-Schmetterlinge der Erde, Supplement, **2**: 29. Holik, 1939, Mitt. münch. ent. Ges., **29**: 67. — Zepče, Bosnien (nördlich von Sarajewo).

ssp. **onobrychis** Denis & Schiffermüller, 1775, Ankündigung eines systematischen Werkes von den Schmetterlingen der Wienergegend, p. 45; 1776, Systematisches Verzeichniss der Schmetterlinge der Wienergegend, p. 45. Hübner, 1796, Sammlung europäischer Schmetterlinge, **2**: 14, pl. 5, fig. 28; 1806, ibidem, Der Ziefer, p. 84. Esper, 1797, Die Schmetterlinge, Supplement, **2** (2): 22, pl. 42, figs. 2-4. Denis & Schiffermüller, 1801, Systematisches Verzeichniss von den Schmetterlingen der Wiener Gegend, **1**: 40. Ochsenheimer, 1808, Die Schmetterlinge von Europa, **2**: 87. Boisduval, 1828, Essai sur une Monographie des Zygénides, p. 92, pl. 6, fig. 1. Freyer, 1858, Neuere Beiträge zur Schmetterlingskunde, **7**: 66, pl. 637, fig. 2. Wagner, 1919, Ent. Mitt., **8**: 183. Reiss, 1930, in Seitz, Die Gross-Schmetterlinge der Erde, Supplement, **2**: 29, pl. 3d; 1933, ibidem, **2**: 273. Holik, 1936, Lambillionea, **36**, pl. 9, figs. 11-15. Forster & Wohlfahrt, 1958, Die Schmetterlinge Mitteleuropas, **3**: 96, pl. 10, figs. 25, 30. — Bisamberg bei Wien, Nieder-Österreich.

scopolii Rocci, 1921, Atti Soc. ligust. Sci. nat. geogr., **32**: 37.

ab. **azona** Wagner, 1919, Ent. Mitt., **8**: 182, 185. Reiss, 1930, in Seitz, Die Gross-Schmetterlinge der Erde, Supplement, **2**: 30.

ab. **vangeli** Aigner-Abafi, 1906, Ent. Z., **19**: 209.

pseudoberolinensis Burgeff, 1926, in Strand, Lepid. Cat., **33**: 44. Reiss, 1930, in Seitz, Die Gross-Schmetterlinge der Erde, Supplement, **2**: 29.

ab. **dupuyi** Burgeff, 1926, in Strand, Lepid. Cat., **33**: 45. Reiss, 1930, in Seitz, Die Gross-Schmetterlinge der Erde, Supplement, **2**: 29.

dupuyi Holik, 1939, Mitt. münch. ent. Ges., **29**: 184.

ab. **laticincta** Burgeff, 1926, Mitt. münch. ent. Ges., **16**: 60. Reiss, 1930, in Seitz, Die Gross-Schmetterlinge der Erde, Supplement, **2**: 29.

ab. **vellayi** Aigner-Abafi, 1899, Rovart. Lapok., **6**: 103, fig. Holik, 1932, Int. ent. Z., **26**: 77.

ab. **influens** Sterzl, 1919, Z. öst. EntVer., **4**: 12. Holik, 1932, Int. ent. Z., **26**: 27.

ab. **totirubra** Seitz, 1907, Die Gross-Schmetterlinge der Erde, **2**: 30, pl. 8f. Tremewan, 1961, Bull. Brit. Mus. (nat. Hist.) Ent., **10** (7): 264, pl. 52, fig. 9.

ab. **melusina** Burgeff, 1926, in Strand, Lepid. Cat., **33**: 45.

Reiss, 1930, in Seitz, Die Gross-Schmetterlinge der Erde, Supplement, **2**: 29.

ab. **asymetrica** Oberthür, 1909, Études de Lépidoptérologie comparée, **3**, pl. 22, fig. 118. Holik, 1932, Int. ent. Z., **26**: 81. Tremewan, 1961, Bull. Brit. Mus. (nat. Hist.) Ent., **10** (7): 264, pl. 52, fig. 10.

ab. **amoena** Staudinger, 1887, Berl. ent. Z., **31**: 39. Dziurzyński, 1904, Jber. wien. ent. Ver., **14**: 53, pl. 2, fig. 17; 1908, Berl. ent. Z., **53**: 57, pl. 2, fig. 19. Seitz, 1907, Die Gross-Schmetterlinge der Erde, **2**: 30, pl. 8e, f; 1912, ibidem, **2**: 444, pl. 56h. Holik, 1932, Int. ent. Z., **26**: 77.

ab. **horvathi** Aigner-Abafi, 1899, Rovart. Lapok., **6**: 103, fig. Holik, 1932, Int. ent. Z., **26**: 81.

ab. **alba** Dziurzyński, 1907, Jber. wien. ent. Ver., **17**: 88, pl. 2, fig. 24. Burgeff, 1926, Mitt. münch. ent. Ges., **16**: 60. Reiss, 1930, in Seitz, Die Gross-Schmetterlinge der Erde, Supplement, **2**: 29.

ab. **confluens** Dziurzyński, 1907, Jber. wien. ent. Ver., **17**: 87, pl. 2, figs. 18-22. Reiss, 1930, in Seitz, Die Gross-Schmetterlinge der Erde, Supplement, **2**: 29.

ab. **bohatschi** Wagner, 1905, Soc. ent., **20**: 73. Dziurzyński, 1907, Jber. wien. ent. Ver., **17**: 87, pl. 2, fig. 17.

ab. **rhodeophaia** Schawerda, 1909, Verh. zool.-bot. Ges. Wien, **59**: 326.

ab. **dichroma** Hirschke, 1906, Jber. wien. ent. Ver., **16**: 94. Reiss, 1930, in Seitz, Die Gross-Schmetterlinge der Erde, Supplement, **2**: 30.

ab. **drastichi** Hirschke, 1906, Jber. wien. ent. Ver., **16**: 95. Dziurzyński, 1907, Jber. wien. ent. Ver., **17**: 87, pl. 2, fig. 23. Seitz, 1912, Die Gross-Schmetterlinge der Erde, **2**: 444. Reiss, 1930, in Seitz, Die Gross-Schmetterlinge der Erde, Supplement, **2**: 30.

ab. **kautzi** Hirschke, 1909, Int. ent. Z., **3**: 198.

kautzi Dziurzyński, 1910, Jber. wien. ent. Ver., **20**: 9.

kautzi Hirschke, 1910, Jber. wien. ent. Ver., **20**: 11.

ab. **grossi** Hirschke, 1906, Jber. wien. ent. Ver., **16**: 95. Alberti, 1955, Ent. Z., **65**: 89-91.

ab. **flaveola** Esper, 1786, Die Schmetterlinge, **2**: 229, pl. 36, fig. 1. Hübner, 1796, Sammlung europäischer Schmetterlinge, **2**, pl. 3, fig. 14; 1806, ibidem, Der Ziefer, p. 84. Seitz, 1907, Die Gross-Schmetterlinge der Erde, **2**: 30, pl. 8d.

luteola Boisduval, 1828, Essai sur une Monographie des Zygénides, pl. 6, fig. 2; 1829, Monographie des Zygénides, Errata et Addenda, p. 2.

ab. **nigra** Reiss, 1926, Int. ent. Z., **20**: 217. Alberti, 1955, Ent. Z., **65**: 89-91. Reiss & Tremewan, 1964, Ent. Rec., **76**: 132.

nigra Reiss, 1926, Die Zygaenen Deutschlands, p. 36, pl. 1, fig.
totanigra Reiss, 1930, in Seitz, Die Gross-Schmetterlinge der Erde, Supplement, **2**: 30, pl. 3d. Reiss & Tremewan, 1964, Ent. Rec., **76**: 132.

ssp. **leonhardi** Reiss, 1921, Int. ent. Z., **15**: 38; 1930, in Seitz, Die Gross-Schmetterlinge der Erde, Supplement, **2**: 28, pl. 3a. Holik, 1939, Mitt. münch. ent. Ges., **29**: 203. — Kapellenberg bei Kronstadt, Siebenbürgen, Transsylvanien, 600-800 m.
ab. **dealbata** Holik, 1939, Mitt. münch. ent. Ges., **29**: 204.
ab. **securigera** Holik, 1939, Mitt. münch. ent. Ges., **29**: 204.

ssp. **interposita** Burgeff, 1926, Mitt. münch. ent. Ges., **16**: 60. Reiss, 1930, in Seitz, Die Gross-Schmetterlinge der Erde, Supplement, **2**: 30. — Luftenberg bei Linz (Donau), Ober-Österreich.

ssp. **carniolica** Scopoli, 1763, Entomologia carniolica, p. 189, fig. 478. Seitz, 1907, Die Gross-Schmetterlinge der Erde, **2**: 29, pl. 8d. Wagner, 1919, Ent. Mitt., **8**: 178; 1922, Soc. ent., **37**: 42. Stauder, 1922, Soc. ent., **37**: 1, 6, 9, 46. Reiss, 1930, in Seitz, Die Gross-Schmetterlinge der Erde, Supplement, **2**: 27. Holik, 1936, Lambillionea, **36**: 108, 127, pl. 8, figs. 1-5; 1939, Mitt. münch. ent. Ges., **29**: 61. Alberti, 1958, Mitt. zool. Mus. Berl., **34**: 306. Forster & Wohlfahrt, 1958, Die Schmetterlinge Mitteleuropas, **3**: 95, pl. 10, figs. 22, 27. — Carniola (Krain) [St. Katherina, Oberkrain].
nigra Dziurzyński, 1909, Berl. ent. Z., **53**: 251.
transiensnigra Dziurzyński, 1909, Berl. ent. Z., **53**: 251.
ab. **cingulata** Burgeff, 1926, in Strand, Lepid. Cat., **33**: 38. Reiss, 1930, in Seitz, Die Gross-Schmetterlinge der Erde, Supplement, **2**: 27.
ab. **posterolineata** Holik, 1939, Mitt. münch. ent. Ges., **29**: 62.
ab. **octonotata** Burgeff, 1926, Mitt. münch. ent. Ges., **16**: 49. Reiss, 1930, in Seitz, Die Gross-Schmetterlinge der Erde, Supplement, **2**: 27.

ssp. **gottscheeina** Burgeff, 1926, Mitt. münch. ent. Ges., **16**: 50. Reiss, 1930, in Seitz, Die Gross-Schmetterlinge der Erde, Supplement, **2**: 27. Holik, 1939, Mitt. münch. ent. Ges., **29**: 62. — Umgebung von Gottschee in Krain.

ssp. **carinthiae** Holik, 1939, Mitt. münch. ent. Ges., **29**: 64. — Ulrichsberg bei Klagenfurt, Kärnten, Österreich.
ab. **pseudoberolinensis** Burgeff, 1926, Mitt. münch. ent. Ges., **16**: 49. Reiss, 1930, in Seitz, Die Gross-Schmetterlinge der Erde, Supplement, **2**: 27.
ab. **dealbata** Holik, 1939, Mitt. münch. ent. Ges., **29**: 64.
ab. **flaveola** Holik, 1939, Mitt. münch. ent. Ges., **29**: 65.

ssp. **styriaca** Holik, 1939, Mitt. münch. ent. Ges., **29**: 66. — Graz, Steiermark, Österreich.

ssp. **syrmica** Holik, 1939, Mitt. münch. ent. Ges., **29**: 183. — Beočin, Fruška Gora, Slavonien.

Tremewan, 1961, Bull. Brit. Mus. (nat. Hist.) Ent., **10** (7): 307, pl. 57, fig. 23.

ab. **azona** Holik, 1939, Mitt. münch. ent. Ges., **29**: 183.

ab. **dupuyi** Burgeff, 1926, Mitt. münch. ent. Ges., **16**: 49. Reiss, 1930, in Seitz, Die Gross-Schmetterlinge der Erde, Supplement, **2**: 27.

ab. **dealbata** Holik, 1939, Mitt. münch. ent. Ges., **29**: 183.

ab. **securigera** Holik, 1939, Mitt. münch. ent. Ges., **29**: 183.

ssp. **microhistria** Holik, 1939, Mitt. münch. ent. Ges., **29**: 174. — Tarnova, Görz, Norditalien.

ssp. **histria** Burgeff, 1926, Mitt. münch. ent. Ges., **16**: 49. Reiss, 1930, in Seitz, Die Gross-Schmetterlinge der Erde, Supplement, **2**: 27. Holik, 1939, Mitt. münch. ent. Ges., **29**: 175. — Triest, Istrien, Norditalien.

ab. **pseudoleonhardi** Holik, 1939, Mitt. münch. ent. Ges., **29**: 177, 178.

ab. **dupuyi** Holik, 1939, Mitt. münch. ent. Ges., **29**: 181.

ssp. **gradiscana** Stauder, 1922, Soc. ent., **37**: 46. Reiss, 1930, in Seitz, Die Gross-Schmetterlinge der Erde, Supplement, **2**: 29. Holik, 1939, Mitt. münch. ent. Ges., **29**: 175. Tremewan, 1966, Ent. Rec., **78**: 32, pl. 1, fig. 3. — Sdraussina bei Gradisca, Litorale Illyricum.

ab. **rubrothoracalis** Stauder, 1922, Soc. ent., **37**: 46. Reiss, 1930, in Seitz, Die Gross-Schmetterlinge der Erde, Supplement, **2**: 29.

ssp. **croatica** Reiss, 1941, Mitt. münch. ent. Ges., **31**: 998, pl. 34, figs. C5, A6. — Fucine, Kroatisches Litoral.

ssp. **siciliana** Reiss, 1921, Int. ent. Z., **15**: 39; 1930, in Seitz, Die Gross-Schmetterlinge der Erde, Supplement, **2**: 29, pl. 3c. — Taormina, Sizilien, Süditalien.

sicilica Ragusa, 1924, Boll. Lab. Zool. Portici, **18**: 94. Tremewan, 1961, Bull. Brit. Mus. (nat. Hist.) Ent., **10** (7): 265, pl. 52, fig. 11.

ssp. **calabricola** Verity, 1946, Redia, **31**: 67 (nomen novum for *calabrica* Turati). — Reggio, Calabria, Italia.

calabrica Turati, 1913, Atti Soc. ital. Sci. nat., **51**: 338 (preoccupied by **calabrica** Calberla, 1895, ssp. of **transalpina** Esper). Reiss, 1930, in Seitz, Die Gross-Schmetterlinge der Erde, Supplement, **2**: 29, pl. 3c.

ab. **cingulata** Turati, 1913, Atti Soc. ital. Sci. nat., **51**: 338. Reiss, 1930, in Seitz, Die Gross-Schmetterlinge der Erde, Supplement, **2**: 29.

ab. **intermedia** Turati, 1913, Atti Soc. ital. Sci. nat., **51**: 338. Reiss, 1930, in Seitz, Die Gross-Schmetterlinge der Erde, Supplement, **2**: 29.

ssp. **aspromontica** Reiss, 1941, Z. wien. EntVer., **26**: 60. ab. **apennina** Reiss, 1941, Z. wien. EntVer., **26**: 61.	Sinopoli (Aspromonte), Süditalien, 500 m.
ssp. **amanda** Reiss, 1921, Int. ent. Z., **15**: 20; 1930, in Seitz, Die Gross-Schmetterlinge der Erde, Supplement, **2**: 29, pl. 3c.	Subiaco, Majella, Gran Sasso, Abruzzen, Italien.
ssp. **dulcis** Burgeff, 1926, Mitt. münch. ent. Ges., **16**: 60. Reiss, 1930, in Seitz, Die Gross-Schmetterlinge der Erde, Supplement, **2**: 29. ab. **confluens** Burgeff, 1926, in Strand, Lepid. Cat., **33**: 43. Reiss, 1930, in Seitz, Die Gross-Schmetterlinge der Erde, Supplement, **2**: 29. ab. **suffusa** Turati, 1913, Atti Soc. ital. Sci. nat., **51**: 339. Holik, 1932, Int. ent. Z., **26**: 81. ab. **amoena** Burgeff, 1926, Mitt. münch. ent. Ges., **16**: 60. Reiss, 1930, in Seitz, Die Gross-Schmetterlinge der Erde, Supplement, **2**: 29.	Mte. Sirente; Mte. Velino, Italien, 1500-2000 m.
ssp. **formiacola** Reiss & Tremewan, 1964, Ent. Rec., **76**: 132 (nomen novum for *magnaustralis* Verity, 1946 [as *formidacola* Reiss & Tremewan (partim)]). *magnaustralis* Verity, 1946, Redia, **31**: 66 (preoccupied by **magnaustralis** Verity, 1926, ssp. of **trifolii** Esper). Querci, 1947, Ent. Rec., **59**: 97. Reiss & Tremewan, 1964, Ent. Rec., **76**: 132.	Formia presso Lazio, Italia.
ssp. **eminens** Verity, 1946, Redia, **31**: 64.	Firenzuola, 500 m.; Covigliaio, 870 m.; Paluzzuolo di Romagna, 700 m., Italia.
ssp. **incerta** Rocci, 1914, Atti Soc. ligust. Sci. nat. geogr., **25**: 226; 1919, ibidem, **30**: 65, 76, pl. 4, figs. 9b, c, d, 10b. Verity, 1920, Boll. Lab. Zool. Portici, **14**: 40. Burgeff, 1926, Mitt. münch. ent. Ges., **16**: 55. Reiss, 1930, in Seitz, Die Gross-Schmetterlinge der Erde, Supplement, **2**: 28, pl. 3b. Verity, 1946, Redia, **31**: 62. ab. **canuta** Rocci, 1914, Atti Soc. ligust. Sci. nat. geogr., **25**: 225. Reiss, 1930, in Seitz, Die Gross-Schmetterlinge der Erde, Supplement, **2**: 28. ab. **apennina** Turati, 1884, Boll. Soc. ent. ital., **16**: 71. Dziurzyński, 1904, Jber. wien. ent. Ver., **14**: 54, pl. 2, fig. 18. Seitz, 1907, Die Gross-Schmetterlinge der Erde, **2**: 30, pl. 8g.	Appennino ligure; Appennino toscana; Italia [Monte Maggio, Reopasso].

Reiss, 1930, in Seitz, Die Gross-Schmetterlinge der Erde, Supplement, **2**: 28. Verity, 1946, Redia, **31**: 65.

pseudoapenina Rocci, 1919, Atti Soc. ligust. Sci. nat. geogr., **30**: 80, pl. 4, fig. 11c.

wiskotti Calberla, 1887, Iris, **1**: 146.

ab. **quadrisignata** Rocci, 1914, Atti Soc. ligust. Sci. nat. geogr., **25**: 223.

ab. **trisignata** Rocci, 1914, Atti Soc. ligust. Sci. nat. geogr., **25**: 223.

ab. **intermedia** Rocci, 1914, Atti Soc. ligust. Sci. nat. geogr., **25**: 224.

ab. **decollata** Rocci, 1914, Atti Soc. ligust. Sci. nat. geogr., **25**: 225.

ab. **quinquesignata** Rocci, 1914, Atti Soc. ligust. Sci. nat. geogr., **25**: 224.

ssp. **livornica** Burgeff, 1926, Mitt. münch. ent. Ges., **16**: 59. Reiss, 1930, in Seitz, Die Gross-Schmetterlinge der Erde, Supplement, **2**: 29. Burgeff, 1950, Portug. acta biol., (A) Goldschmidt: 697, figs. 4b. — Aus den Macchien zwischen Livorno und Pisa, Italien.

ab. **cingulata** Burgeff, 1926, in Strand, Lepid. Cat., **33**: 43. Reiss, 1930, in Seitz, Die Gross-Schmetterlinge der Erde, Supplement, **2**: 29.

ab. **pseudoberolinensis** Burgeff, 1926, Mitt. münch. ent. Ges., **16**: 59. Reiss, 1930, in Seitz, Die Gross-Schmetterlinge der Erde, Supplement, **2**: 29.

ab. **apennina** Burgeff, 1926, Mitt. münch. ent. Ges., **16**: 59. Reiss, 1930, in Seitz, Die Gross-Schmetterlinge der Erde, Supplement, **2**: 29.

ssp. **florentina** Verity, 1920, Boll. Lab. Zool. Portici, **14**: 41. Reiss, 1930, in Seitz, Die Gross-Schmetterlinge der Erde, Supplement, **2**: 29, pl. 3c. Querci, 1947, Ent. Rec., **59**: 98. Burgeff, 1950, Portug. acta biol., (A) Goldschmidt: 697, figs. 4a. — Pian di Mugnone, presso Firenze, Italia.

ssp. **roccii** Verity, 1920, Boll. Lab. Zool. Portici, **14**: 42. Seitz, 1907, Die Gross-Schmetterlinge der Erde, **2**: 30, pl. 8g (as *apennina* Turati). Burgeff, 1914, Mitt. münch. ent. Ges., **5**: 54, pl. 3, figs. 81-83, 89-91 (as *apennina* Turati). Reiss, 1930, in Seitz, Die Gross-Schmetterlinge der Erde, Supplement, **2**: 28, pl. 3b. Holik, 1936, Lambillionea, **36**, pl. 8, figs. 16-20. Burgeff, 1950, Portug. acta biol., (A) Goldschmidt: 697, figs. 4c. — Genova, Italia.

ab. **cingulata** Dziurzyński, 1904, Jber. wien. ent. Ver., **14**: 54, pl. 2, fig. 19. Reiss, 1930, in Seitz, Die Gross-Schmetterlinge der Erde, Supplement, **2**: 29.

cingulata Burgeff, 1914, Mitt. münch. ent. Ges., **5**: 54, pl. 2, fig. 178.

ab. **pseudocarniolica** Rocci, 1914, Atti Soc. ligust. Sci. nat. geogr., **24**: 116. Reiss, 1930, in Seitz, Die Gross-Schmetterlinge der Erde, Supplement, **2**: 29.
pseudocarniolica Rocci, 1914, Soc. ent., **29**: 42.
pseudohedysari Burgeff, 1914, Mitt. münch. ent. Ges., **5**: 54, pl. 3, figs. 98, 105.
ab. **berolinoides** Turati, 1913, Atti Soc. ital. Sci. nat., **51**: 338. Seitz, 1907, Die Gross-Schmetterlinge der Erde, **2**: 30, pl. 8g (as *berolinensis* Staudinger).
pseudoberolinensis Burgeff, 1914, Mitt. münch. ent. Ges., **5**: 54, pl. 3, figs. 97, 104.
dealbata Rocci, 1914, Atti Soc. ligust. Sci. nat. geogr., **24**: 116.
dealbata Rocci, 1914, Soc. ent., **29**: 42.
ab. **dupuyi** Burgeff, 1914, Mitt. münch. ent. Ges., **5**: 54, pl. 2, fig. 176, pl. 3, fig. 96. Reiss, 1930, in Seitz, Die Gross-Schmetterlinge der Erde, Supplement, **2**: 29, pl. 3b.
ab. **nigrocincta** Rocci, 1914, Atti Soc. ligust. Sci. nat. geogr., **24**: 116. Reiss, 1930, in Seitz, Die Gross-Schmetterlinge der Erde, Supplement, **2**: 29, pl. 3b.
nigrocincta Rocci, 1914, Soc. ent., **29**: 42.
nigrosupposita Burgeff, 1914, Mitt. münch. ent. Ges., **5**: 54, pl. 2, fig. 177, pl. 3, fig. 95.
genovensis Reiss, 1914, Int. ent. Z., **8**: 46.
ab. **parvipuncta** Rocci, 1914, Atti Soc. ligust. Sci. nat. geogr., **25**: 225.
ab. **nigricans** Burgeff, 1914, Mitt. münch. ent. Ges., **5**: 54, pl. 3, figs. 86, 87, 93, 94. Reiss, 1930, in Seitz, Die Gross-Schmetterlinge der Erde, Supplement, **2**: 29.
ab. **depauperata** Turati, 1913, Atti Soc. ital. Sci. nat., **51**: 338. Reiss, 1930, in Seitz, Die Gross-Schmetterlinge der Erde, Supplement, **2**: 29, pl. 3c.
paupera Burgeff, 1914, Mitt. münch. ent. Ges., **5**: 54, pl. 3, fig. 88.
deleta Rocci, 1919, Atti Soc. ligust. Sci. nat. geogr., **30**: 71, pl. 4, fig. 7c.
ab. **stoechadoides** Turati, 1913, Atti Soc. ital. Sci. nat., **51**: 338. Reiss, 1930, in Seitz, Die Gross-Schmetterlinge der Erde, Supplement, **2**: 29.
nigrescens Rocci, 1914, Atti Soc. ligust. Sci. nat. geogr., **24**: 116.
nigrescens Rocci, 1914, Soc. ent., **29**: 42.
ornata Burgeff, 1914, Mitt. münch. ent. Ges., **5**: 54, pl. 2, fig. 180, pl. 3, figs. 84, 92.
ab. **octonotata** Turati, 1913, Atti Soc. ital. Sci. nat., **51**: 338. Reiss, 1930, in Seitz, Die Gross-Schmetterlinge der Erde, Supplement, **2**: 29.

octornata Reiss, 1914, Int. ent. Z., **8**: 46.
prolifera Burgeff, 1914, Mitt. münch. ent. Ges., **5**: 54, pl. 2, fig. 178, pl. 3, fig. 83.
ab. **laticlavia** Burgeff, 1914, Mitt. münch. ent. Ges., **5**: 54, pl. 2, fig. 179, pl. 3, fig. 85. Reiss, 1930, in Seitz, Die Gross-Schmetterlinge der Erde, Supplement, **2**: 29.
ab. **canuta** Rocci, 1914, Atti Soc. ligust. Sci. nat. geogr., **25**: 225. Reiss, 1930, in Seitz, Die Gross-Schmetterlinge der Erde, Supplement, **2**: 29.
ab. **cuprea** Turati, 1913, Atti Soc. ital. Sci. nat., **51**: 338. Reiss, 1930, in Seitz, Die Gross-Schmetterlinge der Erde, Supplement, **2**: 29.
ab. **monosignata** Turati, 1913, Atti Soc. ital. Sci. nat., **51**: 338.
ab. **bissignata** Turati, 1913, Atti Soc. ital. Sci. nat., **51**: 338.
ab. **posterolineata** Rocci, 1914, Atti Soc. ligust. Sci. nat. geogr., **24**: 116.
posterolineata Rocci, 1914, Soc. ent., **29**: 42.
ab. **bitincta** Rocci, 1914, Atti Soc. ligust. Sci. nat. geogr., **25**: 225.
ab. **incompleta** Rocci, 1914, Atti Soc. ligust. Sci. nat. geogr., **24**: 116.
incompleta Rocci, 1914, Soc. ent., **29**: 42.
ab. **minima** Rocci, 1914, Atti Soc. ligust. Sci. nat. geogr., **24**: 116.
minima Rocci, 1914, Soc. ent., **29**: 42.
ab. **bicolor** Rocci, 1914, Atti Soc. ligust. Sci. nat. geogr., **24**: 116.
bicolor Rocci, 1914, Soc. ent., **29**: 42.
ab. **carnea** Rocci, 1919, Atti Soc. ligust. Sci. nat. geogr., **30**: 75.

ssp. **padana** Rocci, 1921, Atti Soc. ligust. Sci. nat. geogr., **32**: 38. Reiss, 1930, in Seitz, Die Gross-Schmetterlinge der Erde, Supplement, **2**: 29. Verity, 1946, Redia, **31**: 63. — Parma; Reggio; basso bacino del Po, Italia.

ssp. **notissima** Rocci, 1941, Boll. Ist. Ent. Univ. Bologna, **13**: 127, pl. 3, figs. 23-26. — Collina di Torino, Piedmont, Italia, 600-650 m.
microhedysari Verity, 1946, Redia, **31**: 61.
ab. **virginea** Allioni, 1766, Misc. Taurin., **3**: 192.
ab. **biconjuncta** Rocci, 1941, Boll. Ist. Ent. Univ. Bologna, **13**: 127.
ab. **ragonoti** Gianelli, 1902, Zool. Anz., **25**: 509. Seitz, 1907, Die Gross-Schmetterlinge der Erde, **2**: 30, pl. 8f.
f. t. **autumnalis** Rocci, 1919, Atti Soc. ligust. Sci. nat. geogr., **30**: 65, pl. 4, fig. 7e (as *antumnalis*). Reiss, 1930, in Seitz, Die Gross-Schmetterlinge der Erde, Supplement, **2**: 29. — Colline del Po, Torino, Italia.

ssp. **diniensis** Herrich-Schäffer, 1852, Systematische Bearbeitung der Schmetterlinge von Europa, **6**: 46; 1851, ibidem, **2**, pl. 16, figs. 111, 112 (non-binominal). Oberthür, 1896, Études d'Entomologie, **20**: 52, pl. 7, figs. 118, 121. Seitz, 1907, Die Gross- — Digne, Basses-Alpes, Frankreich.

Schmetterlinge der Erde, **2**: 30, pl. 8e. Oberthür, 1909, Études de Lépidoptérologie comparée, **3**, pl. 22, figs. 115-117; 1910, ibidem, **4**: 631. Wagner, 1919, Ent. Mitt., **8**: 182. Burgeff, 1926, Mitt. münch. ent. Ges., **16**: 53. Reiss, 1930, in Seitz, Die Gross-Schmetterlinge der Erde, Supplement, **2**: 28. Holik, 1936, Lambillionea, **36**, pl. 9, figs. 16, 17. Burgeff, 1950, Portug. acta biol., (A) Goldschmidt: 697, figs. 4i. Dufay, 1966, Bull. mens. Soc. linn. Lyon, **35**: 68.

ab. **hedysaroides** Turati, 1913, Atti Soc. ital. Sci. nat., **51**: 337. Reiss, 1930, in Seitz, Die Gross-Schmetterlinge der Erde, Supplement, **2**: 28.

azona Wagner, 1919, Ent. Mitt., **8**: 182.

ab. **dupuyi** Oberthür, 1909, Études de Lépidoptérologie comparée, **3**, pl. 22, figs. 110-112; 1910, ibidem, **4**: 632. Seitz, 1912, Die Gross-Schmetterlinge der Erde, **2**: 444. Reiss, 1930, in Seitz, Die Gross-Schmetterlinge der Erde, Supplement, **2**: 28. Tremewan, 1961, Bull. Brit. Mus. (nat. Hist.) Ent., **10** (7): 263, pl. 52, fig. 5.

ab. **melusina** Oberthür, 1909, Études de Lépidoptérologie comparée, **3**, pl. 22, figs. 119, 120; 1910, ibidem, **4**: 632. Reiss, 1930, in Seitz, Die Gross-Schmetterlinge der Erde, Supplement, **2**: 28. Holik, 1932, Int. ent. Z., **26**: 82. Tremewan, 1961, Bull. Brit. Mus. (nat. Hist.) Ent., **10** (7): 263, pl. 52, fig. 4.

ssp. **dinioides** Burgeff, 1926, Mitt. münch. ent. Ges., **16**: 53. Reiss, 1930, in Seitz, Die Gross-Schmetterlinge der Erde, Supplement, **2**: 28. Burgeff, 1950, Portug. acta biol., (A) Goldschmidt: 697, figs. 4h.	St. Martin-de-Vésubie, Alpes-Maritimes, Frankreich, 1500 m.
ssp. **magdalenae** Abeille, 1909, Mém. Soc. linn. Provence, **1**: 16. Holik, 1938, Lambillionea, **38**: 170.	St. Baume, Var, France.
ssp. **larchensis** Holik, 1938, Lambillionea, **38**: 169, pl. 8, figs. 18-20.	Larche, Basses-Alpes, 1700-1800 m.
ssp. **miniorubens** Dujardin, 1956, Bull. mens. Soc. linn. Lyon, **25**: 261.	Thorenc, Col de Bleyne, Alpes-Maritimes, France, 1450 m.
ssp. **gessa** Burgeff, 1950, Portug. acta biol., (A) Goldschmidt: 699, fig. 4e.	Oberlauf der Gessa, Roaschia, Entraque, Meeralpen, Norditalien, 1000 m.
ssp. **tanara** Burgeff, 1950, Portug. acta biol., (A) Goldschmidt: 699, fig. 4d.	Viozene, Tanaro Fluss, Meeralpen, Norditalien, 1400-1500 m.

ssp. **subincerta** Rocci, 1941, Boll. Ist. Ent. Univ. Bologna, **13**: 127, pl. 3, figs. 19-22.	Fabrosa Soprana, Alpi Marittime, Piedmont, Italia, 850-950 m.
ssp. **roja** Burgeff, 1950, Portug. acta biol., (A) Goldschmidt: 699, fig. 4f.	Rojatal, südlich des Tenda, Meeralpen, Norditalien.
ssp. **moraulti** Holik, 1938, Lambillionea, **38**: 166, pl. 8, figs. 11-17. ab. **securigera** Holik, 1938, Lambillionea, **38**: 166. ab. **pseudoleonhardi** Holik, 1938, Lambillionea, **38**: 165. ab. **dealbata** Holik, 1938, Lambillionea, **38**: 164.	Le Lautaret, Hautes-Alpes, France, 2000 m.
ssp. **riparia** Rocci, 1941, Boll. Ist. Ent. Univ. Bologna, **13**: 127, pl. 3, figs. 27-32.	Meana; Ulzio; Salabertano; Valle di Susa, Piedmont, Italia, 800-1000 m.
ssp. **piemonticola** Reiss, 1941, Mitt. münch. ent. Ges., **31**: 997, pl. 34, figs. B4, C4, A5, B5.	Col de Sestriere, Piemont, Italien, 1600-1900 m.
ssp. **savoia** Holik, 1938, Lambillionea, **38**: 163, pl. 8, figs. 6-10.	Crévin (Pied du Salève), Haute-Savoie, France, 500 m.
ssp. **menaggia** Przegendza, 1932, Ent. Z., **46**: 113, figs. 17-20. Reiss, 1933, in Seitz, Die Gross-Schmetterlinge der Erde, Supplement, **2**: 272; 1950, Jber. naturf. Ges. Graubünden, **82**: 110. ab. **pseudoleonhardi** Reiss, 1950, Jber. naturf. Ges. Graubünden, **82**: 110.	Menaggio, Comer See, Norditalien, 200 m.
ssp. **marinensis** Przegendza & Prack, 1943, Ent. Z., **56**: 241, figs. 1, 2. ab. **cingulata** Przegendza & Prack, 1943, Ent. Z., **56**: 243. ab. **pseudoroccii** Przegendza & Prack, 1943, Ent. Z., **56**: 243, fig. 7. ab. **parvipuncta** Przegendza & Prack, 1943, Ent. Z., **56**: 243, fig. 6. ab. **dichroma** Przegendza & Prack, 1943, Ent. Z., **56**: 243, fig. 3. ab. **amoenoides** Przegendza & Prack, 1943, Ent. Z., **56**: 243, fig. 4. ab. **amoena** Przegendza & Prack, 1943, Ent. Z., **56**: 243, fig. 5.	Mte. Titano, San Marino, Norditalien, 750 m.

ssp. **hedysari** Hübner, 1796, Sammlung europäischer Schmetterlinge, **2**: 15, pl. 5, fig. 29, pl. 6, fig. 36; 1806, ibidem, Der Ziefer, p. 83. Seitz, 1907, Die Gross-Schmetterlinge der Erde, **2**: 30, pl. 8e. Wagner, 1919, Ent. Mitt., **8**: 180. Rocci, 1921, Atti Soc. ligust. Sci. nat. geogr., **32**: 38. Burgeff, 1926, Mitt. münch. ent. Ges., **16**: 50. Reiss, 1930, in Seitz, Die Gross-Schmetterlinge der Erde, Supplement, **2**: 27. Holik, 1936, Lambillionea, **36**, pl. 8, figs. 6-10. Reiss, 1941, Mitt. münch. ent. Ges., **31**, pl. 34, figs. C3, A4; 1950, Jber. naturf. Ges. Graubünden, **82**: 109, figs. 14, 15. Burgeff, 1950, Portug. acta biol., (A) Goldschmidt: 697, figs. 4g. — Südliche Alpen von Piemont bis Südtirol [Bozen].

ab. **cingulata** Burgeff, 1926, Mitt. münch. ent. Ges., **16**: 50. Reiss, 1930, in Seitz, Die Gross-Schmetterlinge der Erde, Supplement, **2**: 27.

ab. **pseudoberolinensis** Burgeff, 1926, Mitt. münch. ent. Ges., **16**: 50. Oberthür, 1896, Études d'Entomologie, **20**, pl. 7, fig. 117. Reiss, 1930, in Seitz, Die Gross-Schmetterlinge der Erde, Supplement, **2**: 27.

ab. **apenina** Oberthür, 1910, Études de Lépidoptérologie comparée, **4**: 633; 1896, Études d'Entomologie, **20**, pl. 7, fig. 114. Tremewan, 1961, Bull. Brit. Mus. (nat. Hist.) Ent., **10** (7): 264.

pseudoapenina Rocci, 1921, Atti Soc. ligust. Sci. nat. geogr., **32**: 39.

apennina Burgeff, 1926, in Strand, Lepid. Cat., **33**: 39; 1926, Mitt. münch. ent. Ges., **16**: 50. Reiss, 1930, in Seitz, Die Gross-Schmetterlinge der Erde, Supplement, **2**: 27.

ab. **quinquemaculata** Vorbrodt, 1913, in Vorbrodt & Müller-Rutz, Die Schmetterlinge der Schweiz, **2**: 281, fig. 39.

ab. **octonotata** Burgeff, 1926, in Strand, Lepid. Cat., **33**: 39; 1926, Mitt. münch. ent. Ges., **16**: 50. Reiss, 1930, in Seitz, Die Gross-Schmetterlinge der Erde, Supplement, **2**: 27.

ab. **dupuyi** Reiss, 1950, Jber. naturf. Ges. Graubünden, **82**: 109, fig. 16.

ab. **nigrocincta** Reiss, 1950, Jber. naturf. Ges. Graubünden, **82**: 109.

ssp. **mendolensis** Dannehl, 1929, Ent. Z., **43**: 41. Reiss, 1930, in Seitz, Die Gross-Schmetterlinge der Erde, Supplement, **2**: 27. — Höhen der Mendel, Südtirol.

ssp. **anzascana** Verity, 1930, Mem. Soc. ent. ital., **9**: 22. Reiss, 1933, in Seitz, Die Gross-Schmetterlinge der Erde, Supplement, **2**: 272. — Vanzone, Monte Rosa, Valle Anzasca, Italia, 700 m.

ssp. **rhaeticola** Burgeff, 1926, Mitt. münch. ent. Ges., **16**: 50. Reiss, 1930, in Seitz, Die Gross-Schmetterlinge der Erde, — Filisur, Graubünden, Schweiz, 1000 m.

Supplement, **2**: 27; 1950, Jber. naturf. Ges. Graubünden, **82**: 108.

ab. **flaveola** Reiss, 1933, in Seitz, Die Gross-Schmetterlinge der Erde, Supplement, **2**: 272.

ssp. **valesiae** Burgeff, 1926, Mitt. münch. ent. Ges., **16**: 51. Oberthür, 1910, Études de Lépidoptérologie comparée, **4**: 629. Reiss, 1930, in Seitz, Die Gross-Schmetterlinge der Erde, Supplement, **2**: 27. Holik, 1936, Lambillionea, **36**, pl. 9, figs. 1-5. Martigny-Ville, Wallis, Schweiz.

ab. **cingulata** Vorbrodt, 1913, in Vorbrodt & Müller-Rutz, Die Schmetterlinge der Schweiz, **2**: 279.

ab. **pseudoberolinensis** Burgeff, 1926, Mitt. münch. ent. Ges., **16**: 51. Reiss, 1930, in Seitz, Die Gross-Schmetterlinge der Erde, Supplement, **2**: 27.

ab. **laticincta** Burgeff, 1926, Mitt. münch. ent. Ges., **16**: 51. Reiss, 1930, in Seitz, Die Gross-Schmetterlinge der Erde, Supplement, **2**: 27. Holik, 1932, Int. ent. Z., **26**: 81.

ab. **quinquemaculata** Vorbrodt, 1913, in Vorbrodt & Müller-Rutz, Die Schmetterlinge der Schweiz, **2**: 281, fig. 39.

ab. **parvimaculata** Vorbrodt, 1913, in Vorbrodt & Müller-Rutz, Die Schmetterlinge der Schweiz, **2**: 281, fig. 32.

ab. **securigera** Burgeff, 1926, Mitt. münch. ent. Ges., **16**: 52. Reiss, 1930, in Seitz, Die Gross-Schmetterlinge der Erde, Supplement, **2**: 28.

ab. **costalielongata** Vorbrodt, 1913, in Vorbrodt & Müller-Rutz, Die Schmetterlinge der Schweiz, **2**: 280.

ab. **basiconfluens** Vorbrodt, 1913, in Vorbrodt & Müller-Rutz, Die Schmetterlinge der Schweiz, **2**: 280.

ab. **medioconfluens** Vorbrodt, 1913, in Vorbrodt & Müller-Rutz, Die Schmetterlinge der Schweiz, **2**: 280.

ab. **costaliconfluens** Vorbrodt, 1913, in Vorbrodt & Müller-Rutz, Die Schmetterlinge der Schweiz, **2**: 280, fig. 18.

ab. **parallela** Vorbrodt, 1913, in Vorbrodt & Müller-Rutz, Die Schmetterlinge der Schweiz, **2**: 281, fig. 20.

ab. **omniconfluens** Vorbrodt, 1913, in Vorbrodt & Müller-Rutz, Die Schmetterlinge der Schweiz, **2**: 281, fig. 25.

ab. **confluens** Burgeff, 1926, Mitt. münch. ent. Ges., **16**: 52. Reiss, 1930, in Seitz, Die Gross-Schmetterlinge der Erde, Supplement, **2**: 27.

ab. **sinistrotricolor** Lacreuze, 1945, Mitt. schweiz. ent. Ges., **19**: 254.

ab. **tricolor** Oberthür, 1904, Études de Lépidoptérologie comparée, **1**: 52, pl. 3, fig. 30. Holik, 1932, Int. ent. Z., **26**: 81. Tremewan, 1961, Bull. Brit. Mus. (nat. Hist.) Ent., **10** (7): 264, pl. 52, fig. 7.

ab. **jurassica** Blachier, 1905, Bull. Soc. ent. Fr., p. 52; 1906, Ann. Soc. ent. Fr., **75**: 21, pl. 2, fig. 1.

ab. **weileritricolor** Oberthür, 1904, Études de Lépidoptérologie comparée, **1**: 52, pl. 3, fig. 32. Tremewan, 1961, Bull. Brit. Mus. (nat. Hist.) Ent., **10** (7): 264, pl. 52, fig. 8.

ab. **amoena** Burgeff, 1926, in Strand, Lepid. Cat., **33**: 39; 1926, Mitt. münch. ent. Ges., **16**: 51. Reiss, 1930, in Seitz, Die Gross-Schmetterlinge der Erde, Supplement, **2**: 27.

ab. **flava** Vorbrodt, 1913, in Vorbrodt & Müller-Rutz, Die Schmetterlinge der Schweiz, **2**: 279.

ab. **incingulata** Vorbrodt, 1913, in Vorbrodt & Müller-Rutz, Die Schmetterlinge der Schweiz, **2**: 279.

ssp. **modesta** Burgeff, 1914, Mitt. münch. ent. Ges., **5**: 57, pl. 3, figs. 99, 100, 106, 107. Esper, 1779, Die Schmetterlinge, **2**, pl. 17, figs. 4a, b (as *caffra* Linné); 1781, ibidem, **2**: 152 (as *caffra* Linné). Wagner, 1919, Ent. Mitt., **8**: 185. Reiss, 1930, in Seitz, Die Gross-Schmetterlinge der Erde, Supplement, **2**: 28. Haaf, 1952, Veröff. zool. Staatssamml. Münch., **2**: 141, 152, 154, 157, pl. 11 (as *carniolica* Scopoli). Bergmann, 1953, Die Grossschmetterlinge Mitteldeutschlands, **3**: 34, pl. 66, figs. A1-A4, A7, B1-B3, B6, B7. Forster & Wohlfahrt, 1958, Die Schmetterlinge Mitteleuropas, **3**: 95, pl. 10, figs. 23, 28. Heuser & Jöst, 1959, Mitt. Pollichia, (3) **6**: 125.

Mittleres und unteres Rheintal von Mainz bis Coblenz, Deutschland,

ab. **pseudoleonhardi** Reiss, 1941, Mitt. münch. ent. Ges., **31**: 999, pl. 34, fig. A8. Bergmann, 1953, Die Grossschmetterlinge Mitteldeutschlands, **3**, pl. 66, figs. A5, A6.

ab. **securigera** Reiss, 1941, Mitt. münch. ent. Ges., **31**: 998, pl. 34, fig. A7.

ab. **paradoxa** Burgeff, 1914, Mitt. münch. ent. Ges., **5**: 59. Reiss, 1930, in Seitz, Die Gross-Schmetterlinge der Erde, Supplement, **2**: 28; 1941, Mitt. münch. ent. Ges., **31**: 999, pl. 34, fig. C7.

ab. **klapalecki** Burgeff, 1926, in Strand, Lepid. Cat., **33**: 40.

ab. **drastichi** Reiss, 1941, Mitt. münch. ent. Ges., **31**: 998.

ab. **eximia** Heyn, 1913, Ent. Z., **27**: 41, fig. Holik, 1932, Int. ent. Z., **26**: 81.

ab. **albomarginata** Spuler, 1906, in Hofmann, Die Schmetterlinge Europas, **2**: 164. Holik, 1932, Int. ent. Z., **26**: 80.

ab. **confluens** Burgeff, 1926, in Strand, Lepid. Cat., **33**: 40.

ab. **bicolor** Burgeff, 1906, Ent. Z., **20**: 161, fig. 3.

ab. **weileri** Staudinger, 1887, Berl. ent. Z., **31**: 39. Reiss, 1933, in Seitz, Die Gross-Schmetterlinge der Erde, Supplement, **2**: 272; 1941, Mitt. münch. ent. Ges., **31**, pl. 34, fig. B7.

ab. **carnea** Spuler, 1906, in Hofmann, Die Schmetterlinge Europas, **2**: 164.

ab. **flaveola** Burgeff, 1914, Mitt. münch. ent. Ges., **5**: 59, pl. 2, fig. 181, pl. 3, fig. 101. Reiss, 1930, in Seitz, Die Gross-Schmetterlinge der Erde, Supplement, **2**: 28.

ssp. **media** Reiss, 1918, Int. ent. Z., **11**: 201. Wagner, 1919, Ent. Mitt., **8**: 186. Reiss, 1926, Die Zygaenen Deutschlands, p. 35, pl. 2, fig. (as *modesta* Burgeff); 1937, in Schneider, Jh. Ver. vaterl. Naturk. Württemb., **93**: 127 (as *modesta* Burgeff); 1949, Entomon, **1**: 171 (as *modesta* Burgeff). — Weil der Stadt, Neuffen, Württemberg, Deutschland.

ab. **pseudoleonhardi** Reiss, 1949, Entomon, **1**: 171.

ab. **securigera** Reiss, 1949, Entomon, **1**: 171.

ab. **dupuyi** Reiss, 1964, Coridon, (A) **6**: 4.

ab. **philamoena** Reiss, 1918, Int. ent. Z., **11**: 202. Holik, 1932, Int. ent. Z., **26**: 81.

ab. **griseola** Reiss, 1941, Mitt. münch. ent. Ges., **31**: 999.

ab. **meteora** Reiss, 1918, Int. ent. Z., **11**: 202; 1926, Die Zygaenen Deutschlands, p. 36, pl. 1, fig.; 1930, in Seitz, Die Gross-Schmetterlinge der Erde, Supplement, **2**: 28. Holik, 1932, Int. ent. Z., **26**: 82.

ab. **flavicornis** Reiss, 1949, Entomon, **1**: 172.

ab. **detschi** Oberthür, 1910, Études de Lépidoptérologie comparée, **4**: 636, pl. 51, fig. 442. Tremewan, 1961, Bull. Brit. Mus. (nat. Hist.) Ent., **10** (7): 263, pl. 52, fig. 6.

ssp. **diluviicola** Burgeff, 1926, Mitt. münch. ent. Ges., **16**: 54. Reiss, 1930, in Seitz, Die Gross-Schmetterlinge der Erde, Supplement, **2**: 28. — Ihringen und Bickensohl, Kaiserstuhl, Baden, Westdeutschland.

ab. **amoena** Burgeff, 1914, Mitt. münch. ent. Ges., **5**: 59.

ssp. **albarracina** Staudinger, 1887, Berl. ent. Z., **31**: 41. Seitz, 1907, Die Gross-Schmetterlinge der Erde, **2**: 30. Wagner, 1919, Ent. Mitt., **8**: 183. Reiss, 1921, Int. ent. Z., **15**: 39. Burgeff, 1926, Mitt. münch. ent. Ges., **16**: 53. Reiss, 1930, in Seitz, Die Gross-Schmetterlinge der Erde, Supplement, **2**: 28, pl. 3a. — Albarracin, Aragonien, Spanien.

ssp. **sagarraiana** Reiss & Tremewan, 1964, Ent. Rec., **76**: 132 (nomen novum for *catalonica* de Sagarra). — Sorribes, Gosol, Valle de Arán, Andorra, España.

catalonica de Sagarra, 1940, VI Congr. int. Ent., Madrid, p. 392 (preoccupied by **catalonica** de Sagarra, 1924, ssp. of **hilaris** Ochsenheimer). Reiss & Tremewan, 1964, Ent. Rec., **76**: 132.

ssp. **descimonti** Lucas, 1959, Bull. mens. Soc. linn. Lyon, **28**: 303. — Aussonne près Gavarnie, Haute-Garonne (Pyrénées), France.

ssp. **duponti** Rocci, 1921, Atti Soc. ligust. Sci. nat. geogr., **32**: 41. Dupont, 1900, Bull. Soc. Sci. nat. Elbeuf, **18**: 74; 1925, ibidem, **43**: 136; 1927, ibidem, **45**: 74. Reiss, 1930, in Seitz, Die Gross-Schmetterlinge der Erde, Supplement, **2**: 28. Holik, 1936, Lambillionea, **36**: 43, pl. 9, figs. 6-10. — Normandie, France [Pont-de-l'Arche, Eure].

ab. **minor** Rocci, 1919, Atti Soc. ligust. Sci. nat. geogr., **30**: 65.

ssp. **gaumaisiensis** Holik, 1936, Lambillionea, **36**: 182. Derenne, 1935, Amat. Papillons, **7**: 169. Reiss & Tremewan, 1964, Ent. Rec., **76**: 132. — Torgny, Belgique.

torgniensis Lambillion, 1909, Rev. Soc. ent. namur., **9**: 75 (infrasubspecific). Reiss & Tremewan, 1964, Ent. Rec., **76**: 132.

azona Holik, 1936, Lambillionea, **36**: 180.

ab. **cingulata** Holik, 1936, Lambillionea, **36**: 181.

ab. **dissociata** Lambillion, 1909, Rev. Soc. ent. namur., **9**: 75.

ab. **albilunaris** Lambillion, 1909, Rev. Soc. ent. namur., **9**: 76.

ab. **adunata** Lambillion, 1909, Rev. Soc. ent. namur., **9**: 75.

ab. **rubricostata** Lambillion, 1909, Rev. Soc. ent. namur., **9**: 76.

ab. **faustoides** Lambillion, 1909, Rev. Soc. ent. namur., **9**: 75.

ab. **flavicostata** Lambillion, 1909, Rev. Soc. ent. namur., **9**: 76. Holik, 1932, Int. ent. Z., **26**: 80.

ssp. **parisiensis** Holik, 1938, Lambillionea, **38**: 159, pl. 8, figs. 1-5. Tremewan, 1961, Bull. Brit. Mus. (nat. Hist.) Ent., **10** (7): 307, pl. 57, fig. 22. — Lardy, Seine-et-Oise, France.

ab. **flavicornis** Reiss, 1964, Coridon, (A) **6**: 5.

ab. **inversa** Caruel, 1939, Lambillionea, **39**: 122.

ab. **rubrocostata** Holik, 1938, Lambillionea, **38**: 160.

ab. **eximia** Holik, 1938, Lambillionea, **38**: 159.

ab. **amoena** Reiss, 1964, Coridon, (A) **6**: 5.

ssp. **viridis** Przegendza, 1932, Ent. Z., **46**: 113, figs. 25-28. Holik & Reiss, 1932, in Holik, Iris, **46**: 121, pl. 2, figs. 7-12. Reiss, 1933, in Seitz, Die Gross-Schmetterlinge der Erde, Supplement, **2**: 273. Holik & Sheljuzhko, 1956, Mitt. münch. ent. Ges., **46**: 199. — Bei Kijev, Ukraine.

ab. **azona** Holik & Reiss, 1932, in Holik, Iris, **46**: 121. Reiss, 1933, in Seitz, Die Gross-Schmetterlinge der Erde, Supplement, **2**: 273.

ab. **pseudoberolinensis** Holik & Reiss, 1932, in Holik, Iris, **46**: 121. Reiss, 1933, in Seitz, Die Gross-Schmetterlinge der Erde, Supplement, **2**: 273.

ab. **securigera** Holik & Reiss, 1932, in Holik, Iris, **46**: 121. Reiss, 1933, in Seitz, Die Gross-Schmetterlinge der Erde, Supplement, **2**: 273.

ab. **crassimaculata** Holik & Reiss, 1932, in Holik, Iris, **46**: 121. Reiss, 1933, in Seitz, Die Gross-Schmetterlinge der Erde, Supplement, **2**: 273.

ab. **confluens** Holik & Reiss, 1932, in Holik, Iris, **46**: 121. Reiss, 1933, in Seitz, Die Gross-Schmetterlinge der Erde, Supplement, **2**: 273.

ab. **amoena** Holik & Reiss, 1932, in Holik, Iris, **46**: 121. Reiss,

1933, in Seitz, Die Gross-Schmetterlinge der Erde, Supplement, **2**: 273.

ssp. **ludmilae** Obraztsov, 1936, Folia Zool. Hydrobiol., Riga, **9**: 36. Holik & Sheljuzhko, 1956, Mitt. münch. ent. Ges., **46**: 201.

ab. **latocingulata** Obraztsov, 1936, Folia Zool. Hydrobiol., Riga, **9**: 36.

Dolinskaja, Park Vessjolaja Bokovenjka, Cherson, Russland.

ssp. **bessarabica** Holik & Sheljuzhko, 1956, Mitt. münch. ent. Ges., **46**: 198.

Tshobrutshi, Distrikt Akkerman, Bessarabien.

ssp. **tyrasica** Holik, 1939, Ann. Mus. zool. Polon., **12**: 61, pl. 2, figs. 65-68.

Mündungsgebiet des Seret in den Dniestr (= Tyras), (Dobrowlany bei Zaleszezyki), Süd-Podolien.

ssp. **subviridis** Holik, 1932, Iris, **46**: 120, pl. 2, figs. 1-6 (as *subiridis*, corrected to **subviridis**, p. 192). Reiss, 1933, in Seitz, Die Gross-Schmetterlinge der Erde, Supplement, **2**: 273. Holik, 1939, Ann. Mus. zool. Polon., **12**: 57, pl. 2, figs. 62-64.

Lackie bei Zloczow (Lysa Gora), Südost-Polen ohne Süd-Podolien.

ssp. **transviridis** Holik, 1939, Ann. Mus. zool. Polon., **12**: 54.

ab. **privata** Dąbrowski, 1965, Acta zool. cracov., **10**: 100, pl. 8, fig. 2.

Pińczów an der Nida, 30 km. nördlich. von Kazimierza, Wielka, Polen.

ssp. **marusica** Holik, 1935, Iris, **49**: 21, pl. 1, figs. 40-46.

ab. **deannulata** Holik, 1935, Iris, **49**: 24.

ab. **pseudoleonhardi** Holik, 1935, Iris, **49**: 24.

ab. **pseudoonobrychis** Holik, 1935, Iris, **49**: 24.

ab. **pseudocarniolica** Holik, 1935, Iris, **49**: 24.

ab. **pseudomodesta** Holik, 1935, Iris, **49**: 24.

ab. **laticingulata** Holik, 1935, Iris, **49**: 24.

ab. **weileri** Holik, 1935, Iris, **49**: 24.

ab. **rosea** Skala, 1913, Verh. naturf. Ver. Brünn, **51**: 211.

Bielkowitzer Tal, Olmütz, Mähren.

ssp. **moravica** Holik, 1939, Sborn. ent. Odd. nár. Mus. Praze, **17**: 45; 1931, Int. ent. Z., **24**: 434; 1935, Iris, **49**, pl. 1, figs. 47-53. Povolný, 1945, Folia ent., Brno, **8**: 80. Gregor & Povolný, 1946, Folia ent., Brno, **9**: 1, figs. 1-9. Povolný & Gregor, 1946, Folia ent., Brno, **9**, Supplement, 12: 63, 97, pl. 3, figs. 1-9. Gregor & Povolný, 1955, Sborn. ent. Odd. nár. Mus. Praze, **30**: 264, 273, pl. 5, fig. 5.

Klentnitz; Hocheck bei Nikolsburg; Pollauer Bergen, Süd-Mähren.

ssp. **ambigua** Holik, 1935, Iris, **49**: 21, pl. 1, figs. 33-39; 1930, Int. ent. Z., **24**: 433. Povolný & Gregor, 1946, Folia ent., Brno, **9**, Supplement, 12: 63, 97. — Kletten bei Zauchtl, Mährisch-Schlesien.

ab. **bicingulata** Dąbrowski, 1965, Acta zool. cracov., **10**: 100.

ab. **deannulata** Holik, 1935, Iris, **49**: 24.

ab. **pseudoonobrychis** Holik, 1935, Iris, **49**: 24.

ab. **pseudocarniolica** Holik, 1935, Iris, **49**: 24.

ab. **pseudomodesta** Holik, 1935, Iris, **49**: 24.

ab. **pseudoleonhardi** Holik, 1935, Iris, **49**: 24.

ab. **laticingulata** Holik, 1935, Iris, **49**: 24.

ab. **costalielongata** Dąbrowski, 1965, Acta zool. cracov., **10**: 101, fig. 125d, pl. 8, fig. 3.

ab. **weileri** Holik, 1935, Iris, **49**: 24.

ssp. **bohemica** Komárek, 1958, Folia ent. Hung., **11**: 116; 1952, Acta Soc. ent. Bohem. (Čsl.), **49**: 146; 1954, ibidem, **51**: 197, figs. Gregor & Povolný, 1955, Sborn. ent. Odd. nár. Mus. Praze, **30**: 264, 273, pl. 5, fig. 6. — Výsoká n. Labem, Czechoslovakia.

ab. **mariae** Reiprich, 1960, Motýle Slovenska, p. 306, 410, 416.

ab. **flavocincta** Komárek, 1952, Acta Soc. ent. Bohem. (Čsl.), **49**: 201, 210, 211.

ab. **klapaleki** Joukl, 1906, Ent. Z., **20**: 20, fig.; 1907, ibidem, **21**: 92, fig. Reiss, 1930, in Seitz, Die Gross-Schmetterlinge der Erde, Supplement, **2**: 29; 1933, ibidem, **2**: 273. Holik, 1932, Int. ent. Z., **26**: 82.

ssp. **lusatica** Holik, 1935, Iris, **49**: 16, pl. 1, figs. 19-25. — Guben, Niederlausitz, Ostdeutschland.

ab. **cingulata** Holik, 1935, Iris, **49**: 18.

ab. **pseudoleonhardi** Holik, 1935, Iris, **49**: 18.

ab. **pseudocarniolica** Holik, 1935, Iris, **49**: 18.

ab. **pseudomodesta** Holik, 1935, Iris, **49**: 18.

ssp. **pinskica** Reiss, 1941, Z. wien. EntVer., **26**: 59. — Ljesnaja (Roknito), Pinsk, Westrussland.

ab. **cingulata** Reiss, 1941, Z. wien. EntVer., **26**: 60.

ssp. **berolinensis** Lederer, 1853, Verh. zool.-bot. Ver. Wien, **2**: 102. Reiss & Tremewan, 1964, Ent. Rec., **76**: 132. — Rüdersdorf bei Berlin, Norddeutschland.

berolinensis Staudinger, 1871, in Staudinger & Wocke, Catalog der Lepidopteren des Europaeischen Faunengebiets, p. 49. Spuler, 1906, in Hofmann, Die Schmetterlinge Europas, **2**: 164, pl. 77, fig. 33a. Seitz, 1907, Die Gross-Schmetterlinge der Erde, **2**: 30, pl. 8g. Wagner, 1919, Ent. Mitt., **8**: 187. Reiss, 1930, in Seitz, Die Gross-Schmetterlinge der Erde, Supplement, **2**: 28, pl. 3a. Holik, 1935, Iris, **49**: 7, pl. 1, figs. 1-6; 1936, Lambillionea, **36**, pl. 8, figs. 11-15. Forster & Wohlfahrt, 1958, Die Schmetterlinge Mitteleuropas, **3**:96, pl. 10, figs. 24, 29. Reiss & Tremewan, 1964, Ent. Rec., **76**: 132.

ab. **cingulata** Burgeff, 1914, Mitt. münch. ent. Ges., **5**: 59. Reiss, 1930, in Seitz, Die Gross-Schmetterlinge der Erde, Supplement, **2**: 28.
ab. **pseudocarniolica** Burgeff, 1914, Mitt. münch. ent. Ges., **5**: 59. Reiss, 1930, in Seitz, Die Gross-Schmetterlinge der Erde, Supplement, **2**: 28.
ab. **pseudocarniolicarosea** Guhn, 1932, Ent. Jb., **41**: 96.
ab. **pseudomodesta** Burgeff, 1914, Mitt. münch. ent. Ges., **5**: 59. Reiss, 1930, in Seitz, Die Gross-Schmetterlinge der Erde, Supplement, **2**: 28.
ab. **pseudomodestarosea** Guhn, 1932, Ent. Jb., **41**: 96.
ab. **pseudoleonhardi** Guhn, 1932, Ent. Jb., **41**: 96. Reiss, 1933, in Seitz, Die Gross-Schmetterlinge der Erde, Supplement, **2**: 273.
ab. **pseudoapennina** Guhn, 1932, Ent. Jb., **41**: 96. Reiss, 1933, in Seitz, Die Gross-Schmetterlinge der Erde, Supplement, **2**: 273.
ab. **rubricosta** Guhn, 1932, Ent. Jb., **41**: 96. Reiss, 1933, in Seitz, Die Gross-Schmetterlinge der Erde, Supplement, **2**: 273.
ab. **paupera** Guhn, 1932, Ent. Jb., **41**: 96. Reiss, 1933, in Seitz, Die Gross-Schmetterlinge der Erde, Supplement, **2**: 273.
ab. **medioconfluens** Holik, 1935, Iris, **49**: 10.

ssp. **verrina** Burgeff, 1926, Mitt. münch. ent. Ges., **16**: 54. Reiss, 1930, in Seitz, Die Gross-Schmetterlinge der Erde, Supplement, **2**: 28. Holik, 1935, Iris, **49**: 11, pl. 1, figs. 13-18. — Eberswalde am Finowkanal, Kreis Oberbarnim, Norddeutschland.
ab. **pseudoonobrychis** Hannemann, 1931, Int. ent. Z., **24**: 400.
ab. **pseudomodesta** Hannemann, 1931, Int. ent. Z., **24**: 400.
ab. **parallela** Holik, 1935, Iris, **49**: 12.
ab. **medioconfluens** Holik, 1935, Iris, **49**: 12.

ssp. **violascens** Holik, 1935, Iris, **49**: 18, pl. 1, figs. 26-32. — Umgebung von Breslau, Ostdeutschland.
ab. **cingulata** Holik, 1935, Iris, **49**: 19.
ab. **pseudomodesta** Holik, 1935, Iris, **49**: 19.
ab. **pseudocarniolica** Holik, 1935, Iris, **49**: 19.
ab. **securigera** Holik, 1935, Iris, **49**: 18.
ab. **nigrosupposita** Holik, 1935, Iris, **49**: 19.

ssp. **pommerana** Holik, 1935, Iris, **49**: 13, pl. 1, figs. 7-12. — Finkenwalde, Hökendorf usw. bei Stettin, Pommern, Norddeutschland.

Biology

Abeille, 1909, Mém. Soc. linn. Provence, **1**: 16-19. Burgeff, 1912, Z. wiss. InsektBiol., **8**: 186; 1921, Mitt. münch. ent. Ges., **11**: 54; 1950, Portug. acta biol., (A) Goldschmidt:

663-728; 1951, Biol. Zbl., **70**: 1-23. Dąbrowski, 1961, Przegląd Zoologiczny, **5**: 42-47, figs. 1-6; 1963, Folia biol., Kraków, **11**: 339-346, figs. 3a-3l, pl. 1, figs. A1-A10, B1-B8, C1-C10, D1-D8. Dorfmeister, 1854, Verh. zool.-bot. Ver. Wien, **4**: 479. Döring, 1955, Zur Morphologie der Schmetterlingseier, p. 120, pl. 18, fig. 256. Esper, 1797, Die Schmetterlinge, Supplement, **2** (2): 30, pl. 44, figs. 5-8. Freyer, 1858, Neuere Beiträge zur Schmetterlingskunde, **7**: 66, pl. 637, fig. 2. Hepp, 1927, Ent. Z., **41**: 288. Holik, 1937, Lambillionea, **37**: 15-24, 32-45, 80-91; 1938, ibidem, **38**: 51-58, 79-88, 95-102; 1938, Ent. Rdsch., **55**: 320-323, 331-333; 1946, Rev. franç. Lépid., **10**: 250-261, 273-280; 1953, Ent. Z., **63**: 23. Hrubý, 1964, Prodromus Lepidopter Slovenska, p. 478. Kiefer, 1934, Int. ent. Z., **27**: 521-524. Koch, 1955, Wir Bestimmen Schmetterlinge, **2**: 60, 61, pl. 1, fig. 12, pl. 15, fig. 12. Ochsenheimer, 1808, Die Schmetterlinge von Europa, **2**: 87. Reiss, 1926, Die Zygaenen Deutschlands, p. 7; 1930, in Seitz, Die Gross-Schmetterlinge der Erde, Supplement, **2**: 31; 1958, Z. wien. ent. Ges., **43**: 158. Roüast, 1883, Catalogue des Chenilles européennes connues, p. 23. Sarlet, 1964, Mém. Soc. r. ent. Belg., **29**: 10. Spuler, 1910, in Hofmann, Die Raupen der Schmetterlinge Europas, pl. 10, figs. 8a, b. Schwingenschuss, 1939, Ent. Z., **53**: 96. Turati, 1913, Atti Soc. ital. Sci. nat., **51**: 339. Tutt, 1898, Ent. Rec., **10**: 45. Weiser, 1951, Folia ent., Brno, **14**: 130.

SECTION 5

haematina Kollar

Distribution: South-west Iran.

haematina Kollar, 1850, Denkschr. Akad. Wiss. Wien, **1**: 53. Seitz, 1907, Die Gross-Schmetterlinge der Erde, **2**: 27. Reiss, 1930, in Seitz, Die Gross-Schmetterlinge der Erde, Supplement, **2**: 19, pl. 2e; 1933, ibidem, **2**: 258; 1933, Int. ent. Z., **26**: 478; 1938, Ent. Rdsch., **55**: 252, figs. a3, a4. Holik & Sheljuzhko, 1955, Mitt. münch. ent. Ges., **44/45**: 65. Alberti, 1958, Mitt. zool. Mus. Berl., **34**: 340. — Bei Shiraz, Südwestiran.

ab. **reducta** Reiss, 1938, Ent. Rdsch., **55**: 254.

ab. **rubroabdominalis** Reiss, 1938, Ent. Rdsch., **55**: 254, fig. b1.

ab. **rubroconfluens** Reiss, 1938, Ent. Rdsch., **55**: 254.

ab. **rubroperfecta** Reiss, 1938, Ent. Rdsch., **55**: 253.

ab. **flava** Reiss, 1938, Ent. Rdsch., **55**: 253.

ab. **flavoabdominalis** Reiss, 1938, Ent. Rdsch., **55**: 254.

ab. **flavoconfluens** Reiss, 1938, Ent. Rdsch., **55**: 254, fig. b2.

ab. **flavoperfecta** Reiss, 1938, Ent. Rdsch., **55**: 253.

Biology

Burgeff, 1950, Portug. acta biol., (A) Goldschmidt: 663-728. Holik, 1938, Ent. Rdsch., **55**: 349-354, 382-384; 1946, Rev. franç., Lépid., **10**: 250-261, 273-280. Reiss, 1958, Z. wien. ent. Ges., **43**: 160.

SECTION 6

exulans Reiner & Hohenwarth

Distribution: Siberia, Balkans, Abruzzi, Pyrenees, Alps, Scotland, Scandinavia, Lapland.

ssp. **exsiliens** Staudinger, 1881, Stettin. ent. Ztg., **42**: 398. Seitz, 1907, Die Gross-Schmetterlinge der Erde, **2**: 24, pl. 6c. Reiss, 1930, in Seitz, Die Gross-Schmetterlinge der Erde, Supplement, **2**: 12; 1933, ibidem, **2**: 253; 1935, Int. ent. Z., **29**: 141, fig. Holik & Sheljuzhko, 1955, Mitt. münch. ent. Ges., **44/45**: 119. — Ala Tau, Tarbagatai-Gebirge, Sibirien.

ssp. **sajana** Burgeff, 1926, Mitt. münch. ent. Ges., **16**: 25. Reiss, 1930, in Seitz, Die Gross-Schmetterlinge der Erde, Supplement, **2**: 12, pl. 1k; 1933, ibidem, **2**: 253. Holik & Sheljuzhko, 1955, Mitt. münch. ent. Ges., **44/45**: 121. — Munku Sardyk, Sajan-Gebirge, Sibirien, 2500 m.; Schawyr, Tannuola or., Mongolei.

ssp. **montenegrina** Burgeff, 1926, Mitt. münch. ent. Ges., **16**: 24. Schawerda, 1915, Verh. zool.-bot. Ges. Wien, **65**: (89) (as *apfelbecki* Rebel). Reiss, 1930, in Seitz, Die Gross-Schmetterlinge der Erde, Supplement, **2**: 12. Holik, 1937, Mitt. münch. ent. Ges., **27**: 9. — Volujak, Montenegrinisch-herzegowinische Grenze, 2000 m.

ssp. **apfelbecki** Rebel, 1910, Verh. zool.-bot. Ges. Wien, **60**: (4), fig. 2. Reiss, 1930, in Seitz, Die Gross-Schmetterlinge der Erde, Supplement, **2**: 12. Holik, 1937, Mitt. münch. ent. Ges., **27**: 9. Daniel, 1957, Acta Mus. maced. Sci. nat., **4**: 211; 1964, Prirod. Muz. Scopje, no. 2: 14. — Schar Dagh (Ljubeten), Albanien.

albania Bang-Haas, 1926, Novitates Macrolepidopterologicae, **1**: 105 (in error).

ssp. **apennina** Rebel, 1910, Verh. zool.-bot. Ges. Wien., **60**: (5). Reiss & Tremewan, 1964, Ent. Rec., **76**: 133. — Gran Sasso, Abruzzen, Italien.

abruzzina Burgeff, 1926, Mitt. münch. ent. Ges., **16**: 25. Reiss, 1930, in Seitz, Die Gross-Schmetterlinge der Erde, Supplement, **2**: 12, pl. 1k. Reiss & Tremewan, 1964, Ent. Rec., **76**: 133.

ab. **striata** Burgeff, 1914, Mitt. münch. ent. Ges., **5**: 45. Reiss, 1930, in Seitz, Die Gross-Schmetterlinge der Erde, Supplement, **2**: 12.

ssp. **pyrenaica** Burgeff, 1926, Mitt. münch. ent. Ges., **16**: 24. — Mt. Louis

Reiss, 1930, in Seitz, Die Gross-Schmetterlinge der Erde, Supplement, **2**: 12. — und Mt. Canigou, 2500 m., Ostpyrenäen, Frankreich.

ssp. **exulans** Reiner & Hohenwarth, 1792, Botanische Reisen, p. 265, pl. 6, fig. 2. Hübner, 1796, Sammlung europäischer Schmetterlinge, **2**: 13, pl. 2, fig. 12. Esper, 1797, Die Schmetterlinge, Supplement, **2** (2): 17, pl. 41, figs. 1, 2. Hübner, [July 1803]-[15th November 1806], Sammlung europäischer Schmetterlinge, **2**, pl. 20, fig. 101; 1806, ibidem, Der Ziefer, p. 81. Boisduval, 1828, Essai sur une Monographie des Zygénides, p. 47, pl. 3, fig. 3; 1834, Icones historique des Lépidoptères nouveaux ou peu connus, **2**: 54, pl. 54, figs. 4, 5. Duponchel, 1835, in Godart & Duponchel, Histoire naturelle des Lépidoptères ou Papillons de France, Supplement, **2**: 57, pl. 5, figs. 5a, b. Freyer, 1839, Neuere Beiträge zur Schmetterlingskunde, **3**: 13, pl. 200, fig. 2. Seitz, 1907, Die Gross-Schmetterlinge der Erde, **2**: 24, pl. 6c; 1912, ibidem, **2**: 443. Oberthür, 1910, Études de Lépidoptérologie comparée, **4**: 480. Burgeff, 1926, Mitt. münch. ent. Ges., **16**: 23. Reiss, 1930, in Seitz, Die Gross-Schmetterlinge der Erde, Supplement, **2**: 11, 12; 1933, ibidem, **2**: 253. Daniel, 1932, in Osthelder, Die Schmetterlinge Südbayerns, **1**: 567. Verity, 1946, Redia, **31**: 58. Reiss, 1950, Jber. naturf. Ges. Graubünden, **82**: 104, fig. 7. Haaf, 1952, Veröff. zool. Staatssamml. Münch., **2**: 152, 157, pl. 14. Alberti, 1958, Mitt. zool. Mus. Berl., **34**: 316. Forster & Wohlfahrt, 1958, Die Schmetterlinge Mitteleuropas, **3**: 92, pl. 9, figs. 41, 42. — Oberhalb der Pasterze, Gross-Glocknergebiet, Österreich, 2400 m.

ab. **clara** Tutt, 1894, Ent. Rec., **5**: 266. Seitz, 1912, Die Gross-Schmetterlinge der Erde, **2**: 443. Tremewan, 1961, Bull. Brit. Mus. (nat. Hist.) Ent., **10** (7): 265.

ab. **pallida** Tutt, 1897, Ent. Rec., **9**: 14. Seitz, 1912, Die Gross-Schmetterlinge der Erde, **2**: 443. Reiss, 1950, Jber. naturf. Ges. Graubünden, **82**: 105, fig. 8. Tremewan, 1961, Bull. Brit. Mus. (nat. Hist.) Ent., **10** (7): 265, pl. 52, fig. 13.

ab. **exilioides** Burgeff, 1926, Mitt. münch. ent. Ges., **16**: 24. Reiss, 1930, in Seitz, Die Gross-Schmetterlinge der Erde, Supplement, **2**: 12.

ab. **flavicornis** Reiss, 1950, Jber. naturf. Ges. Graubünden, **82**: 105.

ab. **crassimaculata** Vorbrodt, 1913, in Vorbrodt & Müller-Rutz, Die Schmetterlinge der Schweiz, **2**: 257, fig. 7.

ab. **analielongata** Vorbrodt, 1931, Mitt. schweiz. ent. Ges., **14**: 380. Reiss, 1933, in Seitz, Die Gross-Schmetterlinge der Erde, Supplement, **2**: 253.

ab. **costalielongata** Vorbrodt, 1913, in Vorbrodt & Müller-Rutz, Die Schmetterlinge der Schweiz, **2**: 257, fig. 8. Reiss, 1933, in Seitz, Die Gross-Schmetterlinge der Erde, Supplement, **2**: 253.
ab. **apicalielongata** Vorbrodt, 1931, Mitt. schweiz. ent. Ges., **14**: 380. Reiss, 1933, in Seitz, Die Gross-Schmetterlinge der Erde, Supplement, **2**: 253.
ab. **medioconfluens** Vorbrodt, 1913, in Vorbrodt & Müller-Rutz, Die Schmetterlinge der Schweiz, **2**: 257.
ab. **costaliconfluens** Vorbrodt, 1913, in Vorbrodt & Müller-Rutz, Die Schmetterlinge der Schweiz, **2**: 258.
ab. **analiconfluens** Vorbrodt, 1913, in Vorbrodt & Müller-Rutz, Die Schmetterlinge der Schweiz, **2**: 258.
ab. **semistriata** Tutt, 1908, Ent. Rec., **20**: 274.
ab. **confluens** Dziurzyński, 1902, Iris, **15**: 336.
confluens Oberthür, 1910, Études de Lépidoptérologie comparée, **4**: 482. Tremewan, 1961, Bull. Brit. Mus. (nat. Hist.) Ent., **10** (7): 266, pl. 52, fig. 16.
pseudoscabiosae Hoffmann, 1911, Int. ent. Z., **5**: 186.
parallela Vorbrodt, 1913, in Vorbrodt & Müller-Rutz, Die Schmetterlinge der Schweiz, **2**: 258.
ab. **omniconfluens** Vorbrodt, 1913, in Vorbrodt & Müller-Rutz, Die Schmetterlinge der Schweiz, **2**: 258, fig. 27.
ab. **fulva** Spuler, 1906, in Hofmann, Die Schmetterlinge Europas, **2**: 157. Seitz, 1907, Die Gross-Schmetterlinge der Erde, **2**: 24.

ssp. **seekaarensis** Koch, 1940, Z. wien. EntVer., **25**: 123. — Tauernhöhe, Seekaarhaus, Österreich, 1850 m.

ssp. **bourgognei** Le Charles, 1942, Bull. Soc. ent. Fr., **47**: 180. — Madone de Fénestre, Haute-Vésubie, France.

ssp. **nivicola** Le Charles, 1942, Bull. Soc. ent. Fr., **47**: 180, pl. 1, col. I, fig. (as *nicicola*, corrected to **nivicola** in text of pl. 1, p. 180). — Refuge de Carro, Savoie, France, 2700 m.

ssp. **altaretensis** Le Charles, 1942, Bull. Soc. ent. Fr., **47**: 178, pl. 1, cols. A, B, C, figs. Holik, 1936, Lambillionea, **36**: 46. — Le Lautaret, Hautes-Alpes, France, 1800-2200 m.
ab. **minor** Tutt, 1899, A Natural History of the British Lepidoptera, **1**: 449. Seitz, 1912, Die Gross-Schmetterlinge der Erde, **2**: 443. Tremewan, 1961, Bull. Brit. Mus. (nat. Hist.) Ent., **10** (7): 267, pl. 52, fig. 18.
ab. **flavilinea** Tutt, 1894, Ent. Rec., **5**: 267. Reiss, 1930, in Seitz, Die Gross-Schmetterlinge der Erde, Supplement, **2**: 12.

Tremewan, 1961, Bull. Brit. Mus. (nat. Hist.) Ent., **10** (7): 267, pl. 52, fig. 19.

ab. **striata** Tutt, 1896, Proc. ent. Soc. Lond., p. xli. Seitz, 1912, Die Gross-Schmetterlinge der Erde, **2**: 443. Reiss, 1930, in Seitz, Die Gross-Schmetterlinge der Erde, Supplement, **2**: 12. Tremewan, 1961, Bull. Brit. Mus. (nat. Hist.) Ent., **10** (7): 266, pl. 52, fig. 15.

ab. **pulchra** Tutt, 1899, A Natural History of the British Lepidoptera, **1**: 448. Seitz, 1912, Die Gross-Schmetterlinge der Erde, **2**: 443. Tremewan, 1961, Bull. Brit. Mus. (nat. Hist.) Ent., **10** (7): 266, pl. 52, fig. 17.

ab. **flava** Oberthür, 1896, Études d'Entomologie, **20**: 43, pl. 8, fig. 141. Tremewan, 1961, Bull. Brit. Mus. (nat. Hist). Ent., **10** (7): 267, pl. 52, fig. 20.

ssp. **alpigena** Le Charles, 1942, Bull. Soc. ent. Fr., **47**: 179, pl. 1, cols. D, E, F, G, figs.

Bonneval-sur-Arc, Savoie, France.

ab. **latemarginata** Le Charles, 1942, Bull. Soc. ent. Fr., **47**: 179, pl. 1, col. H, fig.

ab. **basaliconfluens** Le Charles, 1942, Bull. Soc. ent. Fr., **47**: 179, pl. 1, col. D, fig.

ssp. **stenostigma** Dujardin, 1956, Bull. mens. Soc. linn. Lyon, **25**: 259.

Montagne de l'Authion, Alpes-Maritimes, France, 2000 m.

ssp. **subochracea** White, 1871/72, Scot. Nat., **1**: 175. South, 1908, The Moths of the British Isles, **2**: 335, pl. 146, fig. 3. James, 1912, Ent. Rec., **24**: 253. Burgeff, 1926, Mitt. münch. ent. Ges., **16**: 24. Reiss, 1930, in Seitz, Die Gross-Schmetterlinge der Erde, Supplement, **2**: 12, pl. 1k. Ford, 1955, Moths, pl. 27, fig. 8. Showler, 1955, Ent. Rec., **67**: 316. Tremewan, 1958, Ent. Gaz., **9**: 189; 1960, ibidem, **11**: 187; 1961, Bull. Brit. Mus. (nat. Hist.) Ent., **10** (7): 266, pl. 52, fig. 14; 1961, Coridon, (A) **1**: 2, pl. C1, fig. 6. South, 1961, The Moths of the British Isles, **2**: 329, pl. 129, fig. 3.

Near Braemar, Aberdeen, Scotland, 2000-2830 ft.

rubbedaria Tutt, 1894, Trans. ent. Soc. Lond., p. xxvii. Tremewan, 1961, Bull. Brit. Mus. (nat. Hist.) Ent., **10** (7): 266.

ssp. **vanadis** Dalman, 1816, K. svenska VetenskAkad. Handl., p. 223. Seitz, 1907, Die Gross-Schmetterlinge der Erde, **2**: 24, pl. 6c. Nordström & Wahlgren, 1941, Svenska Fjäriler, **2**: 326, pl. 46, fig. 3. Feichtenberger, 1965, Z. wien. ent. Ges., **50**: 86, 89, 101.

Lappland.

ab. **dilatata** Burgeff, 1906, Ent. Z., **20**: 154. Seitz, 1907, Die Gross-Schmetterlinge der Erde, **2**: 24.

dilata Burgeff, 1926, in Strand, Lepid. Cat., **33**: 15. Reiss, 1930, in Seitz, Die Gross-Schmetterlinge der Erde, Suppl., **2**: 12.

ab. **confluens** Strand, 1901, Nyt. Mag. Naturv., **39**: 52. Seitz, 1907, Die Gross-Schmetterlinge der Erde, **2**: 24; 1912, ibidem, **2**: 443. Reiss, 1930, in Seitz, Die Gross-Schmetterlinge der Erde, Supplement, **2**: 12.

ssp. **polaris** Holik, 1935, Ent. Tidskr., **56**: 47. Holik & Sheljuzhko, 1955, Mitt. münch. ent. Ges., **44/45**: 119. Rybatschij Halbinsel, Murman Küste, 70° n. Breite, Tundrenzone, 150 m.

ab. **analiconfluens** Holik, 1935, Ent. Tidskr., **56**: 48.

Biology

Barrett, 1895, The Lepidoptera of the British Islands, **2**: 121, pl. 58, fig. 5b. Buckler, 1886, The Larvae of the British Butterflies and Moths, **2**: 13, pl. 19, figs. 1, 1a-1c. Burgeff, 1912, Z. wiss. InsektBiol., **8**: 124; 1921, Mitt. münch. ent. Ges., **11**: 59; 1926, ibidem, **16**: 23; 1950, Portug. acta biol., (A) Goldschmidt: 663-728; 1951, Biol. Zbl., **70**: 1-23. Dürck, 1924, Ent. Rdsch., **41**: 6, 10. Freyer, 1852, Neuere Beiträge zur Schmetterlingskunde, **6**: 178, pl. 590, fig. 1. Holik, 1937, Lambillionea, **37**: 15-24, 32-45, 80-91; 1938, ibidem, **38**: 51-58, 79-88, 95-102; 1938, Ent. Rdsch., **55**: 349-354, 382-384; 1946, Rev. franç. Lépid., **10**: 250-261, 273-280; 1953, Beitr. Ent. Berlin, **3**: 430; 1953, Ent. Z., **63**: 4. Oberthür, 1910, Études de Lépidoptérologie comparée, **4**: 480. Pictet, 1926, Schweiz. ent. Anz., **5** (3): 2; 1926, ibidem, **5** (5): 3. Reiner & Hohenwarth, 1792, Botanische Reisen, p. 265. Reiss, 1926, Die Zygaenen Deutschlands, p. 7; 1930, in Seitz, Die Gross-Schmetterlinge der Erde, Supplement, **2**: 12; 1958, Z. wien. ent. Ges., **43**: 158, 160. Roüast, 1883, Catalogue des Chenilles européennes connues, p. 22. Seppänen, 1954, Suomen Suurperhostoukkien Ravintokasvit, p. 256. Spuler, 1910, in Hofmann, Die Raupen der Schmetterlinge Europas, pl. 9, figs. 22a, b. Strand, 1919, Arch. Naturgesch., **85A** (4): 23. Tutt, 1894, Ent. Rec., **5**: 258; 1899, A Natural History of the British Lepidoptera, **1**: 449. Vorbrodt, 1913, in Vorbrodt & Müller-Rutz, Die Schmetterlinge der Schweiz, **2**: 258; 1922, Int. ent. Z., **16**: 31.

SECTION 7

ignifera Korb

Distribution: Spain.

ssp. **diezma** Tremewan, 1963, Ent. Rec., **75**: 4, pl. 1, figs. 7, 8. Near Diezma, Granada, Spain, 4000 ft.

ssp. **ignifera** Korb, 1897, Iris, **9**: 349. Dziurzyński, 1904, Jber. wien. ent. Ver., **14**: 53, pl. 2, fig. 16. Spuler, 1906, in Hofmann, Beim Gebirgsdorf

Die Schmetterlinge Europas, **2**: 163, pl. 77, fig. 29. Seitz, 1907, Die Gross-Schmetterlinge der Erde, **2**: 27, pl. 7a; 1912, ibidem, **2**: 443. Reiss, 1930, in Seitz, Die Gross-Schmetterlinge der Erde, Supplement, **2**: 19, pl. 2e; 1933, ibidem, **2**: 258; 1936, Ent. Rdsch., **54**: 60, pl. 2, figs. Querci, 1947, Ent. Rec., **59**: 95. Alberti, 1958, Mitt. zool. Mus. Berl., **34**: 318; 1964, Dtsch. ent. Z. (N.F.), **11**: 386, figs. A7, B7, C7; 388, fig. (distribution map); 391, fig. 2. Bonnin, 1966, Alexanor, **4**: 276, 278. — Huelamo bei Cuenca, Neu-Kastilien, Spanien.

ssp. **dertosensis** de Sagarra, 1940, VI Congr. int. Ent., Madrid, p. 391. — Monte Caro, Puertos de Tortosa, Catalonia, España.

Biology

Burgeff, 1950, Portug. acta biol., (A) Goldschmidt: 663-728. Holik, 1937, Lambillionea, **37**: 15-24, 32-45, 80-91; 1938, Ent. Rdsch., **55**: 349-354, 382-384; 1953, Ent. Z., **63**: 6. Korb, 1897, Iris, **9**: 349. Reiss, 1958, Z. wien. ent. Ges., **43**: 160.

ecki Christoph

Distribution: Northern Iran.

ssp. **ecki** Christoph, 1882, Horae Soc. ent. Ross., **17**: 123; 1885, in Romanoff, Mémoires sur les Lépidoptères, **2**: 202, pl. 13, fig. 1. Seitz, 1907, Die Gross-Schmetterlinge der Erde, **2**: 24, pl. 6c. Reiss, 1933, Int. ent. Z., **26**: 476, figs.; 1933, in Seitz, Die Gross-Schmetterlinge der Erde, Supplement, **2**: 258, pl. 16i, k; 1938, Ent. Rdsch., **55**: 313, fig. b3. Holik & Sheljuzhko, 1955, Mitt. münch. ent. Ges., **44/45**: 156. Alberti, 1958, Mitt. zool. Mus. Berl., **34**: 318; 1964, Dtsch. ent. Z. (N.F.), **11**: 386, figs. A6, B6, C6; 388, fig. (distribution map); 391, fig. 3. — Bei Schakuh, Nordpersien (Nordiran), 9000-10000 ft.

ab. **cingulata** Hirschke, 1906, Jber. wien. ent. Ver., **16**: 95.

ssp. **schwingenschussi** Reiss, 1937, Ent. Rdsch., **54**: 454, fig. c1; 1937, ibidem, **55**: 19, fig. a2. Schwingenschuss, 1939, Ent. Z., **53**: 95. Holik & Sheljuzhko, 1955, Mitt. münch. ent. Ges., **44/45**: 157. — Demavend, Elburs Gebirge, Nordiran, 3200-3700 m.

Biology

Burgeff, 1950, Portug. acta biol., (A) Goldschmidt: 663-728. Holik, 1937, Lambillionea, **37**: 15-24, 32-45, 80-91; 1938, Ent. Rdsch., **55**: 349-354, 382-384; 1953, Ent. Z., **63**: 6.

armena Eversmann

Distribution: Transcaucasia.

ssp. **armena** Eversmann, 1851, Bull. Soc. Nat. Moscou, **24**: 625. Seitz, 1907, Die Gross-Schmetterlinge der Erde, **2**: 27, — Au Sud du Caucase [Abas-Tuman. Georgien].

pl. 7c; 1912, ibidem, **2**: 443. Reiss, 1930, in Seitz, Die Gross-Schmetterlinge der Erde, Supplement, **2**: 19; 1933, ibidem, **2**: 259; 1937, Ent. Rdsch., **54**: 455, fig. d1. Koch, 1939, Mitt. münch. ent. Ges., **29**: 409. Haaf, 1952, Veröff. zool. Staatssamml. Münch., **2**: 152, 155, 157, pl. 13. Holik & Sheljuzhko, 1955, Mitt. münch. ent. Ges., **44/45**: 151. Alberti, 1958, Mitt. zool. Mus. Berl., **34**: 317; 1964, Dtsch. ent. Z., (N.F.), **11**:381, 384, fig. (distribution map); 386, figs. A5, B5, C5; 388, fig. (distribution map); 391, figs. 20-22, 30.

kadenii Lederer, 1864, Wien. ent. Monatschr., **8**: 168, pl. 3, fig. 8. Alberti, 1964, Dtsch. ent. Z. (N.F.), **11**: 391, fig. 19.

ab. **deannulata** Koch, 1939, Mitt. münch. ent. Ges., **29**: 410.

ab. **grisea** Holik & Sheljuzhko, 1955, Mitt. münch. ent. Ges., **44/45**: 154.

ab. **dealbata** Holik & Sheljuzhko, 1955, Mitt. münch. ent. Ges., **44/45**: 154.

ab. **parvimaculata** Koch, 1939, Mitt. münch. ent. Ges., **29**: 410.

ab. **sexmaculata** Koch, 1939, Mitt. münch. ent. Ges., **29**: 410.

ab. **quinquemaculata** Koch, 1939, Mitt. münch. ent. Ges., **29**: 410.

ab. **costalielongata** Holik & Sheljuzhko, 1955, Mitt. münch. ent. Ges., **44/45**: 154.

ab. **analielongata** Holik & Sheljuzhko, 1955, Mitt. münch. ent. Ges., **44/45**: 154.

ab. **costaliconfluens** Holik & Sheljuzhko, 1955, Mitt. münch. ent. Ges., **44/45**: 154.

ab. **medioconfluens** Holik & Sheljuzhko, 1955, Mitt. münch. ent. Ges., **44/45**: 154.

ab. **confluens** Burgeff, 1926, Mitt. münch. ent. Ges., **16**: 38. Reiss, 1930, in Seitz, Die Gross-Schmetterlinge der Erde, Supplement, **2**: 19.

ab. **omniconfluens** Holik & Sheljuzhko, 1955, Mitt. münch. ent. Ges., **44/45**: 154.

ab. **flava** Romanoff, 1884, Mémoires sur les Lépidoptères, **1**: 79, pl. 4, fig. 5.

ssp. **caucasica** Rebel, 1901, in Staudinger & Rebel, Catalog der Lepidopteren des Palaearctischen Faunengebietes, p. 382. Reiss, 1930, in Seitz, Die Gross-Schmetterlinge der Erde, Supplement, **2**: 19; 1933, ibidem, **2**: 259. Holik & Sheljuzhko, 1955, Mitt. münch. ent. Ges., **44/45**: 154. Alberti, 1964, Dtsch. ent. Z. (N.F.), **11**: 383, 391, figs. 16-18, 29.

Alpine Region des südlichen Kaukasus.

alpina Dziurzyński, 1903, Iris, **15**: 337. Seitz, 1912, Die Gross-Schmetterlinge der Erde, **2**: 443. Burgeff, 1914, Mitt. münch. ent. Ges., **5**: 50, pl. 2, figs. 156, 164, pl. 6, figs. 39-43.

apennina Dziurzyński, 1904, Jber. wien. ent. Ver., **14**: 47.

ssp. **dombaiensis** Alberti, 1964, Dtsch. ent. Z. (N.F.), **11**: 383, 391, figs. 7-15, 23-28. — An den südwärts geneigten Hängen des Tschutschurtales, am Fusse des Dombai-Ulgen-Massivs, 2200-2900 m.; Nordhängen des Kluchorpasses, 2100-2400m.; Gebiet Teberda-Dombai, nordwestlichen Kaukasus.

Biology

Burgeff, 1950, Portug. acta biol., (A) Goldschmidt: 663-728. Holik, 1937, Lambillionea, **37**: 15-24, 32-45, 80-91; 1938, Ent. Rdsch., **55**: 349-354, 382-384; 1953, Ent. Z., **63**: 6. Reiss, 1958, Z. wien. ent. Ges., **43**: 158.

loti Denis & Schiffermüller

Distribution: Northern Iran, Asia Minor, Europe to central Spain, Scotland, ?Sweden.

ssp. **suleimanica** Reiss, 1937, Ent. Rdsch., **54**: 453, figs. a1, b1; 1937, ibidem, **55**: 19; 1938, Mitt. münch. ent. Ges., **27**: 164. Holik & Sheljuzhko, 1955, Mitt. münch. ent. Ges., **44/45**: 149. — Tacht i Suleiman, Hasankif, Elburs-Gebirge, Nordiran, 1000 m.

ssp. **antiochena** Staudinger, 1887, Berl. ent. Z., **31**: 34, 35. Seitz, 1907, Die Gross-Schmetterlinge der Erde, **2**: 27, pl. 7d. Reiss, 1930, in Seitz, Die Gross-Schmetterlinge der Erde, Supplement, **2**: 19; 1933, ibidem, **2**: 258. Holik & Sheljuzhko, 1955, Mitt. münch. ent. Ges., **44/45**: 147. — Antiochia, Kleinasien.

ssp. **anatolica** Burgeff, 1926, Mitt. münch. ent. Ges., **16**: 37. Seitz, 1907, Die Gross-Schmetterlinge der Erde, **2**: 27, pl. 7d (as *bitorquata* Ménétriés). Reiss, 1930, in Seitz, Die Gross-Schmetterlinge der Erde, Supplement, **2**: 19. Holik & Sheljuzhko, 1955, Mitt. münch. ent. Ges., **44/45**: 146. — Es-Schehir, Südliches Anatolien, Kleinasien.

ab. **rubrescens** Reiss, 1935, Int. ent. Z., **29**: 149, fig.

ab. **totirubra** Reiss, 1935, Int. ent. Z., **29**: 150, fig.

ssp. **phoenicea** Staudinger, 1887, Berl. ent. Z., **31**: 34, 35. Seitz, 1907, Die Gross-Schmetterlinge der Erde, **2**: 27, pl. 7d. Reiss, 1930, in Seitz, Die Gross-Schmetterlinge der Erde, Supplement, **2**: 19; 1933, ibidem, **2**: 258. Holik & Sheljuzhko, 1955, Mitt. münch. ent. Ges., **44/45**: 144. — Malatia, Taurus, Kleinasien.

ssp. **senilis** Burgeff, 1914, Mitt. münch. ent. Ges., **5**: 48, pl. 2, fig. 161, pl. 6, fig. 46. Reiss, 1930, in Seitz, Die Gross-Schmetterlinge der Erde, Supplement, **2**: 19. Holik, 1942, Mitt. naturw. Inst. Sofia, **15**: 255. Holik & Sheljuzhko, 1955, Mitt. münch. ent. Ges., **44/45**: 145. ab. **achilleaeformis** Holik, 1942, Mitt. naturw. Inst. Sofia, **15**: 255.	Alpine Gebirgszone um Malatia, Kleinasien [Lycaonia, Taurus, 2000 m.].
ssp. **pontica** Holik & Sheljuzhko, 1955, Mitt. münch. ent. Ges., **44/45**: 143. ab. **totirubra** Holik & Sheljuzhko, 1955, Mitt. münch. ent. Ges. **44/45**: 144.	Amasia, Pontus, Kleinasien.
ssp. **suanetica** Holik & Sheljuzhko, 1955, Mitt. münch. ent. Ges., **44/45**: 136.	Mestia, Suanetien, Transkaukasien.
ssp. **suchumensis** Holik & Sheljuzhko, 1955, Mitt. münch. ent. Ges., **44/45**: 135.	Novyj Afon, Suchum, Abchasien, Transkaukasien, Westküste.
ssp. **georgiae** Reiss, 1922, Int. ent. Z., **15**: 174; 1930, in Seitz, Die Gross-Schmetterlinge der Erde, Supplement, **2**: 19, pl. 2d. Koch, 1936, Ent. Z., **50**: 18. Holik & Sheljuzhko, 1955, Mitt. münch. ent. Ges., **44/45**: 136.	Abas-Tuman, Transkaukasien, 800-1000 m.
ssp. **eriwanensis** Reiss, 1935, Int. ent. Z., **29**: 150, figs.; 1930, in Seitz, Die Gross-Schmetterlinge der Erde, Supplement, **2**: 19, pl. 2d (as *bitorquata* Ménétriés). Koch, 1936, Ent. Z., **50**: 18. Holik & Sheljuzhko, 1955, Mitt. münch. ent. Ges., **44/45**: 138. ab. **basimaculata** Holik & Sheljuzhko, 1955, Mitt. münch. ent. Ges., **44/45**: 140.	Erivan, Armenisches Bergland.
ssp. **aktashi** Koch, 1936, Ent. Z., **50**: 20. Holik, 1935, Ent. Z., **49**: 31. Holik & Sheljuzhko, 1955, Mitt. münch. ent. Ges., **44/45**: 142. ab. **cingulata** Reiss, 1935, Int. ent. Z., **29**: 150.	Khashkhash-Dagh bei Aktash, Kars, Westarmenien.
ssp. **bitorquata** Ménétriés, 1832, Catalogue raisonné des Objets de Zoologie, p. 259. Staudinger, 1879, Horae Soc. ent. Ross., **14**: 319. Reiss, 1933, in Seitz, Die Gross-Schmetterlinge der Erde, Supplement, **2**: 258. Koch, 1936, Ent. Z., **50**: 18. Holik & Sheljuzhko, 1955, Mitt. münch. ent. Ges., **44/45**: 131. *laphria* Freyer, 1852, Neuere Beiträge zur Schmetterlingskunde, **6**: 135, pl. 568, fig. 2 (nomen dubium).	Caucase, 3000-6000 ft. (Daghestan).
ssp. **narzanica** Sheljuzhko, 1936, Folia Zool. Hydrobiol., Riga, **9**: 19. Holik & Sheljuzhko, 1955, Mitt. münch. ent. Ges., **44/45**: 132.	Kislovodsk, Narzan-Quelle, Kaukasus.

ssp. **karatshaica** Sheljuzhko, 1936, Folia Zool. Hydrobiol., Riga, **9**: 19. Holik & Sheljuzhko, 1955, Mitt. münch. ent. Ges., **44/45**: 133. Tremewan, 1961, Bull. Brit. Mus. (nat. Hist.) Ent., **10** (7): 308, pl. 57, fig. 24. Alberti, 1964, Dtsch. ent. Z. (N.F.), **11**: 391, figs. 5, 6. — Teberda- und Dzhemagat-Täler, Karatshaj Prov., Teberda- Gebiet, Kaukasus.

ssp. **jagludarensis** Holik & Sheljuzhko, 1955, Mitt. münch. ent. Ges., **44/45**: 140. — Berg Jagludara (Nachitshevan), Zangezur Gebirge, Armenien, 2650-3000 m.
ab. **sexmaculata** Holik & Sheljuzhko, 1955, Mitt. münch. ent. Ges., **44/45**: 141.

ssp. **weidingeri** Reiss, 1939, Ent. Z., **53**: 117. Holik & Sheljuzhko, 1955, Mitt. münch. ent. Ges., **44/45**: 128. — Simferopol, Krim.
ab. **flavogrisea** Reiss, 1939, Ent. Z., **53**: 117.
ab. **acumine** Reiss, 1939, Ent. Z., **53**: 117.

ssp. **balcanica** Reiss, 1922, Int. ent. Z., **15**: 175. Burgeff, 1926, Mitt. münch. ent. Ges., **16**: 36. Reiss, 1930, in Seitz, Die Gross-Schmetterlinge der Erde, Supplement, **2**: 18, pl. 2c. Holik, 1938, Mitt. münch. ent. Ges., **27**: 143. — Korična, Bosnien, 1000-1200 m. [Istria, Bosnien, Herzegowina].

ssp. **caliacrensis** Reiss, 1931, Int. ent. Z., **25**: 97; 1933, in Seitz, Die Gross-Schmetterlinge der Erde, Supplement, **2**: 257. Koch, 1937, Ent. Z., **51**: 19. Popescu-Gorj, 1964, Catalogue de la Collection de Lépidoptères „Prof. A. Ostrogovich", p. 66, pl. 5, fig. 18. — Bei Balcic, Rumänische Silberküste, südliche Dobrudscha.
ab. **cingulata** Reiss, 1931, Int. ent. Z., **25**: 98; 1933, in Seitz, Die Gross-Schmetterlinge der Erde, Supplement, **2**: 257. Popescu-Gorj, 1964, Catalogue de la Collection de Lépidoptères „Prof. A. Ostrogovich", p. 67.
ab. **confluens** Reiss, 1931, Int. ent. Z., **25**: 98; 1933, in Seitz, Die Gross-Schmetterlinge der Erde, Supplement, **2**: 257. Popescu-Gorj, 1964, Catalogue de la Collection de Lépidoptères „Prof. A. Ostrogovich", p. 67.
ab. **flavogrisea** Reiss, 1939, Ent. Z., **53**: 117.

ssp. **islimjensis** Holik, 1938, Mitt. münch. ent. Ges., **27**: 147. — Sliven (Islimje), Bulgarien.
ab. **analiconfluens** Holik, 1938, Mitt. münch. ent. Ges., **27**: 147.

ssp. **winneguthi** Holik, 1938, Mitt. münch. ent. Ges., **27**: 145. Daniel, 1958, Fragmenta Balcanica, **2**: 40; 1964, Prirod. Muz. Scopje, no. 2: 15. — Asan-djurd, Galičica-Planina, Mazedonien.
ab. **analiconfluens** Holik, 1938, Mitt. münch. ent. Ges., **27**: 144.
ab. **rubescens** Holik, 1938, Mitt. münch. ent. Ges., **27**: 144.

ssp. **macedonica** Burgeff, 1926, Mitt. münch. ent. Ges., **16**: 37. Reiss, 1930, in Seitz, Die Gross-Schmetterlinge der Erde, — Bei Veles, Plaquscha-

Supplement, **2**: 18, pl. 2d. Koch, 1937, Ent. Z. **51**: 19. Holik, 1938, Mitt. münch. ent. Ges., **27**: 145. Daniel, 1964, Prirod. Muz. Scopje, no. 2: 14. — Planina und Dojran-See, Mazedonien.

ssp. **hafneri** Holik, 1938, Mitt. münch. ent. Ges., **27**: 142. — Auf den Wiesen an der Save bei Laibach, Krain.

ab. **analiconfluens** Holik, 1938, Mitt. münch. ent. Ges., **27**: 141.

ssp. **syrmiensis** Reiss, 1939, Ent. Z., **53**: 117. — Frusca Gora, Syrmien, 300-600 m.

ssp. **peszerensis** Reiss, 1929, Int. ent. Z., **22**: 358; 1930, in Seitz, Die Gross-Schmetterlinge der Erde, Supplement, **2**: 17. — Budapest, Peszér-Peszéradaracs, Ungarn.

ssp. **transsylvaniae** Burgeff, 1926, Mitt. münch. ent. Ges., **16**: 32. Reiss, 1930, in Seitz, Die Gross-Schmetterlinge der Erde, Supplement, **2**: 18. — Gyergyoszent Miklos, Transsylvanien.

ssp. **owsei** Koch, 1935, Int. ent. Z., **28**: 507. — Blaa-Alm, Aussee, Weststeiermark, 650-800 m.

ssp. **loti** Denis & Schiffermüller, 1775, Ankündigung eines systematischen Werkes von den Schmetterlingen der Wienergegend, p. 45; 1776, Systematisches Verzeichniss der Schmetterlinge der Wienergegend, p. 45; 1801, Systematisches Verzeichniss von den Schmetterlingen der Wiener Gegend, **1**: 38. Dujardin, 1953, Bull. mens. Soc. linn. Lyon, **22**: 245. Tremewan, 1958, Ent. Gaz.. **9**: 189; 1960, ibidem, **11**: 187. Bernardi & Viette, 1960, Bull. mens. Soc. linn. Lyon, **29**: 238. — Bisamberg bei Wien, Österreich.

fulvia Fabricius, 1777, Genera Insectorum, p. 275 (nomen dubium). Rocci, 1937, Redia, **22**: 131. Tremewan, 1958, Ent. Gaz., **9**: 189. Bernardi & Viette, 1960, Bull. mens. Soc. linn. Lyon, **29**: 238. Zimsen, 1964, The Type Material of I. C. Fabricius, p. 531.

vindobonica Reiss, 1958, Z. wien. ent. Ges., **43**: 161 (nomen nudum).

ab. **cingulata** Dziurzyński, 1904, Jber. wien. ent. Ver., **14**: 24.

ab. **pseudozobeli** Reiss, 1964, Coridon, (A) **6**: 5.

ab. **blachieri** Dziurzyński, 1907, Jber. wien. ent. Ver., **17**: 85, pl. 2, fig. 11. Seitz, 1912, Die Gross-Schmetterlinge der Erde, **2**: 443. Reiss, 1930, in Seitz, Die Gross-Schmetterlinge der Erde, Supplement, **2**: 17.

ab. **dziurzynskii** Hirschke, 1906, Jber. wien. ent. Ver., **16**: 92. Dziurzyński, 1907, Jber. wien. ent. Ver., **17**: 85, pl. 2, fig. 13; 1908, Berl. ent. Z., **53**: 49, pl. 2, fig. 17.

ab. **confluens** Dziurzyński, 1902, Iris, **15**: 336; 1904, Jber. wien. ent. Ver., **14**: 48, pl. 2, fig. 5.

ab. **flava** Dziurzyński, 1902, Iris, **15**: 337; 1908, Berl. ent. Z., **53**: 48, pl. 2, fig. 14. Seitz, 1907, Die Gross-Schmetterlinge der Erde, **2**: 27, pl. 7d (as *flava* Romanoff). Reiss, 1930, in Seitz, Die Gross-Schmetterlinge der Erde, Supplement, **2**: 17.

ab. **grisescens** Reiss, 1964, Coridon, (A) **6**: 5.

ab. **brunnea** Dziurzyński, 1902, Iris, **15**: 336; 1903, Jber. wien. ent. Ver., **13**: 40; 1908, Berl. ent. Z., **53**: 48, pl. 2, fig. 15.

f. loc. **leinfesti** Reiss, 1939, Ent. Z., **53**: 116. — Bisamberg bei Wien, (Franzosengräben).

ab. **flavopraetexta** Reiss, 1939, Ent. Z., **53**: 117.

ab. **pseudozobel i**Reiss, 1939, Ent. Z., **53**: 117.

ab. **flavogrisea** Reiss, 1939, Ent. Z., **53**: 117.

ssp. **auchensis** Koch, 1941, Mitt. münch. ent. Ges., **31**: 567. — Zliechov, Slatina im Galgoczer Berge, Czechoslovakei.

ab. **cingulata** Reiss, 1964, Coridon, (A) **6**: 6.

ab. **pseudozobeli** Reiss, 1964, Coridon, (A) **6**: 5.

ab. **flavogrisea** Reiss, 1964, Coridon, (A) **6**: 6.

ab. **latemarginata** Reiss, 1964, Coridon, (A) **6**: 6.

ab. **quinquemaculata** Reiss, 1964, Coridon, (A) **6**: 5, fig. 2.

ab. **confluens** Reiss, 1964, Coridon, (A) **6**: 5.

ssp. **lodomerica** Holik, 1939, Ann. Mus. zool. Polon., **12**: 47, pl. 2, figs. 53-56, 59, 60. Holik & Sheljuzhko, 1955, Mitt. münch. ent. Ges., **44/45**: 125. Alberti & Soffner, 1962, Mitt. münch. ent. Ges., **52**: 174. — Potylicz, Holosko, Südostpolen.

ab. **parvimaculata** Holik, 1939, Ann. Mus. zool. Polon., **12**: 48.

ab. **flavopraetexta** Holik, 1939, Ann. Mus. zool. Polon., **12**: 48.

ssp. **obraztsovi** Holik & Sheljuzhko, 1955, Mitt. münch. ent. Ges., **44/45**: 128. Obraztsov, 1936, Folia Zool. Hydrobiol., Riga, **9**: 36 (as *stauderi* Holik & Reiss). — Park von Vesjolaja Bokovenjka, Cherson, Russland.

ssp. **stauderi** Holik & Reiss, 1932, in Holik, Iris, **46**: 118, pl. 1, figs. 16-19. Stauder, 1924, Int. ent. Z., **18**: 53. Reiss, 1933, in Seitz, Die Gross-Schmetterlinge der Erde, Supplement, **2**: 257. Holik & Sheljuzhko, 1955, Mitt. münch. ent. Ges., **44/45**: 125. — Kirillovskije ovragi, Kijev, Ukraine.

ab. **cingulata** Holik & Reiss, 1932, in Holik, Iris, **46**: 119. Reiss, 1933, in Seitz, Die Gross-Schmetterlinge der Erde, Supplement, **2**: 257.

ab. **rubrianata** Holik & Reiss, 1932, in Holik, Iris, **46**: 119. Reiss, 1933, in Seitz, Die Gross-Schmetterlinge der Erde, Supplement, **2**: 257.

ab. **flavopraetexta** Holik & Reiss, 1932, in Holik, Iris, **46**: 119. Reiss, 1933, in Seitz, Die Gross-Schmetterlinge der Erde, Supplement, **2**: 257.

ab. **basimaculata** Sheljuzhko, 1941, Acta Mus. zool. Kijev, **1**: 68, 95, 101.

ab. **confluens** Holik & Reiss, 1932, in Holik, Iris, **46**: 119. Reiss, 1933, in Seitz, Die Gross-Schmetterlinge der Erde, Supplement, **2**: 257.

ab. **flava** Holik & Reiss, 1932, in Holik, Iris, **46**: 119. Reiss, 1933, in Seitz, Die Gross-Schmetterlinge der Erde, Supplement, **2**: 257.

ssp. **austrosilesia** Przegendza & Prack, 1942, Ent. Z., **56**: 160, figs. 1-6.

austrosilesia Przegendza & Prack, 1943, Ent. Z., **56**: 243 (nomen nudum).

Tul, Kreis Teschen, Schlesien, 600 m.

ssp. **beraunensis** Reiss, 1922, Int. ent. Z., **16**: 84; 1926, Die Zygaenen Deutschlands, p. 31, pl. 2, fig. Holik, 1929, Int. ent. Z., **23**: 3 (footnote 9). Reiss, 1930, in Seitz, Die Gross-Schmetterlinge der Erde, Supplement, **2**: 17. Holik, 1939, Sborn. ent. Odd. nár. Mus. Praze, **17**: 42. Povolný, 1945, Folia ent., Brno, **8**: 77. Povolný & Gregor, 1946, Folia ent., Brno, **9**, Supplement, 12: 19, 94. Gregor & Povolný, 1955, Sborn. ent. Odd. nár. Mus. Praze, **30**: 260, 272, pl. 4, figs. 5, 6.

Karlstein, Radotin, Mittelböhmen.

ab. **grisea** Reiss, 1922, Int. ent. Z., **16**: 84.

ssp. **niesiolowskii** Holik, 1939, Ann. Mus. zool. Polon., **12**: 45, pl. 2, figs. 47-52. Povolný & Gregor, 1946, Folia ent., Brno, **9**, Supplement, 12: 19, 94.

Ojców, nördlich von Kraków, Südwestpolen.

ssp. **arragonensis** Staudinger, 1887, Berl. ent. Z., **31**: 34. Seitz, 1907, Die Gross-Schmetterlinge der Erde, **2**: 27. Reiss, 1930, in Seitz, Die Gross-Schmetterlinge der Erde, Supplement, **2**: 17, pl. 2b; 1933, ibidem, **2**: 257; 1936, Ent. Rdsch., **54**: 59. Koch, 1938, Ent. Z., **51**: 398.

Albarracin, Aragonien, Spanien, 1800 m.

ab. **cingulata** Koch, 1938, Ent. Z., **51**: 399.

ab. **rubrescens** Reiss, 1936, Ent. Rdsch., **54**: 60.

ssp. **avilensis** Koch, 1948, Eos, Madr., **24**: 323.

Hoyos del Espino, Sierra de Gredos, Avila, España, 1400 m.

ssp. **soriacola** Tremewan & Manley, 1965, Ent. Rec., **77**: 7.

Abejar, Soria, Spain, 3300 ft.

ssp. **pardoi** Agenjo, 1953, Graellsia, **11**: 2. Tremewan, 1961, Ent. Rec., **73**: 3; 1963, ibidem, **75**: 5. Tremewan & Manley, 1965, Ent. Rec., **77**: 8.

Pesués, Prov. Santander, España, 14 m.

ssp. **tristis** Oberthür, 1884, Études d'Entomologie, **8**: 29; 1896, ibidem, **20**: 54, pl. 7, fig. 126. Seitz, 1907, Die Gross-Schmetterlinge der Erde, **2**: 27, pl. 7d. Oberthür, 1910, Études de

Cauterets, Hautes-Pyrénées, France.

Lépidoptérologie comparée, **4**: 464. Tremewan, 1961, Bull. Brit. Mus. (nat. Hist.) Ent., **10** (7): 269, pl. 52, fig. 26.

ab. **brunnea** Oberthür, 1910, Études de Lépidoptérologie comparée, **4**: 464. Reiss, 1930, in Seitz, Die Gross-Schmetterlinge der Erde, Supplement, **2**: 17. Tremewan, 1961, Bull. Brit. Mus. (nat. Hist.) Ent., **10** (7): 269, pl. 52, fig. 27.

ssp. **interwagneri** Dujardin, 1965, Entomops, Nice, no. 2: 36, fig. — Route de Cabris à St. Vallier, 700 m.; Guillaumes, env. 700 m.; Valberg, route de Guillaumes, 1520 m.; Auron, 1500-1600 m., Alpes-Maritimes, France.

ssp. **wagneri** Millière, 1885, Bull. Soc. ent. Fr., p. xcii. Spuler, 1906, in Hofmann, Die Schmetterlinge Europas, **2**: 156, pl. 77, fig. 10. Seitz, 1907, Die Gross-Schmetterlinge der Erde, **2**: 21; 1912, ibidem, **2**: 441. Oberthür, 1910, Études de Lépidoptérologie comparée, **4**: 469. Burgeff, 1926, Mitt. münch. ent. Ges., **16**: 34. Reiss, 1930, in Seitz, Die Gross-Schmetterlinge der Erde, Supplement, **2**: 17. Weber, 1939, Mitt. schweiz. ent. Ges., **17**: 532. Burgeff, 1950, Portug. acta biol., (A) Goldschmidt: 692, figs. 3d; 1951, Biol. Zbl., **70**: 13, figs. 9d. Loritz, 1961, Bull. Soc. ent. Mulhouse, p. 83-102, fig. — La Turbie, Alpes-Maritimes, France méridionale.

ab. **achilloides** Wagner, 1905, Soc. ent., **19**: 149. Spuler, 1906, in Hofmann, Die Schmetterlinge Europas, **2**: 156, pl. 77, fig. 10a. Seitz, 1907, Die Gross-Schmetterlinge der Erde, **2**: 21. Oberthür, 1910, Études de Lépidoptérologie comparée, **4**: 469, 471. Reiss, 1933, in Seitz, Die Gross-Schmetterlinge der Erde, Supplement, **2**: 257.

ab. **flavocincta** Reiss, 1953, Ent. Z., **63**: 93.

ab. **demarginata** Reiss, 1953, Ent. Z., **63**: 93.

ab. **subcaerulea** Millière, 1886, Ann. Soc. ent. Fr., (6) **6**: 7. Oberthür, 1910, Études de Lépidoptérologie comparée, **4**: 471. Reiss, 1930, in Seitz, Die Gross-Schmetterlinge der Erde, Supplement, **2**: 17.

nigra Dziurzyński, 1906, Ent. Z., **19**: 185 (nomen nudum). Seitz, 1907, Die Gross-Schmetterlinge der Erde, **2**: 21.

ab. **giesekingi** Wagner, 1905, Soc. ent., **19**: 149. Spuler, 1906, in Hofmann, Die Schmetterlinge Europas, **2**: 156, pl. 77, fig. 10b. Seitz, 1907, Die Gross-Schmetterlinge der Erde, **2**:

21. Oberthür, 1910, Études de Lépidoptérologie comparée, **4**: 471. Reiss, 1933, in Seitz, Die Gross-Schmetterlinge der Erde, Supplement, **2**: 257.

wagneri Millière, 1886, Ann. Soc. ent. Fr., (6) **6**: 6, pl. 1, figs. 6, 7 (preoccupied by **wagneri** Millière, 1885, ssp. of **loti** Denis & Schiffermüller).

ab. **paupera** Reiss, 1953, Ent. Z., **63**: 92.

ab. **sexmacula** Dziurzyński, 1908, Berl. ent. Z., **53**: 22. Seitz, 1912, Die Gross-Schmetterlinge der Erde, **2**: 441. Reiss, 1930, in Seitz, Die Gross-Schmetterlinge der Erde, Supplement, **2**: 17.

quadrimaculata Oberthür, 1910, Études de Lépidoptérologie comparée, **4**: 471. Tremewan, 1961, Bull. Brit. Mus. (nat. Hist.) Ent., **10** (7): 268, pl. 52, fig. 25.

ab. **confluens** Dziurzyński, 1908, Berl. ent. Z., **53**: 6, 23.

ssp. **osthelderi** Burgeff, 1926, Mitt. münch. ent. Ges., **16**: 35. Reiss, 1930, in Seitz, Die Gross-Schmetterlinge der Erde, Supplement, **2**: 17, pl. 2c. Burgeff, 1950, Portug. acta biol., (A) Goldschmidt: 692, figs. 3c; 1951, Biol. Zbl., **70**: 13, figs. 9c. — Alassio, Litoral Liguriens, Norditalien.

ssp. **vozea** Przegendza, 1933, Ent. Z., **47**: 27, figs. 1-3. Reiss, 1933, in Seitz, Die Gross-Schmetterlinge der Erde, Supplement, **2**: 257. — Voze bei Spotorno, Riviera di Ponente, Norditalien, 300 m.

ssp. **ligustica** Rocci, 1913, Soc. ent., **28**: 56. Reiss, 1930, in Seitz, Die Gross-Schmetterlinge der Erde, Supplement, **2**: 17, pl. 2c. Burgeff, 1950, Portug. acta biol., (A) Goldschmidt: 692, figs. 3b; 1951, Biol. Zbl., **70**: 13, figs. 9b; 1951, ibidem, **70**: 17, fig. 12. — Genova, Liguria, Italia.

ligustina Burgeff, 1914, Mitt. münch. ent. Ges., **5**: 48, 77, pl. 2, figs. 155, 163, pl. 5, figs. 32-38.

ab. **parva** Rocci, 1915, Atti Soc. ligust. Sci. nat. geogr., **25**: 116, pl. 1, figs. 6d, 7d. Reiss, 1930, in Seitz, Die Gross-Schmetterlinge der Erde, Supplement, **2**: 18.

ab. **decollata** Rocci, 1915, Atti Soc. ligust. Sci. nat. geogr., **25**: 115. Reiss, 1930, in Seitz, Die Gross-Schmetterlinge der Erde, Supplement, **2**: 18.

ab. **pseudoachilleae** Rocci, 1915, Atti Soc. ligust. Sci. nat. geogr., **25**: 112, pl. 1, figs. 3a, 10b, 11b, 1c. Reiss, 1930, in Seitz, Die Gross-Schmetterlinge der Erde, Supplement, **2**: 18.

ab. **divisa** Rocci, 1913, Soc. ent., **28**: 56; 1915, Atti Soc. ligust. Sci. nat. geogr., **25**: 114, pl. 1, figs. 2d, 3d. Reiss, 1930, in Seitz, Die Gross-Schmetterlinge der Erde, Supplement, **2**: 18.

ab. **pseudocynarae** Rocci, 1913, Soc. ent., **28**: 56; 1915, Atti

Soc. ligust. Sci. nat. geogr., **25**: 113, pl. 1, figs. 7a, 8a, 6c, 7c. Reiss, 1930, in Seitz, Die Gross-Schmetterlinge der Erde, Supplement, **2**: 18.

ab. **pseudowagneri** Rocci, 1913, Soc. ent., **28**: 56; 1915, Atti Soc. ligust. Sci. nat. geogr., **25**: 113, pl. 1, figs. 9a, 10a, 11a, 9c. Reiss, 1930, in Seitz, Die Gross-Schmetterlinge der Erde, Supplement, **2**: 18.

ab. **latomarginata** Rocci, 1915, Atti Soc. ligust. Sci. nat. geogr., **25**: 112, pl. 1, figs. 6b, 10c, 11c. Reiss & Tremewan, 1964, Ent. Rec., **76**: 133.

latemarginata Burgeff, 1926, Mitt. münch. ent. Ges., **16**: 36. Reiss, 1930, in Seitz, Die Gross-Schmetterlinge der Erde, Supplement, **2**: 18. Reiss & Tremewan, 1964, Ent. Rec., **76**: 133.

ab. **parvipuncta** Rocci, 1915, Atti Soc. ligust. Sci. nat. geogr., **25**: 115. Reiss, 1930, in Seitz, Die Gross-Schmetterlinge der Erde, Supplement, **2**: 18.

ab. **paupera** Rocci, 1918, Atti Soc. ligust. Sci. nat. geogr., **28**: 147, pl. 3, fig. 3b. Reiss, 1930, in Seitz, Die Gross-Schmetterlinge der Erde, Supplement, **2**: 18.

ab. **diaphana** Rocci, 1915, Atti Soc. ligust. Sci. nat. geogr., **25**: 115. Reiss & Tremewan, 1964, Ent. Rec., **76**: 133.

translucens Burgeff, 1926, Mitt. münch. ent. Ges., **16**: 36. Reiss, 1930, in Seitz, Die Gross-Schmetterlinge der Erde, Supplement, **2**: 18. Reiss & Tremewan, 1964, Ent. Rec., **76**: 133.

ab. **nigrocincta** Rocci, 1921, Atti Soc. ligust. Sci. nat. geogr., **32**: 33.

ab. **flavocincta** Rocci, 1915, Atti Soc. ligust. Sci. nat. geogr., **25**: 115. Reiss, 1930, in Seitz, Die Gross-Schmetterlinge der Erde, Supplement, **2**: 18.

ab. **flavoinspersa** Rocci, 1915, Atti Soc. ligust. Sci. nat. geogr., **25**: 115. Reiss, 1930, in Seitz, Die Gross-Schmetterlinge der Erde, Supplement, **2**: 18.

ab. **conjuncta** Rocci, 1915, Atti Soc. ligust. Sci. nat. geogr., **25**: 114. Reiss, 1930, in Seitz, Die Gross-Schmetterlinge der Erde, Supplement, **2**: 18.

ab. **semiconfluens** Rocci, 1914, Atti Soc. ligust. Sci. nat. geogr., **24**: 197.

ab. **confluens** Rocci, 1913, Soc. ent., **28**: 56.

ab. **anomala** Rocci, 1915, Atti Soc. ligust. Sci. nat. geogr., **25**: 116, pl. 1, fig. 5d.

ab. **rosea** Rocci, 1914, Atti Soc. ligust. Sci. nat. geogr., **24**: 113.

rosea Rocci, 1914, Soc. ent., **29**: 41.

ab. **aurantiaca** Rocci, 1915, Atti Soc. ligust. Sci. nat. geogr., **25**: 115.

ab. **flavescens** Rocci, 1914, Atti Soc. ligust. Sci. nat. geogr., **24**: 113.

flavescens Rocci, 1914, Soc. ent., **29**: 41.

ab. **flava** Hebsacker, 1914, Ent. Z., **28**: 50. Reiss, 1930, in Seitz, Die Gross-Schmetterlinge der Erde, Supplement, **2**: 18.

ab. **fumosa** Rocci, 1916, Atti Soc. ligust. Sci. nat. geogr., **27**: 30.

ssp. **libarnica** Rocci, 1941, Boll. Ist. Ent. Univ. Bologna, **13**: 115, pl. 2, figs. 12, 13. — Arquata Scrivia, Liguria, Italia, 300-400 m.

ssp. **conserta** Rocci, 1941, Boll. Ist. Ent. Univ. Bologna, **13**: 113, pl. 2, figs. 8, 9. — Villarfocchiardo (Valle di Susa), 800 m.; Collina di Torino, 600 m.; Italia.

ssp. **oblita** Rocci, 1941, Boll. Ist. Ent. Univ. Bologna, **13**: 113, pl. 2, figs. 10, 11. — Meana; Salabertano; Valle di Susa, Italia, 1200 m.

ssp. **roccai** Rocci, 1937, Redia, **22**: 133. — Sappada, Alpi Carniche, Italia, 1200 m.

ssp. **propinqua** Rocci, 1926, Boll. Soc. ent. ital., **58**: 72. Reiss, 1930, in Seitz, Die Gross-Schmetterlinge der Erde, Supplement, **2**: 18. — Mte. Alpesisa, Appennino ligure, Italia, 600 m.

ab. **cingulata** Turati, 1923, Atti Soc. ital. Sci. nat., **62**: 44.

ab. **pseudoligustica** Rocci, 1926, Boll. Soc. ent. ital., **58**: 73. Reiss, 1930, in Seitz, Die Gross-Schmetterlinge der Erde, Supplement, **2**: 18.

ssp. **cicaleti** Verity, 1930, Mem. Soc. ent. ital., **9**: 19. Oberthür, 1910, Études de Lépidoptérologie comparée, **4**: 467 (as *triptolemus* Hübner). Verity, 1930, Mem. Soc. ent. ital., **9**: 19 (as *triptolemus* Hübner [partim]). Reiss, 1930, in Seitz, Die Gross-Schmetterlinge der Erde, Supplement, **2**: 18, pl. 2d (as *triptolemus* Hübner); 1933, ibidem, **2**: 257. Burgeff, 1950, Portug. acta biol., (A) Goldschmidt: 692, figs. 3a (as *triptolemus* Hübner); 1951, Biol. Zbl., **70**: 13, figs. 9a (as *triptolemus* Hübner). — Pian di Mugnone, Firenze, Toskana, Italia.

ab. **flavopraetexta** Burgeff, 1926, Mitt. münch. ent. Ges., **16**: 33.

ssp. **tuscamodica** Verity, 1930, Mem. Soc. ent. ital., **9**: 19. Reiss, 1933, in Seitz, Die Gross-Schmetterlinge der Erde, Supplement, **2**: 257. — Monte Conca, Toskana, Italia, 400 m.

ab. **acumine** Verity, 1930, Mem. Soc. ent. ital., **9**: 20. Reiss,

1933, in Seitz, Die Gross-Schmetterlinge der Erde, Supplement, **2**: 256.

ab. **crasseunco** Verity, 1930, Mem. Soc. ent. ital., **9**: 20. Reiss, 1933, in Seitz, Die Gross-Schmetterlinge der Erde, Supplement, **2**: 256.

ab. **uncoflabello** Verity, 1930, Mem. Soc. ent. ital., **9**: 20. Reiss, 1933, in Seitz, Die Gross-Schmetterlinge der Erde, Supplement, **2**: 256.

ab. **tenueunco** Verity, 1930, Mem. Soc. ent. ital., **9**: 20. Reiss, 1933, in Seitz, Die Gross-Schmetterlinge der Erde, Supplement, **2**: 256.

ssp. **planorum** Rocci, 1937, Redia, **22**: 132.

Abbiategrasso e Vigevano, Ticino (Santa Maria del Bosco), Lombardia, Italia.

ssp. **aspera** Burgeff, 1926, Mitt. münch. ent. Ges., **16**: 33. Reiss, 1930, in Seitz, Die Gross-Schmetterlinge der Erde, Supplement, **2**: 18.

Mte. Sirente und Mte. Velino, Abruzzen, Italien, 1500-2000 m.

ab. **elongata** Verity, 1916, Boll. Soc. ent. ital., **47**: 73.

ab. **emirubra** Verity, 1916, Boll. Soc. ent. ital., **47**: 73.

f. t. **italicaaestivalis** Oberthür, 1910, Études de Lépidoptérologie comparée, **4**: 662. Reiss, 1930, in Seitz, Die Gross-Schmetterlinge der Erde, Supplement, **2**: 18 (as *aestivalis*). Tremewan, 1961, Bull. Brit. Mus. (nat. Hist.) Ent., **10** (7): 269, pl. 52, fig. 28.

Roccaroso et Palena, Italie méridionale.

ssp. **verityana** Burgeff, 1926, Mitt. münch. ent. Ges., **16**: 33. Verity, 1916, Boll. Soc. ent. ital., **47**: 73 (as *bellis* Hübner). Reiss, 1930, in Seitz, Die Gross-Schmetterlinge der Erde, Supplement, **2**: 18.

Ascoli Piceno, Mte. Sibillini, Italien, 1200-1700 m.

ssp. **ruberrima** Verity, 1920, Boll. Lab. Zool. Portici, **14**: 37. Reiss & Tremewan, 1964, Ent. Rec., **76**: 133.

Mte. Mainarde, Prov. Caserta, Italia.

maximerubra Burgeff, 1926, Mitt. münch. ent. Ges., **16**: 33. Reiss, 1930, in Seitz, Die Gross-Schmetterlinge der Erde, Supplement, **2**: 18. Reiss & Tremewan, 1964, Ent. Rec., **76**: 133.

ssp. **restricta** Stauder, 1915, Z. wiss. InsektBiol., **11**: 71. Reiss, 1930, in Seitz, Die Gross-Schmetterlinge der Erde, Supplement, **2**: 18. Tremewan, 1966, Ent. Rec., **78**: 32, pl. 1, fig. 4.

Mt. Faito, Sorrentinische Halbinsel, Italien, 1000 m.

punctachilleae Stauder, 1929, Ent. Z., **43**: 30. Tremewan, 1961, Bull. Brit. Mus. (nat. Hist.) Ent., **10** (7): 269, pl. 52, fig. 30.

punctmeliloti Stauder, 1929, Ent. Z., **43**: 31. Tremewan, 1961, Bull. Brit. Mus. (nat. Hist.) Ent., **10** (7): 270, pl. 52, fig. 31.

melilorestricta Stauder, 1929, Ent. Z., **43**: 31. Tremewan, 1961, Bull. Brit. Mus. (nat. Hist.) Ent., **10** (7): 270, pl. 52, fig. 32.

ssp. **hyalowagneri** Burgeff, 1950, Portug. acta biol., (A) Goldschmidt: 694. — St. Martin-Vésubie, Alpes-Maritimes, Frankreich, 1600-1700 m.

ssp. **tendina** Burgeff, 1950, Portug. acta biol., (A) Goldschmidt: 694. — Col di Tenda, Meeralpen, Norditalien, 800-1500 m.

ssp. **achillalpina** Burgeff, 1926, Mitt. münch. ent. Ges., **16**: 34 (nomen novum for *alpina* Oberthür). Boisduval, 1834, Icones historique des Lépidoptères nouveaux ou peu connus, **2**: 49, pl. 53, figs. 6, 7. Reiss, 1930, in Seitz, Die Gross-Schmetterlinge der Erde, Supplement, **2**: 17, pl. 2b. Burgeff, 1950, Portug. acta biol., (A) Goldschmidt: 692, figs. 3e; 1951, Biol. Zbl., **70**: 13, figs. 9e. Dufay, 1966, Bull. mens. Soc. linn. Lyon, **35**: 71. — Digne, Basses-Alpes, Frankreich.

alpina Oberthür, 1910, Études de Lépidoptérologie comparée, **4**: 466 (preoccupied by **alpina** Boisduval, 1834, ssp. of **transalpina** Esper). Tremewan, 1961, Bull. Brit. Mus. (nat. Hist.) Ent., **10** (7): 268, pl. 52, fig. 23.

ab. **carnea** de Saussure, 1914, Bull. Soc. lépid. Genève, **3**: 80, pl. 1, fig. 4.

ssp. **sanctabalmica** Dujardin, 1965, Entomops, Nice, no. 2: 37, fig. — Mazaugues; Sainte-Baume 700 m.; Plan-d'Aups, 615 m.; Var, France.

ssp. **janthina** Boisduval, 1828, Essai sur une Monographie des Zygénides, p. 45, pl. 8, fig. 7; 1834, Icones historique des Lépidoptères nouveaux ou peu connus, **2**: 51, pl. 53, fig. 8. Duponchel, 1835, in Godart & Duponchel, Histoire naturelle des Lépidoptères ou Papillons de France, Supplement, **2**: 143, pl. 12, fig. 7. Oberthür, 1910, Études de Lépidoptérologie comparée, **4**: 467. Reiss, 1930, in Seitz, Die Gross-Schmetterlinge der Erde, Suppl., **2**: 17 (as ab.). Tremewan, 1961, Bull. Brit. Mus. (nat. Hist.) Ent., **10** (7): 268, pl. 52, fig. 24. — Montagnes sousalpines du Bourg d'Oysans et les Alpes de la Provence, France.

ssp. **mesotaenia** Dujardin, 1956, Bull. mens. Soc. linn. Lyon, **25**: 258. Reiss, 1958, Bull. Soc. ent. Mulhouse, p. 53. — Le Rouret, route de Grasse, Alpes-Maritimes, France, 0-1000 m.

ssp. **oreotropha** Dujardin, 1965, Entomops, Nice, no. 2: 35, fig. — Esteng, le

ssp. **oreotropha** *(continued)* — haut Var, Alpes-Maritimes, France, 1500 m.

ssp. **hypsichorica** Dujardin, 1965, Entomops, Nice, no. 2: 34, fig. — Col du Lautaret, Rif. Blanc, 1900 m.; Villar d'Arène, 1800 m., Hautes-Alpes, France.

ssp. **praeclara** Burgeff, 1926, Mitt. münch. ent. Ges., **16**: 32. Reiss, 1930, in Seitz, Die Gross-Schmetterlinge der Erde, Supplement, **2**: 18, pl. 2c. Koch, 1935, Int. ent. Z., **28**: 504. Reiss, 1950, Jber. naturf. Ges. Graubünden, **82**: 108, figs. 12, 13. — Etschtal, Eisacktal und Sarcatal bei Bozen, Südtirol.

sexmaculata Vorbrodt & Müller-Rutz, 1917, Mitt. schweiz. ent. Ges., **12**: 497.

sexmaculata Vorbrodt, 1931, Mitt. schweiz. ent. Ges., **14**: 380. Reiss, 1933, in Seitz, Die Gross-Schmetterlinge der Erde, Supplement, **2**: 256.

ab. **apicalielongata** Vorbrodt, 1913, in Vorbrodt & Müller-Rutz, Die Schmetterlinge der Schweiz, **2**: 255, fig. 10.

ab. **analielongata** Vorbrodt, 1913, in Vorbrodt & Müller-Rutz, Die Schmetterlinge der Schweiz, **2**: 255.

ab. **analiconfluens** Vorbrodt, 1913, in Vorbrodt & Müller-Rutz, Die Schmetterlinge der Schweiz, **2**: 255, fig. 19.

ab. **parallela** Vorbrodt, 1913, in Vorbrodt & Müller-Rutz, Die Schmetterlinge der Schweiz, **2**: 255.

parallela Vorbrodt & Müller-Rutz, 1917, Mitt. schweiz. ent. Ges., **12**: 497.

parallela Vorbrodt, 1921, Mitt. schweiz. ent. Ges., **13**: 203.

ssp. **triptolemus** Hübner, [July 1803]-[15th November 1806], Sammlung europäischer Schmetterlinge, **2**, pl. 20, figs. 96. 97; 1806, ibidem, Der Ziefer, p. 78. Reiss, 1933, in Seitz, Die Gross-Schmetterlinge der Erde, Supplement, **2**: 257. Koch, 1935, Int. ent. Z., **28**: 505. Reiss, 1950, Jber. naturf. Ges. Graubünden, **82**: 107. — Südtirol, Vorberge der Alpen.

ab. **latemarginata** Reiss, 1950, Jber. naturf. Ges. Graubünden, **82**: 107.

ab. **confluens** Burgeff, 1926, in Strand, Lepid. Cat., **33**: 24. Reiss, 1930, in Seitz, Die Gross-Schmetterlinge der Erde, Supplement, **2**: 18.

ssp. **castellana** Stauder, 1929, Ent. Z., **43**: 79. Reiss, 1930, in Seitz, Die Gross-Schmetterlinge der Erde, Supplement, **2**: 17. — Seiseralpe, Südtirol, 1500 m.

Koch, 1935, Int. ent. Z., **28**: 504. Tremewan, 1961, Bull. Brit. Mus. (nat. Hist.) Ent., **10** (7): 269, pl. 52, fig. 29.

ssp. **ladiniae** Reiss, 1953, Z. wien. ent. Ges., **38**: 265, pl. 18, figs. B3, C3, A4, B4. Col. Pralongia, Ladinia, Ost-Dolomiten, 2000-2100 m.

ab. **pseudoachilleae** Reiss, 1953, Z. wien. ent. Ges., **38**: 266, pl. 18, fig. C5.

ab. **parvipuncta** Reiss, 1953, Z. wien. ent. Ges., **38**: 266, pl. 18, fig. A5.

ab. **latemarginata** Reiss, 1953, Z. wien. ent. Ges., **38**: 266, pl. 18, fig. B5.

ab. **demarginata** Reiss, 1953, Z. wien. ent. Ges., **38**: 266, pl. 18, fig. C4.

ab. **flavopraetexta** Reiss, 1953, Z. wien. ent. Ges., **38**: 265, pl. 18, figs. B6, C6.

ab. **conjuncta** Reiss, 1953, Z. wien. ent. Ges., **38**: 265, pl. 18, fig. A6.

ssp. **alpestris** Burgeff, 1914, Mitt. münch. ent. Ges., **5**: 47, 48, pl. 2, figs. 154, 162, pl. 5, figs. 28-31. Reiss, 1930, in Seitz, Die Gross-Schmetterlinge der Erde, Supplement, **2**: 17. Daniel, 1932, in Osthelder, Die Schmetterlinge Südbayerns, **1**: 566. Koch, 1935, Int. ent. Z., **28**: 504. Reiss, 1950, Jber. naturf. Ges. Graubünden, **82**: 105, figs. 9, 10. Saas-Fee im Wallis (Bergün, Engadin); Schweiz.

ab. **locheri** Vorbrodt & Müller-Rutz, 1917, Mitt. schweiz. ent. Ges., **12**: 496.

ab. **pseudoachilleae** Reiss, 1950, Jber. naturf. Ges. Graubünden, **82**: 106.

ab. **flavocincta** Reiss, 1950, Jber. naturf. Ges. Graubünden, **82**: 106.

ab. **flavoinspersa** Reiss, 1950, Jber. naturf. Ges. Graubünden, **82**: 106.

ab. **latemarginata** Reiss, 1950, Jber. naturf. Ges. Graubünden, **82**: 107.

ab. **quadrimaculata** Vorbrodt, 1913, in Vorbrodt & Müller-Rutz, Die Schmetterlinge der Schweiz, **2**: 256, fig. 40.

ab. **pseudowagneri** Reiss, 1950, Jber. naturf. Ges. Graubünden, **82**: 106.

ab. **costalielongata** Vorbrodt, 1913, in Vorbrodt & Müller-Rutz, Die Schmetterlinge der Schweiz, **2**: 255.

ab. **costaliconfluens** Vorbrodt, 1913, in Vorbrodt & Müller-Rutz, Die Schmetterlinge der Schweiz, **2**: 255.

ab. **basiconfluens** Vorbrodt, 1913, in Vorbrodt & Müller-Rutz, Die Schmetterlinge der Schweiz, **2**: 255.

ab. **apicalimaculata** Vorbrodt, 1913, in Vorbrodt & Müller-Rutz, Die Schmetterlinge der Schweiz, **2**: 256.

ab. **gramanni** Vorbrodt, 1914, in Vorbrodt & Müller-Rutz, Die Schmetterlinge der Schweiz, (Nachtrag), **2**: 647.

ab. **flava** Reiss, 1950, Jber. naturf. Ges. Graubünden, **82**: 107, fig. 11.

ssp. **rhenicola** Reiss, 1950, Jber. naturf. Ges. Graubünden, **82**: 107.

Landquart, Churer Rheintal, Schweiz, 500-800 m.

ssp. **rhingauiana** Burgeff, 1926, Mitt. münch. ent. Ges., **16**: 31. Reiss, 1930, in Seitz, Die Gross-Schmetterlinge der Erde, Supplement, **2**: 17. Heuser & Jöst, 1959, Mitt. Pollichia, (3) **6**: 124.

Geisenheim, Rheingau, West-deutschland.

ssp. **rhenana** Reiss, 1922, Int. ent. Z., **16**: 84; 1926, Die Zygaenen Deutschlands, p. 31, pl. 2, fig.; 1930, in Seitz, Die Gross-Schmetterlinge der Erde, Supplement, **2**: 17, pl. 2b. Derenne, 1934, Amat. Papillons, p. 148.

ab. **grisea** Reiss, 1922, Int. ent. Z., **16**: 84.

Kaiserstuhl, Baden.

ssp. **jurassina** Burgeff, 1926, Mitt. münch. ent. Ges., **16**: 31. Reiss, 1930, in Seitz, Die Gross-Schmetterlinge der Erde, Supplement, **2**: 17; 1937, in Schneider, Jh. Ver. vaterl. Naturk. Württemb., **93**: 126; 1949, Entomon, **1**: 170.

ab. **flavopraetexta** Burgeff, 1926, Mitt. münch. ent. Ges., **16**: 32. Reiss, 1930, in Seitz, Die Gross-Schmetterlinge der Erde, Supplement, **2**: 17.

ab. **quinquepuncta** Reiss, 1939, Ent. Z., **53**: 116.

ab. **aurantiaca** Reiss, 1939, Ent. Z., **53**: 116.

ab. **brunnea** Reiss, 1964, Coridon, (A) **6**: 5.

Spaichingen; Tuttlingen; Klingenstein, usw., Schwäbische Alb, Württem-berg,Südwest-deutschland.

ssp. **leptoderma** Dujardin, 1965, Entomops, Nice, no. 2: 34, fig.

ab. **minuta** Lambillion, 1909, Rev. Soc. ent. namur., **9**: 68.

ab. **semicingulata** Lambillion, 1909, Rev. Soc. ent. namur., **9**: 68.

ab. **connexa** Lambillion, 1909, Rev. Soc. ent. namur., **9**: 68.

ab. **juncta** Dufrane, 1936, Bull. (Ann.) Soc. ent. Belg., **76**: 125.

Torgny; Vir-ton; Barvaux; Hockai, Belgique.

ssp. **miniacea** Oberthür, 1910, Études de Lépidoptérologie comparée, **4**: 462. Reiss, 1930, in Seitz, Die Gross-Schmetterlinge der Erde, Supplement, **2**: 17, pl. 2b. Tremewan, 1961, Bull. Brit. Mus. (nat. Hist.) Ent., **10** (7): 267, pl. 52, fig. 21.

ab. **interligata** Pionneau, 1939, Échange, **55**: 26.

ab. **confluens** École Bordelaise, 1933, Act. Soc. linn. Bordeaux, **85**: 138.

ab. **flava** Oberthür, 1896, Études d'Entomologie, **20**: 43, pl. 8, fig. 140. Reiss, 1930, in Seitz, Die Gross-Schmetterlinge der Erde, Supplement, **2**: 17. Tremewan, 1961, Bull. Brit. Mus. (nat. Hist.) Ent., **10** (7): 267, pl. 52, fig. 22.

Dompierre-sur-Mer, Charente-In-férieure, France.

ssp. **hypochlora** Dujardin, 1964, Entomops, Nice, no. 1: 22, fig.

Meyrueis, 700 m.;

ssp. **hypochlora** *(continued)*

Causse-Méjean, Meyrueis, 800 m., Lozère; Peyreleau (Causse Noir), Aveyron, 800 m., Massif Central, France.

ssp. **achilleae** Esper, 1781, Die Schmetterlinge, **2**: 189, pl. 25, figs. 1a, 1b. Ochsenheimer, 1808, Die Schmetterlinge von Europa, **2**: 30. Spuler, 1906, in Hofmann, Die Schmetterlinge Europas, **2**: 156, pl. 75, fig. 45a, pl. 77, fig. 9. Seitz, 1907, Die Gross-Schmetterlinge der Erde, Supplement, **2**: 27, pl. 7c. Oberthür, 1910, Études de Lépidoptérologie comparée, **4**: 461. Burgeff, 1926, Mitt. münch. ent. Ges., **16**: 29. Reiss, 1930, in Seitz, Die Gross-Schmetterlinge der Erde, Supplement, **2**: 16; 1933, ibidem, **2**: 256. Haaf, 1952, Veröff. zool. Staatssamml. Münch., **2**: 152, 155, 157, pl. 13. Bergmann, 1953, Die Grossschmetterlinge Mitteldeutschlands, **3**: 26, pl. 64, figs. D1, D2, D3. Forster & Wohlfahrt, 1958, Die Schmetterlinge Mitteleuropas, **3**: 93, pl. 10, figs. 11, 16. Alberti, 1958, Mitt. zool. Mus. Berl., **34**: 318.

Uffenheim, Franken, Mitteldeutschland [Maintal, Gambach bis Würzburg].

lunata Heydenreich, 1851, Lepidopterorum Europaeorum Catalogus Methodicus, p. 21 (nomen nudum).

ab. **augsburga** Burgeff, 1926, Mitt. münch. ent. Ges., **16**: 86. Reiss, 1930, in Seitz, Die Gross-Schmetterlinge der Erde, Supplement, **2**: 16.

ab. **grisea** Reiss, 1922, Int. ent. Z., **16**: 84. Reiss & Tremewan, 1964, Ent. Rec., **76**: 133.

flavogrisea Burgeff, 1926, in Strand, Lepid. Cat., **33**: 21. Reiss, 1930, in Seitz, Die Gross-Schmetterlinge der Erde, Supplement, **2**: 17. Reiss & Tremewan, 1964, Ent. Rec., **76**: 133.

ab. **amsteinii** Scheven, 1782, in Fuessly, Neues Magazin für die Liebhaber der Entomologie, **1**: 54. Fuessly, 1778, Magazin für die Liebhaber der Entomologie, **1**, pl. 1, fig. 4.

bellis Borkhausen, 1789, Systematische Beschreibung der Europäischen Schmetterlinge, **2**: 122. Fuessly, 1778, Magazin für die Liebhaber der Entomologie, **1**: 127, pl. 1, fig. 4. Borkhausen, 1793, Rheinisches Magazin, **1**: 639. Hübner, 1796, Sammlung europäischer Schmetterlinge, **2** (1), pl. 2, fig. 10. Reiss, 1930, in Seitz, Die Gross-Schmetterlinge der Erde, Supplement, **2**: 16.

bellidis Hübner, 1806, Sammlung europäischer Schmetterlinge, Der Ziefer, p. 78.

ab. **confluens** Burgeff, 1914, Mitt. münch. ent. Ges., **5**: 47.

ab. **bellisconfluens** Spuler, 1906, in Hofmann, Die Schmetterlinge Europas, **2**: 156.

ab. **rubrescens** Reiss, 1922, Int. ent. Z., **16**: 84; 1930, in Seitz, Die Gross-Schmetterlinge der Erde, Supplement, **2**: 16.

ab. **rosea** Reiss, 1922, Int. ent. Z., **16**: 84; 1926, Die Zygaenen Deutschlands, p. 33, pl. 1, fig.

ab. **fulva** Spuler, 1906, in Hofmann, Die Schmetterlinge Europas, **2**: 156.

ab. **nigerrima** Lonitz, 1914/24, Jber. Ges. Fr. Naturw. Gera, **57/67**: 82.

ssp. **scotica** Rowland-Brown, 1919, Entomologist, **52**: 225. Cockayne, 1908, Ent. Rec., **20**: 73. Tutt, 1908, Ent. Rec., **20**: 73. South, 1908, The Moths of the British Isles, **2**: 335, pl. 1, fig. 2. Reid, 1919, Entomologist, **52**: 188. James, 1932, Entomologist, **65**: 224. Woodbridge, 1934, Entomologist, **67**: 238. Cockayne, 1951, Ent. Rec., **63**: 143. Ford, 1955, Moths, pl. 27, fig. 9. Heslop-Harrison, 1955, Ent. Rec., **67**: 177. Tremewan, 1958, Ent. Gaz., **9**: 190; 1961, Coridon, (A) **1**: 2, pl. C1, fig. 7. South, 1961, The Moths of the British Isles, **2**: 330, pl. 129, fig. 4. De Worms, 1962, Entomologist, **95**: 102. Hyde, 1964, Animals, **3** (6): 162-164, fig. 6 (as *purpuralis* Brünnich). — Argyll, West Scotland [Morven; Isle of Mull].

caledoniae Verity, 1930, Mem. Soc. ent. ital., **9**: 21. Reiss, 1933, in Seitz, Die Gross-Schmetterlinge der Erde, Supplement, **2**: 256. Cockayne, 1951, Ent. Rec., **63**: 143.

caledonica Reiss, 1931, Int. ent. Z., **25**: 341; 1931, ibidem, **25**: 359, figs.; 1933, in Seitz, Die Gross-Schmetterlinge der Erde, Supplement, **2**: 256. Cockayne, 1951, Ent. Rec., **63**: 143.

ssp. **zobeli** Reiss, 1921, Int. ent. Z., **15**: 118; 1926, Die Zygaenen Deutschlands, p. 31, pl. 2, fig.; 1930, in Seitz, Die Gross-Schmetterlinge der Erde, Supplement, **2**: 17, pl. 2b. Kuserau, 1942, Ent. Z., **56**: 47. Forster & Wohlfahrt, 1958, Die Schmetterlinge Mitteleuropas, **3**: 93, pl. 10, figs. 12, 17. — Osterode, Ostpreussen [nördliches Mittelpolen].

ab. **grisea** Reiss, 1922, Int. ent. Z., **16**: 84.

ab. **parvimaculata** Holik, 1939, Ann. Mus. zool. Polon., **12**: 43, pl. 2, fig. 43 (as *parvomaculata*, corrected to **parvimaculata** in legend to plate).

ssp. **loquayi** Reiss, 1939, Ent. Z., **53**: 115. Kuserau, 1942, Ent. Z., **56**: 47. — Umgebung von Wiese, Kreis, Lübben, Spreewald, Norddeutschland.

ab. **pseudotristis** Guhn, 1932, Ent. Jb., **41**: 95.

ab. **pseudozobeli** Reiss, 1939, Ent. Z., **53**: 115.

ab. **flavopraetexta** Reiss, 1939, Ent. Z., **53**: 115.

ab. **confluens** Reiss, 1939, Ent. Z., **53**: 115; 1941, Mitt. münch. ent. Ges., **31**: 995.

ssp. **ruefferi** Reiss, 1939, Ent. Z., **53**: 115 (as *rüfferi*). Kuserau, 1942, Ent. Z., **56**: 48. — Finkenwalde bei Stettin, Pommern, Norddeutschland.

ssp. **sueciae** Reiss, 1939, Ent. Z., **53**: 116. Nordström & Wahlgren, 1941, Svenska Fjärilar, **2**: 326. Kuserau, 1942, Ent. Z., **56**: 48. Holik, 1942, Ent. Z., **55**: 236. — ? Umgebung von Stockholm, Schweden.

Biology

Abeille, 1909, Mém. Soc. linn. Provence, **1**: 12. Burgeff, 1912, Z. wiss. InsektBiol., **8**: 123; 1921, Mitt. münch. ent. Ges., **11**: 56; 1950, Portug. acta biol., (A) Goldschmidt: 663-728; 1951, Biol. Zbl., **70**: 1-23. Dorfmeister, 1854, Verh. zool.-bot. Ver. Wien, **4**: 477. Döring, 1955, Zur Morphologie der Schmetterlingseier, p. 120, pl. 17, fig. 253. Holik, 1937, Lambillionea, **37**: 15-24, 32-45, 80-91; 1938, ibidem, **38**: 51-58, 79-88, 95-102; 1938, Ent. Rdsch., **55**: 320-323, 331-333; 1938, ibidem, **55**: 349-354, 382-384; 1946, Rev. franç. Lépid., **10**: 250-261, 273-280; 1953, Ent. Z., **63**: 5. Hrubý, 1964, Prodromus Lepidopter Slovenska, p. 477. Kiefer, 1933, Int. ent. Z., **27**: 252-256; 1934, ibidem, **27**: 521-524. Koch, 1955, Wir Bestimmen Schmetterlinge, **2**: 60, 61, pl. 1, fig. 10, pl. 15, fig. 10. Millière, 1885, Bull. Soc. ent. Fr., p. xcii; 1886, Ann. Soc. ent. Fr., (6) **6**, pl. 1, figs. 3-7. Reiss, 1926, Die Zygaenen Deutschlands, p. 7; 1930, in Seitz, Die Gross-Schmetterlinge der Erde, Supplement, **2**: 19; 1953, Ent. Z., **63**: 92; 1958, Z. wien. ent. Ges., **43**: 158. Roüast, 1883, Catalogue des Chenilles européennes connues, p. 22. Sarlet, 1964, Mém. Soc. r. ent. Belg., **29**: 6. Spuler, 1910, in Hofmann, Die Raupen der Schmetterlinge Europas, pl. 9, fig. 21.

Subg. **Zygaena** Fabricius

Zygaena Fabricius, 1775, Systema Entomologiae, p. 550. Type-species: **Sphinx filipendulae** Linné, 1758, by subsequent designation, Latreille, 1810, Considérations générales, p. 441.

Anthrocera Scopoli, 1777, Introductio ad Historiam naturalem, **10**: 414. Type-species: **Sphinx filipendulae** Linné, 1758, by subsequent designation, Westwood, 1840, Synopsis of the Genera of British Insects, p. 89.

Eutychia Hübner, [1819], Verzeichniss bekannter Schmettlinge [sic], p. 117. Type-species: **Sphinx rhadamanthus** Esper, 1793, by subsequent designation, Tremewan, 1961, Ent. Rec., **73**: 202.

Anthilaria Hübner, [1819], Verzeichniss bekannter Schmettlinge [sic], p. 117. Type-species: **Sphinx lavandulae** Esper, 1783, by subsequent designation, Tremewan, 1961, Ent. Rec., **73**: 202.

Aeacis Hübner, [1819], Verzeichniss bekannter Schmettlinge [sic], p. 117. Type-species: **Sphinx ephialtes** Linné, 1767, by monotypy.

Thermophila Hübner, [1819], Verzeichniss bekannter Schmettlinge [sic], p. 117. Type-species: **Sphinx viciae** Denis & Schiffermüller, 1775, by subsequent designation, Holik & Sheljuzhko, 1957, Mitt. münch. ent. Ges., **47**: 144.

Silvicola Burgeff, 1926, in Strand, Lepid. Cat., **33**: 10. Type-species: **Zygaena chaos** Burgeff, 1926 (= **Zygaena mana** Kirby, 1892), by subsequent designation, Tremewan, 1961, Ent. Rec., **73**: 203.

Peristygia Burgeff, 1926, in Strand, Lepid. Cat., **33**: 25. Type-species: **Zygaena anthyllidis** Boisduval, 1828, by subsequent designation, Tremewan, 1961, Ent. Rec., **73**: 202.

Polymorpha Burgeff, 1926, in Strand, Lepid. Cat., **33**: 65 (preoccupied by **Polymorpha** Soldani, 1791, Testaceogr., **1** (2): 114 [Foraminifera]). Type-species: **Sphinx transalpina** Esper, 1782, by subsequent designation, Tremewan, 1961, Ent. Rec., **73**: 202.

Biezankoia Strand, 1936, Folia Zool. Hydrobiol., Riga, **9**: 167 (nomen novum for *Polymorpha* Burgeff). Type-species: **Sphinx transalpina** Esper, 1782, by subsequent designation, Tremewan, 1961, Ent. Rec., **73**: 202.

Libania Holik & Sheljuzhko, 1956, Mitt. münch. ent. Ges., **46**: 94. Type-species: **Zygaena graslini** Lederer, 1855, by original designation, Holik & Sheljuzhko, 1956, loc. cit.

Usgenta Holik & Sheljuzhko, 1956, Mitt. münch. ent. Ges., **46**: 237. Type-species: **Zygaena huguenini** Staudinger, 1887, by original designation, Holik & Sheljuzhko, 1956, loc. cit.

Huebneriana Holik & Sheljuzhko, 1957, Mitt. münch. ent. Ges., **47**: 144. Type-species: **Sphinx lonicerae** Scheven, 1777, by original designation, Holik & Sheljuzhko, 1957, loc. cit.

Burgeffia Holik & Sheljuzhko, 1958, Mitt. münch. ent. Ges., **48**: 229 (nomen novum for *Polymorpha* Burgeff). Type-species: **Sphinx transalpina** Esper, 1782, by subsequent designation, Tremewan, 1961, Ent. Rec., **73**: 202.

SECTION 1

huguenini Staudinger

Distribution: Fergana.

huguenini Staudinger, 1887, Stettin. ent. Ztg., **48**: 73. Seitz, — Usgent; Margelan; Fergana, Zentralasien.

1907, Die Gross-Schmetterlinge der Erde, **2**: 26, pl. 6i. Reiss, 1933, Int. ent. Z., **26**: 499, figs.; 1933, in Seitz, Die Gross-Schmetterlinge der Erde, Supplement, **2**: 276. Holik & Sheljuzhko, 1956, Mitt. münch. ent. Ges., **46**: 237. Alberti, 1958, Mitt. zool. Mus. Berl., **34**: 338.

Biology

Holik, 1937, Lambillionea, **37**: 15-24, 32-45, 80-91; 1953, Ent. Z., **63**: 25. Reiss, 1958, Z. wien. ent. Ges., **43**: 161.

SECTION 2

graslini Lederer

Distribution: Syria, Mesopotamia, Asia Minor.

ssp. **graslini** Lederer, 1855, Verh. zool.-bot. Ver. Wien, **5**: 197, pl. 2, figs. 3, 4. Staudinger, 1879, Horae Soc. ent. Ross., **14**: 323. Oberthür, 1896, Études d'Entomologie, **20**: 46, pl. 7, fig. 127. Seitz, 1907, Die Gross-Schmetterlinge der Erde, **2**: 25, pl. 6f. Reiss, 1930, in Seitz, Die Gross-Schmetterlinge der Erde, Supplement, **2**: 20, pl. 2f; 1933, ibidem, **2**: 259; 1932, Int. ent. Z., **26**: 271, figs. Haaf, 1952, Veröff. zool. Staatssamml. Münch., **2**: 152, 157, pl. 14. Holik & Sheljuzhko, 1956, Mitt. münch. ent. Ges., **46**: 93, 97. Alberti, 1958, Mitt. zool. Mus. Berl., **34**: 337. — Beirut, Libanon.

ab. **confluens** Oberthür, 1896, Études d'Entomologie, **20**: 46, pl. 7, fig. 128. Dziurzyński, 1903, Iris, **15**: 337; 1903, Jber. wien. ent. Ver., **13**: 40. Seitz, 1907, Die Gross-Schmetterlinge der Erde, **2**: 24, pl. 6f. Reiss, 1932, Int. ent. Z., **26**: 272; 1933, in Seitz, Die Gross-Schmetterlinge der Erde, Supplement, **2**: 259. Tremewan, 1961, Bull. Brit. Mus. (nat. Hist.) Ent., **10** (7): 271, pl. 52, fig. 33.

ssp. **pfeifferi** Reiss, 1932, Int. ent. Z., **26**: 273, figs.; 1933, in Seitz, Die Gross-Schmetterlinge der Erde, Supplement, **2**: 259, pl. 16i. Holik & Sheljuzhko, 1956, Mitt. münch. ent. Ges., **46**: 98. — Bscharre, Libanon, 1600-1850 m.

ssp. **kulzeri** Reiss, 1932, Int. ent. Z., **26**: 273, figs.; 1933, in Seitz, Die Gross-Schmetterlinge der Erde, Supplement, **2**: 259. Holik & Sheljuzhko, 1956, Mitt. münch. ent. Ges., **46**: 98. — Zebdani, Antilibanon, 1100 m.

ssp. **rebeliana** Reiss & Tremewan, 1964, Ent. Rec., **76**: 133 (nomen novum for *rebeli* Reiss). — Yüksek Dagh, Amanus Gebirge, Syrien.

rebeli Reiss, 1932, Int. ent. Z., **26**: 275, figs. (preoccupied by **rebeli** Drenowski, 1928, ssp. of **purpuralis** Brünnich); 1933, in Seitz, Die Gross-Schmetterlinge der Erde, Supplement, **2**: 260. Holik & Sheljuzhko, 1956, Mitt. münch. ent. Ges., **46**: 96. Reiss & Tremewan, 1964. Ent. Rec., **76**: 133.

ab. **sexmaculata** Reiss, 1932, Int. ent. Z., **26**: 274, fig.; 1933, in Seitz, Die Gross-Schmetterlinge der Erde, Supplement, **2**: 260.

ab. **confluens** Reiss, 1932, Int. ent. Z., **26**: 274; 1933, in Seitz, Die Gross-Schmetterlinge der Erde, Supplement, **2**: 260.

ssp. **maraschensis** Reiss, 1935, Int. ent. Z., **29**: 151. Holik & Sheljuzhko, 1956, Mitt. münch. ent. Ges., **46**: 95. — Büyük-Dere, Marasch, Kleinasien, 600-900 m.

ab. **pseudorebeli** Reiss, 1935, Int. ent. Z., **29**: 151.

Biology

Holik, 1937, Lambillionea, **37**: 15-24, 32-45, 80-91; 1938, Ent. Rdsch., **55**: 320-323, 331-333; 1953, Ent. Z., **63**: 15. Lederer, 1855, Verh. zool.-bot. Ver. Wien, **5**: 197, pl. 2, fig. 4. Reiss, 1933, in Seitz, Die Gross-Schmetterlinge der Erde, Supplement, **2**: 260; 1958, Z. wien. ent. Ges., **43**: 161.

SECTION 3

anthyllidis Boisduval

Distribution: Central Pyrenees.

anthyllidis Boisduval, 1828, Essai sur une Monographie des Zygénides, p. 78, pl. 4, fig. 8; 1829, Europaeorum Lepidopterorum Index Methodicus, Errata et Addenda, p. 3; 1834, Icones historique des Lépidoptères nouveaux ou peu connus, **2**: 69, pl. 55, fig. 7. Duponchel, 1835, in Godart & Duponchel, Histoire naturelle des Lépidoptères ou Papillons de France, Supplement, **2**: 76, pl. 7, fig. 1. Freyer, 1845, Neuere Beiträge zur Schmetterlingskunde, **5**: 27, pl. 398, fig. 3 (as *anthillidis*). Herrich-Schäffer, 1843, Systematische Bearbeitung der Schmetterlinge von Europa, **2**, pl. 1, fig. 4; 1846, ibidem, **2**: 40. Oberthür, 1884, Études d'Entomologie, **8**: 30, pl. 1, figs. 14-17. Chrétien, 1893, Naturaliste, **15**: 10. Caradja, 1894, Iris, **6**: 192. Spuler, 1906, in Hofmann, Die Schmetterlinge Europas, **2**: 157, pl. 75, fig. 47, pl. 77, fig. 12. Seitz, 1907, Die Gross-Schmetterlinge der Erde, **2**: 22, pl. 5e; 1912, ibidem, **2**: 442. Oberthür, 1910, Études de Lépidoptérologie comparée, **4**: 474. Reiss, 1930, in Seitz, Die Gross-Schmetterlinge der Erde, Supplement, **2**: 19, pl. 2e; 1933, ibidem, **2**: 259. Haaf, 1952, Veröff. zool. Staatssamml. Münch., **2**: 152, 157, pl. 14. Holik, 1953, NachrBl. bayer. Ent., **2**: 47. Alberti, 1958, Mitt. zool. Mus. Berl., **34**: 317. Tremewan, 1961, Bull. Brit. Mus. (nat. Hist.) Ent., **10** (7): 271, pl. 52, fig. 34, pl. 61, figs. 5, 6. Alberti, 1964, Dtsch. ent. Z. (N.F.), **11**: 386, figs. A3, B3, C3; 388, fig. 3 (distribution map); 391, fig. 1. — Environs de Barèges, Hautes-Pyrénées, France.

erebus Meigen, 1830, Systematische Beschreibung der Europäischen Schmetterlinge, **2**: 90, pl. 59, fig. 4.

ab. **conjuncta** Caradja, 1894, Iris, **6**: 192.
conjuncta Dziurzyński, 1908, Berl. ent. Z., **53**: 30. Seitz, 1912, Die Gross-Schmetterlinge der Erde, **2**: 442.
ab. **nconfluens** Le Charles, 1927, Encycl. ent., (B) 3, Lepidoptera, **2**: 151, pl. 9, fig. 8.

Biology

Burgeff, 1950, Portug. acta biol., (A) Goldschmidt: 663-728. Caradja, 1894, Iris, **6**: 192. Holik, 1937, Lambillionea, **37**: 15-24, 32-45, 80-91; 1946, Rev. franç. Lépid., **10**: 250-261, 273-280; 1953, Ent. Z., **63**: 14. Oberthür, 1884, Études d'Entomologie, **8**: 30, pl. 1, fig. 17. Reiss, 1958, Z. wien. ent. Ges., **43**: 161. Seitz, 1907, Die Gross-Schmetterlinge der Erde, **2**: 22.

SECTION 4

rhadamanthus Esper
Distribution: Portugal, Spain, southern France.
ssp. **algarbiensis** Christ, 1889, Mitt. schweiz. ent. Ges., **8**: 101. Oberthür, 1896, Études d'Entomologie, **20**: 47, pl. 7, fig. 107. Spuler, 1906, in Hofmann, Die Schmetterlinge Europas, **2**: 162. Seitz, 1907, Die Gross-Schmetterlinge der Erde, **2**: 26, pl. 6h. Reiss. 1933, in Seitz, Die Gross-Schmetterlinge der Erde, Supplement, **2**: 260; 1936, Ent. Rdsch., **54**: 92, fig. — Faro, Algarve, Südportugal.
roederi Staudinger, 1892, Iris, **4**: 247.
ssp. **alfacarensis** Reiss, 1922, Int. ent. Z., **15**: 176; 1930, in Seitz, Die Gross-Schmetterlinge der Erde, Supplement, **2**: 20, pl. 2f. Agenjo, 1952, Fáunula Lepidopterológica Almeriense, p. 159. Schmidt-Koehl, 1965, Ent. Z., **75**: 282. — Sierra de Alfacar, Granada, Südspanien.
ab. **flava** Oberthür, 1909, Études de Lépidoptérologie comparée, **3**, pl. 29, fig. 184; 1910, ibidem, **4**: 587. Seitz, 1912, Die Gross-Schmetterlinge der Erde, **2**: 443. Reiss, 1930, in Seitz, Die Gross-Schmetterlinge der Erde, Supplement, **2**: 20. Tremewan, 1961, Bull. Brit. Mus. (nat. Hist.) Ent., **10** (7): 272, pl. 53, fig. 3.
ssp. **caroniana** Reiss, 1965, Ent. Z.. **75**: 68, figs. — Villajoyosa, Alicante, Südostspanien, 150 m.
ab. **pseudoalfacarensis** Reiss, 1965, Ent. Z., **75**: 68.

ssp. **barcina** Verity, 1920, Ent. Rec., **32**: 161. Reiss, 1930, in Seitz, Die Gross-Schmetterlinge der Erde, Supplement, **2**: 20, pl. 2f. — Barcelona, Catalonia, Spain.
ab. **pseudorhadamanthus** Burgeff, 1914, Mitt. münch. ent. Ges., **5**: 60. Reiss, 1930, in Seitz, Die Gross-Schmetterlinge der Erde, Supplement, **2**: 20.

ab. **kiesenwetteri** Herrich-Schäffer, 1852, Systematische Bearbeitung der Schmetterlinge von Europa, **6**: 46. Oberthür, 1896, Études d'Entomologie, **20**: 47. Spuler, 1906, in Hofmann, Die Schmetterlinge Europas, **2**: 162. Seitz, 1907, Die Gross-Schmetterlinge der Erde, **2**: 26, pl. 6h. Oberthür, 1910, Études de Lépidoptérologie comparée, **4**: 590.

staechadis Boisduval, 1834, Icones historique des Lépidoptères nouveaux ou peu connus, **2**: 71, pl. 55, fig. 4 (preoccupied by **stoechadis** Borkhausen, 1793, ssp. of **filipendulae** Linné). Tremewan, 1961, Bull. Brit. Mus. (nat. Hist.) Ent., **10** (7): 273, pl. 53, fig. 4.

kiesenwetterii Herrich-Schäffer, 1851, Systematische Bearbeitung der Schmetterlinge von Europa, **2**, pl. 14, fig. 96 (non-binominal).

latecincta Navás, 1924, Mus. barcin. Sci. nat. Op., **4** (10): 38.

ab. **quinquemaculata** Burgeff, 1926, in Strand, Lepid. Cat., **33**: 28. Reiss, 1930, in Seitz, Die Gross-Schmetterlinge der Erde, Supplement, **2**: 20.

ab. **obscura** Reiss, 1930, in Seitz, Die Gross-Schmetterlinge der Erde, Supplement, **2**: 20, pl. 2g.

ssp. **aragonia** Tremewan, 1961, Ent. Rec., **73**: 4; 1962, ibidem. **74**: 127, pl. 2, figs. 6, 7. — Albarracin, Teruel, Aragon, Spain.

ssp. **gredosica** Reiss, 1936, Ent. Rdsch., **54**: 60, pl. 2, figs. — Cebreros, Sierra de Gredos, Spanien.

ssp. **rasura** Agenjo, 1948, Eos, Madr., **24**: 392. Tremewan & Manley, 1965, Ent. Rec., **77**: 8. — Villasur de Herreros, Burgos, España, 1040 m.

ab. **lambra** Agenjo, 1948, Eos, Madr., **24**: 394.

ssp. **manleyi** Tremewan, 1961, Ent. Rec., **73**: 4; 1963, ibidem, **75**: 6, pl. 1, figs. 9, 10. — La Pena, Huesca, Spain, 2400 ft.

ab. **acingulata** Tremewan, 1961, Ent. Rec., **73**: 4; 1963, ibidem, **75**: 6, pl. 1, fig. 11.

ssp. **pyrenaea** Verity, 1920, Ent. Rec., **32**: 161. Oberthür, 1910, Études de Lépidoptérologie comparée, **4**: 590. Reiss, 1930, in Seitz, Die Gross-Schmetterlinge der Erde, Supplement, **2**: 20. — La Traucada d'Ambouilla, Pyrénées-Orientales, France.

ssp. **azurea** Burgeff, 1914, Mitt. münch. ent. Ges., **5**: 60, pl. 2, fig. 158, pl. 6, figs. 75, 76. Oberthür, 1909, Études de Lépidoptérologie comparée, **3**, pl. 29, fig. 188; 1910, ibidem, **4**: 593. Reiss, 1930, in Seitz, Die Gross-Schmetterlinge der Erde, Supplement, **2**: 21. — Vence, Seealpen, Südfrankreich.

ab. **pseudogrisea** Reiss, 1953, Ent. Z., **63**: 78.

ab. **apicaliconfluens** Reiss, 1953, Ent. Z., **63**: 78.

ab. **confluens** Reiss, 1953, Ent. Z., **63**: 78.

ab. **pseudostygia** Burgeff, 1914, Mitt. münch. ent. Ges., **5**: 60. Reiss, 1930, in Seitz, Die Gross-Schmetterlinge der Erde, Supplement, **2**: 21.

ssp. **oxytropiferens** Verity, 1920, Ent. Rec., **32**: 161. Oberthür, 1910, Études de Lépidoptérologie comparée, **4**: 592.

Castillon near Sospel, Alpes-Maritimes, France.

ab. **albovittata** Verity, 1920, Ent. Rec., **32**: 161. Reiss, 1930, in Seitz, Die Gross-Schmetterlinge der Erde, Supplement, **2**: 21.

ab. **pseudostygia** Reiss, 1953, Ent. Z., **63**: 102.

ssp. **mesolopha** Dujardin, 1956, Bull. mens. Soc. linn. Lyon, **25**: 262.

Les Vallettes près de Tourrettes-sur-Loup, Alpes-Maritimes, France.

ssp. **ochsi** Dujardin, 1956, Bull. mens. Soc. linn. Lyon, **25**: 262.

Collongues, Vallée de l'Esteron, Alpes-Maritimes, France, 700 m.

ssp. **stygia** Burgeff, 1914, Mitt. münch. ent. Ges., **5**: 60, pl. 2, fig. 166, pl. 6, figs. 77, 78. Reiss, 1930, in Seitz, Die Gross-Schmetterlinge der Erde, Supplement, **2**: 21, pl. 2g. Burgeff, 1950, Portug. acta biol., (A) Goldschmidt: 673, figs. 2a; 1951, Biol. Zbl., **70**: 10, figs. 7a. Loritz, 1961, Bull. Soc. ent. Mulhouse, p. 83-102, fig.; 1962, ibidem, p. 17-26, figs. Burgeff, 1965, Nachr. Akad. Wiss. Göttingen, **2**, mat.-phys. Kl., no. 1: 8, figs. 4a.

Bordighera; San Remo, usw., Litoral der Seealpen. Italienische und französische Riviera.

ab. **pseudoazurea** Burgeff, 1914, Mitt. münch. ent. Ges., **5**: 60. Reiss, 1930, in Seitz, Die Gross-Schmetterlinge der Erde, Supplement, **2**: 21.

ab. **quinquemaculata** Oberthür, 1910, Études de Lépidoptérologie comparée, **4**: 595; 1909, ibidem, **3**, pl. 29, fig. 183 (as *kiesenwetteri* Herrich-Schäffer). Reiss, 1930, in Seitz, Die Gross-Schmetterlinge der Erde, Supplement, **2**: 21. Tremewan, 1961, Bull. Brit. Mus. (nat. Hist.) Ent., **10** (7): 271, pl. 52, fig. 35.

ab. **obscura** Oberthür, 1910, Études de Lépidoptérologie comparée, **4**: 595; 1909, ibidem, **3**, pl. 29, fig. 181 (as *kiesenwetteri* Herrich-Schäffer). Reiss, 1930, in Seitz, Die Gross-Schmetterlinge der Erde, Supplement, **2**: 21. Tremewan, 1961, Bull. Brit. Mus. (nat. Hist.) Ent., **10** (7): 271, pl. 52, fig. 36.

ssp. **azureoides** Reiss, 1953, Ent. Z., **63**: 102.

Col de l'Ablé, Seealpen, Frankreich, 1200 m.

ssp. **grisea** Oberthür, 1909, Études de Lépidoptérologie comparée, 3, pl. 29, fig. 187; 1910, ibidem, **4**: 587, 591. Reiss, 1930, in Seitz, Die Gross-Schmetterlinge der Erde, Supplement, **2**: 21, pl. 2g. Haaf, 1952, Veröff. zool. Staatssamml. Münch., **2**: 152, 154, 156, pl. 10 (as *rhadamanthus* Esper). Tremewan, 1961, Bull. Brit. Mus. (nat. Hist.) Ent., **10** (7): 272, pl. 53, fig. 1. Dufay, 1966, Bull. mens. Soc. linn. Lyon, **35**: 69. — Digne, Basses-Alpes, France.

cingulata Oberthür, 1909, Études de Lépidoptérologie comparée, 3, pl. 29, fig. 187. Tremewan, 1961, Bull. Brit. Mus. (nat. Hist.) Ent., **10** (7): 272.

ab. **minicolor** Schawerda, 1929, Z. öst. EntVer., **14**: 120.

ab. **pseudokiesenwetteri** Burgeff, 1926, in Strand, Lepid. Cat., **33**: 28.

ab. **gueneei** Oberthür, 1909, Études de Lépidoptérologie comparée, 3, pl. 29, figs. 185, 186 (with reference to Guenée, 1870, Ann. Soc. ent. Fr., (4) **10**: 20, pl. 7, fig. 12); 1910, ibidem, **4**: 587. Seitz, 1912, Die Gross-Schmetterlinge der Erde, **2**: 443. Reiss, 1930, in Seitz, Die Gross-Schmetterlinge der Erde, Supplement, **2**: 20, 21. Tremewan, 1961, Bull. Brit. Mus. (nat. Hist.) Ent., **10** (7): 272, pl. 53, fig. 2.

confluens Heinrich, 1923, Dtsch. ent. Z., (Beitrag), p. 120.

ab. **subconfluens** Dufrane, 1936, Bull. (Ann.) Soc. ent. Belg., **76**: 124.

ssp. **cleui** Dujardin, 1956, Bull. mens. Soc. linn. Lyon, **25**: 261. — St. Privat, Ardèche, France.

ssp. **rhadamanthus** Esper, 1793, Die Schmetterlinge, Supplement, **2** (2): 13, pl. 40, figs. 1, 2. Hübner, 1796, Sammlung europäischer Schmetterlinge, **2**: 13, pl. 4, fig. 23; 1806, ibidem, Der Ziefer, p. 79. Ochsenheimer, 1808, Die Schmetterlinge von Europa, **2**: 86. Boisduval, 1828, Essai sur une Monographie des Zygénides, p. 91, pl. 5, fig. 8. Herrich-Schäffer, 1844, Systematische Bearbeitung der Schmetterlinge von Europa, **2**, pl. 3, figs. 21, 22; 1846, ibidem, **2**: 43. Oberthür, 1896, Études d'Entomologie, **20**: 47. Spuler, 1906, in Hofmann, Die Schmetterlinge Europas, **2**: 162, pl. 72, fig. 8, pl. 77, fig. 25. Seitz, 1907, Die Gross-Schmetterlinge der Erde, **2**: 26, pl. 6g. Oberthür, 1910, Études de Lépidoptérologie comparée, **4**: 586. Verity, 1920, Ent. Rec., **32**: 158. Reiss, 1930, in Seitz, Die Gross-Schmetterlinge der Erde, Suppl., **2**: 20. Alberti, 1958, Mitt. zool. Mus. Berl., **34**: 312. — Nîmes, Languedoc Südfrankreich.

ab. **cingulata** Lederer, 1852, Verh. zool.-bot. Ver. Wien, **2**: 72. Seitz, 1907, Die Gross-Schmetterlinge der Erde, **2**: 26, pl. 6h.

ab. **confluens** Oberthür, 1910, Études de Lépidoptérologie comparée, **4**: 592.

Biology

Abeille, 1909, Mém. Soc. linn. Provence, **1**: 4-6. Boisduval, Rambur & Graslin, 1832, Collection iconographique et historique des Chenilles, pl. 4, figs. 1-3. Burgeff, 1950, Portug. acta biol., (A) Goldschmidt: 663-728; 1951, Biol. Zbl., **70**: 1-23. Holik, 1937, Lambillionea, **37**: 15-24, 32-45, 80-91; 1938, ibidem, **38**: 51-58, 79-88, 95-102; 1938, Ent. Rdsch., **55**: 320-323, 331-333; 1946, Rev. franç. Lépid., **10**: 250-261, 273-280; 1953, Ent. Z., **63**: 14. Reiss, 1958, Z. wien. ent. Ges., **43**: 161. Roüast, 1883, Catalogue des Chenilles européennes connues, p. 22. Seitz, 1907, Die Gross-Schmetterlinge der Erde, **2**: 26. Spuler, 1910, in Hofmann, Die Raupen der Schmetterlinge Europas, pl. 10, figs. 5a, b, pl. 50, fig. 24.

oxytropis Boisduval

Distribution: Sicily, Italy, north to Liguria and Piedmont.

ssp. **quercii** Verity, 1920, Ent. Rec., **32**: 160. Reiss, 1930, in Seitz, Die Gross-Schmetterlinge der Erde, Supplement, **2**: 20. — Mt. Cuccitiello near S. Martino delle Scale, Palermo, Sicily, 2000 ft.

insulicola Stauder, 1928, Lepid. Rdsch., **2**: 77. Reiss, 1930, in Seitz, Die Gross-Schmetterlinge der Erde, Supplement, **2**: 20. Tremewan, 1962, Ent. Rec., **74**: 127, pl. 2, fig. 8.

ssp. **bruttiensis** Dujardin, 1964, Entomops, Nice, no. 1: 21, fig. — Lago Ampollino (massif de la Sila), Calabria, Italie Sud, 1200 m.

ssp. **laterubra** Verity, 1920, Ent. Rec., **32**: 160; 1920, Boll. Lab. Zool. Portici, **14**: 40. Reiss, 1930, in Seitz, Die Gross-Schmetterlinge der Erde, Supplement, **2**: 20. Haaf, 1952, Veröff. zool. Staatssamml. Münch., **2**: 152, 154, 156, pl. 10 (as *oxytropis* Boisduval). — Villalatina, Vallegrande, Mainarde Mts., Caserta, S. Italy, 1500 ft.

ab. **posticeflaveola** Stauder, 1922, Iris, **36**: 43.

ab. **octonotata** Burgeff, 1926, Mitt. münch. ent. Ges., **16**: 39. Reiss, 1930, in Seitz, Die Gross-Schmetterlinge der Erde, Supplement, **2**: 20.

ab. **reissi** Stauder, 1922, Iris, **36**: 43. Reiss, 1930, in Seitz, Die Gross-Schmetterlinge der Erde, Supplement, **2**: 20.

ab. **corsioides** Burgeff, 1926, Mitt. münch. ent. Ges., **16**: 39. Reiss, 1930, in Seitz, Die Gross-Schmetterlinge der Erde, Supplement, **2**: 20.

ab. **phlebomelas** Stauder, 1922, Iris, **36**: 43. Reiss, 1930, in Seitz, Die Gross-Schmetterlinge der Erde, Supplement, **2**: 20.

ab. **irregularis** Stauder, 1922, Iris, **36**: 43. Reiss, 1930, in Seitz, Die Gross-Schmetterlinge der Erde, Supplement, **2**: 20.

ab. **rubescens** Burgeff, 1926, Mitt. münch. ent. Ges., **16**: 39.

Reiss, 1930, in Seitz, Die Gross-Schmetterlinge der Erde, Supplement, **2**: 20.

ab. **garibaldina** Stauder, 1922, Iris, **36**: 43. Reiss, 1930, in Seitz, Die Gross-Schmetterlinge der Erde, Supplement, **2**: 20.

ssp. **sibyllina** Verity, 1916, Boll. Soc. ent. ital., **47**: 77. Rocci, 1918, Atti Soc. ligust. Sci. nat. geogr., **28**: 153, pl. 3, figs. 10d, 11a, b, c. Verity, 1920, Ent. Rec., **32**: 160. Reiss, 1930, in Seitz, Die Gross-Schmetterlinge der Erde, Supplement, **2**: 20, pl. 2f. — Bolognola, Sibillini Mts., Marche, Italia, 2700 ft.

ab. **tricingulata** Verity, 1916, Boll. Soc. ent. ital., **47**: 77. Reiss, 1930, in Seitz, Die Gross-Schmetterlinge der Erde, Supplement, **2**: 20.

ab. **divisa** Verity, 1916, Boll. Soc. ent. ital., **47**: 77.

ab. **gueneeiformis** Verity, 1916, Boll. Soc. ent. ital., **47**: 78.

ssp. **lucania** Dujardin, 1964, Entomops, Nice, no. 1: 22, fig. — Taverna (Ruoti), Lucania, Italie Sud, 1000 m.

ssp. **blanda** Dujardin, 1964, Entomops, Nice, no. 1: 21, fig. — Lugagnano Val d'Arda, Emilia, Italie, ca. 300 m.

ssp. **pumila** Verity, 1920, Ent. Rec., **32**: 160. Reiss, 1930, in Seitz, Die Gross-Schmetterlinge der Erde, Supplement, **2**: 20. — Futa Pass, Traversa, Tuscany, Italy, 2700 ft.

ssp. **oxytropis** Boisduval, 1828, Essai sur une Monographie des Zygénides, p. 89, pl. 5, fig. 7. Freyer, 1833, Neuere Beiträge zur Schmetterlingskunde, **1**: 28, pl. 14, fig. 2. Boisduval, 1834, Icones historique des Lépidoptères nouveaux ou peu connus, **2**: 70, pl. 55, fig. 3. Duponchel, 1835, in Godart & Duponchel, Histoire naturelle des Lépidoptères ou Papillons de France, Supplement, **2**: 80, pl. 7, fig. 3. Herrich-Schäffer, 1844, Systematische Bearbeitung der Schmetterlinge von Europa, **2**, pl. 3, figs. 19, 20; 1846, ibidem, **2**: 43. Perlini, 1905, Forme di Lepidotteri esclusivamente Italiane, p. 55, pl. 2, fig. 13. Spuler, 1906, in Hofmann, Die Schmetterlinge Europas, **2**: 163, pl. 72, fig. 9, pl. 77, fig. 26. Seitz, 1907, Die Gross-Schmetterlinge der Erde, **2**: 25, pl. 6f. Oberthür, 1910, Études de Lépidoptérologie comparée, **4**: 595. Querci, 1912, in Oberthür, Études de Lépidoptérologie comparée, **6**: 154, 174. Verity, 1920, Ent. Rec., **32**: 158. Reiss, 1930, in Seitz, Die Gross-Schmetterlinge der Erde, Supplement, **2**: 19, pl. 2e. Burgeff, 1950, Portug. acta biol., (A) Goldschmidt: 673, figs. 2d; 1951, Biol. Zbl., **70**: 10, figs. 7d. Alberti, 1958, Mitt. zool. Mus. Berl., **34**: 312. Tremewan, 1961, Bull. Brit. — M. Passerini aux environs de Florence, Toscane, Italie.

Mus. (nat. Hist.) Ent., **10** (7): 273, pl. 53, fig. 5, pl. 62, figs. 1, 2.

ab. **minima** Rocci, 1914, Atti Soc. ligust. Sci. nat. geogr., **25**: 223; 1918, ibidem, **28**: 153, pl. 3, fig. 11d.

ab. **cingulata** Zickert, 1905, Ent. Z., **19**: 117. Spuler, 1906, in Hofmann, Die Schmetterlinge Europas, **2**: 163. Seitz, 1907, Die Gross-Schmetterlinge der Erde, **2**: 25.

ab. **conspicua** Rocci, 1916, Atti Soc. ligust. Sci. nat. geogr., **27**: 31; 1918, ibidem, **28**: 151, pl. 3, figs. 9c, 9d, 10a.

ab. **separata** Rocci, 1914, Atti Soc. ligust. Sci. nat. geogr., **25**: 222; 1918, ibidem, **28**: 152.

ab. **corsicoides** Stauder, 1915, Z. wiss. InsektBiol., **11**: 137. Reiss, 1930, in Seitz, Die Gross-Schmetterlinge der Erde, Supplement, **2**: 19.

ab. **disjuncta** Rocci, 1914, Atti Soc. ligust. Sci. nat. geogr., **25**: 222; 1918, ibidem, **28**: 152, pl. 3, fig. 11e.

ab. **coniuncta** Spuler, 1906, in Hofmann, Die Schmetterlinge Europas, **2**: 163. Seitz, 1912, Die Gross-Schmetterlinge der Erde, **2**: 443. Reiss, 1930, in Seitz, Die Gross-Schmetterlinge der Erde, Supplement, **2**: 19.

ab. **confluens** Zickert, 1905, Ent. Z., **19**: 117. Seitz, 1907, Die Gross-Schmetterlinge der Erde, **2**: 25. Dziurzyński, 1909, Jber. wien. ent. Ver., **19**: 135, pl. 1, fig. 4.

ab. **unita** Rocci, 1914, Atti Soc. ligust. Sci. nat. geogr., **25**: 222; 1918, ibidem, **28**: 153.

ab. **ruberrima** Stauder, 1915, Z. wiss. InsektBiol., **11**: 137. Reiss, 1930, in Seitz, Die Gross-Schmetterlinge der Erde, Supplement, **2**: 19.

ab. **lampadouche** Burgeff, 1914, Mitt. münch. ent. Ges., **5**: 60, pl. 2, fig. 175, pl. 6, fig. 80. Reiss, 1930, in Seitz, Die Gross-Schmetterlinge der Erde, Supplement, **2**: 20.

ab. **rosea** Rocci, 1914, Atti Soc. ligust. Sci. nat. geogr., **25**: 222; 1918, ibidem, **28**: 153.

ab. **aurantiaca** Rocci, 1914, Atti Soc. ligust. Sci. nat. geogr., **25**: 222; 1918, ibidem, **28**: 153.

ssp. **acticola** Burgeff, 1926, Mitt. münch. ent. Ges., **16**: 39. Reiss, 1930, in Seitz, Die Gross-Schmetterlinge der Erde, Supplement, **2**: 20, pl. 2f. Burgeff, 1950, Portug. acta biol., (A) Goldschmidt: 673, figs. 2c; 1951, Biol. Zbl., **70**: 10, figs. 7c; 1965, Nachr. Akad. Wiss. Göttingen, **2**, mat.-phys. Kl., no. 1: 8, figs. 4c.

Alassio; Laigueglia; Porto Maurizio; Italienische Riviera.

ab. **inopinata** Burgeff, 1926, Mitt. münch. ent. Ges., **16**: 40. Reiss, 1930, in Seitz, Die Gross-Schmetterlinge der Erde, Supplement, **2**: 20.

Biology

Arnold, 1919, Mitt. münch. ent. Ges., **9**: 35, figs. Burgeff, 1950, Portug. acta biol., (A) Goldschmidt: 663-728; 1951, Biol. Zbl., **70**: 1-23; 1965, Nachr. Akad. Wiss. Göttingen, **2**, mat.-phys. Kl., no. 1: 8, figs. 4b (**Z. oxytropis acticola** Burgeff ♂ X **Z. rhadamanthus stygia** Burgeff ♀). Holik, 1937, Lambillionea, **37**: 15-24, 32-45, 80-91; 1953, Ent. Z., **63**: 14. Mann, 1859, Wien. ent. Monatschr., **3**: 92. Reiss, 1930, in Seitz, Die Gross-Schmetterlinge der Erde, Supplement, **2**: 20; 1958, Z. wien. ent. Ges., **43**: 161.

theryi de Joannis

Distribution: Algeria.

ssp. **theryi** de Joannis, 1908, Bull. Soc. ent. Fr., p. 203. Seitz, 1912, Die Gross-Schmetterlinge der Erde, **2**: 443. Rothschild, 1917, Novit. zool., **24**: 342. Reiss, 1930, in Seitz, Die Gross-Schmetterlinge der Erde, Supplement, **2**: 21; 1933, ibidem, **2**: 260. Le Charles, 1936, Bull. Soc. ent. Fr., **41**: 48, pl. 1, fig. a. — Philippeville, Constantine, Algérie.

ssp. **nisseni** Rothschild, 1908, Entomologist, **41**: 185; 1917, Novit. zool., **24**: 342, pl. 10, figs. (as *theryi* de Joannis). Reiss, 1930, in Seitz, Die Gross-Schmetterlinge der Erde, Supplement, **2**: 21, pl. 2h (as *theryi* de Joannis); 1937, Ent. Rdsch., **54**: 455, figs. a3, d2 (as *theryi* de Joannis); 1944, Z. wien. ent. Ges., **29**: 189, pl. 35, fig. 23 (as *theryi* de Joannis). Burgeff, 1950, Portug. acta biol., (A) Goldschmidt: 670, figs. b1 (as *theryi* de Joannis); 1951, Biol. Zbl., **70**: 4, figs. 2a (as *theryi* de Joannis). Haaf, 1952, Veröff. zool. Staatssamml. Münch., **2**: 152, 154, 156, pl. 10 (as *theryi* de Joannis). Alberti, 1958, Mitt. zool. Mus. Berl., **34**: 319 (as *theryi* de Joannis). Tremewan, 1961, Bull. Brit. Mus. (nat. Hist.) Ent., **10** (7): 273, pl. 53, fig. 6, pl. 62, figs. 3, 4. — Hamman R'Irha, Algeria.

ssp. **mercyi** Le Charles, 1936, Bull. Soc. ent. Fr., **41**: 48, pl. 1, figs. b, c. — Aïn N'sour, Alger, Algérie, 1000 m.

ab. **trimacula** Le Charles, 1936, Bull. Soc. ent. Fr., **41**: 48, pl. 1, figs. d.

Biology

Burgeff, 1926, Mitt. münch. ent. Ges., **16**: 41; 1950, Portug. acta biol., (A) Goldschmidt: 663-728; 1951, Biol. Zbl., **70**: 1-23. Holik, 1937, Lambillionea, **37**: 15-24, 32-45, 80-91; 1938, Ent. Rdsch., **55**: 320-323, 331-333; 1946, Rev. franç. Lépid., **10**: 250-261, 273-280; 1953, Ent. Z., **63**: 15. Reiss, 1930, in Seitz, Die Gross-Schmetterlinge der Erde, Supplement, **2**: 21; 1958, Z. wien. ent. Ges., **43**: 162. Rothschild, 1917, Novit. zool., **24**: 342, pl. 10.

lavandulae Esper

Distribution: Morocco, Portugal, Spain, southern France to Liguria.

ssp. **michaellae** Rungs & Le Charles, 1943, Bull. Soc. ent. Fr., **48**: 47. Reiss, 1944, Z. wien. ent. Ges., **29**: 72, pl. 35, fig. 24. — Près d'Ifrane, Moyen Atlas, Maroc, 1650 m.

ssp. **izilanica** Reiss, 1944, Z. wien. ent. Ges., **29**: 73, pl. 37, fig. 64. Reisser, 1933, Eos, Madr., **9**: 281 (as *theryi* de Joannis). — Bei Xauen, Rif, Marokko.

ssp. **espunnensis** Reiss, 1922, Int. ent. Z., **15**: 176; 1930, in Seitz, Die Gross-Schmetterlinge der Erde, Supplement, **2**: 21, pl. 2g, h; 1936, Ent. Rdsch., **54**: 92, pl. 2, figs. Agenjo, 1952, Fáunula Lepidopterológica Almeriense, p. 160. — Oberhalb Totana, Sierra de Espuña, Murcia, Südostspanien.

ab. **pseudolavandulae** Reiss, 1922, Int. ent. Z., **15**: 176; 1930, in Seitz, Die Gross-Schmetterlinge der Erde, Suppl., **2**: 21.

ab. **eradiata** Burgeff, 1926, Mitt. münch. ent. Ges., **16**: 40. Reiss, 1930, in Seitz, Die Gross-Schmetterlinge der Erde, Supplement, **2**: 21.

ab. **rubricosta** Reiss, 1963, Stuttgart. Beitr. Naturk., nr. 122: 2.

ssp. **alfacarica** Tremewan, 1961, Ent. Rec., **73**: 5; 1963, ibidem, **75**: 6, pl. 1, figs. 12, 13. — Granada, Sierra de Alfacar, Spain, 3600 ft.

ab. **pseudoespunnensis** Tremewan, 1961, Ent. Rec., **73**: 6; 1963, ibidem, **75**: 6, pl. 1, fig. 14.

ssp. **teruelensis** Reiss, 1936, Ent. Rdsch., **54**: 72, pl. 2, figs. — Umgebung von Albarracin, Teruel, Spanien.

ssp. **oropesica** Reiss, 1965, Ent. Rec., **77**: 86, pl. 1, figs. 3, 4. — Oropesa, Castellon, Spain, 80 m.

ab. **rubricosta** Reiss, 1965, Ent. Rec., **77**: 86.

ssp. **barcelonica** Reiss, 1936, Ent. Rdsch., **54**: 71, pl. 2, figs. Oberthür, 1896, Études d'Entomologie, **20**: 47, pl. 7, fig. 109. Reiss, 1933, in Seitz, Die Gross-Schmetterlinge der Erde, Supplement, **2**: 260 (as *lavandulae* Esper). — Küste bei Barcelona, Catalonien, Spanien.

ab. **octornata** Reiss, 1936, Ent. Rdsch., **54**: 71.

ssp. **huescae** Tremewan, 1963, Ent. Rec., **75**: 6, pl. 1, figs. 15, 16. — Puerto de Santa Barbara, Huesca, Spain, 3300 ft.

ssp. **lecharlesi** Bernardi & Viette, 1959, Entomologiste, **15**: 5. — Villepassans, Aude, France.

ssp. **lavandulae** Esper, 1783, Die Schmetterlinge, **2**: 221, pl. 34, fig. 2. De Villers, 1789, Caroli Linnaei Entomologia, **2**: 114, pl. 4, fig. 20. Ochsenheimer, 1808, Die Schmetterlinge von Europa, **2**: 84. Boisduval, 1828, Essai sur une Monographie — Languedoc, Frankreich [Montpellier].

des Zygénides, p. 83, pl. 5, fig. 4. Herrich-Schäffer, 1844, Systematische Bearbeitung der Schmetterlinge von Europa, **2**, pl. 5, figs. 41, 42; 1846, ibidem, **2**: 48. Spuler, 1906, in Hofmann, Die Schmetterlinge Europas, **2**: 162, pl. 77, fig. 24. Seitz, 1907, Die Gross-Schmetterlinge der Erde, **2**: 25, pl. 6g. Oberthür, 1910, Études de Lépidoptérologie comparée, **4**: 583. Burgeff, 1950, Portug. acta biol., (A) Goldschmidt: 670, figs. b2; 1951, Biol. Zbl., **70**: 4, figs. 2a. Alberti, 1958, Mitt. zool. Mus. Berl., **34**: 319.

lavandulae Fabricius, 1793, Entomologia Systematica, **3** (1): 387. Coquebert, 1799, Illustratio iconographica Insectorum, **1**: 27, pl. 7, figs. 2A, B, C. Zimsen, 1964, The Type Material of I. C. Fabricius, p. 566.

spicae Hübner, 1796, Sammlung europäischer Schmetterlinge, **2**: 17, pl. 4, fig. 25; 1806, ibidem, Der Ziefer, p. 79.

ab. **nigra** Dziurzyński, 1910, Int. ent. Z., **4**: 195.

ab. **pseudoconsobrina** Burgeff, 1926, in Strand, Lepid. Cat., **33**: 29. Reiss, 1930, in Seitz, Die Gross-Schmetterlinge der Erde, Supplement, **2**: 21.

ab. **powelli** Oberthür, 1910, Études de Lépidoptérologie comparée, **4**: 586. Reiss, 1930, in Seitz, Die Gross-Schmetterlinge der Erde, Supplement, **2**: 21. Tremewan, 1961, Bull. Brit. Mus. (nat. Hist.) Ent., **10** (7): 274, pl. 53, fig. 8.

ab. **lutescens** Cockerell, 1889, Entomologist, **22**: 128. Warburg, 1888, Entomologist, **21**: 211. Tremewan, 1966, Ent. Rec., **78**: 33, pl. 1, fig. 5.

ssp. **altalavandulae** Reiss, 1953, Ent. Z., **63**: 102. Loritz, 1961, Bull. Soc. ent. Mulhouse, p. 83-102, fig. — Col de Braus, Seealpen, Frankreich, 1000 m.

ssp. **genoata** Storace, 1956, Boll. Soc. ent. ital., **86**: 140, figs. 5-9. — Genova, Chiappeto, Liguria, Italia.

ab. **staliana** Storace, 1956, Boll. Soc. ent. ital., **86**: 140; 1951, ibidem, **81**: 13. Tremewan, 1965, Ent. Rec., **77**: 88.

ssp. **consobrina** Germar, 1831/35, Fauna Insectorum Europae, **16**, pl. 23. Herrich-Schäffer, 1844, Systematische Bearbeitung der Schmetterlinge von Europa, **2**, pl. 5, fig. 43; 1846, ibidem, **2**: 48. Oberthür, 1896, Études d'Entomologie, **20**: 47, pl. 7, fig. 108. Spuler, 1906, in Hofmann, Die Schmetterlinge Europas, **2**: 162, pl. 72, fig. 7. Seitz, 1907, Die Gross-Schmetterlinge der Erde, **2**: 26, pl. 6g. Reiss, 1930, in Seitz, Die Gross-Schmetterlinge der Erde, Supplement, **2**: 21, pl. 2g. Haaf, 1952, Veröff. zool. Staatssamml. Münch., **2**: 152, 154, 156, pl. 10 (as *lavandulae* Esper). Dufay, 1966, Bull. mens. Soc. linn. Lyon, **35**: 71. — Marseille; Embouchement du Rhône; Hyères; Digne; Südfrankreich.

ab. **quadripuncta** Burgeff, 1926, Mitt. münch. ent. Ges., **16**: 40.

Reiss, 1930, in Seitz, Die Gross-Schmetterlinge der Erde, Supplement, **2**: 21.

ab. **siepii** Oberthür, 1909, Études de Lépidoptérologie comparée, **3**, pl. 28, fig. 162; 1910, ibidem, **4**: 585. Seitz, 1912, Die Gross-Schmetterlinge der Erde, **2**: 442 (as *stoechadis* Borkhausen ab.). Reiss, 1930, in Seitz, Die Gross-Schmetterlinge der Erde, Supplement, **2**: 21. Tremewan, 1961, Bull. Brit. Mus. (nat. Hist.) Ent., **10** (7): 274, pl. 53, fig. 7.

ab. **flavobscura** Le Moult, 1947, Misc. ent., **44**: 64.

Biology

Abeille, 1909, Mém. Soc. linn. Provence, **1**: 1-4. Burgeff, 1950, Portug. acta biol., (A) Goldschmidt: 663-728; 1951, Biol. Zbl., **70**: 1-23. Holik, 1937, Lambillionea, **37**: 15-24, 32-45, 80-91; 1938, ibidem, **38**: 51-58, 79-88, 95-102; 1938, Ent. Rdsch., **55**: 320-323, 331-333; 1946, Rev. franç. Lépid., **10**: 250-261, 273-280; 1953, Ent. Z., **63**: 15. Millière, 1860, Iconographie et Description de Chenilles et Lépidoptères, **1**: 116, pl. 1, figs. 4-7. Reiss, 1958, Z. wien. ent. Ges., **43**: 161. Ribbe, 1909/12, Iris, **23**: 357. Roüast, 1883, Catalogue des Chenilles européennes connues, p. 22. Seitz, 1907, Die Gross-Schmetterlinge der Erde, **2**: 26. Spuler, 1910, in Hofmann, Die Raupen der Schmetterlinge Europas, pl. 10, figs. 4a, b. Tutt, 1898, Ent. Rec., **10**: 91.

SECTION 5

wyatti Reiss & Schulte

Distribution: Afghanistan.

wyatti Reiss & Schulte, 1961, Ent. Z., **71**: 56, figs. 1-4; 1964, ibidem, **74**: 162, figs. 11, 12. — Panjao, West Koh-i-Baba Mts., Afghanistan, 9300 ft.

araratica Staudinger

Distribution: Armenia, Transcaucasia.

araratica Staudinger, 1871, in Staudinger & Wocke, Catalog der Lepidopteren des Europaeischen Faunengebiets, p. 48. Spuler, 1906, in Hofmann, Die Schmetterlinge Europas, **2**: 161. Seitz, 1907, Die Gross-Schmetterlinge der Erde, **2**: 24, pl. 5k. Reiss, 1933, Int. ent. Z., **26**: 503, figs.; 1933, in Seitz, Die Gross-Schmetterlinge der Erde, Supplement, **2**: 278. Holik & Sheljuzhko, 1958, Mitt. münch. ent. Ges., **48**: 249. Alberti, 1958, Mitt. zool. Mus. Berl., **34**: 321. — Ararat, Armenien; Transkaukasien.

ab. **cingulata** Holik & Sheljuzhko, 1958, Mitt. münch. ent. Ges., **48**: 252.

ab. **quinquemaculata** Holik & Sheljuzhko, 1958, Mitt. münch. ent. Ges., **48**: 252.

ab. **flava** Spuler, 1906, in Hofmann, Die Schmetterlinge Europas, **2**: 161. Dziurzyński, 1904, Jber. wien. ent. Ver., **14**: 51 pl. 2, fig. 13 (as *araratica* Staudinger). Sterzl, 1931, Z. Ver. NatBeob., Wien, **6**: 18, pl. 16, fig. 16 (as *araratica* Staudinger).

lederiana Burgeff, 1914, Mitt. münch. ent. Ges., **5**: 61. Lederer, 1870, Bull. (Ann.) Soc. ent. Belg., **13**: 29, pl. 1, fig. 7 (as *stoechadis* Borkhausen var.). Burgeff, 1926, in Strand, Lepid. Cat., **33**: 64 (as *ledereriana*). Reiss, 1930, in Seitz, Die Gross-Schmetterlinge der Erde, Supplement, **2**: 38 (as *lonicerae kindermanni* Oberthür ab.). Holik & Sheljuzhko, 1958, Mitt. münch. ent. Ges., **48**: 251.

ab. **flavaquinquemaculata** Holik & Sheljuzhko, 1958, Mitt. münch. ent. Ges., **48**: 252.

Biology

Holik, 1937, Lambillionea, **37**: 15-24, 32-45, 80-91; 1946, Rev. franç. Lépid., **10**: 250-261, 273-280; 1953, Ent. Z., **63**: 30.

dorycnii Ochsenheimer

Distribution: Northern Iran, Armenia, Asia Minor, Transcaucasia, Caucasus, Crimea.

ssp. **keredjensis** Reiss, 1937, Ent. Rdsch., **55**: 40, figs. b2, b3. Holik & Sheljuzhko, 1958, Mitt. münch. ent. Ges., **48**: 267. — Keredj, Elburs Gebirge, Iran, 1600 m.

ssp. **hasankifensis** Reiss, 1938, Mitt. münch. ent. Ges., **27**: 169. Holik & Sheljuzhko, 1958, Mitt. münch. ent. Ges., **48**: 268. — Hasankif, Särdabtal, Tacht i Suleiman, Elburs Gebirge, Iran, 1000-1400 m.

ssp. **kertshensis** Obraztsov, 1935, Ent. Z., **49**: 54 (nomen novum for *crimea* Burgeff). Holik & Sheljuzhko, 1958, Mitt. münch. ent. Ges., **48**: 258. — Kertsch, Taurien, Krym.

crimea Burgeff, 1926, Mitt. münch. ent. Ges., **16**: 86 (preoccupied by **crymaea** Stauder, 1925, ssp. of **carniolica** Scopoli). Reiss, 1930, in Seitz, Die Gross-Schmetterlinge der Erde, Supplement, **2**: 44, pl. 4l; 1935, Int. ent. Z., **29**: 222. Holik & Sheljuzhko, 1958, Mitt. münch. ent. Ges., **48**: 257.

ssp. **wagneriana** Reiss, 1929, Int. ent. Z., **23**: 151; 1930, in Seitz, Die Gross-Schmetterlinge der Erde, Supplement, **2**: 44; 1935, Int. ent. Z., **29**: 231, fig. Holik & Sheljuzhko, 1958, Mitt. münch. ent. Ges., **48**: 266. — Sultan-Dagh bei Ak-Schehir, Kleinasien, 1300-1700 m.

ssp. **ochtshiensis** Holik & Sheljuzhko, 1958, Mitt. münch. ent. Ges., **48**: 264.	Ochtshi bei Kafan, Zangezur Gebirge, Armenisches Bergland.
ssp. **korbiana** Reiss, 1935, Int. ent. Z., **29**: 230. Koch, 1939, Mitt. münch. ent. Ges., **29**: 414. Haaf, 1952, Veröff. zool. Staatssamml. Münch., **2**: 152, 154, 157, pl. 13 (as *dorycnii* Ochsenheimer). Holik & Sheljuzhko, 1958, Mitt. münch. ent. Ges., **48**: 261.	Achalzich (Chambobel), Transkaukasien.
ab. **quinquemacula** Burgeff, 1926, Mitt. münch. ent. Ges., **16**: 86. Reiss, 1930, in Seitz, Die Gross-Schmetterlinge der Erde, Supplement, **2**: 44.	
ab. **latemarginata** Holik & Sheljuzhko, 1958, Mitt. münch. ent. Ges., **48**: 262.	
ssp. **grusica** Reiss, 1936, Ent. Rdsch., **54**: 103; 1937, ibidem, **55**: 42, fig. c3. Koch, 1939, Mitt. münch. ent. Ges., **29**: 414. Holik & Sheljuzhko, 1958, Mitt. münch. ent. Ges., **48**: 261.	Abastuman, Grusien Berge, Transkaukasien, 1340 m.
ssp. **karabaghensis** Holik & Sheljuzhko, 1958, Mitt. münch. ent. Ges., **48**: 263.	Gadrut und Shusha, Karabagh Gebiet, Transkaukasien.
ssp. **teberdensis** Reiss, 1936, Ent. Rdsch., **54**: 103; 1937, ibidem, **55**: 42, figs. c1, c2. Holik & Sheljuzhko, 1958, Mitt. münch. ent. Ges., **48**: 258.	Teberda-Gebiet, Nordkaukasus.
ssp. **kubana** Holik & Sheljuzhko, 1958, Mitt. münch. ent. Ges., **48**: 258.	Gulkevitshi, Kuban-Gebiet, Nordkaukasus.
ssp. **dorycnii** Ochsenheimer, 1808, Die Schmetterlinge von Europa, **2**: 69. Boisduval, 1828, Essai sur une Monographie des Zygénides, p. 72; 1834, Icones historique des Lépidoptères nouveaux ou peu connus, **2**: 68, pl. 55, fig. 8. Duponchel, 1835, in Godart & Duponchel, Histoire naturelle des Lépidoptères ou Papillons de France, Supplement, **2**: 136, pl. 12, fig. 2 (fig. 3 in error). Herrich-Schäffer, 1844, Systematische Bearbeitung der Schmetterlinge von Europa, **2**, pl. 3, figs. 24, 25; 1846, ibidem, **2**: 39. Freyer, 1839, Neuere Beiträge zur Schmetterlingskunde, **3**: 120, pl. 278, fig. 3. Spuler, 1906, in Hofmann, Die Schmetterlinge Europas, **2**: 161, pl. 72, fig. 4, pl. 77, fig. 22. Seitz, 1907, Die Gross-Schmetterlinge der Erde, **2**: 23, pl. 5d. Reiss, 1935, Int. ent. Z., **29**: 221; 1936, Ent. Rdsch., **54**: 101, figs. Holik & Sheljuzhko, 1958, Mitt. münch. ent. Ges., **48**: 253. Alberti, 1958, Mitt. zool. Mus. Berl., **34**: 321.	Russland [?Vorberge des Kaukasus].

ab. **crocea** Schultz, 1906, Soc. ent., **20**: 170. Dziurzyński, 1908, Berl. ent. Z., **53**: 36, pl. 1, fig. 9. Seitz, 1912, Die Gross-Schmetterlinge der Erde, **2**: 442.

ab. **flava** Reiss, 1935, Int. ent. Z., **29**: 230.

Biology

Holik, 1937, Lambillionea, **37**: 15-24, 32-45, 80-91; 1946, Rev. franç. Lépid., **10**: 250-261, 273-280; 1953, Ent. Z., **63**: 30. Holik & Sheljuzhko, 1958, Mitt. münch. ent. Ges., **48**: 254. Przegendza, 1936, Ent. Rdsch., **53**: 351. Reiss, 1958, Z. wien. ent. Ges., **43**: 161.

senescens Staudinger

Distribution: Asia Minor.

senescens Staudinger, 1887, Berl. ent. Z., **31**: 36. Spuler, 1906, in Hofmann, Die Schmetterlinge Europas, **2**: 161. Seitz, 1907, Die Gross-Schmetterlinge der Erde, **2**: 23, pl. 5e. Burgeff, 1926, Mitt. münch. ent. Ges., **16**: 86. Reiss, 1930, in Seitz, Die Gross-Schmetterlinge der Erde, Supplement, **2**: 44. Holik & Sheljuzhko, 1958, Mitt. münch. ent. Ges., **48**: 265. Alberti, 1958, Mitt. zool. Mus. Berl., **34**: 322. — Marasch und Hadjin im Taurus, Kleinasien.

ab. **sextarubra** Burgeff, 1926, Mitt. münch. ent. Ges., **16**: 86. Reiss, 1930, in Seitz, Die Gross-Schmetterlinge der Erde, Supplement, **2**: 44.

ab. **rubrimacula** Burgeff, 1926, Mitt. münch. ent. Ges., **16**: 86. Reiss, 1930, in Seitz, Die Gross-Schmetterlinge der Erde, Supplement, **2**: 44.

Biology

Holik, 1953, Ent. Z., **63**: 30.

ephialtes Linné

Distribution: Central & southern Europe (west to the Pyrenees), eastern Europe, Balkans, southern Russia.

ssp. **albaflavens** Verity, 1920, Boll. Lab. Zool. Portici, **14**: 39; 1946, Redia, **31**: 81. — Villa Latina, Mte. Mainarde, Caserta, Italia.

ssp. **albarubens** Verity, 1946, Redia, **31**: 79. — Cascine di Firenze, Italia.

albarubens Verity, 1920, Boll. Lab. Zool. Portici, **14**: 39 (nomen nudum).

ssp. **transpadana** Verity, 1946, Redia, **31**: 79. — Soria; Turbigo sul Ticino; Lombardia, Italia.

ssp. **ligus** Verity, 1946, Redia, **31**: 78. — Savona, Liguria di Ponente, Italia.

ssp. **roussilloni** Koch, 1940, Z. wien. EntVer., **25**: 135. Tremewan & Manley, 1965, Ent. Rec., **77**: 8. — Vernet-les-Bains, Ostpyrenäen, Frankreich.

ab. **falcatae** Oberthür, 1910, Études de Lépidoptérologie comparée, **4**: 572; 1896, Études d'Entomologie, **20**: 48, pl. 7, fig. 113 (as *falcatae* Denis & Schiffermüller). Tremewan, 1961, Bull. Brit. Mus. (nat. Hist.) Ent., **10** (7): 274, pl. 53, fig. 10.

ssp. **lurica** Dujardin, 1965, Entomops, Nice, no. 2: 49, fig. Dufay, 1966, Bull. mens. Soc. linn. Lyon, **35**: 71. — Mont Lure (Valbelle, dans la vallée du Jabron), Basses-Alpes, France Sud-Est.

ab. **pallens** Oberthür, 1910, Études de Lépidoptérologie comparée, **4**: 576. Seitz, 1912, Die Gross-Schmetterlinge der Erde, **2**: 443. Tremewan, 1961, Bull. Brit. Mus. (nat. Hist.) Ent., **10** (7): 275, pl. 53, fig. 13.

ssp. **ephialtes** Linné, 1767, Systema Naturae, ed. XII, **1**(2): 806. Scheven, 1777, Der Naturforscher, Halle, **10**: 95, pl. 2, fig. 7. Esper, 1779, Die Schmetterlinge, **2**, pl. 17, fig. 3; 1781, ibidem, **2**: 148. Ochsenheimer, 1808, Die Schmetterlinge von Europa, **2**: 77. Boisduval, 1828, Essai sur une Monographie des Zygénides, p. 85, pl. 5, fig. 5. Speyer & Speyer, 1858, Die geographische Verbreitung der Schmetterlinge Deutschlands und der Schweiz, p. 351. Spuler, 1906, in Hofmann, Die Schmetterlinge Europas, **2**: 161, pl. 72, fig. 5a, pl. 77, fig. 23. Seitz, 1907, Die Gross-Schmetterlinge der Erde, **2**: 23, pl. 5i, k. Oberthür, 1910, Études de Lépidoptérologie comparée, **4**: 566. Vorbrodt, 1913, Mitt. schweiz. ent. Ges., **12**: 165. Sterzl, 1931, Z. Ver. NatBeob.,Wien, **6**: 5, pl. 15, figs, 1, 2. Reiss, 1940, Ent. Z., **54**: 70, 71. Koch, 1940, Z. wien. EntVer. **25**: 123; 1941, ibidem, **26**: 117. Holik, 1952, Ent. NachrBl., Wien, **4**: 16. Alberti, 1958, Mitt. zool. Mus. Berl., **34**: 321. — [Martigny, Wallis, Schweiz].

scheveni Oberthür, 1910, Études de Lépidoptérologie comparée, **4**: 570. Seitz, 1912, Die Gross-Schmetterlinge der Erde, **2**: 442. Tremewan, 1961, Bull. Brit. Mus. (nat. Hist.) Ent., **10**(**7**): 275, pl. 53, fig. 11.

unipunctata Vorbrodt, 1913, in Vorbrodt & Müller-Rutz, Die Schmetterlinge der Schweiz, **2**: 274.

valesiaca Burgeff, 1914, Mitt. münch. ent. Ges., **5**: 69. Reiss, 1930, in Seitz, Die Gross-Schmetterlinge der Erde, Supplement, **2**: 43, pl. 4l.

ab. **herrichschaefferi** Burgeff, 1914, Mitt. münch. ent. Ges., **5**: 66, pl. 4, figs. 148, 149. Reiss, 1930, in Seitz, Die Gross-Schmetterlinge der Erde, Supplement, **2**: 43.

ab. **esperi** Oberthür, 1910, Études de Lépidoptérologie comparée, **4**: 570. Seitz, 1912, Die Gross-Schmetterlinge der Erde, **2**: 442. Tremewan, 1961, Bull. Brit. Mus. (nat. Hist.) Ent., **10**(**7**): 275, pl. 53, fig. 12.

ab. **wullschlegeli** Oberthür, 1909, Études de Lépidoptérologie comparée, 3, pl. 29, fig. 176; 1910, ibidem, **4**: 572. Seitz, 1912, Die Gross-Schmetterlinge der Erde, **2**: 443. Reiss, 1930, in Seitz, Die Gross-Schmetterlinge der Erde, Supplement, **2**: 43, pl. 4i (after Oberthür). Alberti, 1955, Ent. Z., **65**: 89-91. Tremewan, 1961, Bull. Brit. Mus. (nat. Hist.) Ent., **10**(7): 274, pl. 53, fig. 9.

bimaculata Vorbrodt, 1913, in Vorbrodt & Müller-Rutz, Die Schmetterlinge der Schweiz, **2**: 276.

ab. **sophiae** Favre, 1897, in Favre & Wullschlegel, Mitt. schweiz. ent. Ges., **10**: 36. Oberthür, 1904, Études de Lépidoptérologie comparée, **1**: 48, pl. 3, figs. 25-27. Spuler, 1906, in Hofmann, Die Schmetterlinge Europas, **2**: 161. Seitz, 1907, Die Gross-Schmetterlinge der Erde, **2**: 24; 1912, ibidem, **2**: 442. Reiss, 1930, in Seitz, Die Gross-Schmetterlinge der Erde, Supplement, **2**: 43, pl. 4l. Sterzl, 1931, Z. Ver. Nat-Beob., Wien, **6**: 10, pl. 15, fig. 3. Forster & Wohlfahrt, 1958, Die Schmetterlinge Mitteleuropas, **3**: 101, pl. 11, fig. 42.

bipunctata Vorbrodt, 1913, in Vorbrodt & Müller-Rutz, Die Schmetterlinge der Schweiz, **2**: 274.

f. **burgeffi** Bovey, 1950, Arch. Klaus-Stift. VererbForsch., **25**: 36.

f. **pseudomedusa** Burgeff, 1921, Ent. Z., **35**: 40.

unipunctata Vorbrodt, 1913, in Vorbrodt & Müller-Rutz, Die Schmetterlinge der Schweiz, **2**: 275 (preoccupied by *unipunctata* Vorbrodt, 1913, synonym of **ephialtes** Linné, 1767).

quinquemaculata Vorbrodt, 1913, in Vorbrodt & Müller-Rutz, Die Schmetterlinge der Schweiz, **2**: 275 (preoccupied by *quinquemaculata* Vorbrodt, 1913, synonym of **aemilii** Favre, 1897).

f. **aemilii** Favre, 1897, in Favre & Wullschlegel, Mitt. schweiz. ent. Ges., **10**: 36. Spuler, 1906, in Hofmann, Die Schmetterlinge Europas, **2**: 161. Seitz, 1907, Die Gross-Schmetterlinge der Erde, **2**: 24. Sterzl, 1931, Z. Ver. NatBeob., Wien, **6**: 10, pl. 15, fig. 5.

bipunctata Vorbrodt, 1913, in Vorbrodt & Müller-Rutz, Die Schmetterlinge der Schweiz, **2**: 275.

quinquemaculata Vorbrodt, 1913, in Vorbrodt & Müller-Rutz, Die Schmetterlinge der Schweiz, **2**: 275, fig. 6.

f. **mattheyi** Bovey, 1950, Arch. Klaus-Stift. VererbForsch., **25**: 36.

f. **coronilloides** Reiss, 1940, Ent. Z., **54**: 71.

unipunctata Vorbrodt, 1913, in Vorbrodt & Müller-Rutz, Die Schmetterlinge der Schweiz, **2**: 275 (preoccupied by *unipunctata* Vorbrodt, 1913, synonym of **ephialtes** Linné, 1767).

f. **flavobipuncta** Favre, 1903, Supplément à la Faune des

Macrolépidoptères du Valais, p. 9. Reiss, 1930, in Seitz, Die Gross-Schmetterlinge der Erde, Supplement, **2**: 43. Sterzl, 1931, Z. Ver. NatBeob., Wien, **6**: 10, pl. 15, fig. 7a.
bahri Hirschke, 1906, Jber. wien. ent. Ver., **16**: 94. Spuler, 1906, in Hofmann, Die Schmetterlinge Europas, **2**: 162. Seitz, 1907, Die Gross-Schmetterlinge der Erde, **2**: 24.
bipunctata Vorbrodt, 1913, in Vorbrodt & Müller-Rutz, Die Schmetterlinge der Schweiz, **2**: 275.
f. **trigonelloides** Reiss & Tremewan, 1964, Ent. Rec., **76**: 134 (nomen novum for *quinquemaculata* Vorbrodt).
quinquemaculata Vorbrodt, 1913, in Vorbrodt & Müller-Rutz, Die Schmetterlinge der Schweiz, **2**: 276 (preoccupied by *quinquemaculata* Vorbrodt, 1913, synonym of **aemilii** Favre, 1897). Reiss & Tremewan, 1964, Ent. Rec., **76**: 134.
unipunctata Vorbrodt, 1913, in Vorbrodt & Müller-Rutz, Die Schmetterlinge der Schweiz, **2**: 276 (preoccupied by *unipunctata* Vorbrodt, 1913, synonym of **ephialtes** Linné, 1767). Reiss & Tremewan, 1964, Ent. Rec., **76**: 134.
f. **wutzdorffi** Hirschke, 1906, Jber. wien. ent. Ver., **16**: 94. Spuler, 1906, in Hofmann, Die Schmetterlinge Europas, **2**: 162. Seitz, 1907, Die Gross-Schmetterlinge der Erde, **2**: 24. Sterzl, 1931, Z. Ver. NatBeob., Wien, **6**: 10, pl. 15, fig. 9.
bipunctata Vorbrodt, 1913, in Vorbrodt & Müller-Rutz, Die Schmetterlinge der Schweiz, **2**: 276.
f. **peucedanoides** Reiss, 1940, Ent. Z., **54**: 70.
f. **athamanthoides** Reiss, 1940, Ent. Z., **54**: 71.
quinquemaculata Vorbrodt, 1913, in Vorbrodt & Müller-Rutz, Die Schmetterlinge der Schweiz, **2**: 276 (preoccupied by *quinquemaculata* Vorbrodt, 1913, synonym of **aemilii** Favre, 1897).
ab. **diffusa** Burgeff, 1914, Mitt. münch. ent. Ges., **5**:66, pl. 2, fig. 173, pl. 4, fig. 150. Reiss, 1930, in Seitz, Die Gross-Schmetterlinge der Erde, Supplement, **2**: 43.
ssp. **meridiei** Burgeff, 1926, in Strand, Lepid. Cat., **33**: 77 (nomen novum for *meridionalis* Burgeff). Reiss, 1930, in Seitz, Die Gross-Schmetterlinge der Erde, Supplement, **2**: 43, pl. 41 (as *albaflavens* Verity). Verity, 1946, Redia, **31**: 81. Reiss, 1950, Jber. naturf. Ges. Graubünden, **82**: 123, fig. 32. — Bozen; Etschtal und Eisacktal, Südtirol.
meridionalis Burgeff, 1914, Mitt. münch. ent. Ges., **5**: 69 (preoccupied by *meridionalis* Oberthür, 1911 [synonym of **provincialis** Oberthür, 1907, ssp. of **hippocrepidis** Hübner]).
f. **coronilloides** Reiss, 1950, Jber. naturf. Ges. Graubünden, **82**: 124.
ab. **decolorata** Stauder, 1929, Ent. Z., **43**: 80. Tremewan, 1961, Bull. Brit. Mus. (nat. Hist.) Ent., **10**(7): 276, pl. 53, fig. 16.
ssp. **coronillae** Denis & Schiffermüller, 1775, Ankündigung eines — Österreich und Ungarn [Umgebung von Wien].

systematischen Werkes von den Schmetterlingen der Wienergegend, p. 45; 1776, Systematisches Verzeichniss der Schmetterlinge der Wiener Gegend, p. 45. Esper, 1783, Die Schmetterlinge, **2**: 218, pl. 33, fig. 2. Schrank, 1785, in Fuessly, Neues Magazin für die Liebhaber der Entomologie, **2**: 206. Hübner, 1796, Sammlung europäischer Schmetterlinge, **2**: 17, pl. 3, fig. 13; 1806, ibidem, Der Ziefer, p. 83. Denis & Schiffermüller, 1801, Systematisches Verzeichniss von den Schmetterlingen der Wiener Gegend, **1**: 43. Ochsenheimer, 1808, Die Schmetterlinge von Europa, **2**: 79. Werneburg, 1864, Beiträge zur Schmetterlingskunde, **1**: 398. Spuler, 1906, in Hofmann, Die Schmetterlinge Europas, **2**: 162, pl. 77, fig. 23a. Seitz, 1907, Die Gross-Schmetterlinge der Erde, **2**: 24, pl. 6a. Sterzl, 1931, Z. Ver. NatBeob., Wien, **6**: 10, pl. 15, fig. 6a. Reiss, 1940, Ent. Z., **54**: 75. Holik, 1952, Ent. Nachr-Bl., Wien, **4**: 69.

sexmaculata Burgeff, 1921, Ent. Z., **35**: 39.

ab. **costaflavabipuncta** Le Charles, 1927, Encycl. ent., (B) 3, Lepidoptera, **2**: 152, pl. 9, fig. 10.

ab. **schawerdae** Dziurzyński, 1913, Jber. wien. ent. Ver., **23**: 215.

f. **schaefferi** Scheven, 1777, Der Naturforscher, Halle, **10**: 95. Schäffer, 1766, Icones Insectorum circa Ratisbonam indigenorum, **2**, pl. 165, figs. 3, 4. Fuessly, 1778, Magazin für die Liebhaber der Entomologie, **1**: 122; 1778, ibidem, **1**: 135. Esper, 1781, Die Schmetterlinge, **2**: 147, note c. Borkhausen, 1789, Naturgeschichte der Europäischen Schmetterlinge, **2**: 117.

trigonellae Esper, 1783, Die Schmetterlinge, **2**: 219, pl. 33, figs. 3, 4. Boisduval, 1828, Essai sur une Monographie des Zygénides, p. 87, pl. 5, fig. 6. Spuler, 1906, in Hofmann, Die Schmetterlinge Europas, **2**: 162, pl. 77, fig. 23b. Seitz, 1907, Die Gross-Schmetterlinge der Erde, **2**: 24, pl. 6a. Sterzl, 1931, Z. Ver. NatBeob., Wien, **6**: 10, pl. 15, fig. 8a.

quinquemaculata Burgeff, 1921, Ent. Z., **35**: 39.

f. **falcatae** Denis & Schiffermüller, 1775, Ankündigung eines systematischen Werkes von den Schmetterlingen der Wienergegend, p. 45; 1776, Systematisches Verzeichniss der Schmetterlinge der Wienergegend, p. 45. Hübner, 1796, Sammlung europäischer Schmetterlinge, **2**: 17, pl. 5, fig. 33. Denis & Schiffermüller, 1801, Systematischen Verzeichniss von den Schmetterlingen der Wiener Gegend, **1**: 42 (partim). Hyde, 1963, Animals, 3(6): 162-164, fig. 3 (as *trigonellae* Esper).

medusoides Reiss, 1940, Ent. Z., **54**: 71.

f. **icterica** Lederer, 1853, Verh. zool.-bot. Ver. Wien, **2**: 72. Schrank, 1785, in Fuessly, Neues Magazin für die Liebhaber der Entomologie, **2**: 207 (as *aeacus* Denis & Schiffermüller). Hübner, 1796, Sammlung europäischer Schmetterlinge, **2**: 17, pl. 3, fig. 18 (as *aeacus* Denis & Schiffermüller). Ochsenheimer, 1808, Die Schmetterlinge von Europa, **2**: 72, 75 (as *aeacus* Denis & Schiffermüller). Dziurzyński, 1904, Jber. wien. ent. Ver., **14**: 51, pl. 2, fig. 10. Spuler, 1906, in Hofmann, Die Schmetterlinge Europas, **2**: 162. Seitz, 1907, Die Gross-Schmetterlinge der Erde, **2**: 24. Sterzl, 1931, Z. Ver. NatBeob., Wien, **6**: 21, pl. 17, fig. 21.

f. **aeacus** Denis & Schiffermüller, 1775, Ankündigung eines systematischen Werkes von den Schmetterlingen der Wienergegend, p. 45; 1776, Systematisches Verzeichniss der Schmetterlinge der Wienergegend, p. 45. Esper, 1783, Die Schmetterlinge, **2**: 217, pl. 33, fig. 1. Denis & Schiffermüller, 1801, Systematischen Verzeichniss von den Schmetterlingen der Wiener Gegend, **1**: 43. Spuler, 1906, in Hofmann, Die Schmetterlinge Europas, **2**: 162, pl. 72, fig. 1b. Sterzl, 1931, Z. Ver. NatBeob., Wien, **6**: 21, pl. 17, fig. 22.

f. **prinzi** Hirschke, 1906, Jber. wien. ent. Ver., **16**: 94. Spuler, 1906, in Hofmann, Die Schmetterlinge Europas, **2**: 162. Seitz, 1907, Die Gross-Schmetterlinge der Erde, **2**: 24, pl. 6b.

f. **aurantiaca** Hirschke, 1904, Jber. wien. ent. Ver., **14**: 57, pl. 2, fig. 11. Spuler, 1906, in Hofmann, Die Schmetterlinge Europas, **2**: 162.

f. **nigroaeacus** Burgeff, 1921, Ent. Z., **35**: 39. Reiss, 1930, in Seitz, Die Gross-Schmetterlinge der Erde, Supplement, **2**: 43. Sterzl, 1931, Z. Ver. NatBeob., Wien, **6**: 21, pl. 17, fig. 24.

ab. **adalberti** Oberthür, 1910, Études de Lépidoptérologie comparée, **4**: 574. Seitz, 1907, Die Gross-Schmetterlinge der Erde, **2**, pl. 6a (as *icterica* Lederer); 1912, ibidem, **2**: 443. Tremewan, 1961, Bull. Brit. Mus. (nat. Hist.) Ent., **10**(7): 276, pl. 53, fig. 15.

ab. **pallida** Oberthür, 1910, Études de Lépidoptérologie comparée, **4**: 576. Seitz, 1912, Die Gross-Schmetterlinge der Erde, **2**: 443. Tremewan, 1961, Bull. Brit. Mus. (nat. Hist.) Ent., **10**(7): 275, pl. 53, fig. 14.

ab. **maureri** Dziurzyński, 1913, Jber. wien. ent. Ver., **23**: 215.

f. **guenneri** Hirschke, 1906, Jber. wien. ent. Ver., **16**: 94 (as *günneri*). Spuler, 1906, in Hofmann, Die Schmetterlinge Europas, **2**: 162. Seitz, 1907, Die Gross-Schmetterlinge der Erde, **2**: 24. Sterzl, 1931, Z. Ver. NatBeob., Wien, **6**: 18, pl. 16, fig. 18.

f. **metzgeri** Hirschke, 1904, Jber. wien. ent. Ver., **14**: 58, pl. 2, fig. 12. Spuler, 1906, in Hofmann, Die Schmetterlinge Europas, **2**: 162. Seitz, 1907, Die Gross-Schmetterlinge der Erde,

2: 24. Sterzl, 1931, Z. Ver. NatBeob., Wien, **6**: 19, pl. 16, fig. 20.

ssp. **pannonica** Holik, 1937, Lambillionea, **37**: 124; 1939, Sborn. ent. Odd. nár. Mus. Praze, **17**: 46. Koch, 1940, Z. wien. Ent-Ver., **25**: 134. Povolńy, 1945, Folia ent., Brno, **8**: 81. Holik, 1948, Mitt. münch. ent. Ges., **34**: 404. Forster & Wohlfahrt, 1958, Die Schmetterlinge Mitteleuropas, **3**: 101, pl. 11, fig. 41. — Murany-Tisovec, Slovaquie.

ab. **paradisiae** Reiprich, 1960, Motýle Slovenska, p. 304, 410, 416.

ssp. **semimixta** Povolný & Gregor, 1946, Folia ent., Brno, **9**, Supplement, 12: 55, 98, figs. — Pavlovské Vrchy; Uval dolnomar; Hády; Moravia.

ssp. **slabyi** Reiss & Tremewan, 1964, Ent. Rec., **76**: 134 (nomen novum for *montana* Slabý). — Bades Brusno (Brusna), Hron, Czechoslovakia, 400 m.

montana Slabý, 1953, Acta Mus. Silesiae, **3** (A): 46, figs. C3, D1 (preoccupied by **montana** Rothschild, 1925, ssp. of **loyselis** Oberthür). Reiss & Tremewan, 1964, Ent. Rec., **76**: 134.

ssp. **vardarica** Daniel, 1957, Acta Mus. maced. Sci. nat., **4**: 218; 1964, Prirod. Muz. Scopje, no. 2: 20. — Vratnica, Sharplanina, Zentral-Mazedonien, 900 m.

ssp. **istoki** Silbernagel, 1944, Z. wien. ent. Ges., **19**: 186. Daniel, 1958, Fragmenta Balcanica, **2**: 44; 1964, Prirod. Muz. Scopje, no. 2: 20. — Istok, Petrina-Planina, Mazedonien.

ab. **rubrimaculata** Silbernagel, 1944, Z. wien. ent. Ges., **29**: 186.

ssp. **chalkidikae** Holik, 1937, Lambillionea, **37**: 127. Koch, 1938, Ent. Z., **51**: 401. Holik, 1948, Mitt. münch. ent. Ges., **34**: 413. — Mont. Athos, Halbinsel Chalkidike, Griechenland.

ab. **immaculata** Koch, 1938, Ent. Z., **51**: 401.

ssp. **tymphrestica** Holik, 1948, Mitt. münch. ent. Ges., **34**: 413. Staudinger, 1870, Horae Soc. ent. Ross., **7**: 104 (as *ephialtes* Linné). Reiss, 1962, Ent. Z., **72**: 223. Daniel, 1964, Prirod. Muz. Scopje, no. 2: 20. — Veluchi (Beluchi) [= Tymphrestos], Griechenland.

ab. **herrichschaefferirubra** Holik, 1948, Mitt. münch. ent. Ges., **34**: 413.

ssp. **taurida** Holik & Sheljuzhko, 1953, in Holik, Ent. NachrBl., Wien, **5**: 6. Holik & Sheljuzhko, 1958, Mitt. münch. ent. Ges., **48**: 248. — Simferopol, Halbinsel Krim.

ab. **analielongata** Holik & Sheljuzhko, 1953, in Holik, Ent. NachrBl., Wien, **5**: 6.

ssp. **medusa** Pallas, 1771, Reise durch verschiedene Provinzen der Russischen Reichs, **1**: 472. Seitz, 1907, Die Gross-Schmet- — Markofka, Kinelfluss, Samara-Gebiet, Russland.

terlinge der Erde, **2**: 23, pl. 5k. Reiss, 1940, Ent. Z., **54**: 76. Holik, 1952, Ent. NachrBl., Wien, **4**: 18.

ssp. **strandi** Obraztsov, 1936, Folia Zool. Hydrobiol., Riga, **9**: 37. Holik, 1937, Lambillionea, **37**: 127. Holik & Sheljuzhko, 1958, Mitt. münch. ent. Ges., **48**: 246. — Nikolajev, unteres Bugtal, Ukraine.

ab. **rubricauda** Burgeff, 1914, Mitt. münch. ent. Ges., **5**: 66 (as *rubricunda*, corrected to **rubricauda** Burgeff, 1914, ibidem, **5**: 78). Reiss, 1930, in Seitz, Die Gross-Schmetterlinge der Erde, Supplement, **2**: 43.

ab. **albobasimaculata** Obraztsov, 1936, Folia Zool. Hydrobiol., Riga, **9**: 38.

ab. **basiunimaculata** Obraztsov, 1936, Folia Zool. Hydrobiol., Riga, **9**: 38.

ab. **coloretincta** Obraztsov, 1936, Folia Zool. Hydrobiol., Riga, **9**: 38.

ab. **ephialtescarnea** Obraztsov, 1936, Folia Zool. Hydrobiol., Riga, **9**: 37.

ab. **sexmacula** Obraztsov, 1936, Folia Zool. Hydrobiol., Riga, **9**: 37.

ssp. **tambovensis** Holik & Sheljuzhko, 1953, in Holik, Ent. NachrBl., Wien, **5**: 9. Holik & Sheljuzhko, 1958, Mitt. münch. ent. Ges., **48**: 238. — Chobotovo Distrikt (Kozlov), Gouv. Tambov, Russland.

ssp. **podolica** Holik, 1932, Iris, **46**: 132, pl. 2, figs. 32-35. Reiss, 1933, in Seitz, Die Gross-Schmetterlinge der Erde, Supplement, **2**: 278. Holik & Sheljuzhko, 1958, Mitt. münch. ent. Ges., **48**: 242. — Rakulowa, Podolien.

ssp. **kiewensis** Reiss, 1932, in Holik, Iris, **46**: 130, pl. 2, fig. 36. Reiss, 1933, in Seitz, Die Gross-Schmetterlinge der Erde, Supplement, **2**: 278. Holik & Sheljuzhko, 1958, Mitt. münch. ent. Ges., **48**: 243. — Umgebung von Kiew, Kirillovskije ovragi, Ukraine.

?*cingulata* Linde, 1894, Mém. Soc. Amis. Sci. nat. Moscou, **86**: 3 (second pagination), (infrasubspecific). Holik & Sheljuzhko, 1958, Mitt. münch. ent. Ges., **48**: 175. Tremewan, 1965, Ent. Rec., **77**: 89.

ssp. **danastriensis** Holik, 1939, Ann. Mus. zool. Polon., **12**: 122, pl. 4, fig. 135, pl. 5, figs. 153-160, 162-164. Holik & Sheljuzhko, 1958, Mitt. münch. ent. Ges., **48**: 240. — Wolczkow, Oberlauf des Dnjestr, Polen.

f. **nigropeucedani** Holik, 1939, Ann. Mus. zool. Polon., **12**: 122.

peucedaninigrescens Holik, 1939, Ann. Mus. zool. Polon., **12**: 136, pl. 5, fig. 161.

f. **nigroathamanthae** Holik, 1939, Ann. Mus. zool. Polon., **12**: 122.

ssp. **retyesati** Holik, 1948, Mitt. münch. ent. Ges., **34**: 415. — Retyesat Gebirge, Transsylvanien.

ab. **costielongata** Holik, 1948, Mitt. münch. ent. Ges., **34**: 415.

ab. **atritella** Hirschke, 1910, Verh. zool.-bot. Ges. Wien, **60**: 416. Reiss, 1930, in Seitz, Die Gross-Schmetterlinge der Erde, Supplement, **2**: 43. Sterzl, 1931, Z. Ver. NatBeob., Wien, **6**: 17, pl. 16, fig. 14.

ssp. **fatrica** Holik, 1937, Lambillionea, **37**: 122. — Strecsno, Monts. Fatra (la Fatra mineure), Slovaquie occidentale.

ssp. **athamanthae** Esper, 1789, Die Schmetterlinge, Supplement, **2**(2): 4, pl. 37, figs. 5, 6. Spuler, 1906, in Hofmann, Die Schmetterlinge Europas, **2**: 162, pl. 77, fig. 23d. Seitz, 1907, Die Gross-Schmetterlinge der Erde, **2**: 24, pl. 6b. Sterzl, 1931, Z. Ver. NatBeob., Wien, **6**: 19, pl. 16, fig. 19. Reiss, 1940, Ent. Z., **54**: 75. Holik, 1953, Ent. NachrBl., Wien, **5**: 29. — Bei Lemberg, Galizien.

veronicae Borkhausen, 1789, Naturgeschichte der Europäischen Schmetterlinge nach systematischer Ordnung, **2**: 162; 1793, Rheinisches Magazin, **1**: 635. Holik, 1936, Ent. Rdsch., **53**: 407.

ssp. **praegracilis** Koch, 1940, Z. wien. EntVer., **25**: 145. — Naleczów, nordwestlich von Lublin, Mittelpolen.

ssp. **styria** Burgeff, 1914, Mitt. münch. ent. Ges., **5**: 68. Reiss, 1930, in Seitz, Die Gross-Schmetterlinge der Erde, Supplement, **2**: 43, pl. 41. — Thörl, Obersteiermark, Österreich.

ssp. **chremisa** Koch, 1940, Z. wien. EntVer., **25**: 135. — Dürnstein bei Krems, Wachau, Österreich.

ssp. **antiqua** Slabý, 1953, Acta Mus. Silesiae, **3** (A): 47, figs. D2, D3, E1-E3, F1-F3, G1-G3, H1-H3. — Giraltovce; Prešov; Remetské; [Tälern des Tisa Gebiets], Czechoslovakia, 200-300 m.

ssp. **galgoczensis** Koch, 1940, Z. wien. Ent Ver., **25**: 146. Slabý, 1953, Acta Mus. Silesiae, **3** (A), pl. 7, figs. A1-A3, B1-B3, C1, C2 (type and forms). — Zliechov (Zsolt), Galgoczer Gebirge (Veterné hole), Slovakei.

ssp. **bohemia** Reiss, 1922, Int. ent. Z., **16**: 83; 1926, Die Zygaenen Deutschlands, p. 25, 27, pl. 2, fig., pl. 1, figs. (forms). Holik, 1929, Int. Ent. Z., **23**: 3 (footnote 10). Reiss, 1930, in Seitz, Die Gross-Schmetterlinge der Erde, Supplement, **2**: — Karlstein bei Prag, Böhmen.

43, pl. 4i, k (type and forms). Forster & Wohlfahrt, 1958, Die Schmetterlinge Mitteleuropas, **3**: 101, pl. 11, figs. 31-40 (type and forms).

ab. **tricingulata** Holik, 1926, Int. ent. Z., **20**: 241. Reiss, 1930, in Seitz, Die Gross-Schmetterlinge der Erde, Supplement, **2**: 43. Sterzl, 1931, Z. Ver. NatBeob., Wien, **6**: 18, pl. 16, fig. 15.

ab. **diffusa** Burgeff, 1914, Mitt. münch. ent. Ges., **5**: 66, pl. 2, fig. 174, pl. 4, figs. 145-147. Reiss, 1930, in Seitz, Die Gross-Schmetterlinge der Erde, Supplement, **2**: 43. Sterzl, 1931, Z. Ver. NatBeob., Wien, **6**: 17, pl. 16, fig. 13.

f. **nigroicterica** Holik, 1919, Z. öst. EntVer., **4**: 111. Reiss, 1930, in Seitz, Die Gross-Schmetterlinge der Erde, Supplement, **2**: 43. Sterzl, 1931, Z. Ver. NatBeob., Wien, **6**: 21, pl. 17, fig. 23.

f. **ephialtoides** Reiss, 1922, Int. ent. Z., **16**: 83; 1930, in Seitz, Die Gross-Schmetterlinge der Erde, Supplement, **2**: 43.

f. **pseudocoronillae** Holik, 1919, Z. öst. EntVer., **4**: 112. Reiss, 1930, in Seitz, Die Gross-Schmetterlinge der Erde, Supplement, **2**: 43.

ssp. **rudolfi** Koch, 1940, Z. wien. EntVer., **25**: 147. — Neutitschein, Sudeten, Mährisch Schlesien.

ssp. **rubens** Verity, 1946, Redia, **31**: 77. — Issy; Lardy; Seine-et-Oise, France.

rubens Verity, 1920, Boll. Lab. Zool. Portici, **14**: 39 (nomen nudum).

parisica Reiss, 1959, Bull. Soc. ent. Mulhouse, p. 57. Bernardi & Viette, 1960, Bull. Soc. ent. Mulhouse, p. 31.

ssp. **peucedani** Esper, 1781, Die Schmetterlinge, **2**: 191, pl. 25, fig. 2a. Hübner, 1796, Sammlung europäischer Schmetterlinge, **2**: 17, pl. 16, figs. 75, 76. Ochsenheimer, 1808, Die Schmetterlinge von Europa, **2**: 70. Boisduval, 1828, Essai sur une Monographie des Zygénides, p. 68, pl. 4, fig. 6. Herrich-Schäffer, 1844, Systematische Bearbeitung der Schmetterlinge von Europa, **2**, pl. 7, figs. 52, 53 (as *hippocrepidis* Hübner); 1846, ibidem, **2**: 39. Spuler, 1906, in Hofmann, Die Schmetterlinge Europas, **2**: 162, pl. 77, fig. 23c. Seitz, 1907, Die Gross-Schmetterlinge der Erde, **2**: 24, pl. 6b. Sterzl, 1931, Z. Ver. NatBeob., Wien, **6**: 18, pl. 16, fig. 17. Reiss, 1937, in Schneider, Jh. Ver. vaterl. Naturk. Württemb., **93**: 129 (as *borealis* Burgeff [partim]); 1940, Ent. Z., **54**: 71; 1949, Entomon, **1**: 175. Haaf, 1952, Veröff. zool. Staatssamml. Münch., **2**: 152, 154, 157, pl. 13 (as *ephialtes* Linné). Holik, 1953, Ent. NachrBl., Wien, **5**: 7. Bergmann, 1953, Die Grossschmetterlinge Mitteldeutschlands, **3**: 62, pl. 69, figs. C1, C2, — Erlangen; Uffenheim in Franken [Mittel- und Süddeutschland].

C4-C6. Heuser & Jöst, 1959, Mitt. Pollichia, (3) **6**: 128.
ab. **acingulata** Francke, 1920, Z. öst. EntVer, **5**: 4. Reiss, 1930, in Seitz, Die Gross-Schmetterlinge der Erde, Supplement, **2**: 43.
f. **prinzi** Reiss, 1964, Coridon, (A) **6**: 6.
f. **medusoides** Reiss, 1949, Entomon, **1**: 175.
f. **ephialtoides** Reiss, 1949, Entomon, **1**: 175.
f. **athamanthoides** Reiss, 1949, Entomon, **1**: 175.
f. **intermedia** Spuler, 1906, in Hofmann, Die Schmetterlinge Europas, **2**: 162.
f. **semipuncta** Ziegler, 1911, Int. ent. Z., **5**: 139.
ssp. **borealis** Burgeff, 1914, Mitt. münch. ent. Ges., **5**: 68. Reiss, 1926, Die Zygaenen Deutschlands, p. 25, 26 (partim), pl. 2, fig.; 1930, in Seitz, Die Gross-Schmetterlinge der Erde, Supplement, **2**: 43. — Magdeburg; Berlin; Norddeutschland.
ab. **extrema** Reiss, 1927, Int. ent. Z., **21**: 289; 1930, in Seitz, Die Gross-Schmetterlinge der Erde, Supplement, **2**: 43.
f. **athamanthoides** Reiss, 1940, Ent. Z., **54**: 71.
ssp. **baltica** Holik, 1937, Lambillionea, **37**: 122; 1939, Ann. Mus. zool. Polon., **12**: 112, pl. 4, figs. 123-126. Reiss, 1940, Ent. Z., **54**: 74. — Osterode, Prusse occidentale.
[The following forms and aberrations were described by Dryja from specimens obtained by crossing different subspecies and forms during experimental breeding. As these forms and aberrations cannot be placed under any one particular subspecies, they are listed below in the same order as they were described].
Dryja, Antoni, 1959, Badania nad Polimorfizmem Kraśnika Zmiennego (*Zygaena ephialtes* L.), pp. 1-404, pls. I-VII, Warszawa, 1959.
f. **pseudotrigonellae** Dryja, 1959, loc. cit., p. 90, pl. 5, fig. 42.
f. **nigroprinzi** Dryja, 1959, loc. cit., p. 158, pl. 3, fig. 7.
f. **nigroaurantiaca** Dryja, 1959, loc. cit., p. 158, pl. 5, fig. 57.
f. **prinzidecedens** Dryja, 1959, loc. cit., p. 158.
f. **aurantiacadecedens** Dryja, 1959, loc. cit., p. 158, pl. 5, fig. 54.
f. **ictericadecedens** Dryja, 1959, loc. cit., p. 160, pl. 5, fig. 8.
f. **aeacusdecedens** Dryja, 1959, loc. cit., p. 160, pl. 5, fig. 40.
f. **rubromutabilis** Dryja, 1959, loc. cit., p. 161, pl. 5, fig. 5.
f. **flavomutabilis** Dryja, 1959, loc. cit., p. 161, pl. 5, fig. 7.
f. **aurantiacomutabilis** Dryja, 1959, loc. cit., p. 161, pl. 7, ser. 6, fig. 3.
f. **bicolorata** Dryja, 1959, loc. cit., p. 162, pl. 7, ser. 6, fig. 4.
f. **parvulus** Dryja, 1959, loc. cit., p. 163, pl. 7, ser. 5, figs. 5-10.
f. **rubrofucosa** Dryja, 1959, loc. cit., p. 163, 185, pl. 7, ser. 5, fig. 2.
ab. **flavofucosa** Dryja, 1959, loc. cit., p. 185, pl. 7, ser. 5, fig. 3.

ab. **aurantiacofucosa** Dryja, 1959, loc. cit., p. 185, pl. 6, ser. 2, fig. 2.
ab. **rubropuncta** Dryja, 1959, loc. cit., p. 195, pl. 7, ser. 6, fig. 1.
ab. **flavopuncta** Dryja, 1959, loc. cit., p. 195, pl. 7, ser. 6, fig. 2.
ab. **flavomaculosa** Dryja, 1959, loc. cit., p. 197, pl. 6, ser. 2, fig. 4.
ab. **duplex** Dryja, 1959, loc. cit., p. 203, pl. 6, ser. 2, fig. 7.
ab. **flavoduplex** Dryja, 1959, loc. cit., p. 203, pl. 6, ser. 3, fig. 8.
ab. **bimutata** Dryja, 1959, loc. cit., p. 203, pl. 6, ser. 2, fig. 9.
ab. **flavobimutata** Dryja, 1959, loc. cit., p. 203, pl. 6, ser. 2, fig. 10.
ab. **bimaculosa** Dryja, 1959, loc. cit., p. 203, pl. 6, ser. 2, fig. 11.
ab. **flavobimaculosa** Dryja, 1959, loc. cit., p. 203, pl. 6, ser. 2, fig. 12.
ab. **triplex** Dryja, 1959, loc. cit., p. 204, pl. 6, ser. 2, fig. 13.
ab. **flavotriplex** Dryja, 1959, loc. cit., p. 204, pl. 6, ser. 2, fig. 14.
ab. **magnimacula** Dryja, 1959, loc. cit., p. 221, pl. 6, ser. 1, figs. 8-11.
ab. **parvimacula** Dryja, 1959, loc. cit., p. 221, pl. 6, ser. 1, figs, 12, 13.
ab. **basiunimacula** Dryja, 1959, loc. cit., p. 226, pl. 6, ser. 1, figs. 4-7.
ab. **basibicolora** Dryja, 1959, loc. cit., p. 231, pl. 7, ser. 6, figs. 5, 6.
ab. **basiimmaculata** Dryja, 1959, loc. cit., p. 231, pl. 6, ser. 1, figs. 1, 2.
ab. **basidepuncta** Dryja, 1959, loc. cit., p. 231, pl. 6, ser. 1, fig. 3.
ab. **pallidopeucedani** Dryja, 1959, loc. cit., p. 234, pl. 6, ser. 3, fig. 9.
ab. **pallidoicterica** Dryja, 1959, loc. cit., p. 234, pl. 6, ser. 3, fig. 11.
ab. **pallidoephialtes** Dryja, 1959, loc. cit., p. 234, pl. 6, ser. 3, fig. 12.
ab. **pallidocoronillae** Dryja, 1959, loc. cit., p. 234.
ab. **violaceopeucedani** Dryja, 1959, loc. cit., p. 236, pl. 6, ser. 3, fig. 3.
ab. **violaceoicterica** Dryja, 1959, loc. cit., p. 236, pl. 6, ser. 3, fig. 5.
ab. **violaceoephialtes** Dryja, 1959, loc. cit., p. 236, pl. 6, ser. 3, fig. 7.
ab. **violaceocoronillae** Dryja, 1959, loc. cit., p. 236, pl. 6, ser. 3, fig. 8.
ab. **violaceometzgeri** Dryja, 1959, loc. cit., p. 236.
ab. **violaceomedusa** Dryja, 1959, loc. cit., p. 236.
ab. **semipeucedani** Dryja, 1959, loc. cit., p. 243, pl. 7, ser. 6, fig. 8.

ab. **semiguenneri** Dryja, 1959, loc. cit., p. 243, pl. 7, ser. 6, fig. 9.
ab. **violaceotriplex** Dryja, 1959, loc. cit., p. 363, pl. 7, ser. 6, fig. 10.
ab. **flavobipuncta** Dryja, 1959, loc. cit., p. 390, pl. 6, ser. 2, fig. 6.
ab. **pallidoaeacus** Dryja, 1959, loc. cit., p. 391, pl. 6, ser. 3, fig. 10.
ab. **pallidomedusa** Dryja, 1959, loc. cit., p. 391, pl. 7, ser. 4, fig. 10.
ab. **pallidonigroaeacus** Dryja, 1959, loc. cit., p. 391, pl. 7, ser. 4, fig. 11.
ab. **pallidotrigonellae** Dryja, 1959, loc. cit., p. 391, pl. 6, ser. 3, fig. 13.
ab. **violaceoaeacus** Dryja, 1959, loc. cit., p. 391, pl. 6, ser. 3, fig. 6.
ab. **violaceoprinzi** Dryja, 1959, loc. cit., p. 391, pl. 6, ser. 3, fig. 4.

Biology

Boisduval, Rambur & Graslin, 1832, Collection iconographique et historique des Chenilles, pl. 2, figs. 1-3. Bovey, 1941, Rev. suisse Zool., **48**: 1-90, figs. 1-16, pl. 1, figs. 1-48; 1942, Arch. Klaus-Stift. VererbForsch., **17**: 432; 1948, ibidem, **23**: 499-502; 1966, Rev. suisse Zool., **73**: 193-218, figs. 1-12. Burgeff, 1912, Z. wiss. InsektBiol., **8**: 185, 186, 198; 1921, Ent. Z., **35**: 21-22, 26, 31, 35, 38-40; 1950, Portug. acta biol., (A) Goldschmidt: 663-728. De Lattin, 1952, Verh. dtsch. zool. Ges., p. 452-460, figs. 1-5. Dorfmeister, 1854, Verh. zool.-bot. Ver. Wien, **4**: 480. Dryja, 1959, Badania nad Polimorfizmem Kraśnika Zmiennego (*Zygaena ephialtes* L.), p. 1-404, pls. 1-7. Esper, 1781, Die Schmetterlinge, **2**: 191, pl. 25, figs. 2β, 2γ, 2δ. Ford, 1964, Ecological Genetics, p. 132-134. Guhn, 1932, Ent. Jb., **41**: 93. Haneld, 1903, Berl. ent. Z., **48**: 26. Hepp, 1927, Ent. Z., **41**: 288. Holik, 1910, Int. ent. Z., **4**: 135; 1919, Z. öst. EntVer., **4**: 103-104, 111-113; 1937, Lambillionea, **37**: 15-24, 32-45, 80-91; 1938, ibidem, **38**: 51-58, 79-88, 95-102; 1938, Ent. Rdsch., **55**: 320-323, 331-333; 1946, Rev. franç. Lépid., **10**: 250-261, 273-280; 1953, Ent. Z., **63**: 29. Hrubý, 1964, Prodromus Lepidopter Slovenska, p. 481. Kiefer, 1934, Int. ent. Z., **27**: 521-524. Koch, 1955, Wir Bestimmen Schmetterlinge, **2**: 64, 65, pl. 1, figs. 20, 20a, 20b, pl. 15, fig. 20. Ochsenheimer, 1808, Die Schmetterlinge von Europa, **2**: 70, 77. Povolný & Pijáček, 1949, Sborn. přír. Společ. Mor. Ostravé, **10**: 400-410, figs. Reichl, 1958, Z. wien. ent. Ges., **43**: 250-265, figs. 1, 2 (distribution maps); 1959, ibidem, **44**: 50-64, fig. (distribution map). Reiss, 1926, Die Zygaenen Deutschlands, p. 7; 1930, in Seitz, Die Gross-Schmetterlinge der Erde, Supplement, **2**: 44; 1958, Z. wien. ent. Ges., **43**: 161. Roüast, 1883, Catalogue des Chenilles européennes connues, p. 22. Seitz, 1907, Die Gross-Schmetterlinge der Erde, **2**: 24. Spuler, 1910, in Hofmann, Die Raupen der Schmetterlinge Europas, pl. 10, figs. 3a, b.

transalpina Esper

Distribution: Italy (excluding Sicily), Adriatic coast, Tyrol, Alps, southern Germany.

ssp. **calabrica** Calberla, 1895, Iris, **8**: 226. Perlini, 1905, Forme di Lepidotteri esclusivamente Italiane, p. 55, pl. 3, figs. 14, 15. Spuler, 1906, in Hofmann, Die Schmetterlinge Europas, **2**: 161, pl. 77, fig. 21b. Seitz, 1907, Die Gross-Schmetterlinge der Erde, **2**: 23, pl. 5i. Burgeff, 1926, Mitt. münch. ent. Ges., **16**: 78. — Calabrien, Süditalien.

spicae Staudinger, 1894, Iris, **7**: 254 (preoccupied by *spicae* Hübner, 1796, synonym of **lavandulae** Esper, 1783).

ab. **hexamacula** Turati, 1910, Boll. Lab. Zool. Portici, **4**: 162.

ab. **rhodomelas** Turati, 1910, Boll. Lab. Zool. Portici, **4**: 162.

ab. **tertiaedeleta** Stauder, 1922, Iris, **36**: 42.

ab. **tripicta** Stauder, 1929, Ent. Z., **43**: 132. Reiss, 1930, in Seitz, Die Gross-Schmetterlinge der Erde, Supplement, **2**: 39.

ab. **nanina** Stauder, 1929, Ent. Z., **43**: 133.

ab. **bipuncta** Stauder, 1929, Ent. Z., **43**: 132. Reiss, 1930, in Seitz, Die Gross-Schmetterlinge der Erde, Supplement, **2**: 39.

ab. **bichroma** Stauder, 1929, Ent. Z., **43**: 132.

ab. **aristocratica** Stauder, 1922, Iris, **36**: 42.

ab. **mediodefecta** Stauder, 1922, Iris, **36**: 42.

ab. **albicincta** Burgeff, 1914, Mitt. münch. ent. Ges., **5**: 63, pl. 4, fig. 142. Reiss, 1930, in Seitz, Die Gross-Schmetterlinge der Erde, Supplement, **2**: 39.

hispana Verity, 1920, Ent. Rec., **32**: 31. Rocci, 1935, Boll. Soc. ent. ital., **67**: 163; 1938, Redia, **24**: 175.

ab. **zickerti** Burgeff, 1914, Mitt. münch. ent. Ges., **5**: 63.

ssp. **sorrentina** Staudinger, 1894, Iris, **7**: 254. Calberla, 1895, Iris, **8**: 223. Spuler, 1906, in Hofmann, Die Schmetterlinge Europas, **2**: 161, pl. 77, fig. 21c. Seitz, 1907, Die Gross-Schmetterlinge der Erde, **2**: 23, pl. 5i. Verity, 1920, Ent. Rec., **32**: 31. Rocci, 1938, Redia, **24**: 170. — Halbinsel von Sorrent, Campanien, Umgebung von Neapel, Italien.

rubromixta Stauder, 1921, N. Beitr. syst. Insektenk., **2**: 31. Tremewan, 1966, Ent. Rec., **78**: 33, pl. 1, fig. 6.

azurea Stauder, 1923, Ent. Anz., **3**: 21.

ab. **viridescens** Stauder, 1923, Ent. Anz., **3**: 21.

ab. **chrysomelas** Stauder, 1923, Ent. Anz., **3**: 21.
ab. **sexmacula** Dziurzyński, 1908, Berl. ent. Z., **53**:5, 34. Seitz, 1912, Die Gross-Schmetterlinge der Erde, **2**: 442. Reiss, 1930, in Seitz, Die Gross-Schmetterlinge der Erde, Supplement, **2**: 39.
ab. **holiki** Stauder, 1922, Iris, **36**: 42.
ab. **cynariformis** Stauder, 1921, N. Beitr. syst. Insektenk., **2**: 31.
ab. **galvagnii** Stauder, 1915, Iris, **29**: 32.
ab. **gramanni** Stauder, 1915, Z. wiss. InsektBiol., **11**: 135, pl. 2, fig. 21. Reiss, 1930, in Seitz, Die Gross-Schmetterlinge der Erde, Supplement, **2**: 39.
ab. **sheljuzhkoi** Stauder, 1915, Z. wiss. InsektBiol.,**11**: 136.
ab. **spoliata** Stauder, 1915, Z. wiss. InsektBiol., **11**: 136.
ab. **evanescens** Stauder, 1921, N. Beitr. syst. Insektenk., **2**: 31.
ab. **crassimaculata** Stauder, 1921, N. Beitr. syst. Insektenk., **2**: 31.
ab. **heptamacula** Stauder, 1921, N. Beitr. syst. Insektenk., **2**: 31.
ab. **quadrupla** Stauder, 1921, N. Beitr. syst. Insektenk., **2**: 31.
ab. **depauperata** Burgeff, 1926, in Strand, Lepid. Cat., **33**: 68. Reiss, 1930, in Seitz, Die Gross-Schmetterlinge der Erde, Supplement, **2**: 39.
ab. **pseudomaritima** Turati, 1910, Boll. Lab. Zool. Portici, **4**: 161.
pseudomaritima Burgeff, 1926, Mitt. münch. ent. Ges., **16**: 77. Reiss, 1930, in Seitz, Die Gross-Schmetterlinge der Erde, Supplement, **2**: 39.
ab. **pseudolitorea** Burgeff, 1926, Mitt. münch. ent. Ges., **16**: 78. Reiss, 1930, in Seitz, Die Gross-Schmetterlinge der Erde, Supplement, **2**: 39.
ab. **obscurissima** Cannaviello, 1903, Rev. Soc. ent. namur., p. 45.
pseudocalabrica Burgeff, 1926, in Strand, Lepid. Cat., **33**: 68. Reiss, 1930, in Seitz, Die Gross-Schmetterlinge der Erde, Supplement, **2**: 39.
ab. **posticebipuncta** Stauder, 1922, Iris, **36**: 42. Reiss, 1930, in Seitz, Die Gross-Schmetterlinge der Erde, Supplement, **2**: 39.
ab. **posticetripuncta** Stauder, 1922, Iris, **36**: 42. Reiss, 1930, in Seitz, Die Gross-Schmetterlinge der Erde, Supplement, **2**: 39.
ab. **roseopicta** Turati, 1910, Boll. Lab. Zool. Portici, **4**: 162. Reiss, 1930, in Seitz, Die Gross-Schmetterlinge der Erde, Supplement, **2**: 39.
ab. **albinotica** Stauder, 1915, Z. wiss. InsektBiol., **11**: 135. Tremewan, 1966, Ent. Rec. **78**: 33, pl. 1, fig. 7.
ab. **flavoalbescens** Stauder, 1915, Z. wiss. InsektBiol., **11**: 136.

ab. **flavescens** Turati, 1910, Boll. Lab. Zool. Portici, **4**: 162. Reiss, 1930, in Seitz, Die Gross-Schmetterlinge der Erde, Supplement, **2**: 39.

ab. **aurantiaca** Oberthür, 1911, Études de Lépidoptérologie comparée, **5**(1): 208, pl. 62, fig. 583. Tremewan, 1961, Bull. Brit. Mus. (nat. Hist.) Ent., **10**(7): 278, pl. 53, fig. 23.

ab. **carnea** Stauder, 1915, Z. wiss. InsektBiol., **11**: 136.

ab. **adflata** Turati, 1910, Boll. Lab. Zool. Portici, **4**: 161.

ab. **aureomaculata** Stauder, 1921, N. Beitr. syst. Insektenk., **2**: 30. Tremewan, 1966, Ent. Rec., **78**: 33, pl. 1, fig. 8.

ab. **flavomixta** Stauder, 1921, N. Beitr. syst. Insektenk., **2**: 31. Tremewan, 1966, Ent. Rec., **78**: 34, pl. 1, fig. 9.

ab. **pentachroma** Stauder, 1921, N. Beitr. syst. Insektenk., **2**: 31.

ab. **ochraceomaculata** Stauder, 1921, N. Beitr. syst. Insektenk., **2**: 31. Tremewan, 1966, Ent. Rec., **78**: 34, pl. 1, fig. 10.

ab. **xanthographa** Germar, 1837/38, Fauna Insectorum Europae, **16**, pl. 22. Herrich-Schäffer, 1844, Systematische Bearbeitung der Schmetterlinge von Europa, **2**, pl. 5, fig. 40; 1846, ibidem, **2**: 48. Reiss, 1930, in Seitz, Die Gross-Schmetterlinge der Erde, Supplement, **2**: 39.

boisduvalii Heydenreich, 1851, Lepidopterorum Europaeorum Catalogus Methodicus, p. 22 (nomen nudum). Oberthür, 1896, Études d'Entomologie, **20**: 44, pl. 7, fig. 111. Dziurzyński, 1904, Jber. wien. ent. Ver., **14**: 50, pl. 2, fig. 9. Zickert, 1904, Ent. Z., **18**: 9, figs. a, b. Seitz, 1907, Die Gross-Schmetterlinge der Erde, **2**: 23, pl. 5i. Oberthür, 1911, Études de Lépidoptérologie comparée, **5**(1), pl. 62, figs. 580, 582.

boisduvalii Lederer, 1853, Verh. zool.-bot. Ver. Wien. **2**: 71, 97.

ab. **zickerti** Hoffmann, 1904, Ent. Z., **18**: 9. Seitz, 1907, Die Gross-Schmetterlinge der Erde, **2**: 23.

zickerti Burgeff, 1926, in Strand, Lepid. Cat., **33**: 68. Reiss, 1930, in Seitz, Die Gross-Schmetterlinge der Erde, Supplement, **2**: 39.

ab. **sexmaculata** Dziurzyński, 1908, Berl. ent. Z., **53**: 35.

ab. **verityi** Stauder, 1921, N. Beitr. syst. Insektenk., **2**: 30. Reiss, 1930, in Seitz, Die Gross-Schmetterlinge der Erde, Supplement, **2**: 39.

ab. **centripunctata** Stauder, 1921, N. Beitr. syst. Insektenk., **2**: 30. Reiss, 1930, in Seitz, Die Gross-Schmetterlinge der Erde, Supplement, **2**: 39.

ab. **radiatula** Stauder, 1921, N. Beitr. syst. Insektenk., **2**: 30. Reiss, 1930, in Seitz, Die Gross-Schmetterlinge der Erde, Supplement, **2**: 39.

ab. **nigroinspersa** Burgeff, 1926, Mitt. münch. ent. Ges., **16**:

77. Reiss, 1930, in Seitz, Die Gross-Schmetterlinge der Erde, Supplement, **2**: 39.

ssp. **collina** Burgeff, 1926, Mitt. münch. ent. Ges., **16**: 77. Reiss, 1930, in Seitz, Die Gross-Schmetterlinge der Erde, Supplement, **2**: 39. Rocci, 1938, Redia, **24**: 168. — Formia, Provinz Caserta, Italien.

italica Dziurzyński, 1904, Jber. wien. ent.Ver., **14**: 50 (infrasubspecific). Seitz, 1907, Die Gross-Schmetterlinge der Erde, **2**: 23, pl. 5h; 1912, ibidem, **2**: 442.

ab. **pseudosorrentina** Turati, 1910, Boll. Lab. Zool. Portici, **4**: 161. Oberthür, 1911, Études de Lépidoptérologie comparée, **5** (1): 207, pl. 62, fig. 564. Reiss, 1930, in Seitz, Die Gross-Schmetterlinge der Erde, Supplement, **2**: 39.

ab. **impar** Oberthür, 1911, Études de Lépidoptérologie comparée, **5** (1): 206, pl. 62, figs. 569, 571. Tremewan, 1961, Bull. Brit. Mus. (nat. Hist.) Ent., **10** (7): 277, pl. 53, fig. 22.

ab. **depauperata** Turati, 1910, Boll. Lab. Zool. Portici, **4**: 162. Reiss, 1930, in Seitz, Die Gross-Schmetterlinge der Erde, Supplement, **2**: 39.

ab. **annulata** Turati, 1910, Boll. Lab. Zool. Portici, **4**: 161. Reiss, 1930, in Seitz, Die Gross-Schmetterlinge der Erde, Supplement, **2**: 39.

ab. **burgeffi** Rocci, 1938, Redia, **24**: 169.

ab. **lutea** Turati, 1910, Boll. Lab. Zool. Portici, **4**: 161.

ssp. **albana** Burgeff, 1926, Mitt. münch. ent. Ges., **16**: 78. Reiss, 1930, in Seitz, Die Gross-Schmetterlinge der Erde, Supplement, **2**: 39. Rocci, 1938, Redia, **24**: 169. — Bei Albano, Albanergebirge, Italien, 300 m.

ssp. **tenuissima** Burgeff, 1914, Mitt. münch. ent. Ges., **5**: 63. Perlini, 1905, Forme di Lepidotteri esclusivamente Italiane, p. 55, pl. 2, fig. 12 (as *boisduvalii* Heydenreich). Reiss, 1930, in Seitz, Die Gross-Schmetterlinge der Erde, Supplement, **2**: 39. Rocci, 1938, Redia, **24**: 139. — Sarracinesco; Monte Gennaro, 1200 m., Italien.

ssp. **sabina** Rocci, 1938, Redia, **24**: 140. — Monte Gennaro, 800 m., Fiuggi, Italia.

ssp. **latina** Verity, 1920, Ent. Rec., **32**: 31; 1920, Boll. Lab. Zool. Portici, **14**: 40. Querci, 1920, Ent. Rec., **32**: 27, 28. Burgeff, 1926, Mitt. münch. ent. Ges., **16**: 78. Reiss, 1930, in Seitz, Die Gross-Schmetterlinge der Erde, Supplement, **2**: 39, pl. 4d, e. Rocci, 1938, Redia, **24**: 175. — Villalatina, Mainarde mountains, Caserta, Italy.

ssp. **tilaventa** Holik, 1935, Z. öst. EntVer., **20**: 62. Rocci, 1938, Redia, **24**: 131. Forster & Wohlfahrt, 1958, Die Schmetterlinge Mitteleuropas, **3**: 99, pl. 11, fig. 27. Reichl, 1962, Biol. Glasnik, **15**: 141. — Hügel am Flusslauf des Tagliamento, Friaul, Norditialen

rubra Holik, 1935, Z. öst. EntVer., **20**: 62.

ab. **sorrentinaeformisrubra** Holik, 1935, Z. öst. EntVer., **20**: 62.
ab. **quinquemacularubra** Reiss, 1963, Stuttgart. Beitr. Naturk., nr. 122: 3.
ab. **flava** Holik, 1935, Z. öst. EntVer., **20**: 62. Forster & Wohlfahrt, 1958, Die Schmetterlinge Mitteleuropas, **3**, pl. 11, fig. 28.
ab. **sorrentinaeformisflava** Holik, 1935, Z. öst. EntVer., **20**: 62.
ab. **quinquemaculaflava** Reiss, 1963, Stuttgart. Beitr. Naturk., nr. 122: 3, figs. 7, 8.
ab. **fumosaflava** Reiss, 1963, Stuttgart. Beitr. Naturk., nr. 122: 2, figs. 5, 6.

ssp. **mauriae** Rocci, 1936, Boll. Soc. ent. ital., **68**: 45; 1938, Redia, **24**: 129. — Passo della Mauria, Tagliamento, Italia, 800-1200 m.

ssp. **emendata** Verity, 1916, Boll. Soc. ent. ital., **47**: 76; 1920, Ent. Rec., **32**: 30. Rocci, 1918, Atti Soc. ligust. Sci. nat. geogr., **28**, pl. 2, figs. 1a-1c, 2a. Burgeff, 1926, Mitt. münch. ent. Ges., **16**: 72. Reiss, 1930, in Seitz, Die Gross-Schmetterlinge der Erde, Supplement, **2**: 38, pl. 4d. Rocci, 1938, Redia, **24**: 145. — Colle Torri presso Macerata, Marche, Italia.
emendataeformis Rocci, 1918, Atti Soc. ligust. Sci. nat. geogr., **28**: 127.
ab. **flava** Sicher, 1906, Jber. wien. ent. Ver., **16**: 91, pl. 1, fig. 9. Reiss, 1930, in Seitz, Die Gross-Schmetterlinge der Erde, Supplement, **2**: 39, pl. 4d.
f. t. **autumnalis** Verity, 1916, Boll. Soc. ent. ital., **47**: 76. — Sibillini, Italia.
ab. **anticeconjuncta** Verity, 1916, Boll. Soc. ent. ital., **47**: 77.

ssp. **altitudinaria** Turati, 1910, Boll. Lab. Zool. Portici, **4**: 161. Oberthür, 1911, Études de Lépidoptérologie comparée, **5** (1): 209. Rocci, 1916, Atti Soc. ligust. Sci. nat. geogr., **27**: 23; 1918, ibidem, **28**: 130, pl. 2, figs. 7d, 8b. Verity, 1920, Ent. Rec., **32**: 30. Reiss, 1930, in Seitz, Die Gross-Schmetterlinge der Erde, Supplement, **2**: 40, pl. 4e. — Gran Sasso, Alte valli intermedie dell'Abruzzo, Italia, 1500-1800 m.
vernalis Costantini, 1916, Atti Soc. Nat. Mat. Modena, **49**: 29.
altitudinariaeformis Rocci, 1918, Atti Soc. ligust. Sci. nat. geogr., **28**: 127.
subalticoloides Rocci, 1918, Atti Soc. ligust. Sci. nat. geogr., **28**: 131.
ab. **cingulata** Turati, 1923, Atti Soc. ital. Sci. nat., **62**: 44. Reiss, 1930, in Seitz, Die Gross-Schmetterlinge der Erde, Supplement, **2**: 40.
ab. **privata** Turati, 1910, Boll. Lab. Zool. Portici, **4**: 161.
privata Costantini, 1916, Atti Soc. Nat. Mat. Modena, **49**: 29. Reiss, 1930, in Seitz, Die Gross-Schmetterlinge der Erde, Supplement, **2**: 40.

ab. **depauperata** Costantini, 1916, Atti Soc. Nat. Mat. Modena, **49**: 29. Reiss, 1930, in Seitz, Die Gross-Schmetterlinge der Erde, Supplement, **2**: 40.

ab. **amplomarginata** Rocci, 1916, Atti Soc. ligust. Sci. nat. geogr., **27**: 27; 1918, ibidem, **28**: 133, pl. 2, figs. 8a, 8b, 11a-11d, pl. 3, fig. 8b. Reiss, 1930, in Seitz, Die Gross-Schmetterlinge der Erde, Supplement, **2**: 40.

ab. **sorrentinaeformis** Rocci, 1918, Atti Soc. ligust. Sci. nat. geogr., **28**: 133, pl. 2, fig. 10d, pl. 3, figs. 7d, 8a. Reiss, 1930, in Seitz, Die Gross-Schmetterlinge der Erde, Supplement, **2**: 40.

f. t. **aestivalis** Oberthür, 1910, Études de Lépidoptérologie comparée, **4**: 663. Tremewan, 1961, Bull. Brit. Mus. (nat. Hist.) Ent., **10** (7): 278, pl. 53, fig. 24. — Roccaroso et Palena, Italie.

minima Costantini, 1916, Atti Soc. Nat. Mat. Modena, **49**: 29.

ssp. **nigraltudinaria** Rocci, 1937, Redia, **22**: 142; 1938, ibidem, **24**: 138. — Mte. Autore, Appennino, Italia, 1200 m.

ab. **nigerrima** Rocci, 1937, Redia, **22**: 142.

ab. **pseudosorrentinaeformis** Rocci, 1937, Redia, **22**: 142.

ab. **flava** Rocci, 1937, Redia, **22**: 142.

ssp. **promunturii** Burgeff, 1921, Mitt. münch. ent. Ges., **11**: 54; 1926, ibidem, **16**: 75. Reiss, 1930, in Seitz, Die Gross-Schmetterlinge der Erde, Supplement, **2**: 39. Rocci, 1938, Redia, **24**: 160. Burgeff, 1951, Biol. Zbl., **70**: 18, figs. 13. Haaf, 1952, Veröff. zool. Staatssamml. Münch., **2**: 152, 154, 157, pl. 13 (as *maritima* Oberthür). — Capo Mele, Laigueglia, Italienische Riviera.

ab. **anthrax** Burgeff, 1926, Mitt. münch. ent. Ges., **16**: 75. Reiss, 1930, in Seitz, Die Gross-Schmetterlinge der Erde, Supplement, **2**: 39.

ssp. **pinguis** Burgeff, 1926, Mitt. münch. ent. Ges., **16**: 76. Rocci, 1916, Atti Soc. ligust. Sci. nat. geogr., **27**: 6 (as *maritima* Oberthür); 1918, ibidem, **28**: 120, pl. 2, fig. 1d (as *maritima* Oberthür). Reiss, 1930, in Seitz, Die Gross-Schmetterlinge der Erde, Supplement, **2**: 39. Rocci, 1938, Redia, **24**: 162. — Ligurischer Appenin um Genua (nicht unmittelbar am Meer), Italien.

ab. **incompleta** Rocci, 1914, Atti Soc. ligust. Sci. nat. geogr., **25**: 221.

ab. **pseudosorrentina** Burgeff, 1926, Mitt. münch. ent. Ges., **16**: 76. Reiss, 1930, in Seitz, Die Gross-Schmetterlinge der Erde, Supplement, **2**: 39.

ab. **parva** Rocci, 1914, Atti Soc. ligust. Sci. nat. geogr., **24**: 115; 1918, Atti Soc. ligust. Sci. nat. geogr., **28**: 157, pl. 3, fig. 4a.

parva Rocci, 1914, Soc. ent., **29**: 42.

ab. **parvimaculata** Rocci, 1914, Atti Soc. ligust. Sci. nat. geogr., **24**: 115. Reiss, 1930, in Seitz, Die Gross-Schmetterlinge der Erde, Supplement, **2**: 39.

parvimaculata Rocci, 1914, Soc. ent., **29**: 42.
ab. **decirclata** Rocci, 1914, Atti Soc. ligust. Sci. nat. geogr., **25**: 221.
ab. **apicalis** Rocci, 1914, Atti Soc. ligust. Sci. nat. geogr., **24**: 115; 1918, Atti Soc. ligust. Sci. nat. geogr., **28**: 157, pl. 3, fig. 5c.
apicalis Rocci, 1914, Soc. ent., **29**: 41.
ab. **circumscripta** Turati, 1919, Atti Soc. ital. Sci. nat., **58**: 175. Reiss, 1930, in Seitz, Die Gross-Schmetterlinge der Erde, Supplement, **2**: 39.
ab. **amplimacula** Rocci, 1914, Atti Soc. ligust. Sci. nat. geogr., **25**: 221.
amplomaculata Rocci, 1916, Atti Soc. ligust. Sci. nat. geogr., **27**: 12; 1918, ibidem, **28**: 157, pl. 3, fig. 4b. Reiss, 1930, in Seitz, Die Gross-Schmetterlinge der Erde, Supplement, **2**: 39.
ab. **aurantiaca** Rocci, 1914, Atti Soc. ligust. Sci. nat. geogr., **24**: 115.
aurantiaca Rocci, 1914, Soc. ent., **29**: 42.
ab. **diffusa** Rocci, 1914, Atti Soc. ligust. Sci. nat. geogr., **24**: 115; 1918, Atti Soc. ligust. Sci. nat. geogr., **28**: 157, pl. 3, fig. 5d. Reiss, 1930, in Seitz, Die Gross-Schmetterlinge der Erde, Supplement, **2**: 39.
diffusa Rocci, 1914, Soc. ent., **29**: 42.
ab. **rosea** Rocci, 1916, Atti Soc. ligust. Sci. nat. geogr., **27**: 14.
f. t. **postpinguis** Rocci, 1938, Redia, **24**: 163. — Riviere ligure in località un po distanti dal mare, Italia.

ssp. **intermedia** Rocci, 1914, Atti Soc. ligust. Sci. nat. geogr., **24**: 115; 1916, Atti Soc. ligust. Sci. nat. geogr., **27**: 15; 1918, ibidem, **28**: 125, pl. 2, figs. 6a-6d. Verity, 1920, Ent. Rec., **32**: 30. Rocci, 1938, Redia, **24**: 152. Reiss & Tremewan, 1964, Ent. Rec., **76**: 134. — Val Scrivia da Romo; Val Bisagno, Appennino, 400-600 m., Italia.
intermedia Rocci, 1914, Soc. ent., **29**: 42.
cisapennina Rocci, 1918, Atti Soc. ligust. Sci. nat. geogr., **28**: 127.
paramaritima Rocci, 1918, Atti Soc. ligust. Sci. nat. geogr., **28**: 127.
pseudoemendata Rocci, 1918, Atti Soc. ligust. Sci. nat. geogr., **28**: 128.
maritimoides Rocci, 1918, Atti Soc. ligust. Sci. nat. geogr., **28**: 128.
vulgaris Rocci, 1918, Atti Soc. ligust. Sci. nat. geogr., **28**: 128.
paralticolaria Rocci, 1918, Atti Soc. ligust. Sci. nat. geogr., **28**: 128.

interjacens Burgeff, 1926, Mitt. münch. ent. Ges., **16**: 74. Reiss, 1930, in Seitz, Die Gross-Schmetterlinge der Erde, Supplement, **2**: 39. Reiss & Tremewan, 1964, Ent. Rec., **76**: 134.

ab. **sorrentinoides** Rocci, 1918, Atti Soc. ligust. Sci. nat. geogr., **28**: 130.

ab. **quinqueguttata** Rocci, 1914, Atti Soc. ligust. Sci. nat. geogr., **25**: 221. Reiss, 1930, in Seitz, Die Gross-Schmetterlinge der Erde, Supplement, **2**: 39.

f. t. **pinguisintermedia** Rocci, 1941, Boll. Ist. Ent. Univ. Bologna, **13**: 119. — Appennino ligure-piemontese.

ssp. **subalticola** Rocci, 1918, Atti Soc. ligust. Sci. nat. geogr., **28**: 127 (nomen novum for *intermedia* Costantini). — Appennino ligure, Italia, 800-1000 m.

intermedia Costantini, 1916, Atti Soc. Nat. Mat. Modena, **49**: 29 (preoccupied by **intermedia** Rocci, 1914, ssp. of **transalpina** Esper).

vernalis Costantini, 1916, Atti Soc. Nat. Mat. Modena, **49**: 29 (infrasubspecific). Rocci, 1938, Redia, **24**: 156.

pseudointermedia Rocci, 1918, Atti Soc. ligust. Sci. nat. geogr., **28**: 131, pl. 2, figs. 9b, 10a-10c (infrasubspecific). Reiss, 1930, in Seitz, Die Gross-Schmetterlinge der Erde, Supplement, **2**: 39. Rocci, 1938, Redia, **24**: 137.

pseudointermedia Burgeff, 1926, Mitt. münch. ent. Ges., **16**: 74.

ab. **privata** Costantini, 1916, Atti Soc. Nat. Mat. Modena, **49**: 29. Reiss, 1930, in Seitz, Die Gross-Schmetterlinge der Erde, Supplement, **2**: 39.

ab. **depauperata** Costantini, 1916, Atti Soc. Nat. Mat. Modena, **49**: 29. Reiss, 1930, in Seitz, Die Gross-Schmetterlinge der Erde, Supplement, **2**: 39.

ab. **emendataeformis** Rocci, 1938, Redia, **24**: 157.

f. t. **minima** Costantini, 1916, Atti Soc. Nat. Mat. Modena, **49**: 29. — Appennino ligure, Italia, 800-1000 m.

ssp. **tigulii** Rocci, 1938, Redia, **24**: 161. — Mte. Portofino, Liguria, Italia, 250 m.

ssp. **macromaritima** Rocci, 1936, Boll. Soc. ent. ital., **68**: 46; 1938, Redia, **24**: 168. — Livorno, Italia.

ssp. **sorrentinaeformis** Rocci, 1938, Redia, **24**: 139. — Mte. Mosca, Appennino lucchese, Italia, 1300 m.

ab. **obscurata** Rocci, 1938, Redia, **24**: 139.

ssp. **sorrentinoides** Rocci, 1938, Redia, **24**: 155. — Avellano, Appennino marchigiano, Italia, 600 m.

ssp. **litorea** Burgeff, 1926, Mitt. münch. ent. Ges., **16**: 76 (nomen novum for *transiens* Rocci). Reiss, 1930, in Seitz, Die Gross-Schmetterlinge der Erde, Supplement, **2**: 39, pl. 4d. — Küstengebiet um Genua, Norditalien.

transiens Rocci, 1913, Soc. ent., **28**: 56 (preoccupied by **transiens** Staudinger, 1887, ssp. of **carniolica** Scopoli); 1914, Atti Soc. ligust. Sci. nat. geogr., **25**: 221; 1916, ibidem, **27**: 18; 1918, ibidem, **28**, pl. 2, figs. 3b-3d, 4a. Verity, 1920, Ent. Rec., **32**: 32. Rocci, 1938, Redia, **24**: 165.

ab. **apicalis** Rocci, 1914, Atti Soc. ligust. Sci. nat. geogr., **24**: 115.

apicalis Rocci, 1914, Soc. ent., **29**: 41.

ab. **pseudocalabrica** Rocci, 1916, Atti Soc. ligust. Sci. nat. geogr., **27**: 22. Reiss, 1930, in Seitz, Die Gross-Schmetterlinge der Erde, Supplement, **2**: 39.

ab. **pseudostoechadis** Rocci, 1914, Atti Soc. ligust. Sci. nat. geogr., **24**: 115.

pseudostoechadis Rocci, 1914, Soc. ent., **29**: 42.

ab. **pentasignata** Rocci, 1916, Atti Soc. ligust. Sci. nat. geogr., **27**: 22.

ab. **diffusa** Rocci, 1914, Atti Soc. ligust. Sci. nat. geogr., **24**: 115.

diffusa Rocci, 1914, Soc. ent., **29**: 42.

ab. **parva** Rocci, 1914, Atti Soc. ligust. Sci. nat. geogr., **24**: 115.

parva Rocci, 1914, Soc. ent., **29**: 42.

ab. **depuncta** Turati, 1910, Boll. Lab. Zool. Portici, **4**: 162.

ab. **decimaculata** Rocci, 1913, Soc. ent., **28**: 56; 1918, Atti Soc. ligust. Sci. nat. geogr., **28**: 157, pl. 3, fig. 4d.

ab. **parvimaculata** Rocci, 1914, Atti Soc. ligust. Sci. nat. geogr., **24**: 115; 1918, Atti Soc. ligust. Sci. nat. geogr., **28**: 157, pl. 3, fig. 4c.

parvimaculata Rocci, 1914, Soc. ent., **29**: 42.

ab. **undecimaculata** Rocci, 1913, Soc. ent., **28**: 56.

ab. **aurantiaca** Rocci, 1914, Atti Soc. ligust. Sci. nat. geogr., **24**: 115.

aurantiaca Rocci, 1914, Soc. ent., **29**: 42.

f. t. **macrotransiens** Rocci, 1938, Redia, **24**: 166. — Genova, Italia.

ssp. **dominatrix** Rocci, 1941, Boll. Ist. Ent. Univ. Bologna, **13**: 118, pl. 2, figs. 20-22. — Collina di Torino, Piedmont, Italia, 700 m.

ssp. **subemendata** Rocci, 1936, Boll. Soc. ent. ital., **68**: 45; 1938, Redia, **24**: 147. — Boschi di Mirafiori, Torino, Italia.

ssp. **maritima** Oberthür, 1898, Bull. Soc. ent. Fr., p. 22; 1907, — La Turbie, Nice, Alpes-Maritimes, France.

Ann. Soc. ent. Fr., **76**: 39. Spuler, 1906, in Hofmann, Die Schmetterlinge Europas, **2**: 161 (partim). Seitz, 1907, Die Gross-Schmetterlinge der Erde, **2**: 23, pl. 5h. Oberthür, 1909, Études de Lépidoptérologie comparée, **3**, pl. 30, figs. 190, 191. Burgeff, 1926, Mitt. münch. ent. Ges., **16**: 74. Rocci, 1935, Mem. Soc. ent. ital., **14**: 47-58, figs.; 1938, Redia, **24**: 143. Tremewan, 1961, Bull. Brit. Mus. (nat. Hist.) Ent., **10** (7): 276, pl. 53, fig. 17.

ab. **pseudopromunturii** Reiss, 1953, Ent. Z., **63**: 95.

ab. **trimaculata** Oberthür, 1909, Études de Lépidoptérologie comparée, **3**, pl. 30, fig. 189. Tremewan, 1961, Bull. Brit. Mus. (nat. Hist.) Ent., **10** (7): 277, pl. 53, fig. 19. Reiss & Tremewan, 1964, Ent. Rec., **76**: 134.

trimacula Reiss, 1930, in Seitz, Die Gross-Schmetterlinge der Erde, Supplement, **2**: 39. Reiss & Tremewan, 1964, Ent. Rec., **76**: 134.

ab. **depauperata** Reiss, 1953, Ent. Z., **63**: 95.

ab. **infuscata** Reiss, 1958, Bull. Soc. ent. Mulhouse, p. 51.

ab. **albicinctus** Reiss, 1958, Bull. Soc. ent. Mulhouse, p. 51.

f. t. **postmaritima** Dujardin, 1956, Bull. mens. Soc. linn. Lyon, **25**: 258. — Nice, quartier de la Conque, Alpes-Maritimes, France.

ssp. **altamaritima** Dujardin, 1956, Bull. mens. Soc. linn. Lyon, **25**: 257. — La Colmiane (Valdeblore), St. Martin-de-Vésubie, Alpes-Maritimes, France, 1500 m.

ssp. **helvetica** Rothschild & Bethune-Baker, 1920, Proc. ent. Soc. Lond., p. xlix. Tremewan, 1961, Bull. Brit. Mus. (nat. Hist.) Ent., **10** (7): 276, pl. 53, fig. 18. — Moulinet, Alpes-Maritimes, France.

ssp. **transalpina** Esper, 1781, Die Schmetterlinge, **2**: 142; 1779, ibidem, **2**, pl. 16, fig. f (as *filipendulae* Linné). Perlini, 1905, Forme di Lepidotteri esclusivamente Italiane, p. 54, pl. 6, fig. 15. Spuler, 1906, in Hofmann, Die Schmetterlinge Europas, **2**: 160, pl. 72, fig. 3b. Seitz, 1907, Die Gross-Schmetterlinge der Erde, **2**: 23, pl. 5g. Oberthür, 1910, Études de Lépidoptérologie comparée, **4**: 578; 1911, ibidem, **5** (1): 198. Querci, 1912, in Oberthür, Études de Lépidoptérologie comparée, **6**: 151, 171. Turati, 1912, in Oberthür, Études de Lépidoptérologie comparée, **6**: 180. Verity, 1916, Boll. Soc. ent. ital., **47**: 75. Reiss, 1930, in Seitz, Die Gross-Schmetterlinge der Erde, Supplement, **2**: 38. Rocci, 1935, Mem. Soc. ent. ital., **14**: 47-58, figs.; 1938, Redia, **24**: 113. Verity, 1946, — Italienische Voralpen [Umgebung des Lago Maggiore].

Redia, **31**: 69. Alberti, 1956, Dtsch. ent. Z., (N.F.), **3**: 91, figs.; 1956, Z. wien. ent. Ges., **41**: 231-239, figs.; 1958, Bull. Soc. ent. Mulhouse, p. 1-9, figs.; 1958, Mitt. zool. Mus. Berl., **34**: 322. Forster & Wohlfahrt, 1958, Die Schmetterlinge Mitteleuropas, **3**: 99, pl. 11, figs. 3, 8, 13.

ratisbonica Scheven, 1782, in Fuessly, Neues Magazin für die Liebhaber der Entomologie, **1**: 54. Fuessly, 1778, Magazin für die Liebhaber der Entomologie, **1**: 126; 1778, ibidem, **1**: 139, pl. 1, fig. 3.

ssp. **redempta** Rocci, 1937, Redia, **22**: 139; 1938, ibidem, **24**: 150. — Garniga (Mte Bondone), Trento, Italia, 700-800 m.

ssp. **hilfi** Reiss, 1922, Int. ent. Z., **15**: 176. Burgeff, 1926, Mitt. münch. ent. Ges., **16**: 78. Reiss, 1930, in Seitz, Die Gross-Schmetterlinge der Erde, Supplement, **2**: 40, pl. 4e. Rocci, 1938, Redia, **24**: 130. Holik, 1942, Ent. Z., **55**: 237; 1948, Mitt. münch. ent. Ges., **34**: 399. — Fuzine, Litoral Kroatiens; Istrien.

ab. **sticheli** Stauder, 1915, Iris, **29**: 32.

ab. **paulae** Naufock, 1913, Boll. Soc. adriat. Sci. nat., **27**: 103.

ssp. **athicaria** Burgeff, 1926, Mitt. münch. ent. Ges., **16**: 79. Reiss, 1930, in Seitz, Die Gross-Schmetterlinge der Erde, Supplement, **2**: 40. Rocci, 1938, Redia, **24**: 151. — Umgebung von Bozen, Etschtal, Südtirol.

costazzina Stauder, 1929, Ent. Z., **43**: 81. Tremewan, 1961, Bull. Brit. Mus. (nat. Hist.) Ent., **10** (7): 278, pl. 53, fig. 25.

korbi Reiss, 1930, in Seitz, Die Gross-Schmetterlinge der Erde, Supplement, **2**: 42, pl. 4g; 1935, Int. ent. Z., **29**: 229. Holik, 1942, Ent. Z., **55**: 236. Holik & Sheljuzhko, 1958, Mitt. münch. ent. Ges., **48**: 234. — [?Nordkaukasus].

ab. **cingulata** Burgeff, 1914, Mitt. münch. ent. Ges., **5**: 63. Reiss, 1930, in Seitz, Die Gross-Schmetterlinge der Erde, Supplement, **2**: 40.

ab. **hyalina** Stauder, 1929, Ent. Z., **43**: 132.

ssp. **nambinica** Rocci, 1936, Boll. Soc. ent. ital., **68**: 43; 1938, Redia, **24**: 128. — Lago Nambino (Madonna di Campiglio), Italia, 1800-2000 m.

ssp. **misurinae** Rocci, 1936, Boll. Soc. ent. ital., **68**: 43; 1938, Redia, **24**: 128. — Misurina, Trentino, Italia, 1800 m.

ssp. **subjugi** Rocci, 1936, Boll. Soc. ent. ital., **68**: 44; 1938, Redia, **24**: 128. — Bormio, Val Venosta, Italia, 1500 m.

ssp. **tonalensis** Rocci, 1936, Boll. Soc. ent. ital., **68**: 43; 1938, Redia, **24**: 128. — Ponte di Legno, Passo del Tonale, Val Camonica, Italia, 1500 m.

ssp. **holikiana** Rocci, 1937, Redia, **22**: 140; 1938, ibidem, **24**: 129. — Zwiselstein (Zwieselstein), Tirolo, 1500 m.

ssp. **carentaniae** Rocci, 1937, Redia, **22**: 141; 1938, ibidem, **24**: 130. Holik, 1948, Mitt. münch. ent. Ges., **34**: 395. — Ulrichsberg sopra Klagenfurt, Carinzia.
ab. **pseudohilfi** Rocci, 1937, Redia, **22**: 141.

ssp. **poschiavica** Reiss, 1950, Jber. naturf. Ges. Graubünden, **82**: 123, fig. 31. — Campocologno, Puschlav, Graubünden, Schweiz, 600 m.
ab. **parvimaculata** Vorbrodt, 1913, in Vorbrodt & Müller-Rutz, Die Schmetterlinge der Schweiz, **2**: 273-274, figs. 29-31.

ssp. **alpina** Boisduval, 1834, Icones historique des Lépidoptères nouveaux ou peu connus, **2**: 66, pl. 53, fig. 9. Oberthür, 1907, Ann. Soc. ent. Fr., **76**: 39; 1909, Études de Lépidoptérologie comparée, **3**, pl. 30, figs. 194-196; 1911, ibidem, **5** (1): 211. Burgeff, 1926, Mitt. münch. ent. Ges., **16**: 79. Heinrich, 1926, Int. ent. Z., **20**: 202, figs. 1, 2. Reiss, 1930, in Seitz, Die Gross-Schmetterlinge der Erde, Supplement, **2**: 40, pl. 4e. Rocci, 1938, Redia, **24**: 125. Reiss, 1941, Mitt. münch. ent. Ges., **31**, pl. 34, fig. B9. Tremewan, 1961, Bull. Brit. Mus. (nat. Hist.) Ent., **10** (7): 277. — Grenoble, Isère, France.
ferulae Lederer, 1853, Verh. zool.-bot. Ver. Wien, **2**: 71, 96.
pseudoalpina Turati, 1910, Boll. Lab. Zool. Portici, **4**: 161. Rocci, 1938, Redia, **24**: 148.
alpicola Verity, 1920, Ent. Rec., **32**: 29.
ab. **cingulata** Burgeff, 1914, Mitt. münch. ent. Ges., **5**: 63.
ab. **reducta** Turati, 1910, Boll. Lab. Zool. Portici, **4**: 161. Reiss, 1930, in Seitz, Die Gross-Schmetterlinge der Erde, Supplement, **2**: 40.
quinquemaculata Vorbrodt, 1913, in Vorbrodt & Müller-Rutz, Die Schmetterlinge der Schweiz, **2**: 274.
ab. **basiconfluens** Vorbrodt, 1913, in Vorbrodt & Müller-Rutz, Die Schmetterlinge der Schweiz, **2**: 273, fig. 11.
ab. **medioconfluens** Vorbrodt, 1913, in Vorbrodt & Müller-Rutz, Die Schmetterlinge der Schweiz, **2**: 273, fig. 12.
ab. **apicaliconfluenta** Reiss, 1964, Coridon, (A) **6**: 6, fig. 3.
ab. **costaliconfluens** Vorbrodt, 1913, in Vorbrodt & Müller-Rutz, Die Schmetterlinge der Schweiz, **2**: 273.

costimaculata Rocci, 1921, Atti Soc. ligust. Sci. nat. geogr., **32**: 36.

ab. **costalielongata** Vorbrodt, 1913, in Vorbrodt & Müller-Rutz, Die Schmetterlinge der Schweiz, **2**: 273.

anticeconjuncta Rocci, 1921, Atti Soc. ligust. Sci. nat. geogr., **32**: 36.

ab. **rubrocosta** Rocci, 1921, Atti Soc. ligust. Sci. nat. geogr., **32**: 36.

ab. **rubrostriata** Rocci, 1921, Atti Soc. ligust. Sci. nat. geogr., **32**: 36.

ab. **rubropicta** Rocci, 1938, Redia, **24**: 127.

ab. **grisescens** Oberthür, 1920, Études de Lépidoptérologie comparée, **17**: 62, pl. 507, fig. 4248. Tremewan, 1961, Bull. Brit. Mus. (nat. Hist.) Ent., **10** (7): 277, pl. 53, fig. 21.

ab. **avellanea** Prohaska, 1915, Int. ent. Z., **9**: 51.

ab. **incarnata** Vorbrodt, 1921, Mitt. schweiz. ent. Ges., **13**: 204.

ab. **flava** Oberthür, 1907, Ann. Soc. ent. Fr., **76**: 39; 1909, Études de Lépidoptérologie comparée, **3**, pl. 30, fig. 200. Reiss, 1930, in Seitz, Die Gross-Schmetterlinge der Erde, Supplement, **2**: 40. Tremewan, 1961, Bull. Brit. Mus. (nat. Hist.) Ent., **10** (7): 277, pl. 53, fig. 20.

ssp. **segusterica** Dujardin, 1965, Entomops, Nice, no. 2: 50, fig. Dufay, 1966, Bull. mens. Soc. linn. Lyon, **35**: 72.	St. Auban, Basses-Alpes, France, 450 m.
ssp. **dufayi** Dujardin, 1965, Entomops, Nice, no. 2: 51, fig. Dufay, 1966, Bull. mens. Soc. linn. Lyon, **35**: 72.	Meyrueis, Lozère; Col Saint-Jean, Drôme, 1150 m.; St. Genis-Laval Obs., Rhône; Col du Rousset; Col de Crémonne, Haute-Drôme, France.
ssp. **astragalpina** Rocci, 1938, Redia, **24**: 120.	Kaiserstuhl, Valle del Reno, Svizzera [Schweiz], 1200 m.
ssp. **megastragali** Rocci, 1938, Redia, **24**: 119.	Reuchnet, Bözingen, Jura, Svizzera [Schweiz].
ssp. **victrix** Rocci, 1941, Boll. Ist. Ent. Univ. Bologna, **13**: 117, pl. 2, figs. 23-25.	Meana (Valle di Susa), Piedmont, Italia, 800-1000 m.

ssp. **piemontiae** Reiss, 1941, Mitt. münch. ent. Ges., **31**: 1001, pl. 34, fig. C9. — Cesana Claviere, Piemont, Italien, 1500-1900 m.

ssp. **frigidalpina** Rocci, 1941, Boll. Ist. Ent. Univ. Bologna, **13**: 116, pl. 2, figs. 14-17. — Forno Alpia Graie, Piedmont, Italia, 1300 m.

ssp. **lacustris** Rocci, 1938, Redia, **24**: 158. — Laveno, Como, Italia.

ssp. **hartigi** Rocci, 1936, Boll. Soc. ent. ital., **68**: 42. Hartig, 1930, Studi trentini, **11** (8) (3): 244, figs. 15-18. Rocci, 1938, Redia, **24**: 122. — Lajon, Valle dell' Isarco, Trentino, Italia, 900 m.

ssp. **jugi** Burgeff, 1926, Mitt. münch. ent. Ges., **16**: 80 (nomen novum for *altissima* Burgeff). Reiss, 1930, in Seitz, Die Gross-Schmetterlinge der Erde, Supplement, **2**: 40, pl. 4e. Rocci, 1938, Redia, **24**: 127. Reiss, 1941, Mitt. münch. ent. Ges., **31**: 1001, pl. 34, fig. A10; 1950, Jber. naturf. Ges. Graubünden, **82**: 122, fig. 30. — Stilfser Joch, Ortlergebiet, Österreich, 1800-2400 m.

altissima Burgeff, 1914, Mitt. münch. ent. Ges., **5**: 63, pl. 4, figs. 138-141 (preoccupied by **altissima** Burgeff, 1914, ssp. of **sogdiana** Erschoff). Reiss, 1926, Die Zygaenen Deutschlands, p. 25.

ssp. **splugena** Burgeff, 1926, Mitt. münch. ent. Ges., **16**: 80. Reiss, 1930, in Seitz, Die Gross-Schmetterlinge der Erde, Supplement, **2**: 40, pl. 4f. Rocci, 1938, Redia, **24**: 124. Reiss, 1950, Jber. naturf. Ges. Graubünden, **82**: 119, fig. 28. Burgeff, 1965, Nachr. Akad. Wiss. Göttingen, **2**, mat.-phys. Kl., no. 14: 198, fig. 11, nos. 1-6, fig. 21 (distribution map). — Bei San Giacomo im Mesoccotal; Dorf Splügen, Schweiz.

splugena Vorbrodt, 1925, Mitt. schweiz. ent. Ges., **13**: 464 (nomen nudum).

ab. **cingulata** Vorbrodt, 1925, Mitt. schweiz. ent. Ges., **13**: 464.

ab. **flavinrubra** Burgeff, 1926, Mitt. münch. ent. Ges., **16**: 81. Reiss, 1930, in Seitz, Die Gross-Schmetterlinge der Erde, Supplement, **2**: 41.

ssp. **rhodani** Burgeff, 1965, Nachr. Akad. Wiss. Göttingen, **2**, mat.-phys. Kl., no. 14: 198, fig. 12, nos. 1-6. — Von Fiesch bis über Sierre, ober Rhonetal, Schweiz.

ssp. **jurassoboica** Burgeff, 1926, Mitt. münch. ent. Ges., **16**: 81. Reiss, 1930, in Seitz, Die Gross-Schmetterlinge der Erde, Supplement, **2**: 41. Daniel, 1932, in Osthelder, Die Schmetterlinge Südbayerns, **1**: 575, pl. 21, figs. 34, 35. Rocci, 1938, Redia, **24**: 121. — Bei Bad Reichenhall; in der Ramsau; Brandkopf bei Berchtesgaden, 1100 m.; Königseegebiet, Südbayern, Süddeutschland.

ssp. **nantuatium** Verity, 1946, Redia, **31**: 76. — Bex; Lavey (Vaud), Svizzero [Schweiz].

ssp. **rhaetiaemixta** Reiss, 1950, Jber. naturf. Ges. Graubünden, **82**: 120. — Bergün, Albulatal, Schweiz, 1400-1500 m.
ab. **pallens** Ziegler, 1911, Int. ent. Z., **5**: 139.
ab. **apicaliconfluens** Vorbrodt, 1913, in Vorbrodt & Müller-Rutz, Die Schmetterlinge der Schweiz, **2**: 273, fig. 13.

ssp. **relicta** Reiss, 1950, Jber. naturf. Ges. Graubünden, **82**: 121, fig. 29. — Pontresina, Schafberg, Schweiz, 2200 m.

ssp. **glockneriana** Leinfest, 1949, Wien. ent. Rdsch., **1**: 11. — Heiligenblut, Grossglocknergebiet, Kärnten, Österreich, 1350 m.
ab. **cingulata** Leinfest, 1949, Wien. ent. Rdsch., **1**: 12.

ssp. **gulsensis** Daniel, 1954, Z. wien. ent. Ges., **39**: 68, pl. 2, figs. 25-30. — Gulsenburg, Oberes Murtal, Steiermark, Österreich, 600-900 m.

ssp. **angelicotransalpina** Daniel, 1954, Z. wien. ent. Ges., **39**: 60, pl. 3, figs. 37-50, 53-60; 1955, NachrBl. bayer. Ent., **4**: 49-56. Forster & Wohlfahrt, 1958, Die Schmetterlinge Mitteleuropas, **3**: 100, pl. 11, figs. 17, 22. — Puxberg bei Teufenbach, Oberes Murtal, Steiermark, Österreich.

ssp. **osthelderiana** Reiss, 1941, Mitt. münch. ent. Ges., **31**: 1003, pl. 34, figs. A11, B11. Daniel, 1932, in Osthelder, Die Schmetterlinge Südbayerns, **1**, pl. 21, figs. 13-18 (as *alpina* Boisduval). Reiss, 1950, Jber. naturf. Ges. Graubünden, **82**: 122. Forster & Wohlfahrt, 1958, Die Schmetterlinge Mitteleuropas, **3**: 99, pl. 11, figs. 4, 9, 14. — Umgebung von Kochel, Bayr. Alpen, Süddeutschland.
ab. **paradoxa** Michalk, 1931, Int. ent. Z., **25**: 261-262, fig. Reiss, 1933, in Seitz, Die Gross-Schmetterlinge der Erde, Supplement, **2**: 277.
ab. **apicaliconfluens** Reiss, 1941, Mitt. münch. ent. Ges., **31**: 1003.

ssp. **bavarica** Burgeff, 1921, Mitt. münch. ent. Ges., **11**: 102. Reiss, 1926, Die Zygaenen Deutschlands, p. 24. Reiss & Tremewan, 1964, Ent. Rec., **76**: 134. — Umgebung der Osterseen bei Seeshaupt und bei Bernried, süd- und südwestlich des Starnberger Sees, Südbayern, Süddeutschland.
boica Burgeff, 1926, Mitt. münch. ent. Ges., **16**: 80. Reiss, 1930, in Seitz, Die Gross-Schmetterlinge der Erde, Supplement, **2**: 40. Daniel, 1932, in Osthelder, Die Schmetterlinge Südbayerns, **1**: 572, pl. 21, figs. 9-12. Rocci, 1938, Redia, **24**: 119.

Reiss, 1941, Mitt. münch. ent. Ges., **31**: 1002, pl. 34, figs. B10, C10; 1949, Entomon, **1**: 173. Reiss & Tremewan, 1964, Ent. Rec., **76**: 134.

ssp. **boicophila** Reiss, 1941, Mitt. münch. ent. Ges., **31**: 1004. Daniel, 1932, in Osthelder, Die Schmetterlinge Südbayerns, **1**, pl. 21, figs. 6-8 (as *boica* Burgeff).

Falkenstein bei Pfronten, Südbayern, Süddeutschland.

ssp. **astragali** Borkhausen, 1793, Rheinisches Magazin, **1**: 631. Freyer, 1845, Neuere Beiträge zur Schmetterlingskunde, **5**: 117, pl. 452. Seitz, 1907, Die Gross-Schmetterlinge der Erde, **2**: 23, pl. 5g, h. Burgeff, 1914, Mitt. münch. ent. Ges., **5**: 64, pl. 2, figs. 188, 189, pl. 4, figs. 131-135. Reiss, 1926, Die Zygaenen Deutschlands, p. 23, 24, pl. 2, fig.; 1930, in Seitz, Die Gross-Schmetterlinge der Erde, Supplement, **2**: 40, pl. 4e. Forster & Wohlfahrt, 1958, Die Schmetterlinge Mitteleuropas, **3**: 99, pl. 11, figs. 5, 10, 15. Heuser & Jöst, 1959, Mitt. Pollichia, (3) **6**: 127.

Mittleres Rheintal und Nahetal, Westdeutschland [Gonsenheim bei Mainz und Geisenheim].

stultulus Heydenreich, 1851, Lepidopterorum Europaeorum Catalogus Methodicus, p. 22 (nomen nudum).

hopfferi Heydenreich, 1851, Lepidopterorum Europaeorum Catalogus Methodicus, p. 22 (nomen nudum).

hopfferi Lederer, 1853, Verh. zool.-bot. Ver. Wien, **2**: 71.

ab. **confluens** Spuler, 1906, in Hofmann, Die Schmetterlinge Europas, **2**: 161.

ab. **omniconfluens** Vorbrodt, 1913, in Vorbrodt & Müller-Rutz, Die Schmetterlinge der Schweiz, **2**: 273, fig. 24. Reiss, 1930, in Seitz, Die Gross-Schmetterlinge der Erde, Supplement, **2**: 40.

ab. **fulva** Spuler, 1906, in Hofmann, Die Schmetterlinge Europas, **2**: 161.

ab. **flava** Spuler, 1906, in Hofmann, Die Schmetterlinge Europas, **2**: 161. Freyer, 1845, Neuere Beiträge zur Schmetterlingskunde, **5**: 44 (as *hippocrepidis* Hübner ab.).

hybr. **flammula** Burgeff, 1914, Mitt. münch. ent. Ges., **5**: 66, pl. 2, fig. 190, pl. 4, figs. 136, 137 (**transalpina astragali** Borkhausen ♂ X **hippocrepidis jurassica** Burgeff ♀). Reiss, 1926, Die Zygaenen Deutschlands, p. 25. Holik, 1933, Iris, **47**: 18.

Biology

Abeille, 1909, Mém. Soc. linn. Provence, **1**: 12-16. Alberti, 1956, Z. wien. ent. Ges., **41**: 231-239; 1958, Bull. Soc. ent. Mulhouse, p. 1-9. Boisduval, Rambur & Graslin, 1832, Collection iconographique et historique des Chenilles, pl. 5, figs. 3, 4. Burgeff, 1910, Z. wiss. InsektBiol., **6**: 42; 1912, ibidem, **8**: 185, 197, 198; 1921, Mitt. münch. ent. Ges., **11**: 50, 57; 1926, ibidem, **16**: 80; 1950, Portug. acta biol., (A) Goldschmidt: 663-728; 1951, Biol. Zbl., **70**: 1-23; 1965, Nachr. Akad. Wiss. Göttingen, **2**, mat.-phys. Kl., no. 14: 191, figs. Holik, 1937, Lambillionea, **37**: 15-24, 32-45, 80-91; 1938, ibidem, **38**: 51-58, 79-88, 95-102; 1946, Rev. franç. Lépid., **10**: 250-261, 273-280; 1953, Ent. Z., **63**: 28. Kiefer, 1934, Int. ent. Z., **27**: 521-524. Oberthür, 1907, Ann. Soc. ent. Fr., **76**: 40; 1911, Études de Lépidoptérologie comparée, **5** (1), pl. 85, fig. 823. Querci, 1911, in Oberthür, Études de Lépidoptérologie comparée, **5** (1): 223-224. Reichl, 1962, Biol. Glasnik, **15**: 141-156. Reiss, 1926, Die Zygaenen Deutschlands, p. 7; 1930, in Seitz, Die Gross-Schmetterlinge der Erde, Supplement, **2**: 42; 1958, Z. wien. ent. Ges., **43**: 161. Rocci, 1918, Atti Soc. ligust. Sci. nat. geogr., **28**: 119-137. Roüast, 1883, Catalogue des Chenilles européennes connues, p. 22. Seitz, 1907, Die Gross-Schmetterlinge der Erde, **2**: 23. Stauder, 1915, Z. wiss. InsektBiol., **11**: 74-75, 132-137. Sterzl, 1931, Z. Ver. NatBeob., Wien, **6**: 5-7. Tutt, 1898, Ent. Rec., **10**: 45.

hippocrepidis Hübner
Distribution: North and east Spain, France, Germany.

ssp. **centricatalonae** Burgeff, 1926, Mitt. münch. ent. Ges., **16**: 82. Reiss, 1930, in Seitz, Die Gross-Schmetterlinge der Erde, Supplement, **2**: 41, pl. 4f. Rocci, 1938, Redia, **24**: 187. — Berg Tagamanent, 50 km. nördlich Barcelona; Mont Serrat, Spanien.

ssp. **philippsi** Romei, 1927, Ent. Rec., **39**: 109. Reiss, 1929, Int. ent. Z., **22**: 358; 1930, in Seitz, Die Gross-Schmetterlinge der Erde, Supplement, **2**: 41, pl. 4g; 1936, Ent. Rdsch., **54**: 91, pl. 2, fig. Rocci, 1938, Redia, **24**: 187. — Serrania de Cuenca, Castile, Spain.

ssp. **rupicola** Rocci, May, 1936, Boll. Soc. ent. ital., **68**: 41; 1938, Redia, **24**: 187. Tremewan, 1963, Ent. Rec., **75**: 7; 1963, ibidem, **75**: 253. — Fuente Dé, Picos de Europa, Spagna.

asturiensis Reiss, November, 1936, Ent. Rdsch., **54**: 91, pl. 2, figs. Tremewan, 1963, Ent. Rec., **75**: 7.

ab. **cingulata** Reiss, 1936, Ent. Rdsch., **54**: 91.

ssp. **marujae** Tremewan & Manley, 1965, Ent. Rec., **77**: 9. — Jaca, 2700 ft.; La Pena, 2400 ft.; Huesca, Spain.

ssp. **centripyrenaea** Burgeff, 1926, Mitt. münch. ent. Ges., **16**: 82; 1926, in Strand, Lepid. Cat., **33**: 72. Oberthür, 1909, Études de Lépidoptérologie comparée, **3**, pl. 30, fig. 197 (as *alpina* Boisduval). Reiss, 1930, in Seitz, Die Gross- — Hochpyrenäen, Frankreich.

Schmetterlinge der Erde, Supplement, **2**: 41. Rocci, 1938, Redia, **24**: 185.

ab. **brunnea** Dziurzyński, 1903, Jber. wien. ent. Ver., **13**: 40; 1903, Iris, **15**: 337.

ssp. **occidentalis** Oberthür, 1907, Ann. Soc. ent. Fr., **76**: 41. Dziurzyński, 1908, Berl. ent. Z., **53**: 35, pl. 1, fig. 10. Oberthür, 1909, Études de Lépidoptérologie comparée, **3**, pl. 30, fig. 203. Reiss, 1930, in Seitz, Die Gross-Schmetterlinge der Erde, Supplement, **2**: 41, pl. 4g. Rocci, 1938, Redia, **24**: 188. Tremewan, 1961, Bull. Brit. Mus. (nat. Hist.) Ent., **10** (7): 280, pl. 54, fig. 3. Dompierre-sur-Mer, Charente-Inférieure, France.

ab. **cingulata** Hirschke, 1906, Jber. wien. ent. Ver., **16**: 95. Seitz, 1907, Die Gross-Schmetterlinge der Erde, **2**: 23.

micingulata Oberthür, 1907, Ann. Soc. ent. Fr., **76**: 42; 1909, Études de Lépidoptérologie comparée, **3**, pl. 30, fig. 207. Tremewan, 1961, Bull. Brit. Mus. (nat. Hist.) Ent., **10** (7): 280, pl. 54, fig. 4.

semicingulata Seitz, 1912, Die Gross-Schmetterlinge der Erde, **2**: 442.

ab. **vertebralis** Le Charles, 1927, Encycl. ent., (B) 3, Lepidoptera, **2**: 152, pl. 9, fig. 11. Reiss, 1930, in Seitz, Die Gross-Schmetterlinge der Erde, Supplement, **2**: 41.

ab. **rubricostata** Derenne, 1932, Lambillionea, **32**: 50.

ab. **miltosa** Candèze, 1883, Feuill. jeun. Nat., **13**: 47. Seitz, 1907, Die Gross-Schmetterlinge der Erde, **2**: 23. Oberthür, 1909, Études de Lépidoptérologie comparée, **3**, pl. 30, figs. 204, 205.

pseudomiltosa Pionneau, 1939, Échange, **55**: 28.

ab. **rosea** Oberthür, 1907, Ann. Soc. ent. Fr., **76**: 43. Seitz, 1912, Die Gross-Schmetterlinge der Erde, **2**: 442. Tremewan, 1961, Bull. Brit. Mus. (nat. Hist.) Ent., **10** (7): 280, pl. 54, fig. 6.

ab. **vigei** Oberthür, 1907, Ann. Soc. ent. Fr., **76**: 43; 1909, Études de Lépidoptérologie comparée, **3**, pl. 30, fig. 201. Tremewan, 1961, Bull. Brit. Mus. (nat. Hist.) Ent., **10** (7): 281, pl. 54, fig. 7.

flava École Bordelaise, 1933, Act. Soc. linn. Bordeaux, **85**: 141.

ab. **argentalis** Lucas, 1958, Bull. mens. Soc. linn. Lyon, **27**: 67.

f. t. **postoccidentalis** Rocci, 1938, Redia, **24**: 189. Oberthür, 1909, Études de Lépidoptérologie comparée, **3**, pl. 30, fig. 206 (as *occidentalis* Oberthür). Charente-Inférieure, France.

ab. **pallidior** Oberthür, 1907, Ann. Soc. ent. Fr., **76**: 43; 1909, Études de Lépidoptérologie comparée, **3**, pl. 30, fig. 202. Seitz, 1912, Die Gross-Schmetterlinge der Erde, **2**: 442. Tremewan, 1961, Bull. Brit. Mus. (nat. Hist.) Ent., **10** (7): 280, pl. 54, fig. 5.

ssp. **centralis** Oberthür, 1907, Ann. Soc. ent. Fr., **76**: 40. Dupont, 1900, Bull. Soc. Sci. nat. Elbeuf, **18**: 71 (as *hippocrepidis* Hübner). Oberthür, 1911, Études de Lépidoptérologie comparée, **5** (1): 212. Dupont, 1925, Bull. Soc. Sci. nat. Elbeuf, **43**: 133; 1927, ibidem, **45**: 73. Reiss, 1930, in Seitz, Die Gross-Schmetterlinge der Erde, Supplement, **2**: 41, pl. 4g. Derenne, 1935, Amat. Papillons, **7**: 172. Rocci, 1938, Redia, **24**: 183. Tremewan, 1961, Bull. Brit. Mus. (nat. Hist.) Ent., **10** (7): 279, pl. 53, fig. 26. — Lardy, Seine-et-Oise, France.

ab. **rubricinctata** Lambillion, 1909, Rev. Soc. ent. namur., **9**: 72.

rubricingulata Lambillion, 1909, Rev. Soc. ent. namur., **9**: 72.

ab. **aldini** Lucas, 1920, Bull. Soc. ent. Fr., p. 230.

ab. **flava** Oberthür, 1896, Études d'Entomologie, **20**: 44, pl. 8, fig. 144; 1907, Ann. Soc. ent. Fr., **76**: 41. Reiss, 1930, in Seitz, Die Gross-Schmetterlinge der Erde, Supplement, **2**: 41. Tremewan, 1961, Bull. Brit. Mus. (nat. Hist.) Ent., **10** (7): 279, pl. 53, fig. 28.

ab. **nigricans** Oberthür, 1896, Études d'Entomologie, **20**: 44, pl. 8, fig. 146; 1907, Ann. Soc. ent. Fr., **76**: 41. Seitz, 1907, Die Gross-Schmetterlinge der Erde, **2**: 23. Tremewan, 1961, Bull. Brit. Mus. (nat. Hist.) Ent., **10** (7): 279, pl. 53, fig. 27.

ssp. **curtisi** Tremewan, 1961, Ent. Rec., **73**: 139; 1962, ibidem, **74**: 128, pl. 2, fig. 9. Dufay, 1966, Bull. mens. Soc. linn. Lyon, **35**: 72. — Dieulefit, Drôme, France.

ab. **cingulata** Tremewan, 1961, Ent. Rec., **73**: 140; 1962, ibidem, **74**: 128, pl. 2, fig. 11.

ab. **miniacens** Tremewan, 1961, Ent. Rec., **73**: 140; 1962, ibidem, **74**: 128, pl. 2, fig. 10.

ssp. **provincialis** Oberthür, 1907, Ann. Soc. ent. Fr., **76**: 45; 1909, Études de Lépidoptérologie comparée, **3**, pl. 30, figs. 192, 193. Reiss, 1930, in Seitz, Die Gross-Schmetterlinge der Erde, Supplement, **2**: 41. Rocci, 1938, Redia, **24**: 190. Tremewan, 1961, Bull. Brit. Mus. (nat. Hist.) Ent., **10** (7): 279, pl. 54, fig. 1. — Montrieux près Méounes, Var, France.

meridionalis Oberthür, 1911, Études de Lépidoptérologie comparée, **5** (1): 213. Tremewan, 1961, Bull. Brit. Mus. (nat. Hist.) Ent., **10** (7): 279, pl. 54, fig. 2.

f. t. **aestivoprovincialis** Burgeff, 1926, Mitt. münch. ent. Ges., **16**: 82. Reiss, 1930, in Seitz, Die Gross-Schmetterlinge der Erde, Supplement, **2**: 41. Rocci, 1938, Redia, **24**: 159, 190. — Umgebung von Marseille, Frankreich.

ssp. **jurassica** Burgeff, 1914, Mitt. münch. ent. Ges., **5**: 65, pl. 2, figs. 186, 187, pl. 4, figs. 124-128. Reiss, 1926, Die Zygaenen Deutschlands, p. 23, pl. 2, fig. Reiss & Tremewan, 1964, Ent. Rec., **76**: 134. — Geislingen an der Steige, Schwäbische Alb, Württemberg;

jurassicola Burgeff, 1926, Mitt. münch. ent. Ges., **16**: 82. Reiss, 1930, in Seitz, Die Gross-Schmetterlinge der Erde, Supplement, **2**: 41, pl. 4f; 1937, in Schneider, Jh. Ver. vaterl. Naturk. Württemb., **93**: 128. Rocci, 1938, Redia, **24**: 192. Haaf, 1952, Veröff. zool. Staatssamml. Münch., **2**: 152, 154, 157, pl. 13 (as *hippocrepidis* Hübner). Bergmann, 1953, Die Grossschmetterlinge Mitteldeutschlands, **3**, pl. 68, figs. B3, B4. Reiss & Tremewan, 1964, Ent. Rec., **76**: 135. Burgeff, 1965, Nachr. Akad. Wiss. Göttingen, **2**, mat.-phys. Kl., no. 1: 9, figs. 6c. — Eichstätt, Fränkischer Jura, Bayern, Süddeutschland.

ab. **alpinoides** Reiss, 1922, Int. ent. Z., **16**: 83; 1926, Die Zygaenen Deutschlands, p. 24, pl. 2, fig.; 1930, in Seitz, Die Gross-Schmetterlinge der Erde, Supplement, **2**: 41.

ab. **quinquemaculata** Reiss, 1964, Coridon, (A) **6**: 6, fig. 4.

ab. **apicaliconfluens** Reiss, 1964, Coridon, (A) **6**: 6.

ab. **totirubra** Reiss, 1920, Int. ent. Z., **14**: 118; 1926, Die Zygaenen Deutschlands, p. 24, pl. 1, fig. (as *totarubra*); 1930, in Seitz, Die Gross-Schmetterlinge der Erde, Supplement, **2**: 41, pl. 4f.

ab. **rubrobrunneata** Przegendza, 1926, Ent. Z., **40**: 345, pl. 4, fig. 2.

ab. **brunneata** Przegendza, 1926, Ent. Z., **40**: 345, pl. 4, figs. 4, 5. Reiss, 1930, in Seitz, Die Gross-Schmetterlinge der Erde, Supplement, **2**: 41.

ab. **flaveola** Kaufmann, 1909, Ent. Z., **23**: 121.

ab. **flava** Kaufmann, 1909, Ent. Z., **23**: 121. Reiss, 1926, Die Zygaenen Deutschlands, p. 24, pl. 1, fig. (as *flava* Dziurzyński); 1930, in Seitz, Die Gross-Schmetterlinge der Erde, Supplement, **2**: 41, pl. 4f.

ab. **albobrunneata** Przegendza, 1926, Ent. Z., **40**: 345, pl. 4, fig. 6.

ssp. **hippocrepidis** Hübner, [1796]-[24th December 1799], Sammlung europäischer Schmetterlinge, **2**, pl. 17, fig. 83; 1806, ibidem, Der Ziefer, p. 79. Ochsenheimer, 1808, Die Schmetterlinge von Europa, **2**: 63. Herrich-Schäffer, 1844, Systematische Bearbeitung der Schmetterlinge von Europa, **2**, pl. 7, figs. 54, 55; 1846, ibidem, **2**: 41. Speyer & Speyer, 1858, Die geographische Verbreitung der Schmetterlinge Deutschlands und der Schweiz, p. 349. Spuler, 1906, in Hofmann, Die Schmetterlinge Europas, **2**: 161, pl. 77, fig. 21a (as *astragali* Borkhausen). Burgeff, 1914, Mitt. münch. ent. Ges., **5**: 65, pl. 2, figs. 184, 185, pl. 4, figs. 118-121. Reiss, 1926, Die Zygaenen Deutschlands, p. 23, 24; 1930, in Seitz, Die Gross-Schmetterlinge der Erde, Supplement, **2**: 41. Rocci, 1935, Mem. Soc. ent. ital., **14**: 47-58, figs. Reiss, 1937, in Schneider, Jh. Ver. vaterl. Naturk. Württemb., **93**: 128. Rocci, 1938, — Bei Jena, Thüringen [Kalkberge Mittel- und Süddeutschlands ohne Schwäb. Alb und Fränkischer Jura].

Redia, **24**: 181. Reiss, 1949, Entomon, **1**: 173. Bergmann, 1953, Die Grossschmetterlinge Mitteldeutschlands, **3**: 54, pl. 68, figs. A1, A2, C1-C4. Alberti, 1956, Z. wien. ent. Ges., **41**: 231-239, figs.; 1958, Bull. Soc. ent. Mulhouse, p. 1-9, figs.; 1958, Mitt. zool. Mus. Berl., **34**: 323. Forster & Wohlfahrt, 1958, Die Schmetterlinge Mitteleuropas, **3**: 99, pl. 11, figs. 20, 25, 30.

ab. **cingulata** Burgeff, 1914, Mitt. münch. ent. Ges., **5**: 65. Reiss, 1930, in Seitz, Die Gross-Schmetterlinge der Erde, Supplement, **2**: 41.

ab. **hippocrepis** Hübner, [July 1803]-[21st December 1806], Sammlung europäischer Schmetterlinge, **2**, pl. 22, fig. 105.

huebneri Burgeff, 1914, Mitt. münch. ent. Ges., **5**: 65. Reiss, 1926, Die Zygaenen Deutschlands, p. 24; 1930, in Seitz, Die Gross-Schmetterlinge der Erde, Supplement, **2**: 41.

ab. **rubescens** Burgeff, 1926, Mitt. münch. ent. Ges., **16**: 81. Reiss, 1930, in Seitz, Die Gross-Schmetterlinge der Erde, Supplement, **2**: 41.

ab. **flava** Burgeff, 1926, in Strand, Lepid. Cat., **33**: 71. Reiss, 1930, in Seitz, Die Gross-Schmetterlinge der Erde, Supplement, **2**: 41.

ssp. **allgaviana** Burgeff, 1926, Mitt. münch. ent. Ges., **16**: 82. Reiss, 1930, in Seitz, Die Gross-Schmetterlinge der Erde, Supplement, **2**: 41. Rocci, 1938, Redia, **24**: 193.

Gerstruben und Obersdorf, Allgäu, Südbayern, Süddeutschland.

ab. **cingulata** Burgeff, 1926, Mitt. münch. ent. Ges.. **16**: 82. Reiss, 1930, in Seitz, Die Gross-Schmetterlinge der Erde, Supplement, **2**: 41. Daniel, 1932, in Osthelder, Die Schmetterlinge Südbayerns, **1**: 574, pl. 21, figs. 1-5.

Biology

Alberti, 1956, Z. wien. ent. Ges., **41**: 231-239; 1958, Bull. Soc. ent. Mulhouse, p. 1-9. Boisduval, Rambur & Graslin, 1832, Collection iconographique et historique des Chenilles, pl. 1, figs. 4-7. Burgeff, 1912, Z. wiss. InsektBiol., **8**: 185; 1950, Portug. acta biol., (A) Goldschmidt: 663-728; 1965, Nachr. Akad. Wiss. Göttingen, **2**, mat.-phys. Kl., no. 14: 191, figs. Freyer, 1833, Neuere Beiträge zur Schmetterlingskunde, **1**: 157, pl. 86, figs. 2, 3; 1845, ibidem, **5**: 117, pl. 452. Holik, 1937, Lambillionea, **37**: 15-24, 32-45, 80-91; 1938, ibidem, **38**: 51-58, 79-88, 95-102; 1946, Rev. franç. Lépid., **10**: 250-261, 273-280; 1953, Ent. Z., **63**: 28. Kiefer, 1933, Int. ent. Z., **27**: 252-256. Koch, 1955, Wir Bestimmen Schmetterlinge, **2**: 62, 63, pl. 1, fig. 17, pl. 15, fig. 17 (as *transalpina* Esper). Oberthür, 1907, Ann. Soc. ent. Fr., **76**: 44. Reiss, 1926, Die Zygaenen Deutschlands, p. 7 (as *transalpina* Esper); 1930, in Seitz. Die Gross-Schmetterlinge der Erde,

Supplement, **2**: 42; 1958, Z. wien. ent. Ges., **43**: 161. Spuler, 1910, in Hofmann, Die Raupen der Schmetterlinge Europas, pl. 10, figs. 2a, b, c.

angelicae Ochsenheimer

Distribution: Macedonia, Bosnia, Herzegowina, Bukowina, Hungary, Ukraine, Poland, Czechoslovakia, Krain, Styria, Austria, Bavaria, Saxony, East Prussia.

ssp. **herzegowinensis** Reiss, 1922, Int. ent. Z., **16**: 66; 1930, in Seitz, Die Gross-Schmetterlinge der Erde, Supplement, **2**: 42, pl. 4h. Koch, 1938, Z. öst. EntVer., **23**: 17. Holik, 1948, Mitt. münch. ent. Ges., **34**: 391.

balcani Burgeff, 1926, Mitt. münch. ent. Ges., **16**: 85. Koch, 1938, Z. öst. EntVer., **23**: 17. Holik, 1948, Mitt. münch. ent. Ges., **34**: 390. Daniel, 1958, Fragmenta Balcanica, **2**: 43; 1964, Prirod. Muz. Scopje, no. 2: 19.

Ubli (Ublö), Herzegowina (montenegrinische Grenze) [Bosnien, Korična; Mazedonien, bei Hudowa].

ssp. **ternovanensis** Koch, 1938, Z. öst. EntVer., **23**: 17. Holik, 1948, Mitt. münch. ent. Ges., **34**: 389. Reichl, 1962, Biol. Glasnik, **15**: 142.

ab. **sexmaculata** Holik, 1948, Mitt. münch. ent. Ges., **34**: 390.

ab. **ternoflava** Koch, 1938, Z. öst. EntVer., **23**: 18. Reiss, 1930, in Seitz, Die Gross-Schmetterlinge der Erde, Supplement, **2**: 42, pl. 4h (as *doleschalli* Rühl).

Ternovaner-Wald, 800-900 m. [Auf dem Tschaun, Julisch-Venetien].

ssp. **sheljuzhkoiana** Holik & Reiss, 1932, in Holik, Iris, **46**: 127, pl. 2, figs. 21-24. Reiss, 1933, in Seitz, Die Gross-Schmetterlinge der Erde, Supplement, **2**: 277. Holik & Sheljuzhko, 1958, Mitt. münch. ent. Ges., **48**: 232. Alberti & Soffner, 1962, Mitt. münch. ent. Ges., **52**: 175 (as *zamoscensis* Koch).

Butsha; Kijev, Kirillovskije ovragi, Ukraine.

ab. **privata** Holik & Reiss, 1932, in Holik, Iris, **46**: 128. Reiss, 1933, in Seitz, Die Gross-Schmetterlinge der Erde, Supplement, **2**: 278.

ab. **nigroinspersa** Holik & Sheljuzhko, 1958, Mitt. münch. ent. Ges., **48**: 232.

ab. **costalielongata** Holik & Reiss, 1932, in Holik, Iris, **46**: 128. Reiss, 1933, in Seitz, Die Gross-Schmetterlinge der Erde, Supplement, **2**: 278.

ab. **analiconfluens** Sheljuzhko, 1941, Acta Mus. zool. Kijev, **1**: 79, 95, 101.

ab. **confluens** Holik & Reiss, 1932, in Holik, Iris, **46**: 128. Reiss, 1933, in Seitz, Die Gross-Schmetterlinge der Erde, Supplement, **2**: 278.

ab. **crocea** Sheljuzhko, 1941, Acta Mus. zool. Kijev, **1**: 79, 95, 101.

ab. **doleschalli** Holik & Reiss, 1932, in Holik, Iris, **46**: 128. Reiss, 1933, in Seitz, Die Gross-Schmetterlinge der Erde, Supplement, **2**: 277.

ssp. **rentschi** Koch, 1941, Mitt. münch. ent. Ges., **31**: 568. Gregor & Povolný, 1955, Sborn. ent. Odd. nár. Mus. Praze, **30**: 262, 273, pl. 4, fig. 3. — Beim Dorf Lubochna, Grosse Fatra (Velická Fatra), Czechoslovakia.

ab. **hornicekii** Reiprich, 1960, Motýle Slovenska, p. 302, 410, 416, pl. 56, fig. D3.

ssp. **leopoliensis** Holik, 1939, Ann. Mus. zool. Polon., **12**: 106, pl. 4, figs. 136-138. — Rzesna Polska, Holosko, Umgebung von Lemberg, Südostpolen.

ab. **transcarpathina** Hormuzaki, 1902, Soc. ent., **17**: 139. Seitz, 1907, Die Gross-Schmetterlinge der Erde, **2**: 22. Holik & Sheljuzhko, 1958, Mitt. münch. ent. Ges., **48**: 231.

ssp. **zamoscensis** Koch, 1941, Mitt. münch. ent. Ges., **31**: 568. — Zawada bei Zamosc, nordwestlich von Tomaszów Lubelski, Südostpolen.

ssp. **polonica** Holik, 1939, Ann. Mus. zool. Polon., **12**: 104. — Pomiechowo, Nördliches Mittelpolen.

ab. **privata** Holik, 1939, Ann. Mus. zool. Polon., **12**: 104, pl. 4, fig. 139.

ab. **parvimaculata** Holik, 1939, Ann. Mus. zool. Polon., **12**: 104.

ab. **reducta** Dąbrowski, 1965, Acta zool. cracov., **10**: 124, fig. 130a, pl. 8, fig. 8, pl. 11, fig. 7.

ab. **costalielongata** Holik, 1939, Ann. Mus. zool. Polon., **12**: 104, pl. 4, fig. 140.

ssp. **angelicae** Ochsenheimer, 1808, Die Schmetterlinge von Europa, **2**: 67. Hübner, [1808]-[20th June 1813], Sammlung europäischer Schmetterlinge, **2**, pl. 26, figs. 120, 121. Spuler, 1906, in Hofmann, Die Schmetterlinge Europas, **2**: 160, pl. 72, fig. 2, pl. 77, fig. 20. Seitz, 1907, Die Gross-Schmetterlinge der Erde, **2**: 22, pl. 5a. Burgeff, 1926, Mitt. münch. ent. Ges., **16**: 84. Reiss, 1926, Die Zygaenen Deutschlands, p. 18. Koch, 1938, Z. öst. EntVer., **23**: 15. Holik, 1939, Sborn. ent. Odd. nár. Mus. Praze, **17**: 46. Haaf, 1952, Veröff. zool. Staatssamml. Münch., **2**: 152, 154, 157, pl. 13. Alberti, 1956, Dtsch. ent. Z. (N.F.), **3**: 91, figs.; 1956, Z. wien. ent. Ges., **41**: 231-239, figs.; 1958, Bull. Soc. ent. Mulhouse, p. 1-9, figs.; 1958, Mitt. zool. Mus. Berl., **34**: 322. — Dresden Deutschland; Wien, Österreich.

latipennis Herrich-Schäffer, 1852, Systematische Bearbeitung der Schmetterlinge von Europa, **6**: 44; 1851, ibidem, **2**, pl. 15, fig. 105 (non-binominal).

ab. **cingulata** Spuler, 1906, in Hofmann, Die Schmetterlinge Europas, **2**: 160.

cingulata Dziurzyński, 1908, Berl. ent. Z., **53**: 12, 28. Seitz, 1912, Die Gross-Schmetterlinge der Erde, **2**: 442. Reiss, 1926, Die Zygaenen Deutschlands, p. 18.

ab. **subdivisa** Stauder, 1922, Ent. Anz., **2**: 102. Reiss, 1930, in Seitz, Die Gross-Schmetterlinge der Erde, Supplement, **2**: 42.
ab. **sexmacula** Dziurzyński, 1906, Ent. Z., **19**: 185. Seitz, 1907, Die Gross-Schmetterlinge der Erde, **2**: 22. Dziurzyński, 1908, Berl. ent. Z., **53**: 28. Reiss, 1926, Die Zygaenen Deutschlands, p. 18.
ab. **xanthomarginata** Peschke, 1942, Ent. Z., **55**: 234, fig. 3.
ab. **rudolfi** Povolný & Gregor, 1946, Folia ent., Brno, **9**, Supplement, 12: 39, 97, pl. 2, fig. 6b.
ab. **striata** Reiss, 1964, Coridon, (A) **6**: 8, fig. 5.
ab. **confluens** Dziurzyński, 1901, Jber. wien. ent. Ver., **11**: 117, fig.; 1903, Iris, **15**: 337; 1904, Jber. wien. ent. Ver., **14**: 50, pl. 2, fig. 8; 1908, Berl. ent. Z., **53**: 28, pl. 1, fig. 6. Seitz, 1907, Die Gross-Schmetterlinge der Erde, **2**: 22. Reiss, 1926, Die Zygaenen Deutschlands, p. 18.
ab. **carnea** Dziurzyński, 1908, Berl. ent. Z., **53**: 6, 13, 28. Seitz, 1912, Die Gross-Schmetterlinge der Erde, **2**: 442. Reiss, 1926, Die Zygaenen Deutschlands, p. 18.
ab. **doleschalli** Rühl, 1891, Soc. ent., **6**: 105. Dziurzyński, 1908, Berl. ent. Z., **53**: 28, pl. 1, fig. 5. Reiss, 1926, Die Zygaenen Deutschlands, p. 18.
ab. **brunensis** Skala, 1913, Verh. naturf. Ver. Brünn., **51**: 209. Reiss, 1926. Die Zygaenen Deutschlands, p. 19.

ssp. **isaria** Burgeff, 1926, Mitt. münch. ent. Ges., **16**: 85. Reiss, 1930, in Seitz, Die Gross-Schmetterlinge der Erde, Supplement, **2**: 42. Daniel, 1955, NachrBl. bayer. Ent., **4**: 49-56.

Pupplinger Au bei Wolfratshausen im Isartal; bei Deisenhofen südöstlich von München, Südbayern, Süddeutschland.

ssp. **rhatisbonensis** Burgeff, 1914, Mitt. münch. ent. Ges., **5**: 66. Hübner, 1796, Sammlung europäischer Schmetterlinge, **2**, pl. 5, fig. 32 (as *loti* Denis & Schiffermüller). Herrich-Schäffer, 1846, Systematische Bearbeitung der Schmetterlinge von Europa, **2**: 37, 41 (note), (as *angelicae* Ochsenheimer). Reiss, 1926, Die Zygaenen Deutschlands, p. 18, 19, pl. 2, fig.; 1930, in Seitz, Die Gross-Schmetterlinge der Erde, Supplement, **2**: 42, pl. 4i; 1953, Z. wien. ent. Ges., **38**: 141, pl. 8, figs. 8-12. Bergmann, 1953, Die Grossschmetterlinge Mitteldeutschlands, **3**: 62, pl. 68, figs. C7, D7. Daniel, 1955, NachrBl. bayer. Ent., **4**: 49-56. Forster & Wohlfahrt, 1958, Die Schmetterlinge Mitteleuropas, **3**: 100, pl. 11, figs. 18, 23.
ratisponensis Burgeff, 1965, Nachr. Akad. Wiss. Göttingen, **2**, mat.-phys. Kl., no. 1: 10, figs. 6a, fig. 8a.

Umgebung von Regensburg, Fränkischer Jura, Süddeutschland.

ab. **cingulata** Reiss, 1964, Coridon, (A) **6**: 8.

ab. **perdita** Reiss, 1925, Int. ent. Z., **19**: 147.

ab. **paucipuncta** Reiss, 1964, Coridon, (A) **6**: 8.

ab. **pseudoangelicae** Reiss, 1925, Int. ent. Z., **19**: 147; 1930, in Seitz, Die Gross-Schmetterlinge der Erde, Supplement, **2**: 42; 1953, Z. wien. ent. Ges., **38**: 141, pl. 8, fig. 13.

ab. **pseudotransalpina** Daniel, 1954, Z. wien. ent. Ges., **39**: 56, pl. 2, figs. 11-13.

ab. **elegantoides** Reiss, 1922, Int. ent. Z., **16**: 67; 1926, Die Zygaenen Deutschlands, p. 18; 1930, in Seitz, Die Gross-Schmetterlinge der Erde, Supplement, **2**: 42; 1953, Z. wien. ent. Ges., **38**: 141, pl. 8, fig. 14.

ab. **costalielongata** Reiss, 1964, Coridon, (A) **6**: 8, fig. 6.

ab. **confluens** Reiss, 1964, Coridon, (A) **6**: 8.

ab. **dichroma** Reiss, 1925, Int. ent. Z., **19**: 148; 1926, Die Zygaenen Deutschlands, p. 19.

ab. **aurantiaca** Reiss, 1964, Coridon, (A) **6**: 8.

ab. **flava** Reiss, 1935, Int. ent. Z., **28**: 541.

f. loc. **carolimagni** Burgeff, 1965, Nachr. Akad. Wiss. Göttingen, **2**, mat.-phys. Kl., no. 1: 12, figs. 6b (as *carolimangni* [partim]). **Maintal, Fränkischer Jura, Süddeutschland.**

hybr. **angelipina** Ronnicke, 1933, Int. ent. Z., **27**: 65 (**angelicae angelicae** Ochsenheimer ♂ X **transalpina tilaventa** Holik ab. **flava** Holik ♀). Holik, 1933, Iris, **47**: 16. Koch, 1938, Z. öst. EntVer., **23**: 18.

hybr. **angelicojurassica** Przegendza, 1926, Ent. Z., **40**: 295, figs. 3, 3a (**angelicae rhatisbonensis** Burgeff ♂ X **hippocrepidis jurassica** Burgeff ♀). Reiss, 1930, in Seitz, Die Gross-Schmetterlinge der Erde, Supplement, **2**: 43. Holik, 1933, Iris, **47**: 16.

Biology

Alberti, 1956, Z. wien. ent. Ges., **41**: 231-239. Burgeff, 1912, Z. wiss. InsektBiol., **8**: 185, 197, 198; 1926, Mitt. münch. ent. Ges., **16**: 85; 1950, Portug. acta biol., (A) Goldschmidt: 663-728; 1951, Biol. Zbl., **70**: 1-23. Dąbrowski, 1963, Folia biol., Kraków, **11**: 340, 341, figs. 2a, 2b, pl. 1, figs. B10, D10. Dorfmeister, 1854, Verh. zool.-bot. Ver. Wien, **4**: 479. Holik, 1937, Lambillionea, **37**: 15-24, 32-45, 80-91; 1938, Ent. Rdsch., **55**: 320-323, 331-333; 1946, Rev. franç. Lépid., **10**: 250-261, 273-280; 1953, Ent. Z., **63**: 29. Hrubý, 1964, Prodromus Lepidopter Slovenska, p. 480. Koch, 1955, Wir Bestimmen Schmetterlinge, **2**: 64, 65, pl. 1, fig. 19. Ochsenheimer, 1808, Die Schmetterlinge von Europa, **2**: 67. Reiss, 1926, Die Zygaenen Deutschlands, p. 7; 1930, in Seitz, Die Gross-Schmetterlinge der Erde, Supplement, **2**: 43; 1958,

Z. wien. ent. Ges., **43**: 161. Roüast, 1883, Catalogue des Chenilles européennes connues, p. 22.

elegans Burgeff

Distribution: Schwäbische Alb, south-west Germany.

elegans Burgeff, 1913, Mitt. münch. ent. Ges., **4**: 82, pl. 4, fig. 1; 1914, ibidem, **5**: 66, pl. 2, figs. 183, 191, pl. 4, figs. 122, 123, 129, 130. Reiss, 1920, Int. ent. Z., **14**: 21. Burgeff, 1926, Mitt. münch. ent. Ges., **16**: 83. Reiss, 1926, Die Zygaenen Deutschlands, p. 17, pl. 2, fig.; 1930, in Seitz, Die Gross-Schmetterlinge der Erde, Supplement, **2**: 42, pl. 4g, h; 1937, in Schneider, Jh. Ver. vaterl. Naturk. Württemb., **93**: 129. Gremminger, 1943, Ent. Z., **56**: 226. Reiss, 1949, Entomon, **1**: 174. Haaf, 1952, Veröff. zool. Staatssamml. Münch., **2**: 152, 154, 157, pl. 13. Reiss, 1953, Z. wien. ent. Ges., **38**: 132, pl. 8, figs. 1-7. Alberti, 1956, Dtsch. ent. Z., (N.F.), **3**: 91, figs.; 1956, Z. wien. ent. Ges., **41**: 231-239, figs.; 1958, Bull. Soc. ent. Mulhouse, p. 1-9, figs.; 1958, Mitt. zool. Mus. Berl., **34**: 323. Forster & Wohlfahrt, 1958, Die Schmetterlinge Mitteleuropas, **3**: 100, pl. 11, figs. 16, 21, 26. Burgeff, 1965, Nachr. Akad. Wiss. Göttingen, **2**, mat.-phys. Kl., no. **1**: 12, fig. 7 (distribution map). — Geislingen an der Steige, Schwäbische Alb, Württemberg, Südwestdeutschland.

ab. **cingulata** Burgeff, 1913, Mitt. münch. ent. Ges., **4**: 82. Reiss, 1930, in Seitz, Die Gross-Schmetterlinge der Erde, Supplement, **2**: 42.

cingulata Reiss, 1914, Int. ent. Z., **8**: 158; 1920, ibidem, **14**: 22; 1926, Die Zygaenen Deutschlands, p. 17.

ab. **sexmaculata** Reiss, 1920, Int. ent. Z., **14**: 22; 1926, Die Zygaenen Deutschlands, p. 17; 1930, in Seitz, Die Gross-Schmetterlinge der Erde, Supplement, **2**: 42, pl. 4h.

ab. **quinquemaculata** Reiss, 1920, Int. ent. Z., **14**: 22; 1926, Die Zygaenen Deutschlands, p. 17; 1930, in Seitz, Die Gross-Schmetterlinge der Erde, Supplement, **2**: 42.

ab. **splendida** Reiss, 1920, Int. ent. Z., **14**: 22; 1930, in Seitz, Die Gross-Schmetterlinge der Erde, Supplement, **2**: 42, pl. 4h.

confluens Reiss, 1925, Int. ent. Z., **19**: 146; 1926, Die Zygaenen Deutschlands, p. 17, pl. 1, fig.

ab. **burgeffi** Reiss & Tremewan, 1964, Ent. Rec., **76**: 135 (nomen novum for *confluens* Burgeff).

confluens Burgeff, 1926, Mitt. münch. ent. Ges., **16**: 83 (preoccupied by *confluens* Reiss, 1925, synonym of **splendida** Reiss, 1920). Reiss, 1930, in Seitz, Die Gross-Schmetterlinge der Erde, Supplement, **2**: 42.

ab. **extrema** Reiss, 1923, Int. ent. Z., **17**: 6; 1926, Die Zygaenen

Deutschlands, p. 18; 1930, in Seitz, Die Gross-Schmetterlinge der Erde, Supplement, **2**: 42.
ab. **totirubra** Reiss, 1964, Coridon, (A) **6**: 8, fig. 7.
ab. **rosea** Reiss, 1920, Int. ent. Z., **14**: 23; 1926, Die Zygaenen Deutschlands, p. 18.
ab. **dichroma** Reiss, 1920, Int. ent. Z., **14**: 116; 1926, Die Zygaenen Deutschlands, p. 18.
hybr. **burgeffensis** Reiss, 1927, Int. ent. Z., **21**: 289 (**elegans** Burgeff X **hippocrepidis jurassica** Burgeff). Burgeff, 1914, Mitt. münch. ent. Ges., **5**: 66, pl. 4, fig. 123; 1926, ibidem, **16**: 84. Reiss, 1930, in Seitz, Die Gross-Schmetterlinge der Erde, Supplement, **2**: 42; 1949, Entomon, **1**: 174.

Biology

Alberti, 1956, Z. wien. ent. Ges., **41**: 231-239. Burgeff, 1913, Mitt. münch. ent. Ges., **4**: 81; 1950, Portug. acta biol., (A) Goldschmidt: 663-728; 1951, Biol. Zbl., **70**: 1-23. Holik, 1937, Lambillionea, **37**: 15-24, 32-45, 80-91; 1953, Ent. Z., **63**: 29. Koch, 1955, Wir Bestimmen Schmetterlinge, **2**: 62, 63. Reiss, 1926, Die Zygaenen Deutschlands, p. 7, 18; 1930, in Seitz, Die Gross-Schmetterlinge der Erde, Supplement, **2**: 42; 1953, Z. wien. ent. Ges., **38**: 132; 1958, ibidem, **43**: 161.

SECTION 6

persephone Zerny
Distribution: Great Atlas, Morocco.
persephone Zerny, 1934, Z. öst. EntVer., **19**: 29, pl. 5, figs. 1, 2. Reiss, 1937, Ent. Rdsch., **54**: 455, figs. b3, c3; 1944, Z. wien. ent. Ges., **29**: 187, pl. 37, fig. 65. Alberti, 1958, Mitt. zool. Mus. Berl., **34**: 313.

Tizi 'n Tachdirt, Grosser Atlas, Marokko, 3100 m.

Biology

Holik, 1938, Ent. Rdsch., **55**: 320-323, 331-333; 1953, Ent. Z., **63**: 26. Reiss, 1958, Z. wien. ent. Ges., **43**: 162.

SECTION 7

amanica Reiss
Distribution: Northern Syria.
amanica Reiss, 1935, Int. ent. Z., **29**: 191, figs. Dziurzyński, 1908, Berl. ent. Z., **53**: 43 (as *ledereri* Rebel). Oberthür, 1910, Études de Lépidoptérologie comparée, **4**: 564 (as *gurda* Lederer). Reiss, 1930, Int. ent. Z., **23**: 521 (as *cilicica* Burgeff); 1930, in Seitz, Die Gross-Schmetterlinge der Erde, Supple-

Düldül-Dagh, Jeschildere, nördliches Amanus Gebirge, Syrien.

ment, **2**: 32, pl. 3h (as *cilicica* Burgeff); 1933, ibidem, **2**: 275 (as *cilicica* Burgeff); 1953, Z. wien. ent. Ges., **38**: 141, pl. 8, figs. 17, 18. Holik & Sheljuzhko, 1957, Mitt. münch. ent. Ges., **47**: 179 (as *cilicica* Burgeff). Reiss & Tremewan, 1960, Bull. Brit. Mus. (nat. Hist.) Ent., **9**(10): 466, pl. 22, figs. 18, 21, pl. 25, figs. 14, 15.

confluens Dziurzyński, 1910, Int. ent. Z., **4**: 195 (infrasubspecific).

confluens Dziurzyński, 1914, Int. ent. Z., **8**: 34 (infrasubspecific).

ab. **cingulata** Reiss, 1935, Int. ent. Z., **29**: 192, fig.

ab. **sexmaculata** Reiss, 1935, Int. ent. Z., **29**: 192, fig.

ab. **quinquemaculata** Reiss, 1935, Int. ent. Z., **29**: 192, fig.

Biology

Burgeff, 1950, Portug. acta biol., (A) Goldschmidt: 663-728. Holik, 1953, Ent. Z., **63**: 28. Reiss, 1958, Z. wien. ent. Ges., **43**: 163.

laphria Herrich-Schäffer

Distribution: Asia Minor.

ssp. **laphria** Herrich-Schäffer, 1852, Systematische Bearbeitung der Schmetterlinge von Europa, **6**: 44. Dziurzyński, 1908, Berl. ent. Z., **53**: 43 (as *laphria* Freyer). Reiss, 1930, Int. ent. Z., **23**: 524 (as *laphria* Freyer); 1930, in Seitz, Die Gross-Schmetterlinge der Erde, Supplement, **2**: 33, pl. 3h (as *laphria* Freyer); 1935, Int. ent. Z., **29**: 187 (as *laphria* Freyer). Haaf, 1952, Veröff. zool. Staatssamml. Münch., **2**: 152, 154, 157, pl. 12 (as *laphria* Freyer). Holik & Sheljuzhko, 1957, Mitt. münch. ent. Ges., **47**: 174 (as *laphria* Freyer). Reiss & Tremewan, 1960, Bull. Brit. Mus. (nat. Hist.) Ent., **9** (10): 466. — Amasia, Kleinasien.

laphira Herrich-Schäffer, 1851, Systematische Bearbeitung der Schmetterlinge von Europa, **2**, pl. 16, fig. 108 (non-binominal). Reiss & Tremewan, 1960, Bull. Brit. Mus. (nat.Hist.) Ent., **9**(10): 466, pl. 22, fig. 19, pl. 25, figs. 12, 13.

ledereri Rebel, 1901, in Staudinger & Rebel, Catalog der Lepidopteren des Palaearctischen Faunengebietes, p. 385. Seitz, 1907, Die Gross-Schmetterlinge der Erde, **2**: 25, pl. 6e.

cilicica Burgeff, 1926, Mitt. münch. ent. Ges., **16**: 65. Reiss, 1930, Int. ent. Z., **23**: 521. Alberti, 1958, Mitt. zool. Mus. Berl., **34**: 320. Reiss & Tremewan, 1960, Bull. Brit. Mus. (nat. Hist.) Ent., **9** (10): 466.

ssp. **philomelica** Reiss, 1935, Int. ent. Z., **29**: 190, figs.; 1953, Z. wien. ent. Ges., **38**: 141, pl. 8, figs. 15, 16. Holik & Sheljuzhko, 1957, Mitt. münch. ent. Ges., **47**: 179. — Ak-Schehir, Kleinasien, 1000-1500 m.

ab. **sexmaculata** Reiss, 1935, Int. ent. Z., **29**: 190.
ab. **quinquemaculata** Reiss, 1935, Int. ent. Z., **29**: 190, fig.
ab. **confluens** Burgeff, 1914, Mitt. münch. ent. Ges., **5**: 61. Reiss, 1930, in Seitz, Die Gross-Schmetterlinge der Erde, Supplement, **2**: 33.
ab. **totirubra** Reiss, 1935, Int. ent. Z., **29**: 190, fig.

Biology

Burgeff, 1950, Portug. acta biol., (A) Goldschmidt: 663-728. Reiss, 1958, Z. wien. ent. Ges., **43**: 163.

viciae Denis & Schiffermüller
Distribution: Northern Iran, Armenia, Caucasus, Central Asia, Siberia, east and central Europe to the Pyrenees and north Catalonia, Great Britain, Scandinavia.

ssp. **burgeffiana** Reiss, 1930, in Seitz, Die Gross-Schmetterlinge der Erde, Supplement, **2**: 32, pl. 3h; 1933, Int. ent. Z., **26**: 499, figs. Holik & Sheljuzhko, 1957, Mitt. münch. ent. Ges., **47**: 161. — Kuldsar, Nordost-Iran.

ssp. **kasikoparana** Reiss, 1935, Int. ent. Z., **29**: 187, fig. Holik & Sheljuzhko, 1957, Mitt. münch. ent. Ges., **47**: 160. — Kasikoparan, Kulp, Westarmenien, Kleinasien.

ssp. **martirosica** Holik & Sheljuzhko, 1957, Mitt. münch. ent. Ges., **47**: 159. — Martiros, Dorf Azizbekov (Pashalu), Daralagëz Gebirge, 1650-2000 m., Armenisches Bergland, Russisch Armenien.
ab. **deannulata** Holik & Sheljuzhko, 1957, Mitt. münch. ent. Ges., **47**: 159.

ssp. **kotshubeji** Holik & Sheljuzhko, 1957, Mitt. münch. ent. Ges., **47**: 156. — Novyj Afonj, Abchasien, Transkaukasien.

ssp. **tbilisiensis** Reiss, 1935, Int. ent. Z., **29**: 187, figs. Koch, 1939, Mitt. münch. ent. Ges., **29**: 412. Holik & Sheljuzhko, 1957, Mitt. münch. ent. Ges., **47**: 157. — Abas-Tuman, Provinz Tiflis, Georgien, Transkaukasien, 1000 m.

ssp. **teberdina** Holik & Sheljuzhko, 1957, Mitt. münch. ent. Ges., **47**: 154. — Teberdagebiet, Ciskaukasien.

ssp. **digorica** Holik, 1939, Ann. Mus. zool. Polon., **13**: 253, pl. — Karaugom, Nordossetien, Ciskaukasien, 2500 m.

23, figs. 21-24. Holik & Sheljuzhko, 1957, Mitt. münch. ent. Ges., **47**: 155.

ssp. **lesghierica** Holik, 1943, Z. wien. ent. Ges., **28**: 133. Holik & Sheljuzhko, 1957, Mitt. münch. ent. Ges., **47**: 155. — Lesghier, Gunib, Dagestan, Ciskaukasien, ca. 2000 m.

ssp. **tindiensis** Holik, 1943, Z. wien. ent. Ges., **28**: 133. Holik & Sheljuzhko, 1957, Mitt. münch. ent. Ges., **47**: 156. — Tindi im Bogos Gebirge, Dagestan, Ciskaukasien, 1300 m.

ssp. **tarkiana** Holik & Sheljuzhko, 1957, Mitt. münch. ent. Ges., **47**: 155. — Berg Tarki bei Petrovsk (Machatsh-Kala), Dagestan, Ciskaukasien.

ab. **apicalielongata** Holik & Sheljuzhko, 1957, Mitt. münch. ent. Ges., **47**: 155.

ssp. **mongolica** Rebel, 1901, in Staudinger & Rebel, Catalog der Lepidopteren des Palaearctischen Faunengebietes, p. 383. Seitz, 1907, Die Gross-Schmetterlinge der Erde, **2**: 25. Reiss, 1933, in Seitz, Die Gross-Schmetterlinge der Erde, Suppl., **2**: 275. Holik & Sheljuzhko, 1957, Mitt. münch. ent. Ges., **47**: 166. Daniel, 1965, Reichenbachia, **7** (10): 94. — Urga, Mongolei

ssp. **confusa** Staudinger, 1881, Stettin. ent. Ztg., **42**: 398. Seitz, 1907, Die Gross-Schmetterlinge der Erde, **2**: 25, pl. 6d. Reiss, 1933, in Seitz, Die Gross-Schmetterlinge der Erde, Supplement, **2**: 275. Holik & Sheljuzhko, 1957, Mitt. münch. ent. Ges., **47**: 163. — Semiretshje, Dzhungarskij Ala-tau, Zentralasien.

ab. **pseudomeliloti** Reiss, 1933, in Seitz, Die Gross-Schmetterlinge der Erde, Supplement, **2**: 275.

melilotiformis Holik, 1943, Z. wien. ent. Ges., **28**: 134.

ab. **purpuraliformis** Holik, 1943, Z. wien. ent. Ges., **28**: 134.

ab. **totarubra** Holik, 1943, Z. wien. ent. Ges., **28**: 134.

ssp. **intersita** Holik, 1943, Z. wien. ent. Ges., **28**: 134. Holik & Sheljuzhko, 1957, Mitt. münch. ent. Ges., **47**: 163. — Tshimgan, Syr-Darja, Zentralasien, 1500-1600 m.

ssp. **occidosibirica** Holik & Sheljuzhko, 1957, Mitt. münch. ent. Ges., **47**: 165. — Tobolsk; Paninbugor; Dorf Shaposhnikova; Westsibirisches Steppengebiet.

ssp. **dahurica** Boisduval, 1834, Icones historique des Lépidoptères nouveaux ou peu connus, **2**: 57, pl. 54, fig. 7. Dupon- — Daourie, Siberie.

chel, 1835, in Godart & Duponchel, Histoire naturelle des Lépidoptères ou Papillons de France, Supplement, **2**: 134, pl. 12, fig. 3 (fig. 2 is erroneous). Spuler, 1906, in Hofmann, Die Schmetterlinge Europas, **2**: 158 (partim). Seitz, 1907, Die Gross-Schmetterlinge der Erde, **2**: 25, pl. 6d. Reiss, 1933, in Seitz, Die Gross-Schmetterlinge der Erde, Supplement, **2**:275. Holik & Sheljuzhko, 1957, Mitt. münch. ent. Ges., **47**: 166. Tremewan, 1961, Bull. Brit. Mus. (nat. Hist.) Ent., **10**(7): 282, pl. 54, fig. 15, pl. 63, figs. 5, 6.

ssp. **schneideri** Reiss, 1932, Ent. Rdsch., **49**: 166, pl. 1, figs.; 1933, in Seitz, Die Gross-Schmetterlinge der Erde, Supplement, **2**: 274. Holik & Sheljuzhko, 1957, Mitt. münch. ent. Ges., **47**: 153.
Urgunerwald, Kalkanova und Utshaly, 830 m., südlicher Ural (Osthang).

ssp. **bosniensis** Reiss, 1922, Int. ent. Z., **15**: 180; 1930, in Seitz, Die Gross-Schmetterlinge der Erde, Supplement, **2**: 32. Holik, 1943, Mitt. münch. ent. Ges., **33**: 311.
Livno, Korična, Bosnien, 1000 m.

ssp. **menoetius** Burgeff, 1926, Mitt. münch. ent. Ges., **16**: 64. Reiss, 1930, in Seitz, Die Gross-Schmetterlinge der Erde, Supplement, **2**: 32. Holik, 1943, Mitt. münch. ent. Ges., **33**: 310.
Vuciabara bei Gacko, Herzegowina, 1300 m.

ssp. **silbernageli** Reiss, 1943, Z. wien. ent. Ges., **28**: 109; 1962, Ent. Z., **72**: 225. Daniel, 1964, Prirod. Muz. Scopje, no. 2:16.
ab. **pseudostenzi** Reiss, 1962, Ent. Z., **72**: 226.
ab. **medioconfluens** Reiss, 1962, Ent. Z., **72**: 226.
ab. **sexpunctata** Reiss, 1943, Z. wien. ent. Ges., **28**: 110.
Bei Ochrid (Kalkstein), Petrinaplanina, Mazedonien, 1600 m.

ssp. **vardariensis** Holik, 1943, Mitt. münch. ent. Ges., **33**: 312. Daniel, 1958, Fragmenta Balcanica, **2**: 41; 1964, Prirod. Muz. Scopje, no. 2: 16.
Pena-Fluss bei Brodec, Shar-Planina, Mazedonien, 1100 m.

ssp. **caradjai** Holik & Sheljuzhko, 1957, Mitt. münch. ent. Ges., **47**: 150.
Grumazesti; Kl. Neamtu; Slanic; Transsylvanische Ostkarpathen.
ab. **dacica** Caradja, 1893, Iris, **6**: 192. Seitz, 1912, Die Gross-Schmetterlinge der Erde, **2**: 443. Holik, 1943, Mitt. münch. ent. Ges., **33**: 307.
annulata Caradja, 1895, Iris, **8**: 71, 72; 1896, ibidem, **9**: 4. Seitz, 1907, Die Gross-Schmetterlinge der Erde, **2**: 25.
dacica Burgeff, 1914, Mitt. münch. ent. Ges., **5**: 61. Reiss, 1930, in Seitz, Die Gross-Schmetterlinge der Erde, Supplement, **2**: 32; 1933, ibidem, **2**: 274.
ab. **pseudomeliloti** Burgeff, 1926, in Strand, Lepid. Cat., **33**: 50. Reiss, 1930, in Seitz, Die Gross-Schmetterlinge der Erde, Supplement, **2**: 32.
ab. **pseudoitalica** Burgeff, 1926, in Strand, Lepid. Cat., **33**: 50.

Reiss, 1930, in Seitz, Die Gross-Schmetterlinge der Erde, Supplement, **2**: 32.

ssp. **carinthicola** Reiss, 1943, Z. wien. ent. Ges., **28**: 109. — Bei Klagenfurt, Grafenstein, Kärnten, Österreich.

ssp. **sicula** Calberla, 1895, Iris, **8**: 216. Perlini, 1905, Forme di Lepidotteri esclusivamente Italiane, p. 53, pl. 1, fig. 15. Spuler, 1906, in Hofmann, Die Schmetterlinge Europas, **2**: 158. Seitz, 1907, Die Gross-Schmetterlinge der Erde, **2**: 25. Reiss, 1930, in Seitz, Die Gross-Schmetterlinge der Erde, Supplement, **2**: 32, pl. 3h. — Mistretta, Sizilien, 1000 m.

ab. **melilotoides** Ragusa, 1924, Boll. Lab. Zool. Portici, **18**: 93. Tremewan, 1961, Bull. Brit. Mus. (nat. Hist.) Ent., **10**(7): 282, pl. 54, fig. 12.

ab. **cingulata** Ragusa, 1924, Boll. Lab. Zool. Portici, **18**: 87. Tremewan, 1961, Bull. Brit. Mus. (nat. Hist.) Ent., **10**(7): 282, pl. 54, fig. 13.

ssp. **silaecola** Verity, 1930, Mem. Soc. ent. ital., **9**: 23 (nomen novum for *silana* Turati). Burgeff, 1926, Mitt. münch. ent. Ges., **16**: 64 (as *sicula* Calberla). Reiss, 1930, in Seitz, Die Gross-Schmetterlinge der Erde, Supplement, **2**: 32, pl. 3h. — Sila, Calabria, Italia.

silana Turati, 1923, Boll. Soc. ent. ital., **55**: 118 (preoccupied by **silana** Burgeff, 1914, ssp. of **lonicerae** Scheven).

ab. **cingulata** Turati, 1923, Boll. Soc. ent. ital., **55**: 119. Reiss, 1930, in Seitz, Die Gross-Schmetterlinge der Erde, Supplement, **2**: 32.

ab. **charonides** Turati, 1923, Boll. Soc. ent. ital., **55**: 119.

ab. **rubricosta** Turati, 1923, Boll. Soc. ent. ital., **55**: 119. Reiss, 1930, in Seitz, Die Gross-Schmetterlinge der Erde, Supplement, **2**: 32.

ab. **confluens** Turati, 1923, Boll. Soc. ent. ital., **55**: 119. Reiss, 1930, in Seitz, Die Gross-Schmetterlinge der Erde, Supplement, **2**: 32.

ab. **melilocalabra** Stauder, 1915, Z. wiss. InsektBiol., **11**: 73.

ssp. **giussana** Stauder, 1929, Ent. Z., **43**: 79; 1915, Z. wiss. InsektBiol., **11**: 72 (as *teriolensis* Speyer & Speyer). Reiss, 1930, in Seitz, Die Gross-Schmetterlinge der Erde, Supplement, **2**: 32. Tremewan, 1961, Bull. Brit. Mus. (nat. Hist.) Ent., **10** (7): 282, pl. 54, fig. 11. — Sila Giusso, Mte. Faito, Halbinsel Sorrento, Italien, 1000 m.

ssp. **caroni** Holik, 1943, Z. wien. ent. Ges., **28**: 132. — Roccaraso, Abruzzen, Italien, 1250 m.

ab. **cingulata** Holik, 1943, Z. wien. ent. Ges., **28**: 131.

ssp. **italica** Caradja, 1895, Iris, **8**: 71, 72. Seitz, 1907, Die Gross-Schmetterlinge der Erde, **2**: 25, pl. 6d. Dziurzyński, — Litoral des Ligurischen

1908, Berl. ent. Z., **53**: 42, pl. 2, fig. 12. Reiss, 1930, in Seitz, Die Gross-Schmetterlinge der Erde, Supplement, **2**: 32. — Appenins, Norditalien [Umgebung von Genua].

ab. **cingulata** Dziurzyński, 1908, Berl. ent. Z., **53**: 12.

ab. **examaculata** Rocci, 1914, Atti Soc. ligust. Sci. nat. geogr., **25**: 219; 1918, ibidem, **28**: 146. Reiss, 1930, in Seitz, Die Gross-Schmetterlinge der Erde, Supplement, **2**: 32.

incompleta Rocci, 1918, Atti Soc. ligust. Sci. nat. geogr., **28**: 146.

ab. **octornata** Reiss, 1964, Coridon, (A) **6**: 9.

ab. **paupercula** Rocci, 1918, Atti Soc. ligust. Sci. nat. geogr., **28**: 146, pl. 3, fig. 2b. Reiss, 1930, in Seitz, Die Gross-Schmetterlinge der Erde, Supplement, **2**: 32.

ab. **kerleri** Reiss, 1913, Soc. ent., **28**: 76; 1922, Int. ent. Z., **15**: 179; 1926, Die Zygaenen Deutschlands, p. 30, pl. 1, fig.; 1930, in Seitz, Die Gross-Schmetterlinge der Erde, Supplement, **2**: 32, pl. 3g. Alberti, 1955, Ent. Z., **65**: 89-91.

ab. **rubefacta** Rocci, 1916, Atti Soc. ligust. Sci. nat. geogr., **27**: 30; 1918, ibidem, **28**: 146, pl. 3, fig. 2a. Reiss, 1930, in Seitz, Die Gross-Schmetterlinge der Erde, Supplement, **2**: 32.

ab. **nigra** Dziurzyński, 1906, Ent. Z., **19**: 185. Reiss, 1930, in Seitz, Die Gross-Schmetterlinge der Erde, Supplement, **2**: 32.

ab. **biguttata** Rocci, 1918, Atti Soc. ligust. Sci. nat. geogr., **28**: 146. Reiss, 1930, in Seitz, Die Gross-Schmetterlinge der Erde, Supplement, **2**: 32.

ab. **nigerrima** Rocci, 1916, Atti Soc. ligust. Sci. nat. geogr., **27**: 30; 1918, ibidem, **28**: 146. Reiss, 1930, in Seitz, Die Gross-Schmetterlinge der Erde, Supplement, **2**: 32.

ab. **melas** Przegendza, 1932, Ent. Z., **46**: 116, fig. 39. Reiss, 1933, in Seitz, Die Gross-Schmetterlinge der Erde, Supplement, **2**: 274, pl. 16m.

ssp. **silenus** Burgeff, 1926, Mitt. münch. ent. Ges., **16**: 64. Reiss, 1930, in Seitz, Die Gross-Schmetterlinge der Erde, Supplement, **2**: 32. — Maresca, Italien.

ssp. **stentzii** Freyer, 1839, Neuere Beiträge zur Schmetterlingskunde, **3**: 120, pl. 278, fig. 4. Herrich-Schäffer, 1846, Systematische Bearbeitung der Schmetterlinge von Europa, **2**: 36; 1847, ibidem, **2**, pl. 12, figs. 86, 87. Reiss, 1930, in Seitz, Die Gross-Schmetterlinge der Erde, Supplement, **2**: 32. Holik, 1943, Mitt. münch. ent. Ges., **33**: 309. Tremewan, 1961, Bull. Brit. Mus. (nat. Hist.) Ent., **10**(7): 282, pl. 54, fig. 14. — Görz; Pontiebba; Julisch Venetien, Istrien.

ghilianii Ghiliani, 1854, Mem. R. Acad. Torino, (2) **14**: 149 (nomen nudum).

ssp. **teriolensis** Speyer & Speyer, 1858, Die Geographische Verbreitung der Schmetterlinge Deutschlands und der — Bei Meran, Südtirol

Schweiz, **1**: 462. Perlini, 1905, Forme di Lepidotteri esclusivamente Italiane, p. 53, pl. 4, fig. 18. Spuler, 1906, in Hofmann, Die Schmetterlinge Europas, **2**: 158 (partim). Seitz, 1907, Die Gross-Schmetterlinge der Erde, **2**: 25, pl. 6e. Reiss, 1930, in Seitz, Die Gross-Schmetterlinge der Erde, Supplement, **2**: 32; 1950, Jber. naturf. Ges. Graubünden, **82**: 112. [Etschtal; aus dem Misox, Mesoccotal, Schweiz].

ab. **decora** Lederer, 1853, Verh. zool.-bot. Ver. Wien, **2**: 125. Perlini, 1905, Forme di Lepidotteri esclusivamente Italiane, p. 53, pl. 1, fig. 14. Spuler, 1906, in Hofmann, Die Schmetterlinge Europas, **2**, pl. 77, fig. 15b (as *stentzii* Freyer). Reiss, 1930, in Seitz, Die Gross-Schmetterlinge der Erde, Supplement, **2**: 32; 1950, Jber. naturf. Ges. Graubünden, **82**: 112, fig. 21.

stentzii Herrich-Schäffer, 1846, Systematische Bearbeitung der Schmetterlinge von Europa, **2**: 40; 1844, ibidem, **2**, pl. 3, fig. 23 (non-binominal); 1852, ibidem, **6**: 44 (preoccupied). Seitz, 1907, Die Gross-Schmetterlinge der Erde, **2**: 25, pl. 6e.

cingulata Frey, 1884, Mitt. schweiz. ent. Ges., **7**: 14. Tremewan, 1961, Bull. Brit. Mus. (nat. Hist.) Ent., **10**(7): 281, pl. 54, fig. 10.

cingulata Vorbrodt, 1913, in Vorbrodt & Müller-Rutz, Die Schmetterlinge der Schweiz, **2**: 260.

ab. **quinquemaculata** Vorbrodt, 1913, in Vorbrodt & Müller-Rutz, Die Schmetterlinge der Schweiz, **2**: 261. Reiss, 1930, in Seitz, Die Gross-Schmetterlinge der Erde, Suppl., **2**: 32.

ab. **parvimaculata** Vorbrodt, 1913, in Vorbrodt & Müller-Rutz, Die Schmetterlinge der Schweiz, **2**: 261. Reiss, 1933, in Seitz, Die Gross-Schmetterlinge der Erde, Supplement, **2**: 274.

ab. **apicaliconfluens** Vorbrodt, 1913, in Vorbrodt & Müller-Rutz, Die Schmetterlinge der Schweiz, **2**: 260, fig. 14. Reiss, 1933, in Seitz, Die Gross-Schmetterlinge der Erde, Supplement, **2**: 274.

apicaliconfluens Vorbrodt & Müller-Rutz, 1917, Mitt. schweiz. ent. Ges., **12**: 497.

ab. **analielongata** Vorbrodt, 1931, Mitt. schweiz. ent. Ges., **14**: 381. Reiss, 1933, in Seitz, Die Gross-Schmetterlinge der Erde, Supplement, **2**: 274.

ab. **omniconfluens** Vorbrodt, 1931, Mitt. schweiz. ent. Ges., **14**: 381. Reiss, 1933, in Seitz, Die Gross-Schmetterlinge der Erde, Supplement, **2**: 274.

ab. **flava** Reiss, 1950, Jber. naturf. Ges. Graubünden, **82**: 113, fig. 22.

ssp. **charon** Hübner, 1796, Sammlung europäischer Schmetterlinge, **2**: 15, pl. 4, fig. 21; 1806, ibidem, Der Ziefer, p. 81. Ligurische Alpen Süd-

Herrich-Schäffer, 1846, Systematische Bearbeitung der Schmetterlinge von Europa, **2**: 40, pl. 9, figs. 69, 70. Christ, 1880, Mitt. schweiz. ent. Ges., **6**: 40. Perlini, 1905, Forme di Lepidotteri esclusivamente Italiane, p. 52, pl. 4, fig. 17. Reiss, 1930, in Seitz, Die Gross-Schmetterlinge der Erde, Supplement, **2**: 32; 1958, Bull. Soc. ent. Mulhouse, p. 55. Dujardin, 1965, Entomops, Nice, no. 2: 48. — piemonts, Italien.

ssp. **barnabeica** Reiss, 1958, Bull. Soc. ent. Mulhouse, p. 55; 1930, in Seitz, Die Gross-Schmetterlinge der Erde, Supplement, **2**: 32, pl. 3g (as *charon* Hübner). — Col de Vence, St. Barnabé, Alpes-Maritimes, France, 900-1000 m.

barnabeica Reiss, 1958, Z. wien. ent. Ges., **43**: 163 (nomen nudum).

ab. **decora** Reiss, 1958, Bull. Soc. ent. Mulhouse, p. 56.

ab. **sextaseparata** Reiss, 1958, Bull. Soc. ent. Mulhouse, p. 55.

ab. **tenuelimbata** Reiss, 1958, Bull. Soc. ent. Mulhouse, p. 56 (as *tenualimbata*, corrected to **tenuelimbata**, p. 84).

ab. **scabiosaeformis** Le Charles, 1927, Encycl. ent., (B) 3, Lepidoptera, **2**: 152, pl. 9, fig. 9. Reiss, 1930, in Seitz, Die Gross-Schmetterlinge der Erde, Supplement, **2**: 32.

ssp. **dourbensis** Leinfest, 1965, Entomops, Nice, no. 3: 75, fig.; 1963, Bull. Soc. ent. Fr., **68**: 61. Dufay, 1966, Bull. mens. Soc. linn. Lyon, **35**: 71. — Les Dourbes, environ de Digne, Basses-Alpes, France, 1600 m.

ab. **cingulata** Leinfest, 1965, Entomops, Nice, no. 3: 76.

ab. **confluens** Leinfest, 1965, Entomops, Nice, no. 3: 76.

ab. **flava** Leinfest, 1965, Entomops, Nice, no. 3: 76.

ssp. **rhaetica** Burgeff, 1926, Mitt. münch. ent. Ges., **16**: 63 (nomen novum for *alpina* Reiss). Reiss, 1930, in Seitz, Die Gross-Schmetterlinge der Erde, Supplement, **2**: 32; 1950, Jber. naturf. Ges. Graubünden, **82**: 111, fig. 20. — Filisur, Albulatal, Graubünden, Schweiz.

alpina Reiss, 1922, Int. ent. Z., **16**: 67 (preoccupied by **alpina** Boisduval, 1834, ssp. of **transalpina** Esper).

ab. **sexmaculata** Vorbrodt, 1913, in Vorbrodt & Müller-Rutz, Die Schmetterlinge der Schweiz, **2**: 261, fig. 28.

ab. **unimaculata** Vorbrodt, 1913, in Vorbrodt & Müller-Rutz, Die Schmetterlinge der Schweiz, **2**: 261, fig. 42. Reiss, 1930, in Seitz, Die Gross-Schmetterlinge der Erde, Supplement, **2**: 31.

staefae Vorbrodt, 1913, in Vorbrodt & Müller-Rutz, Die Schmetterlinge der Schweiz, **2**: 261.

ab. **basiconfluens** Vorbrodt, 1913, in Vorbrodt & Müller-Rutz, Die Schmetterlinge der Schweiz, **2**: 260.

ab. **analiconfluens** Vorbrodt, 1913, in Vorbrodt & Müller-Rutz, Die Schmetterlinge der Schweiz, **2**: 261.

ab. **costaliconfluens** Vorbrodt, 1913, in Vorbrodt & Müller-Rutz, Die Schmetterlinge der Schweiz, **2**: 261.

ab. **parallela** Vorbrodt, 1914, in Vorbrodt & Müller-Rutz, Die Schmetterlinge der Schweiz, (Nachtrag), **2**: 648.
ab. **omniconfluens** Vorbrodt, 1913, in Vorbrodt & Müller-Rutz, Die Schmetterlinge der Schweiz, **2**: 261, fig. 26.
ab. **funerea** Cornelsen, 1923, Int. ent. Z., **16**: 213. Reiss, 1930, in Seitz, Die Gross-Schmetterlinge der Erde, Supplement, **2**: 32.

ssp. **subglocknerica** Reiss, 1943, Z. wien. ent. Ges., **28**: 108. — Kleine Fleiss, Grossglocknergebiet, Kärnten, Österreich, 1350 m.
pseudofilipendulae Stauder, 1924, Int. ent. Z., **18**: 52 (infrasubspecific).
Tremewan, 1966, Ent. Rec., **78**: 34, pl. 1, fig. 11.
ab. **charonides** Reiss, 1943, Z. wien. ent. Ges., **28**: 108.
ab. **crassimaculata** Reiss, 1943, Z. wien. ent. Ges., **28**: 108.
ab. **pseudomeliloti** Reiss, 1943, Z. wien. ent. Ges., **28**: 108.

ssp. **submontana** Reiss, 1926, Int. ent. Z., **19**: 311; 1926, Die Zygaenen Deutschlands, p. 30. — Bei Oberstdorf, bayrische Alpen, Süddeutschland.

ssp. **viciae** Denis & Schiffermüller, 1775, Ankündigung eines systematischen Werkes von den Schmetterlingen der Wienergegend, p. 45; 1776, Systematisches Verzeichniss der Schmetterlinge der Wienergegend, p. 45. Schrank, 1782, in Fuessly, Neues Magazin für die Liebhaber der Entomologie, **2**: 208. Denis & Schiffermüller, 1801, Systematisches Verzeichniss von den Schmetterlingen der Wiener Gegend, **1**: 56. Dujardin, 1953, Bull. mens. Soc. linn. Lyon, **22**: 246. Tremewan, 1958, Ent. Gaz., **9**: 190; 1960, ibidem, **11**: 188. Bernardi & Viette, 1960, Bull. mens. Soc. linn. Lyon, **29**: 244. Dujardin, 1965, Entomops, Nice, no. 2: 48. — Bei Wien [Bisamberg], Österreich.
ab. **pygmaeana** Schnaider, 1950, Bull. ent. Pologne, **19**: 243.
ab. **pygmaea** Dąbrowski, 1965, Acta zool. cracov., **10**: 93, 113, 115, fig. 29, pl. 8, fig. 5.
ab. **totarubra** Dziurzyński, 1914, Int. ent. Z., **8**: 33. Reiss, 1926, Die Zygaenen Deutschlands, p. 30; 1930, in Seitz, Die Gross-Schmetterlinge der Erde, Supplement, **2**: 31.
ab. **brunnea** Sterzl, 1921, Z. öst. EntVer., **6**: 60.
brunnea Dąbrowski, 1965, Acta zool. cracov., **10**: 115, 171, pl. 10, fig. 18.

ssp. **stiefi** Reiss, 1943, Z. wien. ent. Ges., **28**: 106. Povolný, 1945, Folia ent., Brno, **8**: 78. Gregor & Povolný, 1955, Sborn. ent. Odd. nár. Mus. Praze, **30**: 261, 272, pl. 5, fig. 3. — Bei Olmütz, Wachhübel, Mähren, 300-400 m.
ab. **pygmaeoides** Reiss, 1943, Z. wien. ent. Ges., **28**: 107.
ab. **pseudostentzi** Reiss, 1943, Z. wien. ent. Ges., **28**: 107.
ab. **pseudomeliloti** Reiss, 1943, Z. wien. ent. Ges., **28**: 106.
ab. **sexpunctata** Reiss, 1943, Z. wien. ent. Ges., **28**: 107.
ab. **crassimaculata** Reiss, 1943, Z. wien. ent. Ges., **28**: 107.

ab. **medioconfluens** Reiss, 1943, Z. wien. ent. Ges., **28**: 106.
ab. **analiconfluens** Reiss, 1943, Z. wien. ent. Ges., **28**: 107.
ab. **costaliconfluens** Reiss, 1943, Z. wien. ent. Ges., **28**: 107.
ab. **rubricosta** Reiss, 1943, Z. wien. ent. Ges., **28**: 107.
ab. **carnea** Reiss, 1943, Z. wien. ent. Ges., **28**: 107.
ab. **dichroma** Reiss, 1943, Z. wien. ent. Ges., **28**: 108.

ssp. **meliloti** Esper, 1793, Die Schmetterlinge, Supplement, **2** (2): 10, pl. 39, figs. 1-8. Ochsenheimer, 1808, Die Schmetterlinge von Europa, **2**: 43. Boisduval, 1828, Essai sur une Monographie des Zygénides, p. 51, pl. 3, fig. 5; 1834, Icones historique des Lépidoptères nouveaux ou peu connus, **2**: 56, pl. 54, fig. 6. Duponchel, 1835, in Godart & Duponchel, Histoire naturelle des Lépidoptères ou Papillons de France, Supplement, **2**: 62, pl. 5, fig. 7. Herrich-Schäffer, 1846, Systematische Bearbeitung der Schmetterlinge von Europa, **2**: 35, pl. 11, fig. 78. Speyer & Speyer, 1858, Die geographische Verbreitung der Schmetterlinge Deutschlands und der Schweiz, p. 345. Spuler, 1906, in Hofmann, Die Schmetterlinge Europas, **2**: 158, pl. 75, fig. 49, pl. 77, fig. 15. Seitz, 1907, Die Gross-Schmetterlinge der Erde, **2**: 25, pl. 6d. Oberthür, 1910, Études de Lépidoptérologie comparée, **4**: 484. Reiss, 1926, Die Zygaenen Deutschlands, p. 29, pl. 2, fig. Dufour, 1927, Misc. ent., **30**: 93. Reiss, 1930, in Seitz, Die Gross-Schmetterlinge der Erde, Supplement, **2**: 31, pl. 3g. Derenne, 1934, Amat. Papillons, **7**: 149. Reiss, 1937, in Schneider, Jh. Ver. vaterl. Naturk. Württemb., **93**: 127. Kuserau, 1942, Ent. Z., **56**: 45. Reiss, 1943, Z. wien. ent. Ges., **28**: 105; 1949, Entomon, **1**: 172. Haaf, 1952, Veröff. zool. Staatssamml. Münch., **2**: 152, 154, 157, pl. 12. Bergmann, 1953, Die Grossschmetterlinge Mitteldeutschlands, **3**: 41. Forster & Wohlfahrt, 1958, Die Schmetterlinge Mitteleuropas, **3**: 96, pl. 10, figs. 32, 33, 37. Alberti, 1958, Mitt. zool. Mus. Berl., **34**: 320. Heuser & Jöst, 1959, Mitt. Pollichia, (3) **6**: 125.

Erlangen, Franken [Mittel- und Süddeutschland].

buglossi Boisduval, 1829, Monographie des Zygénides, Errata et Addenda, p. 2; 1834, Icones historique des Lépidoptères nouveaux ou peu connus, **2**: 56. Duponchel, 1835, in Godart & Duponchel, Histoire naturelle des Lépidoptères ou Papillons de France, Supplement, **2**: 138, pl. 12, fig. 4.

heydenreichii Heydenreich, 1851, Lepidopterorum Europaeorum Catalogus Methodicus, p. 21 (nomen nudum).

ab. **pseudostentzii** Burgeff, 1926, in Strand, Lepid. Cat., **33**: 48. Reiss, 1930, in Seitz, Die Gross-Schmetterlinge der Erde, Supplement, **2**: 31.
ab. **melilotella** Lambillion, 1909, Rev. Soc. ent. namur., **9**: 68.
ab. **fimbriata** Burgeff, 1926, Mitt. münch. ent. Ges., **16**: 63.

Reiss, 1930, in Seitz, Die Gross-Schmetterlinge der Erde, Supplement, **2**: 31.

ab. **sexmaculata** Lambillion, 1909, Rev. Soc. ent. namur., **9**: 68.

ab. **obscura** Reiss, 1926, Int. ent. Z., **19**: 310, 311; 1926, Die Zygaenen Deutschlands, p. 29, 30.

ab. **atrata** Metschl, 1925, Int. ent. Z., **19**: 27. Reiss, 1935, Int. ent. Z., **28**: 542.

ab. **conjuncta** Lambillion, 1909, Rev. Soc. ent. namur., **9**: 68.

medioconfluens Vorbrodt, 1913, in Vorbrodt & Müller-Rutz, Die Schmetterlinge der Schweiz, **2**: 260.

ab. **ligata** Lambillion, 1909, Rev. Soc. ent. namur., **9**: 69.

ab. **basimedioconfluens** Vorbrodt, 1913, in Vorbrodt & Müller-Rutz, Die Schmetterlinge der Schweiz, **2**: 260.

ab. **rubescens** Reiss, 1943, Z. wien. ent. Ges., **28**: 105.

ab. **irregularis** Reiss, 1964, Coridon, (A) **6**: 9.

ab. **pseudoconfusa** Burgeff, 1926, in Strand, Lepid. Cat., **33**: 48. Reiss, 1930, in Seitz, Die Gross-Schmetterlinge der Erde, Supplement, **2**: 31.

ab. **flava** Burgeff, 1906, Ent. Z., **20**: 162. Seitz, 1907, Die Gross-Schmetterlinge der Erde, **2**: 25. Reiss, 1930, in Seitz, Die Gross-Schmetterlinge der Erde, Supplement, **2**: 31.

ssp. **hesperina** Dujardin, 1965, Entomops, Nice, no. 2: 48, fig. — Cauterets, Hautes-Pyrénées, France.

ssp. **farriolsi** de Sagarra, 1925, Butll. Inst. catal. Hist. nat., (2) **5**: 273. Reiss, 1930, in Seitz, Die Gross-Schmetterlinge der Erde, Supplement, **2**: 32. — Puigsacalm, Cabrera, Catalonia, España.

ssp. **anglica** Reiss, 1931, Int. ent. Z., **25**: 344; 1931, ibidem, **25**: 359, figs. Tugwell, 1888, Young. Nat., **9**: 53, 99, 131, 174. Briggs, 1888, Young. Nat., **9**: 82, 108, 153, 188. Barrett, 1895, The Lepidoptera of the British Islands, **2**: 124, pl. 59, fig. 1. South, 1908, The Moths of the British Isles, **2**: 336, pl. 146, figs. 4, 5. Reiss, 1933, in Seitz, Die Gross-Schmetterlinge der Erde, Supplement, **2**: 274. Ford, 1955, Moths, p. 144, 229, pl. 26, fig. 14. South, 1961, The Moths of the British Isles, **2**: 330, pl. 129, figs. 5, 6. Tremewan, 1958, Ent. Gaz., **9**: 190; 1960, ibidem, **11**: 188; 1961, Coridon, (A) **1**: 2, pl. C1, fig. 8. — Lyndhurst, New Forest, England.

ab. **sexpunctata** Tutt, 1899, A Natural History of the British Lepidoptera, **1**: 455. Reiss, 1930, in Seitz, Die Gross-Schmetterlinge der Erde, Supplement, **2**: 31. Tremewan, 1961, Bull. Brit. Mus. (nat. Hist.) Ent., **10** (7): 281, pl. 54, fig. 9.

ab. **confluens** Tutt, 1899, A Natural History of the British

Lepidoptera, **1**: 456. Barrett, 1895, The Lepidoptera of the British Islands, **2**: 124, pl. 59, fig. 1a. South, 1908, The Moths of the British Isles, **2**, pl. 148, fig. 1 (as *confusa* Staudinger). Seitz, 1912, Die Gross-Schmetterlinge der Erde, **2**: 443. Reiss, 1930, in Seitz, Die Gross-Schmetterlinge der Erde, Supplement, **2**: 31. Tremewan, 1961, Bull. Brit. Mus. (nat. Hist.) Ent., **10** (7): 281, pl. 54, fig. 8; 1961, Coridon, (A) **1**: 2, pl. C1, fig. 9.

ab. **rubescens** Holik, 1943, Z. wien. ent. Ges., **28**: 132. Barrett, 1895, The Lepidoptera of the British Islands, **2**, pl. 59, fig. 1b.

ssp. **masurica** Reiss, 1939, Ent. Z., **53**: 115. Kuserau, 1942, Ent. Z., **56**: 47. — Rüdzanny, Masurische Seen, Ostpreussen.

ssp. **nigrescens** Reiss, 1921, Int. ent. Z., **15**: 118; 1926, Die Zygaenen Deutschlands, p. 29, pl. 2, fig. Reiss & Tremewan, 1964, Ent. Rec., **76**: 135. — Grünortspitze, Osterode, Ostpreussen.

nigrina Burgeff, 1926, in Strand, Lepid. Cat., **33**: 49. Reiss, 1930, in Seitz, Die Gross-Schmetterlinge der Erde, Supplement, **2**: 31, pl. 3g. Holik, 1939, Ann. Mus. zool. Polon., **12**: 64, pl. 3, figs. 74-76. Nordström & Wahlgren, 1941, Svenska Fjärilar, p. 327, pl. 46, fig. 4. Kuserau, 1942, Ent. Z., **56**: 45. Reiss & Tremewan, 1964, Ent. Rec., **76**: 135.

ab. **pseudostentzii** Reiss, 1930, in Seitz, Die Gross-Schmetterlinge der Erde, Supplement, **2**: 32.

ab. **sexpunctata** Reiss, 1931, Int. ent. Z., **25**: 345; 1933, in Seitz, Die Gross-Schmetterlinge der Erde, Supplement, **2**: 274.

ssp. **engleri** Reiss, 1939, Ent. Z., **53**: 114. Kuserau, 1942, Ent. Z., **56**: 45. — Finkenwalde bei Stettin; Berliner Umgebung; Norddeutschland.

ab. **quadrimaculata** Guhn, 1932, Ent. Jb., **41**: 94. Reiss, 1933, in Seitz, Die Gross-Schmetterlinge der Erde, Supplement, **2**: 274.

ab. **sexpunctata** Reiss, 1939, Ent. Z., **53**: 114.

ab. **rubricosta** Reiss, 1943, Z. wien. ent. Ges., **28**: 105.

ab. **confluens** Guhn, 1932, Ent. Jb., **41**: 94.

ab. **rubescens** Reiss, 1943, Z. wien. ent. Ges., **28**: 105.

ab. **flava** Reiss, 1941, Mitt. münch. ent. Ges., **31**: 1000.

ssp. **estonica** Holik & Sheljuzhko, 1957, Mitt. münch. ent. Ges., **47**: 148. — Reval, Estland.

ssp. **ehnbergii** Reuter, 1893, Acta Soc. Fauna Flora fenn., **9** (6): 22. Spuler, 1906, in Hofmann, Die Schmetterlinge Europas, **2**: 158 (partim). Seitz, 1907, Die Gross-Schmetterlinge der Erde, **2**: 25. Reiss, 1930, in Seitz, Die Gross-Schmetterlinge der Erde, Supplement, **2**: 31; 1933, ibidem, **2**: 274. — Kuhmois, Finnland.

ab. **confluens** Burgeff, 1926, in Strand, Lepid. Cat., **33**: 49.

Reiss, 1930, in Seitz, Die Gross-Schmetterlinge der Erde, Supplement, **2**: 31.

Biology

Barrett, 1895, The Lepidoptera of the British Islands, **2**: 124, pl. 59, fig. 1c. Burgeff, 1912, Z. wiss. InsektBiol., **8**: 124, 197, 198; 1926, Mitt. münch. ent. Ges., **16**: 63, 64; 1950, Portug. acta biol., (A) Goldschmidt: 663-728; 1951, Biol. Zbl., **70**: 1-23. Carpenter, 1937, J. Soc. Brit. Ent., **1**: 178. Döring, 1955, Der Morphologie der Schmetterlingseier, p. 127, pl. 18, fig. 257. Dorfmeister, 1854, Verh. zool.-bot. Ver. Wien, **4**: 478. Esper, 1793, Die Schmetterlinge, Supplement, **2** (2): 10, pl. 39, figs. 4-8. Guhn, 1932, Ent. Jb., **41**: 94. Holik, 1937, Lambillionea, **37**: 15-24, 32-45, 80-91; 1938, ibidem, **38**: 51-58, 79-88, 95-102; 1938, Ent. Rdsch., **55**: 320-323, 331-333; 1953, Ent. Z., **63**: 25. Hrubý, 1964, Prodromus Lepidopter Slovenska, p. 478. Koch, 1955, Wir Bestimmen Schmetterlinge, **2**: 60, 61, pl. 1, fig. 13. Leinfest, 1966, Entomops, Nice, no. 3: 76. Ochsenheimer, 1808, Die Schmetterlinge von Europa, **2**: 43. Nordström & Wahlgren, 1941, Svenska Fjärilar, p. 327, pl. 46, fig. 41. Reiss, 1926, Die Zygaenen Deutschlands, p. 7; 1930, in Seitz, Die Gross-Schmetterlinge der Erde, Supplement, **2**: 32; 1958, Z. wien. ent. Ges., **43**: 161. Roüast, 1883, Catalogue des Chenilles européennes connues, p. 22. Sarlet, 1964, Mém. Soc. r. ent. Belg., **29**: 6. Seitz, 1907, Die Gross-Schmetterlinge der Erde, **2**: 25. Seppänen, 1954, Suomen Suurperhostoukkien Ravintokasvit, p. 256. Spuler, 1910, in Hofmann, Die Raupen der Schmetterlinge Europas, (Nachtrag), pl. 9, fig. 22. Tremewan, 1965, Ent. Gaz., **16**: 119-124. Tutt, 1899, A Natural History of the British Lepidoptera, **1**: 460.

niphona Butler

Distribution: Amurland, North Korea, Japan.

ssp. **christophi** Staudinger, 1887, in Romanoff, Mémoires sur les Lépidoptères, **3**: 173, pl. 8, fig. 9. Dziurzyński, 1908, Berl. ent. Z., **53**: 43, pl. 1, fig. 11 (as *niphona* Butler). Reiss, 1931, Int. ent. Z., **25**: 353, figs.; 1933, in Seitz, Die Gross-Schmetterlinge der Erde, Supplement, **2**: 275. Holik & Sheljuzhko, 1957, Mitt. münch. ent. Ges., **47**: 171. — Bei Raddefka, Amurgebiet.

ab. **quinquemaculata** Reiss, 1931, Int. ent. Z., **25**: 357; 1933, in Seitz, Die Gross-Schmetterlinge der Erde, Supplement, **2**: 275.

ab. **confluens** Reiss, 1931, Int. ent. Z., **25**: 357; 1933, in Seitz, Die Gross-Schmetterlinge der Erde, Supplement, **2**: 275.

ssp. **coreana** Reiss, 1931, Int. ent. Z., **25**: 358, figs.; 1933, in — Sei-Shin, Nord-Korea.

Seitz, Die Gross-Schmetterlinge der Erde, Supplement, **2**: 275. Holik & Sheljuzhko, 1957, Mitt. münch. ent. Ges., **47**: 173.

ab. **pseudochristophi** Reiss, 1931, Int. ent. Z., **25**: 358; 1933, in Seitz, Die Gross-Schmetterlinge der Erde, Suppl., **2**: 275.

ab. **confluens** Reiss, 1931, Int. ent. Z., **25**: 358, figs.; 1933, in Seitz, Die Gross-Schmetterlinge der Erde, Supplement, **2**: 275.

ssp. **niphona** Butler, 1877, Ann. Mag. nat. Hist., (4) **20**: 393; 1878, Illustrations of Typical Specimens of Lepidoptera Heterocera in the Collection of the British Museum, **2**: 4, pl. 21, fig. 9. Leech, 1888, Proc. zool. Soc. Lond., p. 597. Staudinger, 1892, in Romanoff, Mémoires sur les Lépidoptères, **6**: 251. Seitz, 1907, Die Gross-Schmetterlinge der Erde, **2**: 25, pl. 6e. Reiss, 1931, Int. ent. Z., **25**: 353, figs.; 1933, in Seitz, Die Gross-Schmetterlinge der Erde, Supplement, **2**: 275. Holik & Sheljuzhko, 1957, Mitt. münch. ent. Ges., **47**: 168. Esaki et al., 1957, Icones Heterocerorum Japonicorum, **21**: 157, pl. 28, fig. 823. Alberti, 1958, Mitt. zool. Mus. Berl., **34**: 320. Inoue et al., 1959, Iconographia Insectorum Japonicorum, **1**: 228, pl. 163, fig. 13. Tremewan, 1961, Bull. Brit. Mus. (nat. Hist.) Ent., **10** (7): 283, pl. 54, fig. 16, pl. 62, figs. 5, 6. — Yokohama, Japan.

ab. **quinquemaculata** Reiss, 1931, Int. ent. Z., **25**: 357; 1933, in Seitz, Die Gross-Schmetterlinge der Erde, Suppl., **2**: 275.

Biology

Burgeff, 1950, Portug. acta biol., (A) Goldschmidt: 663-728. Holik, 1937, Lambillionea, **37**: 15-24, 32-45, 80-91; 1953, Ent. Z., **63**: 28. Okano, 1949, Bull. Tokohu Ent. Soc., **2**: 9. Tremewan, 1960, Entomologist, **93**: 108.

SECTION 8

gallica Oberthür

Distribution: Basses-Alpes, Hautes-Alpes, Alpes-Maritimes, Vaucluse, France.

ssp. **gallica** Oberthür, 1898, Bull. Soc. ent. Fr., p. 21. Seitz, 1907, Die Gross-Schmetterlinge der Erde, **2**: 19. Oberthür, 1909, Études de Lépidoptérologie comparée, **3**, pl. 28, fig. 173; 1910, ibidem, **4**: 427. Reiss, 1930, in Seitz, Die Gross-Schmetterlinge der Erde, Supplement, **2**: 9, pl. 1h; 1933, ibidem, **2**: 251. Holik, 1936, Lambillionea, **36**: 50. Praviel, 1944, Rev. franç. Lépid., **10**: 146. Haaf, 1952, Veröff. zool. Staatssamml. Münch., **2**: 151, 153, 156, pl. 4. Le Charles, 1953, Rev. franç. Lépid., **14**: 13, figs. 3, 4, pl. 2, fig. 2, pl. 3, — **Environs de Digne, Basses-Alpes, France, 1000 m.**

fig. 1. Verity, 1953, Rev. franç. Lépid., **14**: 50. Reiss, 1953, Z. wien. ent. Ges., **38**: 135, pl. 9, figs. 10, 11. Alberti, 1958, Mitt. zool. Mus. Berl., **34**: 314. Loritz, 1961, Bull. Soc. ent. Mulhouse, p. 83-102, fig. Tremewan, 1961, Bull. Brit. Mus. (nat. Hist.) Ent., **10**(7): 283, pl. 54, fig. 17, pl. 63, figs. 1, 2. Tremewan & Reiss, 1964, Ent. Rec., **76**: 2, 4. Loritz, 1964, Bull. Soc. ent. Mulhouse, p. 51-86, figs. Dufay, 1966, Bull. mens. Soc. linn. Lyon, **35**: 69.

ab. **interrupta** Reiss, 1953, Z. wien. ent. Ges., **38**: 136.

ssp. **frigidagallica** Dujardin, 1956, Bull. mens. Soc. linn. Lyon, **25**: 254. Loritz, 1957, Bull. mens. Soc. linn. Lyon, **26**: 155; 1962, Bull. Soc. ent. Mulhouse, p. 17-26, figs. Tremewan & Reiss, 1964, Ent. Rec., **76**: 4. Loritz, 1964, Bull. Soc. ent. Mulhouse, p. 51-86, figs. Dufay, 1966, Bull. mens. Soc. linn. Lyon, **35**: 69. — Céüze (environs de Gap), Hautes-Alpes, France, 1500 m.

ssp. **bordei** Dujardin, 1965, Entomops, Nice, no. 2: 38, fig. — Caussols, Maison Rouz, env. 1000-1200 m.; Col de Ferrier, 1200 m.; Col de la Sine, 1360 m.; route de Thorenc à Canaux, 1220 m.; Gourdon, route de Caussols; St. Vallier, 700 m.; Alpes-Maritimes, France.

Biology

Burgeff, 1950, Portug. acta biol., (A) Goldschmidt: 663-728. Holik, 1953, Ent. Z., **63**: 4. Loritz, 1957, Bull. mens. Soc. linn. Lyon, **26**: 155-158; 1964, Bull. Soc. ent. Mulhouse, p. 53-54 (footnote). Loritz, Borelly & Tichy, 1959, Bull. Soc. ent. Mulhouse, p. 33-35. Reiss, 1953, Z. wien. ent. Ges., **38**: 136, fig. 3.

giesekingiana Reiss

Distribution: Alpes-Maritimes, France.

giesekingiana Reiss, 1930, in Seitz, Die Gross-Schmetterlinge der Erde, Supplement, **2**: 9, pl. 1h (nomen novum for *interrupta* Boursin); 1933, ibidem, **2**: 251. Holik, 1936, Lambillionea, **36**: 50. Praviel, 1944, Rev. franç. Lépid., **10**: 146. Haaf, 1952, Veröff. zool. Staatssamml. Münch., **2**: 151, 153, 156, — St. Barnabé (Vence à Coursegoules), Alpes-Maritimes, France, 1000 m.

pl. 4 (as *interrupta* Boursin). Reiss, 1953, Ent. Z., **63**: 77; 1953, Z. wien. ent. Ges., **38**: 135, 141, pl. 9, figs. 1-8; 1958, Bull. Soc. ent. Mulhouse, p. 53. Loritz & Fiammengo, 1957, Bull. Soc. ent. Mulhouse, p. 47 (as *gallica* Oberthür). Alberti, 1958, Mitt. zool. Mus. Berl., **34**: 314. Loritz, 1959, Bull. Soc. ent. Mulhouse, p. 10, fig. (as *gallica* Oberthür). Tremewan & Reiss, 1964, Ent. Rec., **76**: 4. Loritz, 1964, Bull. Soc. ent. Mulhouse, p. 51-86, figs. Dujardin, 1965, Entomops, Nice, no. 2: 40.

interrupta Boursin, 1923, Bull. Soc. ent. Fr., p. 68, fig. 1 (infrasubspecific); 1923, Ann. Soc. ent. Fr., **92**: 321 (infrasubspecific). Le Charles, 1953, Rev. franç Lépid., **14**: 14, figs. 5, 6, pl. 2, fig. 3, pl. 3, fig. 2. Tremewan & Reiss, 1964, Ent. Rec., **76**: 5. Dujardin, 1965, Entomops, Nice, no. 2: 39, 40.

ab. **paupera** Reiss, 1953, Z. wien. ent. Ges., **38**: 141, pl. 9, fig. 9.

ab. **confluens** Dujardin, 1956, Bull. mens. Soc. linn. Lyon, **25**: 256.

ab. **salmonea** Dujardin, 1956, Bull. mens. Soc. linn. Lyon, **25**: 256.

ssp. **hemicharis** Dujardin, 1965, Entomops, Nice, no. 2: 41, fig. — Col de Braüs, Alpes-Maritimes, France, 1000 m.

Biology

Burgeff, 1950, Portug. acta biol., (A) Goldschmidt: 663-728. Dujardin, 1965, Entomops, Nice, no. 2: 40, 41. Holik, 1953, Ent. Z., **63**: 4. Loritz, 1964, Bull. Soc. ent. Mulhouse, p. 53-54 (footnote). Reiss, 1953, Z. wien. ent. Ges., **38**: 135; 1958, ibidem, **43**: 181.

nevadensis Rambur

Distribution: Spain, Portugal.

ssp. **nevadensis** Rambur, 1866, Catalogue systématique des Lépidoptères de l'Andalousie, p. 166, pl. 1, fig. 10. Spuler, 1906, in Hofmann, Die Schmetterlinge Europas, **2**: 155, pl. 77, fig. 5c. Seitz, 1907, Die Gross-Schmetterlinge der Erde, **2**: 19, pl. 4e. Oberthür, 1910, Études de Lépidoptérologie comparée, **4**: 440. Reiss, 1930, in Seitz, Die Gross-Schmetterlinge der Erde, Supplement, **2**: 11, pl. 1k; 1933, ibidem, **2**: 251; 1931, Int. ent. Z., **25**: 109, fig. Koch, 1948, Eos, Madr., **24**: 324. Marten, 1956, Ent. Z., **66**: 276. Alberti, 1958, Mitt. zool. Mus. Berl., **34**: 314. Tremewan & Reiss, 1964, Ent. Rec., **76**: 5, 6, fig. 1. — Parties moyennes de la Sierra Nevada, Espagne.

atlantica Le Charles, 1957, Rev. franç. Lépid., **16**: 21, pl. 5, fig. 37 (nomen nudum). Tremewan & Reiss, 1964, Ent. Rec., **76**: 6.

ssp. **dumalis** Marten, 1957, Ent. Z., **67**: 14. Tremewan & Reiss, 1964, Ent. Rec., **76**: 6.

Sierra de los Filabres, oberhalb Baza, Süd-spanien, 1400 m.

ssp. **kricheldorffi** Reiss, 1933, in Seitz, Die Gross-Schmetterlinge der Erde, Supplement, **2**: 252; 1931, Int. ent. Z., **25**: 111, fig. Tremewan & Reiss, 1964, Ent. Rec., **76**: 6.

Umgebung von Guarda, Portugal, 500 m.

ab. **scabiosoides** Reiss, 1931, Int. ent. Z., **25**: 111, fig.; 1933, in Seitz, Die Gross-Schmetterlinge der Erde, Supplement, **2**: 252.

ssp. **guadalupei** Koch, 1948, Eos, Madr., **24**: 326. Tremewan & Reiss, 1964, Ent. Rec., **76**: 6.

Guadalupe, Prov. Cáceres, España, 654 m.

ssp. **schmidti** Reiss, 1931, Int. ent. Z., **25**: 112, figs.; 1933, in Seitz, Die Gross-Schmetterlinge der Erde, Supplement, **2**: 252, pl. 16k. Tremewan, 1961, Bull. Brit. Mus. (nat. Hist.) Ent., **10**(7): 308, pl. 57, fig. 26. Tremewan & Reiss, 1964, Ent. Rec., **76**: 6.

Umgebung von Arenas St. Pedro, Prov. Avila (Sierra de Gredos), Spanien.

ssp. **muda** Marten, 1957, Ent. Z., **67**: 15. Tremewan & Reiss, 1964, Ent. Rec., **76**: 7.

Oberes Teratal, zwischen den Seen Laguna de Yengua und Laguna de Villachica, östlich vom Berg Moncalvo, Prov. Zamora, Spanien, 1300 m.

ssp. **falleriana** Reiss, 1931, Int. ent. Z., **25**: 111, figs.; 1933, in Seitz, Die Gross-Schmetterlinge der Erde, Supplement, **2**: 252. Koch, 1948, Eos, Madr., **24**: 326. Tremewan & Reiss, 1964, Ent. Rec., **76**: 7.

Albarracin, Sierra Noguera, Spanien, 1400-1600 m.

ab. **amplomaculata** Reiss, 1936, Ent. Rdsch., **54**: 29.

ab. **confluens** Reiss, 1936, Ent. Rdsch., **54**: 29.

ssp. **picos** Agenjo, 1953, Graellsia, **11**: 1. Tremewan, 1961, Ent. Rec., **73**: 6; 1963, ibidem, **75**: 8. Tremewan & Reiss, 1964, Ent. Rec., **76**: 7.

Fuente Dé, Camaleño, Santander (Picos de Europa), España, 1001 m.

ssp. **timida** Marten, 1956, Ent. Z., **66**: 287. Tremewan & Reiss, 1964, Ent. Rec., **76**: 7.

agenjoi Le Charles, 1957, Rev. franç. Lépid., **16**: 21, pl. 6,

In den Bergen zwischen den Städten

figs. 39, 40; 1960, Bull. Soc. ent. Fr., **65**: 103. Tremewan & Reiss, 1964, Ent. Rec., **76**: 7. Castellón und Tortosa, Ostspanien, 900 m.

Biology

Holik, 1953, Ent. Z., **63**: 4. Reiss, 1933, in Seitz, Die Gross-Schmetterlinge der Erde, Supplement, **2**: 252.

mana Kirby

Distribution: Transcaucasia, Ciscaucasia.

ssp. **mana** Kirby, 1892, A Synonymic Catalogue of Lepidoptera Heterocera (Moths), p. 64 (nomen novum for *erebus* Staudinger). Holik, 1939, Ent. Rdsch., **56**: 114. Koch, 1939, Mitt. münch. ent. Ges., **29**: 399. Holik, 1941, Ent. Z., **54**: 214. Reiss, 1953, Z. wien. ent. Ges., **38**: 141, pl. 9, figs. 12, 13. Verity, 1953, Rev. franç. Lépid., **14**: 50. Le Charles, 1953, Rev. franç. Lépid., **14**: 13. Holik & Sheljuzhko, 1955, Mitt. münch. ent. Ges., **44/45**: 112. Alberti, 1958, Mitt. zool. Mus. Berl., **34**: 315. Tremewan & Reiss, 1964, Ent. Rec., **76**: 8. Loritz, 1964, Bull. Soc. ent. Mulhouse, p. 51-86, figs. Alberti, 1964, Mitt. münch. ent. Ges., **54**: 262. Adshara Gebiet, Georgien, Transkaukasien.

erebus Staudinger, 1867, Stettin. ent. Ztg., **28**: 101 (preoccupied by *erebus* Meigen, 1830, synonym of **anthyllidis** Boisduval, 1829). Romanoff, 1884, Mémoires sur les Lépidoptères, **1**: 78, pl. 4, fig. 4. Seitz, 1907, Die Gross-Schmetterlinge der Erde, **2**: 19, pl. 4c. Dziurzyński, 1908, Berl. ent. Z., **53**: 17, pl. 1, fig. 2. Reiss, 1935, Int. ent. Z., **29**: 122, fig. Le Charles, 1953, Rev. franç. Lépid., **14**: 13, figs. 1, 2, pl. 2, figs. 1, 4. Verity, 1953, Rev. franç. Lépid., **14**: 50.

erebaea Burgeff, 1926, Mitt. münch. ent. Ges., **16**: 15.

ab. **interrupta** Dziurzyński, 1906, Ent. Z., **19**: 185.

ab. **confluens** Koch, 1939, Mitt. münch. ent. Ges., **29**: 400, 401.

ssp. **chaos** Burgeff, 1926, Mitt. münch. ent. Ges., **16**: 15. Reiss, 1930, in Seitz, Die Gross-Schmetterlinge der Erde, Supplement, **2**: 10, pl. 1h; 1953, Z. wien. ent. Ges., **38**: 141, pl. 8, figs. 12, 13 (as *mana* Kirby). Holik & Sheljuzhko, 1955, Mitt. münch. ent. Ges., **44/45**: 116. Tremewan, 1961, Bull. Brit. Mus. (nat. Hist.) Ent., **10(7)**: 308, pl. 57, fig. 25. Tremewan & Reiss, 1964, Ent. Rec., **76**: 9. Bethania bei Tiflis, Georgien, Transkaukasien.

interrupta Burgeff, 1914, Mitt. münch. ent. Ges., **5**: 45, pl. 5, fig. 18 (infrasubspecific).

ssp. **tarkiensis** Holik & Sheljuzhko, 1955, Mitt. münch. ent. Ges., **44/45**: 115. Tremewan & Reiss, 1964, Ent. Rec., **76**: 9. Berg Tarki bei Petrovsk (Machatsh-Kala), Dagestan, Ciskaukasien.

ab. **divisa** Holik & Sheljuzhko, 1955, Mitt. münch. ent. Ges., **44/45**: 115.

Biology

Burgeff, 1950, Portug. acta biol., (A) Goldschmidt: 663-728. Holik, 1953, Ent. Z., **63**: 4.

rjabovi Holik

Distribution: Armenia.

rjabovi Holik, 1939, Ent. Rdsch., **56**: 115. Koch, 1939, Mitt. münch. ent. Ges., **29**: 399, 403; 1940, Ent. Z., **54**: 199. Holik, 1941, Ent. Z., **54**: 213. Reiss, 1953, Z. wien. ent. Ges., **38**: 141, pl. 9, figs. 15-18. Holik & Sheljuzhko, 1955, Mitt. münch. ent. Ges., **44/45**: 117; 1958, ibidem, **48**: 273. Alberti, 1958, Mitt. zool. Mus. Berl., **34**: 316. Tremewan & Reiss, 1964, Ent. Rec., **76**: 9. Loritz, 1964, Bull. Soc. ent. Mulhouse, p. 51-86, figs. — Daratshitshag im armenischen Bergland, im Eichenwald, 2000 m.

ab. **conjuncta** Holik & Sheljuzhko, 1955, Mitt. münch. ent. Ges., **44/45**: 118.

ab. **scabiosaeformis** Holik, 1941, Ent. Z., **54**: 213.

Biology

Holik, 1953, Ent. Z., **63**: 4.

teberdica Reiss

Distribution: North Caucasus.

teberdica Reiss, 1939, Ent. Z., **53**: 113; 1953, Z. wien. ent. Ges., **38**: 141, pl. 9. fig. 14. Holik & Sheljuzhko, 1955, Mitt. münch. ent. Ges., **44/45**: 114; 1958, ibidem, **48**: 273. Loritz, 1957, Bull. Soc. ent. Mulhouse, p. 58. Alberti, 1958, Mitt. zool. Mus. Berl., **34**: 315. Tremewan & Reiss, 1964, Ent. Rec., **76**: 10. Loritz, 1964, Bull. Soc. ent. Mulhouse, p. 51-86, figs. — Teberda Gebiet, Nord Kaukasus.

romeo Duponchel

Distribution: Sicily, Italy, southern Alps, southern France west to the East Pyrenees.

ssp. **romeo** Duponchel, 1835, in Godart & Duponchel, Histoire naturelle des Lépidoptères ou Papillons de France, Supplement, **2**: 131, pl. 12, fig. 1. Perlini, 1905, Forme di Lepidotteri esclusivamente Italiane, p. 51, pl. 4, fig. 14. Spuler, 1906, in Hofmann, Die Schmetterlinge Europas, **2**: 154. Seitz, 1907, Die Gross-Schmetterlinge der Erde, **2**: 19, pl. 4e. Oberthür, 1910, Études de Lépidoptérologie comparée, **4**: 438. Reiss, 1930, in Seitz, Die Gross-Schmetterlinge der Erde, Supplement, **2**: 11, pl. 1i; 1933, ibidem, **2**: 252; 1935, Int. ent. Z., **29**: 169. Holik, 1935, Int. ent. Z., **29**: 61; 1935, ibidem, **29**: 195; 1944, Iris, **57**: 53. Alberti, 1958, Mitt. zool. Mus. Berl., **34**: 314. Tremewan & Reiss, 1964, Ent. Rec., **76**: 47, fig. 1. — Randazzo (pied de l'Etna), Sicilie.

celeus Herrich-Schäffer, 1846, Systematische Bearbeitung der

Schmetterlinge von Europa, **2**: 38; 1844, ibidem, **2**, pl. 6, figs. 48, 49 (non-binominal).

ab. **analiconjuncta** Burgeff, 1926, Mitt. münch. ent. Ges., **16**: 20. Reiss, 1930, in Seitz, Die Gross-Schmetterlinge der Erde, Supplement, **2**: 11.

ssp. **calberlai** Burgeff, 1926, Mitt. münch. ent. Ges., **16**: 23. Calberla, 1895, Iris, **8**: 209. Reiss, 1930, in Seitz, Die Gross-Schmetterlinge der Erde, Supplement, **2**: 11. Holik, 1944, Iris, **57**: 53. Haaf, 1952, Veröff. zool. Staatssamml. Münch., **2**: 151, 156, pl. 4 (as *neapolitana* Calberla). Tremewan & Reiss, 1964, Ent. Rec., **76**: 47. — Sila; San Fili di Cosenza, Calabrien, Süditalien.

ab. **cingulata** Burgeff, 1914, Mitt. münch. ent. Ges., **5**: 60. Reiss, 1930, in Seitz, Die Gross-Schmetterlinge der Erde, Supplement, **2**: 11.

ssp. **neapolitana** Calberla, 1895, Iris, **8**: 209. Hoffmann, 1904, Ent. Z., **18**: 5, figs. a, b. Perlini, 1905, Forme di Lepidotteri esclusivamente Italiane, p. 51, pl. 4, fig. 15. Spuler, 1906, in Hofmann, Die Schmetterlinge Europas, **2**: 155, pl. 77, fig. 5b. Seitz, 1907, Die Gross-Schmetterlinge der Erde, **2**: 19, pl. 4e. Reiss, 1930, in Seitz, Die Gross-Schmetterlinge der Erde, Supplement, **2**: 11, pl. 1i. Holik, 1944, Iris, **57**: 52. Tremewan & Reiss, 1964, Ent. Rec., **76**: 47. — Auf den Campanien umgebenden Kalkgebirgen, Italien, 500-1000 m.

ab. **analiconjuncta** Burgeff, 1926, Mitt. münch. ent. Ges., **16**: 22. Reiss, 1930, in Seitz, Die Gross-Schmetterlinge der Erde, Supplement, **2**: 11.

ab. **hoffmanni** Zickert, 1903, Ent. Z., **17**: 61. Hoffmann, 1904, Ent. Z., **18**: 5, figs. c, d. Dziurzyński, 1904, Jber. wien. ent. Ver., **14**: 46, pl. 2, fig. 2. Seitz, 1907, Die Gross-Schmetterlinge der Erde, **2**: 19; 1912, ibidem, **2**: 441, pl. 56h.

ab. **nigerrima** Zickert, 1904, Nat. sicil., **17**: 70. Perlini, 1905, Forme di Lepidotteri esclusivamente Italiane, p. 51, pl. 3, fig. 12. Seitz, 1907, Die Gross-Schmetterlinge der Erde, **2**: 19. Reiss, 1933, in Seitz, Die Gross-Schmetterlinge der Erde, Supplement, **2**: 252, Alberti, 1955, Ent. Z., **65**: 89-91.

ab. **flaveola** Zickert, 1904, Nat. sicil., **17**: 69. Seitz, 1907, Die Gross-Schmetterlinge der Erde, **2**: 19.

ssp. **faitocola** Tremewan & Reiss, 1964, Ent. Rec., **76**: 48 (nomen novum for *faitensis* Holik). — Mte. Faito, Sorrento, Italien.

faitensis Holik, 1944, Iris, **57**: 53 (preoccupied by **faitensis** Stauder, 1929, ssp. of **punctum** Ochsenheimer).

ab. **equensis** Stauder, 1915, Z. wiss. InsektBiol., **11**: 71.

ssp. **adumbrata** Burgeff, 1926, Mitt. münch. ent. Ges., **16**: 22. Reiss, 1930, in Seitz, Die Gross-Schmetterlinge der Erde, Supplement, **2**: 11, pl. 1i. Holik, 1944, Iris, **57**: 52. Tremewan & Reiss, 1964, Ent. Rec., **76**: 48. — Mte. Sirente, Abruzzen, Italien, 1500-2000 m.

ab. **absoluta** Dannehl, 1927, Lepid. Rdsch., **1**: 47. Reiss, 1930, in Seitz, Die Gross-Schmetterlinge der Erde, Supplement, **2**: 11. Alberti, 1955, Ent. Z., **65**: 89-91.

ssp. **jalina** Rostagno, 1911, Boll. Soc. Zool. ital., (2) **12**: 106. Holik, 1944, Iris, **57**: 51. Tremewan & Reiss, 1964, Ent. Rec., **76**: 48. — Monti Aurunci, Italia.

minima Turati, 1915, Atti Soc. ital. Sci. nat., **53**: 608. Reiss, 1930, in Seitz, Die Gross-Schmetterlinge der Erde, Supplement, **2**: 11. Holik, 1944, Iris, **57**:51. Tremewan & Reiss, 1964, Ent. Rec., **76**: 48.

ab. **orionjalina** Rostagno, 1911, Boll. Soc. Zool. ital., (2) **12**: 106.

ab. **conjunctajalina** Rostagno, 1911, Boll. Soc. Zool. ital., (2) **12**: 107.

ssp. **romana** Burgeff, 1926, Mitt. münch. ent. Ges., **16**: 21. Reiss, 1930, in Seitz, Die Gross-Schmetterlinge der Erde, Supplement, **2**: 11. Holik, 1944, Iris, **57**: 51. Tremewan & Reiss, 1964, Ent. Rec., **76**: 49. — Albanergebirge und römischen Campagna, Italien.

ab. **scabiosaeformis** Burgeff, 1926, Mitt. münch. ent. Ges., **16**: 21. Reiss, 1930, in Seitz, Die Gross-Schmetterlinge der Erde, Supplement, **2**: 11.

ssp. **orion** Herrich-Schäffer, 1843, Systematische Bearbeitung der Schmetterlinge von Europa, **2**, pl. 1, fig. 3; 1846, ibidem, **2**: 33. Perlini, 1905, Forme di Lepidotteri esclusivamente Italiane, p. 50, pl. 4, fig. 13. Spuler, 1906, in Hofmann, Die Schmetterlinge Europas, **2**: 155, pl. 75, fig. 41a, pl. 77, fig. 5a (partim). Seitz, 1907, Die Gross-Schmetterlinge der Erde, **2**: 19, pl. 4d. Verity, 1920, Boll. Lab. Zool. Portici, **14**: 36. Holik, 1944, Iris, **57**: 49. Tremewan & Reiss, 1964, Ent. Rec., **76**: 49. — Toskana; Marche (Sibillini), Italien.

ab. **transapennina** Calberla, 1895, Iris, **8**: 206. Seitz, 1907, Die Gross-Schmetterlinge der Erde, **2**: 19.

f.t. **aestiva** Burgeff, 1914, Mitt. münch. ent. Ges., **5**: 60, pl. 3, figs. 108-111. Reiss, 1930, in Seitz, Die Gross-Schmetterlinge der Erde, Supplement, **2**: 11. — Cutigliano, Etruskischer Apennin, Italien.

ssp. **megorion** Burgeff, 1926, Mitt. münch. ent. Ges., **16**: 21. Reiss, 1930, in Seitz, Die Gross-Schmetterlinge der Erde, Supplement, **2**: 11. Tremewan & Reiss, 1964, Ent. Rec., **76**: 49. — Pegli; Genua; Litoral der italienischen und französischen Riviera.

ab. **striata** Reiss, 1958, Bull. Soc. ent. Mulhouse, p. 51.

ab. **quinquemacula** Reiss, 1958, Bull. Soc. ent. Mulhouse, p. 52.

ssp. **oriomezon** Dujardin, 1965, Entomops, Nice, no. 2: 37, fig. — Col de Tende, Alpes-Maritimes, France, 1400-1500 m.

ssp. **mecogana** Dujardin, 1965, Entomops, Nice, no. 2: 38, fig.

Saint-Etienne-de-Tinée, 1400 m.; Le Bourguet; Roya, vallée de la Tinée, 1500m.; Isola, Tinée, 1000 m.; le Pra, Alpes-Maritimes, France.

ssp. **loritzi** Reiss, 1958, Bull. Soc. ent. Mulhouse, p. 56. Tremewan & Reiss, 1964, Ent. Rec., **76**: 50.
loritzi Reiss, 1958, Z. wien. ent. Ges., **43**: 182 (nomen nudum).

St. Barnabé, Col de Vence, Alpes-Maritimes, France, 900-1000 m.

ssp. **parvorion** Holik, 1944, Iris, **57**: 48. Tremewan & Reiss, 1964, Ent. Rec., **76**: 50. Dufay, 1966, Bull. mens. Soc. linn. Lyon, **35**: 69.

Digne, Basses-Alpes, Frankreich.

ssp. **subalpina** Calberla, 1895, Iris, **8**: 205. Spuler, 1906, in Hofmann, Die Schmetterlinge Europas, **2**: 154 (partim). Seitz, 1907, Die Gross-Schmetterlinge der Erde, **2**: 19, pl. 4e. Holik, 1944, Iris, **57**: 45. Forster & Wohlfahrt, 1958, Die Schmetterlinge Mitteleuropas, **3**: 92, pl. 10, figs. 5, 10. Tremewan & Reiss, 1964, Ent. Rec., **76**: 51.

Piemont, Italien.

ab. **conjuncta** Calberla, 1895, Iris, **8**: 206. Seitz, 1907, Die Gross-Schmetterlinge der Erde, **2**: 19.

ssp. **freyeri** Lederer, 1853, Verh. zool.-bot. Ver. Wien, **2**: 70, 94 (nomen novum for *triptolemus* Hübner sensu Freyer). Freyer, 1833, Neuere Beiträge zur Schmetterlingskunde, **1**: 28, pl. 14, fig. 4 (as *triptolemus* Hübner). Tremewan & Reiss, 1964, Ent. Rec., **76**: 51.

Südliche Alpentäler der Schweiz und Tirols ohne oberes Etsch- und Eisacktal.

meridionalis Vorbrodt, 1913, in Vorbrodt & Müller-Rutz, Die Schmetterlinge der Schweiz, **2**: 253. Tremewan & Reiss, 1964, Ent. Rec., **76**: 51.

ephemerina Burgeff, 1926, Mitt. münch. ent. Ges., **16**: 20. Reiss, 1930, in Seitz, Die Gross-Schmetterlinge der Erde, Supplement, **2**: 11. Holik, 1944, Iris, **57**: 47. Reiss, 1950, Jber. naturf. Ges. Graubünden, **82**: 103, figs. 5, 6. Tremewan & Reiss, 1964, Ent. Rec., **76**: 51.

ab. **romeiformis** Burgeff, 1926, Mitt. münch. ent. Ges., **16**: 20. Reiss, 1930, in Seitz, Die Gross-Schmetterlinge der Erde, Supplement, **2**: 11.

ab. **latemarginata** Reiss, 1950, Jber. naturf. Ges. Graubünden, **82**: 103.

ab. **divisa** Vorbrodt, 1913, in Vorbrodt & Müller-Rutz, Die Schmetterlinge der Schweiz, **2**: 254, fig. 38.

ab. **parallela** Vorbrodt, 1913, in Vorbrodt & Müller-Rutz, Die Schmetterlinge der Schweiz, **2**: 253.

ssp. **orionides** Burgeff, 1926, Mitt. münch. ent. Ges., **16**: 21. Reiss, 1930, in Seitz, Die Gross-Schmetterlinge der Erde, Supplement, **2**: 11. Holik, 1944, Iris, **57**: 50. Tremewan & Reiss, 1964, Ent. Rec., **76**: 51. — Trient; Sarchetal; Condino; Täler des Adamellogebietes, Italien.

ssp. **lozerica** Holik, 1944, Iris, **57**: 49. Tremewan & Reiss, 1964, Ent. Rec., **76**: 52. — Florac, Lozère, Frankreich.

ssp. **urania** Marten, 1957, Ent. Z., **67**: 218. Tremewan & Reiss, 1964, Ent. Rec., **76**: 52. — Umgebung von Ripoll, spanische Ostpyrenäen, 900-1100 m.

Biology.

Abeille, 1909, Mém. Soc. linn. Provence, **1**: 9-11. Burgeff, 1912, Z. wiss. InsektBiol., **8**: 123; 1950, Portug. acta biol., (A) Goldschmidt: 663-728; 1951, Biol. Zbl., **70**: 1-23. Dufay, 1966, Bull. mens. Soc. linn. Lyon, **35**: 70. Holik, 1938, Lambillionea, **38**: 51-58, 79-88, 95-102; 1953, Ent. Z., **63**: 4. Marten, 1957, Ent. Z., **67**: 217, figs. Reiss, 1958, Z. wien. ent. Ges., **43**: 161.

osterodensis Reiss

Distribution: Urals, Russia, Siberia, Caucasus, Ukraine, east Europe west to the Pyrenees and Aragon, and north to Sweden and Finland.

ssp. **asiatica** Burgeff, 1926, Mitt. münch. ent Ges., **16**: 19. Reiss, 1930, in Seitz, Die Gross-Schmetterlinge der Erde, Supplement, **2**: 10, pl. 1i; 1932, Ent. Rdsch., **49**: 162, pl. 1, figs. Holik, 1939, Rev. franç. Lépid., **9**: 275, pl. 7, figs. 10, 11. Holik & Sheljuzhko, 1955, Mitt. münch. ent. Ges., **44/45**: 104. Tremewan & Reiss, 1964, Ent. Rec., **76**: 53. — Sojmonovsk im nördlichen Teil des Südurals.

ssp. **filipjevi** Holik, 1939, Rev. franç. Lépid., **9**: 276, pl. 7, figs. 12-15. Holik & Sheljuzhko, 1955, Mitt. münch. ent. Ges., **44/45**: 105. Tremewan & Reiss, 1964, Ent. Rec., **76**: 53. — 50 km. südöstlich von Uzjan, Bashkirien.

ab. **divisa** Holik, 1939, Rev. franç. Lépid., **9**: 276, pl. 7, figs. 16, 17.

ssp. **saratovensis** Holik & Sheljuzhko, 1955, Mitt. münch. ent. Ges., **44/45**: 106. Tremewan & Reiss, 1964, Ent. Rec., **76**: 54. — Saratov, Südrussland.

transiens Spuler, 1906, in Hofmann, Die Schmetterlinge Europas, **2**: 155 (preoccupied by **transiens** Staudinger, 1887, ssp. of **carniolica** Scopoli).

ssp. **sibirica** Holik & Sheljuzhko, 1955, Mitt. münch. ent. Ges., **44/45**: 109. Tremewan & Reiss, 1964, Ent. Rec., **76**: 54. — Tobolsk, Westsibirisches Steppengebiet.

ssp. **altaica** Holik & Sheljuzhko, 1955, Mitt. münch. ent. Ges., **44/45**: 111. Tremewan & Reiss, 1964, Ent. Rec., **76**: 54. — Altai, Zentralsibirisches Gebirgsland.

ssp. **kenteina** Burgeff, 1926, Mitt. münch. ent. Ges., **16**: 19. Reiss, 1930, in Seitz, Die Gross-Schmetterlinge der Erde, Supplement, **2**: 10. Holik & Sheljuzhko, 1955, Mitt. münch. ent. Ges., **44/45**: 111. Tremewan & Reiss, 1964, Ent. Rec., **76**: 54. — Nördlich Urga, Kentei Gebirge, Mongolei.

ab. **divisa** Staudinger, 1892, Iris, **5**: 343. Seitz, 1907, Die Gross-Schmetterlinge der Erde, **2**: 19, pl. 4d.

ssp. **caucasi** Burgeff, 1926, Mitt. münch. ent. Ges., **16**: 19 (nomen novum for *caucasica* Spuler). Reiss, 1930, in Seitz, Die Gross-Schmetterlinge der Erde, Supplement, **2**: 10, pl. 4n. Koch, 1939, Mitt. münch. ent. Ges., **29**: 408. Holik & Sheljuzhko, 1955, Mitt. münch. ent. Ges., **44/45**: 107. Tremewan & Reiss, 1964, Ent. Rec., **76**: 54. — Achalzich, Georgien, Transkaukasien.

caucasica Spuler, 1906, in Hofmann, Die Schmetterlinge Europas, **2**: 155 (preoccupied by **caucasica** Rebel, 1901, ssp. of **armena** Eversmann).

ab. **divisa** Koch, 1939, Mitt. münch. ent. Ges., **29**: 408.

ssp. **irpenjensis** Holik & Reiss, 1932, in Holik, Iris, **46**: 114, pl. 1, figs. 12-15. Reiss, 1933, in Seitz, Die Gross-Schmetterlinge der Erde, Supplement, **2**: 253. Holik & Sheljuzhko, 1955, Mitt. münch. ent. Ges., **44/45**: 101. Tremewan & Reiss, 1964, Ent. Rec., **76**: 75. — Irpenj bei Kijev, Waldzone, Nord-Ukraine.

ssp. **praecarpathica** Holik, 1942, Ent. Z., **56**: 198. Tremewan & Reiss, 1964, Ent. Rec., **76**: 77. — Smrkovica, Djumbir Gebiet in den kleinen Karpathen.

ab. **mediointerrupta** Holik, 1942, Ent. Z., **56**: 198.

ab. **analiinterrupta** Holik, 1942, Ent. Z., **56**: 198.

ab. **medioanaliinterrupta** Holik, 1942, Ent. Z., **56**: 198.

ssp. **austrocarpathica** Holik, 1942, Ent. Z., **56**: 198. Tremewan & Reiss, 1964, Ent. Rec., **76**: 77. — Kosów; Kobaki, Nordhang der Ost-Karpathen.

ab. **mediointerrupta** Holik, 1942, Ent. Z., **56**: 198.

ab. **medioanaliinterrupta** Holik, 1942, Ent. Z., **56**: 198.

ssp. **polonia** Przegendza, 1933, Ent. Z., **47**: 27, figs. 4-6. Reiss, 1933, in Seitz, Die Gross-Schmetterlinge der Erde, Supplement, **2**: 253. Holik, 1939, Ann. Mus. zool. Polon., **12**: 27, pl. 5, figs. 171-174. Tremewan & Reiss, 1964, Ent. Rec., **76**: 77. — Szerszeniowce bei Lemberg, Galizien, Polen.

ab. **mediointerrupta** Holik, 1939, Ann. Mus. zool. Polon., **12**: 26.

ssp. **warszawiensis** Holik, 1939, Ann. Mus. zool. Polon., **12**: 26, pl. 1, figs. 31-33. Tremewan & Reiss, 1964, Ent. Rec., **76**: 77. — Pyry bei Warszawa, Polen.

ssp. **budensis** Holik, 1942, Ent. Z., **56**: 197 (as *dubensis*; **budensis** Holik, p. 198); 1943, ibidem, **57**: 45 (corrected to **budensis** Holik). Tremewan & Reiss, 1964, Ent. Rec., **76**: 75. — Budapest, Budakeszi, Ungarn.

ssp. **matrana** Burgeff, 1926, Mitt. münch. ent. Ges., **16**: 18. Reiss, 1930, in Seitz, Die Gross-Schmetterlinge der Erde, Supplement, **2**: 10. Le Charles, 1956, Rev. franç. Lépid., **15**: 14, pl. 5, fig. 3. Forster & Wohlfahrt, 1958, Die Schmetterlinge Mitteleuropas, **3**: 91, pl. 10, figs. 4, 9. Tremewan & Reiss, 1964, Ent. Rec., **76**: 75. — Matra Gebirge, Nord-Ungarn, 500-800 m.

ab. **analiinterrupta** Reiss, 1964, Coridon, (A) **6**: 9.

ab. **divisa** Burgeff, 1926, Mitt. münch. ent. Ges., **16**: 18. Reiss, 1930, in Seitz, Die Gross-Schmetterlinge der Erde, Supplement, **2**: 10.

ab. **quinquemacula** Burgeff, 1926, Mitt. münch. ent. Ges., **16**: 18. Reiss, 1930, in Seitz, Die Gross-Schmetterlinge der Erde, Supplement, **2**: 10.

ssp. **koricnensis** Reiss, 1922, Int. ent. Z., **16**: 66; 1930, in Seitz, Die Gross-Schmetterlinge der Erde, Supplement, **2**: 11, pl. 1i. Holik, 1937, Mitt. münch. ent. Ges., **27**: 7. Tremewan & Reiss, 1964, Ent. Rec., **76**: 76. — Maklen-Pass, Korična, Bosnien.

ssp. **goriziana** Koch, 1937, in Holik, Mitt. münch. ent. Ges., **27**: 7. Tremewan & Reiss, 1964, Ent. Rec., **76**: 76. — Görz, Istrien.

ssp. **ladina** Holik, 1944, Iris, **57**: 44. Forster & Wohlfahrt, 1958, Die Schmetterlinge Mitteleuropas, **3**: 91, pl. 10, figs. 3, 8. Tremewan & Reiss, 1964, Ent. Rec., **76**: 77. — Gröden, Dolomiten, Italien.

ssp. **curvata** Burgeff, 1926, Mitt. münch. ent. Ges., **16**: 17. Reiss, 1930, in Seitz, Die Gross-Schmetterlinge der Erde, Supplement, **2**: 10. Holik, 1937, Mitt. münch. ent. Ges., **27**: 6. Tremewan & Reiss, 1964, Ent. Rec., **76**: 77. — Bruck an der Mur, Thörl, Steiermark, Österreich.

ab. **analiinterrupta** Reiss, 1964, Coridon, (A) **6**: 9.

ab. **quinquemaculata** Reiss, 1964, Coridon, (A) **6**: 9.

ab. **flava** Pieszczek, 1903, Verh. zool.-bot. Ges. Wien, **53**: 570. Dziurzyński, 1904, Jber. wien. ent. Ver., **14**: 46, pl. 2, fig. 3. Seitz, 1907, Die Gross-Schmetterlinge der Erde, **2**: 19. Reiss, 1930, in Seitz, Die Gross-Schmetterlinge der Erde, Supplement, **2**: 10.

ssp. **tenuicurva** Burgeff, 1926, Mitt. münch. ent. Ges., **16**: 18. Reiss, 1930, in Seitz, Die Gross-Schmetterlinge der Erde, Supplement, **2**: 10. Povolný, 1945, Folia ent., Brno, **8**: 77. Gregor & Povolný, 1955, Sborn. ent. Odd. nár. Mus. Praze, **30**: 259, 271, pl. 3, fig. 3. Tremewan & Reiss, 1964, Ent. Rec., **76**: 77. — Neuhütten, Karlstein, Böhmisches Mittelgebirge.

ab. **analiinterrupta** Reiss, 1964, Coridon, (A) **6**: 9.

ab. **interrupta** Reiss, 1922, Int. ent. Z., **16**: 67; 1926, Die Zygaenen Deutschlands, p. 13. Holik, 1929, Int. ent. Z., **23**: 2 (footnote 6).

ab. **confluens** Reiss, 1964, Coridon, (A) **6**: 9.

ssp. **kessleri** Reiss, 1950, Jber. naturf. Ges. Graubünden, **82**: 102, fig. 4. Tremewan & Reiss, 1964, Ent. Rec., **76**: 78. — Albulatal, Bergün, Schweiz, 1300-1400 m.

ab. **mediointerrupta** Reiss, 1950, Jber. naturf. Ges. Graubünden, **82**: 103.

ssp. **validior** Burgeff, 1926, Mitt. münch. ent. Ges., **16**: 17. Reiss, 1930, in Seitz, Die Gross-Schmetterlinge der Erde, Suppl., **2**: 10. Tremewan & Reiss, 1964, Ent. Rec., **76**: 78. — Martigny-Ville, Rhônetal im Wallis, Schweiz.

ab. **mediointerrupta** Vorbrodt, 1913, in Vorbrodt & Müller-Rutz, Die Schmetterlinge der Schweiz, **2**: 253, fig. 35. Reiss, 1930, in Seitz, Die Gross-Schmetterlinge der Erde, Supplement, **2**: 11.

ab. **divisa** Burgeff, 1926, Mitt. münch. ent. Ges., **16**: 17. Reiss, 1930, in Seitz, Die Gross-Schmetterlinge der Erde, Supplement, **2**: 10.

ab. **analiinterrupta** Vorbrodt, 1913, in Vorbrodt & Müller-Rutz, Die Schmetterlinge der Schweiz, **2**: 253, fig. 37. Reiss, 1930, in Seitz, Die Gross-Schmetterlinge der Erde, Supplement, **2**: 10.

ssp. **expansa** Le Charles, 1957, Rev. franç. Lépid., **16**: 20, pl. 6, figs. 9-11; 1960, Bull. soc. ent. Fr., **65**: 103. Tremewan & Reiss, 1964, Ent. Rec., **76**: 78. — Lac de Montrion, Haute-Savoie, France, 1200 m.

ssp. **droiti** Le Charles, 1960, Bull. Soc. ent. Fr., **65**: 103; 1957, Rev. franç. Lépid., **16**: 20. Tremewan & Reiss, 1964, Ent. Rec., **76**: 78. Dufay, 1966, Bull. mens. Soc. linn. Lyon, **35**: 70. — Céüze, Hautes-Alpes, France.

ssp. **schultei** Dujardin, 1956, Bull. mens. Soc. linn. Lyon, **25**: 256. Tremewan & Reiss, 1964, Ent. Rec., **76**: 79. Dufay, 1966, Bull. mens. Soc. linn. Lyon, **35**: 70. — Les Dourbes (près de Digne), Basses-Alpes, France, 1500 m.

ssp. **trimacula** Le Charles, 1957, Rev. franç. Lépid., **16**: 15, pl. 6, fig. 8; 1960, Bull. Soc. ent. Fr., **65**: 103. Tremewan & Reiss, 1964, Ent. Rec., **76**: 80. — Forêt de Sainte-Maure, Indre, France.

ssp. **eupyrenaea** Burgeff, 1926, Mitt. münch. ent. Ges., **16**: 20. Oberthür, 1884, Études d'Entomologie, **8**: 27. Reiss, 1930, in Seitz, Die Gross-Schmetterlinge der Erde, Supplement, **2**: 11. Holik, 1936, Lambillionea, **36**: 49; 1944, Iris, **57**: 54. Koch, 1948, Eos, Madr., **24**: 322. Tremewan & Reiss, 1964, Ent. Rec., **76**: 80. — Vernet-les-Bains; Mt. Canigou, 800-1200 m.; Col de Jou, 1800 m.; Ostpyrenäen, Frankreich.

ssp. **leridana** Marten, 1957, Ent. Z., **67**: 218. Oberthür, 1884, Études d'Entomologie, **8**: 27. Tremewan & Reiss, 1964, Ent. Rec., **76**: 81. — Espot, Prov. Lerida, spanische Pyrenäen, 1000-1100 m.

ssp. **cantabrica** Marten, 1957, Ent. Z., **67**: 217. Reiss, 1931, Int. ent. Z., **25**: 113. Koch, 1948, Eos, Madr., **24**: 322. Tremewan & Reiss, 1964, Ent. Rec., **76**: 81.

Kantabrische Gebirge zwischen der Sierra de Covadonga und dem Massiv der Picos de Europa, Spanien, 500-700 m.

ssp. **valida** Burgeff, 1926, Mitt. münch. ent. Ges., **16**: 17. Reiss, 1926, Die Zygaenen Deutschlands, p. 12, pl. 2, fig. (as *scabiosae* Scheven); 1930, in Seitz, Die Gross-Schmetterlinge der Erde, Supplement, **2**: 10; 1937, in Schneider, Jh. Ver. vaterl. Naturk. Württemb., **93**: 125. Bergmann, 1953, Die Grossschmetterlinge Mitteldeutschlands, **3**: 24, pl. 64, figs. C1-C3. Tremewan & Reiss, 1964, Ent. Rec., **76**: 79.

ab. **divisa** Burgeff, 1926, Mitt. münch. ent. Ges., **16**: 17. Reiss, 1930, in Seitz, Die Gross-Schmetterlinge der Erde, Supplement, **2**: 10.

Klingenstein bei Ulm; Pfullingen; Lautertal bei Herrlingen; Neuffen; Schwäbische Alb, Süddeutschland.

ssp. **lineata** Reiss, 1933, in Seitz, Die Gross-Schmetterlinge der Erde, Supplement, **2**: 253, pl. 16k. Spuler, 1906, in Hofmann, Die Schmetterlinge Europas, **2**: 154 (as *scabiosae* Scheven [partim]). Seitz, 1907, Die Gross-Schmetterlinge der Erde, **2**: 19, pl. 4d (as *scabiosae* Scheven). Reiss, 1935, Int. ent. Z., **29**: 171; 1937, in Schneider, Jh. Ver. vaterl. Naturk. Württemb., **93**: 125; 1949, Entomon, **1**: 170. Haaf, 1952, Veröff. zool. Staatssamml. Münch., **2**: 151, 153, 156, pl. 4 (as *scabiosae* Scheven). Forster & Wohlfahrt, 1958, Die Schmetterlinge Mitteleuropas, **3**: 91, pl. 10, figs. 1, 6 (as *scabiosae* Scheven). Tremewan & Reiss, 1964, Ent. Rec., **76**: 79.

Dollnstein, Fränkischer Jura [Mittel- und Süddeutschland ohne Schwäbische Alb].

ab. **costalielongata** Reiss, 1964, Coridon, (A) **6**: 9.

ab. **confluens** Spuler, 1906, in Hofmann, Die Schmetterlinge Europas, **2**: 154. Reiss, 1930, in Seitz, Die Gross-Schmetterlinge der Erde, Supplement, **2**: 10.

ab. **citrina** Spuler, 1906, in Hofmann, Die Schmetterlinge Europas, **2**: 154. Seitz, 1907, Die Gross-Schmetterlinge der Erde, **2**: 19.

ssp. **hassica** Burgeff, 1926, Mitt. münch. ent. Ges., **16**: 17. Reiss, 1930, in Seitz, Die Gross-Schmetterlinge der Erde, Supplement, **2**: 10. Heuser & Jöst, 1959, Mitt. Pollichia, (3) **6**: 124. Tremewan & Reiss, 1964, Ent. Rec., **76**: 79.

Ingelheim; Heidesheim; Rhein-Tal, Hessen, West-Deutschland.

ssp. **vosegiensis** Le Charles, 1960, Bull. Soc. ent. Fr., **65**: 103 (nomen novum for *vogesiaca* Le Charles). Tremewan & Reiss, 1964, Ent. Rec., **76**: 78.

vogesiaca Le Charles, 1957, Rev. franç. Lépid., **16**: 20, pl. 6,

Nonnenbruch près Cernay; Mulhouse; Haut-Rhin, France.

fig. 6 (preoccupied by **vogesiaca** Przegendza, 1932, ssp. of **trifolii** Esper); 1960, Bull. Soc. ent. Fr., **65**: 103.

ssp. **osterodensis** Reiss, 1921, Int. ent. Z., **15**: 118; 1926, Die Zygaenen Deutschlands, p. 13, pl. 2, fig.; 1930, in Seitz, Die Gross-Schmetterlinge der Erde, Supplement, **2**: 10, pl. 1i. Forster & Wohlfahrt, 1958, Die Schmetterlinge Mitteleuropas, **3**: 91, pl. 10, figs. 2, 7. Alberti, 1958, Mitt. zool. Mus. Berl., **34**: 315 (as *scabiosae* Scheven). Tremewan & Reiss, 1964, Ent. Rec., **76**: 53, 79, fig. 1. Alberti, 1964, Mitt. münch. ent. Ges., **54**: 262. — Grünortspitze, Osterode, Ostpreussen.

ab. **divisa** Reiss, 1930, in Seitz, Die Gross-Schmetterlinge der Erde, Supplement, **2**: 10.

ssp. **masoviensis** Reiss, 1941, Z. wien. EntVer., **26**: 58. Tremewan & Reiss, 1964, Ent. Rec., **76**: 80. — Rüdzanny (Masuren), Ostpreussen.

ssp. **haegeri** Reiss, 1941, Z. wien. EntVer., **26**: 58. Tremewan & Reiss, 1964, Ent. Rec., **76**: 80. — Bublitz (Stadtwald), Ostpommern, Norddeutschland.

Biology

Burgeff, 1912, Z. wiss. InsektBiol., **8**: 123; 1950, Portug. acta biol., (A) Goldschmidt: 663-728; 1951, Biol. Zbl., **70**: 1-23. Dorfmeister, 1854, Verh. zool.-bot. Ver. Wien, **4**: 481. Dufay, 1966, Bull. mens. Soc. linn. Lyon, **35**: 70. Holik, 1937, Lambillionea, **37**: 15-24, 32-45, 80-91; 1938, Ent. Rdsch., **55**: 320-323, 331-333; 1953, Ent. Z., **63**: 3. Hrubý, 1964, Prodromus Lepidopter Slovenska, p. 475. Koch, 1955, Wir Bestimmen Schmetterlinge, **2**: 58, 59, pl. 1, fig. 8, pl. 15, fig. 8. Reiss, 1926, Die Zygaenen Deutschlands, p. 7 (as *scabiosae* Scheven); 1930, in Seitz, Die Gross-Schmetterlinge der Erde, Supplement, **2**: 11; 1958, Z. wien. ent. Ges., **43**: 161. Roüast, 1883, Catalogue des Chenilles européennes connues, p. 22. Seitz, 1907, Die Gross-Schmetterlinge der Erde, **2**: 20. Seppänen, 1954, Suomen Suurperhostoukkien Ravintokasvit, p. 255.

SECTION 9

ramburii Herrich-Schäffer

Distribution: Northern Syria, Asia Minor, Greece.

ssp. **ramburii** Herrich-Schäffer, January, 1861, Neue Schmetterlinge aus Europa und den angrenzenden Ländern, p. 32, figs. 161, 162. Reiss & Tremewan, 1960, Bull. Brit. Mus. (nat. Hist.) Ent., **9** (10): 467. Reiss, 1962, Ent. Z., **72**: 226. — Antiochia, Syrien.

ramburi Lederer, May, 1861, Wien. ent. Monatschr., **5**: 151, pl. 1, fig. 10. Spuler, 1906, in Hofmann, Die Schmetterlinge

Europas, **2**: 160. Seitz, 1907, Die Gross-Schmetterlinge der Erde, **2**: 23, pl. 5f. Oberthür, 1910, Études de Lépidoptérologie comparée, **4**: 564. Reiss, 1930, in Seitz, Die Gross-Schmetterlinge der Erde, Supplement, **2**: 35, pl. 3n. Holik & Sheljuzhko, 1958, Mitt. münch. ent. Ges., **48**: 192. Alberti, 1958, Mitt. zool. Mus. Berl., **34**: 325.

ab. **sexmaculata** Reiss, 1935, Int. ent. Z., **29**: 209.

ab. **quinquemaculata** Reiss, 1935, Int. ent. Z., **29**: 209.

ab. **totirubra** Reiss, 1933, in Seitz, Die Gross-Schmetterlinge der Erde, Supplement, **2**: 276.

omniconfluens Holik & Sheljuzhko, 1958, Mitt. münch. ent. Ges., **48**: 197 (nomen nudum).

ssp. **mersina** Herrich-Schäffer, January, 1861, Neue Schmetterlinge aus Europa und den angrenzenden Ländern, p. 32, fig. 163. Reiss & Tremewan, 1960, Bull. Brit. Mus. (nat. Hist.) Ent., **9** (10): 467. Reiss, 1962, Ent. Z., **72**: 227.

Mersin, Kleinasien.

gurda Lederer, May, 1861, Wien. ent. Monatschr., **5**: 152, pl. 1, fig. 9. Seitz, 1907, Die Gross-Schmetterlinge der Erde, **2**: 23, 31. Reiss, 1930, in Seitz, Die Gross-Schmetterlinge der Erde, Supplement, **2**: 35. Holik & Sheljuzhko, 1958, Mitt. münch. ent. Ges., **48**: 195.

ssp. **rosa** Oberthür, 1909, Études de Lépidoptérologie comparée, **3**, pl. 22, figs. 106, 107; 1910, ibidem, **4**: 565. Reiss, 1930, in Seitz, Die Gross-Schmetterlinge der Erde, Supplement, **2**: 35, pl. 3n (after Oberthür); 1933, ibidem, **2**: 276. Holik & Sheljuzhko, 1958, Mitt. münch. ent. Ges., **48**: 196. Tremewan, 1961, Bull. Brit. Mus. (nat. Hist.) Ent., **10** (7): 292, pl. 55, fig. 16.

Région d'Akbès, Syrie.

ssp. **noacki** Reiss, 1962, Ent. Z., **72**: 226, 228, figs. 1-3.

ab. **basimediounita** Reiss, 1962, Ent. Z., **72**: 228.

ab. **unita** Reiss, 1962, Ent. Z., **72**: 228.

Umgebung von Kalavryta, Peloponnes (Morea), Griechenland, 750 m.

ssp. **helmosica** Reiss, 1962, Ent. Z., **72**: 230.

ab. **medioseparata** Reiss, 1962, Ent. Z., **72**: 230.

ab. **sextaseparata** Reiss, 1962, Ent. Z., **72**: 230.

Helmos (Chelmos), Peloponnes (Morea), Griechenland, 1700 m.

ssp. **europensis** Daniel, 1957, Acta Mus. maced. Sci. nat. **4**: 217. Reiss, 1962, Ent. Z., **72**: 231. Daniel, 1964, Prirod. Muz. Scopje, no. 2: 17.

Doiransee, Stary-Doiran, Süd-Mazedonien, 150-300 m.

Biology

Holik, 1953, Ent. Z., **63**: 28.

filipendulae Linné

Distribution: Syria, Asia Minor, Ciscaucasia, Transcaucasia, Europe including Scandinavia and the British Isles.

ssp. **syriaca** Oberthür, 1896, Études d'Entomologie, **20**: 46, 49, pl. 8, fig. 137; 1910, Études de Lépidoptérologie comparée, **4**: 564. Reiss, 1930, in Seitz, Die Gross-Schmetterlinge der Erde, Supplement, **2**: 35, pl. 3n. Holik & Sheljuzhko, 1958, Mitt. münch. ent. Ges., **48**: 187. Tremewan, 1961, Bull. Brit. Mus. (nat. Hist.) Ent., **10** (7): 292, pl. 55, figs. 18, 19. Région d'Akbès, Syrie.

sexmaculata Oberthür, 1896, Études d'Entomologie, **20**, pl. 8, fig. 137; 1910, Études de Lépidoptérologie comparée, **4**: 564. Tremewan, 1961, Bull. Brit. Mus. (nat. Hist.) Ent., **10** (7): 292.

ab. **quinquemaculata** Oberthür, 1896, Études d'Entomologie, **20**: 49, pl. 8, fig. 136; 1910, Études de Lépidoptérologie comparée, **4**: 564. Tremewan, 1961, Bull. Brit. Mus. (nat. Hist.) Ent., **10** (7): 293, pl. 55, fig. 20.

ab. **confluens** Oberthür, 1896, Études d'Entomologie, **20**: 46, pl. 8, fig. 138. Tremewan, 1961, Bull. Brit. Mus. (nat. Hist.) Ent., **10** (7): 293, pl. 55, fig. 21.

sexmaculataconfluens Oberthür, 1910, Études de Lépidoptérologie comparée, **4**: 564. Tremewan, 1961, Bull. Brit. Mus. (nat. Hist.) Ent., **10** (7): 293.

ssp. **kulpiensis** Reiss, 1935, Int. ent. Z., **29**: 209. Holik & Sheljuzhko, 1958, Mitt. münch. ent. Ges., **48**: 182. Kulp, West Armenien, Kleinasien.

ssp. **zangezurica** Holik & Sheljuzhko, 1958, Mitt. münch. ent. Ges., **48**: 181. Dorf Ochtshi bei Kafan, Zangezur-Gebirge, Nachitshevan, Russisch-Armenien, 2300-2500 m.

ssp. **karsiana** Sheljuzhko, 1936, Folia Zool. Hydrobiol., Riga, **9**: 21. Holik & Sheljuzhko, 1958, Mitt. münch. ent. Ges., **48**: 181. Sarykamysh, Kars-Provinz, West-Armenien, Kleinasien.

ssp. **hadjina** Rebel, 1901, in Staudinger & Rebel, Catalog der Lepidopteren des Palaearctischen Faunengebietes, p. 384. Spuler, 1906, in Hofmann, Die Schmetterlinge Europas, **2**: 159. Seitz, 1907, Die Gross-Schmetterlinge der Erde, **2**: 22, pl. 5c. Reiss, 1935, Int. ent. Z., **29**: 207, fig. Holik & Sheljuzhko, 1958, Mitt. münch. ent. Ges., **48**: 185. Hadjin, Taurus, Kleinasien.

taurica Dziurzyński, 1908, Berl. ent. Z., **53**: 33. Seitz, 1912, Die Gross-Schmetterlinge der Erde, **2**: 442 (as *transalpina* Esper ssp.).

tauriana Burgeff, 1926, in Strand, Lepid. Cat., **33**: 58. Reiss,

1930, in Seitz, Die Gross-Schmetterlinge der Erde, Supplement, **2**: 35, pl. 3o. Holik & Sheljuzhko, 1958, Mitt. münch. ent. Ges., **48**: 186.

ab. **quinquemaculata** Holik & Sheljuzhko, 1958, Mitt. münch. ent. Ges., **48**: 186.

ssp. **anodolitia** Reiss, 1929, Int. ent. Z., **23**: 152; 1930, ibidem, **23**: 523, 525; 1930, in Seitz, Die Gross-Schmetterlinge der Erde, Supplement, **2**: 35, pl. 3n; 1935, Int. ent. Z., **29**: 207. Holik & Sheljuzhko, 1958, Mitt. münch. ent. Ges., **48**: 189. — Ak-Schehir, Kleinasien.

ssp. **akdaghi** Holik & Sheljuzhko, 1958, Mitt. münch. ent. Ges., **48**: 185. — Ak-Dagh, Amasia, Kleinasien.

ssp. **tirabzonica** Koch, 1942, Iris, **56**: 93. — Trapezunt, Schwarzes-Meer-Gebiet, Kleinasien.

ssp. **tiefi** Holik & Sheljuzhko, 1958, Mitt. münch. ent. Ges., **48**: 178. — Novyj Afonj bei Suchum, Abchasien, Transkaukasien.

ssp. **borzhomensis** Sheljuzhko, 1936, Folia Zool. Hydrobiol., Riga, **9**: 21. Koch, 1939, Mitt. münch. ent. Ges., **29**: 413. Holik & Sheljuzhko, 1958, Mitt. münch. ent. Ges., **48**: 179. — Borzhom, Georgien, Transkaukasien.

ssp. **ciscaucasica** Sheljuzhko, 1936, Folia Zool. Hydrobiol., Riga, **9**: 20. Holik & Sheljuzhko, 1958, Mitt. münch. ent. Ges., **48**: 176. Tremewan, 1961, Bull. Brit. Mus. (nat. Hist.) Ent., **10** (7): 309, pl. 57, fig. 28. — Teberda-Gebiet, Ciskaukasien.

ssp. **wojtusiaki** Holik, 1939, Ann. Mus. zool. Polon., **13**: 254, pl. 23, figs. 16-18. Holik & Sheljuzhko, 1958, Mitt. münch. ent. Ges., **48**: 177. Tremewan, 1961, Bull. Brit. Mus. (nat. Hist.) Ent., **10** (7): 309, pl. 57, fig. 29. — Karaugom, Nord-Ossetien, Ciskaukasien, 1800 m.

ab. **quinquemaculata** Holik, 1939, Ann. Mus. zool. Polon., **13**: 255, pl. 23, fig. 19.

ab. **medioconfluens** Holik, 1939, Ann. Mus. zool. Polon., **13**: 255.

ssp. **petsherkensis** Holik & Reiss, 1932, in Holik, Iris, **46**: 126, pl. 2, figs. 13-18. Reiss, 1933, in Seitz, Die Gross-Schmetterlinge der Erde, Supplement, **2**: 276. Holik & Sheljuzhko, 1958, Mitt. münch. ent. Ges., **48**: 172. — Butsha, Umgebung von Kijev, Kirillovskije ovragi, Ukraine.

ab. **medioconfluens** Holik & Reiss, 1932, in Holik, Iris, **46**: 126.

ab. **apicaliconfluens** Holik & Reiss, 1932, in Holik, Iris, **46**: 126.

ssp. **pulchroidea** Holik, 1939, Rev. franç. Lépid., **9**: 279, pl. 7, — Borissovka,

figs. 25, 26. Holik & Sheljuzhko, 1958, Mitt. münch. ent. Ges., **48**: 169. ab. **medioconfluens** Holik, 1939, Rev. franç. Lépid., **9**: 278. ab. **apicaliconfluens** Holik, 1939, Rev. franç. Lépid., **9**: 278. ab. **medioapicaliconfluens** Holik, 1939, Rev. franç. Lépid., **9**: 278, pl. 7, fig. 27. ab. **latelimbata** Holik, 1939, Rev. franç. Lépid., **9**: 278, pl. 7, fig. 28.	50 km. westlich von Bjelgorod, Zentral Russland.
ssp. **tambovana** Holik & Sheljuzhko, 1958, Mitt. münch. ent. Ges., **48**: 169.	Kozlov, Gub. Tambov, Zentral Russland.
ssp. **gemina** Burgeff, 1914, Mitt. münch. ent. Ges., **5**: 61, pl. 3, figs. 115, 116; 1926, ibidem, **16**: 67. Reiss, 1930, in Seitz, Die Gross-Schmetterlinge der Erde, Supplement, **2**: 34, pl. 31; 1936, Ent. Rdsch., **54**: 74. ab. **privata** Burgeff, 1906, Ent. Z., **20**: 162, fig. 9. Seitz, 1907, Die Gross-Schmetterlinge der Erde, **2**: 22 (as *lonicerae* Scheven ab.). Reiss, 1930, in Seitz, Die Gross-Schmetterlinge der Erde, Supplement, **2**: 34.	Sierra Segura, Provinz Murcia, Südspanien.
ssp. **geminoides** Reiss, 1936, Ent. Rdsch., **54**: 75. ab. **paupera** Reiss, 1936, Ent. Rdsch., **54**: 76.	Sierra de Gredos, Spanien, 1600 m.
ssp. **kricheldorffiana** Reiss, 1936, Ent. Rdsch., **54**: 75, pl. 2, fig. *microseeboldi* Verity, 1946, Redia, **31**: 67. Tremewan, 1963, Ent. Rec., **75**: 253. ab. **sexmaculata** Tremewan, 1961, Ent. Rec., **73**: 7; 1962, ibidem, **74**: 128, pl. 2, fig. 12; 1963, ibidem, **75**: 8, pl. 1, fig. 17.	La Liebana, Asturien (Picos de Europa), Spanien.
ssp. **himmighofeni** Burgeff, 1926, Mitt. münch. ent. Ges., **16**: 69. Oberthür, 1910, Études de Lépidoptérologie comparée, **4**: 545 (as *kindermanni* Oberthür [partim]). Reiss, 1930, in Seitz, Die Gross-Schmetterlinge der Erde, Supplement, **2**: 34, pl. 31. Marten, 1956, Ent. Z., **66**: 56. *catalonica* de Sagarra, 1925, Bull. Inst. catal. Hist. nat., (2) **5**: 272 (preoccupied by **catalonica** de Sagarra, 1924, ssp. of **hilaris** Ochsenheimer). *antecosta* Marten, 1956, Ent. Z., **66**: 56. *rustica* Marten, 1956, Ent. Z., **66**: 56.	Barcelona, Katalonien, Spanien.
ssp. **medianera** Marten, 1956, Ent. Z., **66**: 57.	Collsuspina, oberhalb des Ortes Centellas, Provinz Barcelona, Spanien, 800 m.
ssp. **agutangula** Marten, 1956, Ent. Z., **66**: 47.	Bergland nordöstlich von Morella,

ssp. **agutangula** *(continued)* — wo die Provinzen Tarragona und Castellon zusamenstossen, Spanien.

f. loc. **sagitta** Marten, 1956, Ent. Z., **66**: 58. Tremewan, 1963, Ent. Rec., **75**: 253. Tremewan & Manley, 1965, Ent. Rec., **77**: 10. — Penches, Provinz Burgos, Spanien, 700 m.

f. loc. **rutilans** Marten, 1956, Ent. Z., **66**: 58. — Echarren, Provinz Navarra, Spanien, 470 m.

ssp. **trevinca** Marten, 1956, Ent. Z., **66**: 46. — Oberes Tera-Tal, der Pena Trevinca benachbart, Prov. Zamora, Spanien, 1400 m.

ssp. **aleonada** Marten, 1956, Ent. Z., **66**: 46. Tremewan & Manley, 1965, Ent. Rec., **77**: 10. — Oberhalb des Ortes Villar del Ala, Provinz Soria, Spanien, 1200 m.

ssp. **seeboldi** Oberthür, 1910, Études de Lépidoptérologie comparée, **4**: 543. Reiss, 1930, in Seitz, Die Gross-Schmetterlinge der Erde, Supplement, **2**: 34, pl. 3k; 1936, Ent. Rdsch., **54**: 75, pl. 2, fig. Tremewan, 1961, Bull. Brit. Mus. (nat. Hist.) Ent., **10** (7): 292, pl. 55, fig. 15. Tremewan & Manley, 1965, Ent. Rec., **77**: 10. — Bilbao, Espagne septentrionale.

ssp. **altapyrenaica** Le Charles, 1949, Rev. franç. Lépid., **12**: 179. — Cauterets (Port d'Espagne); Gavarnie, 900-1500 m.; Hautes-Pyrénées, France.

ssp. **pyrenes** Verity, 1921, Ent. Rec., **33**: 122. Oberthür, 1910, Études de Lépidoptérologie comparée, **4**: 538 (as *dubia* Staudinger). Reiss, 1930, in Seitz, Die Gross-Schmetterlinge der Erde, Supplement, **2**: 34. — Vernet-les-Bains, Pyrénées-Orientales, France.

quinquemaculata Oberthür, 1910, Études de Lépidoptérologie comparée, **4**: 539 (infrasubspecific); 1909, ibidem, **3**, pl. 28, figs. 169, 171 (as *dubia* Staudinger). Tremewan, 1961, Bull. Brit. Mus. (nat. Hist.) Ent., **10** (7): 288, pl. 55, fig. 5.

ab. **sexmaculata** Oberthür, 1910, Études de Lépidoptérologie comparée, **4**: 539; 1909, ibidem, **3**, pl. 28, fig. 170 (as *dubia* Staudinger). Tremewan, 1961, Bull. Brit. Mus. (nat. Hist.) Ent., **10** (7): 288, pl. 55, fig. 6.

ssp. **gemella** Marten, 1956, Ent. Z., **66**: 41, 45. — Ribas de Fressér, Ostpyrenäen, 900-1400 m.

ssp. **siciliensis** Verity, 1917, Bull. Soc. ent. Fr., p. 223; 1921, Ent. Rec., **33**: 111. Reiss, 1930, in Seitz, Die Gross-Schmetterlinge der Erde, Supplement, **2**: 35, pl. 3m. — Les collines aux environs des Palermo, Sicilie.

ab. **tenuelimbata** Verity, 1921, Ent. Rec., **33**: 147.

ab. **tenuiorelimbata** Verity, 1921, Ent. Rec., **33**: 147.

ab. **tenuissimelimbata** Verity, 1921, Ent. Rec., **33**: 147.

ssp. **calabra** Verity, 1917, Bull. Soc. ent. Fr., p. 223; 1921, Ent. Rec., **33**: 112. Reiss, 1930, in Seitz, Die Gross-Schmetterlinge der Erde, Supplement, **2**: 35. — Piano di Carmelia, Aspromonte, Calabria, Italie, 1200 m.

ab. **pulcherrimaeformis** Verity, 1921, Ent. Rec., **33**: 110. Reiss, 1930, in Seitz, Die Gross-Schmetterlinge der Erde, Supplement, **2**: 35.

ssp. **calabraochsenheimeri** Verity, 1921, Ent. Rec., **33**: 112. — Coastal range, San Fili, Calabria, Italy, 900 m.

ssp. **campaniae** Rebel, 1901, in Staudinger & Rebel, Catalog der Lepidopteren des Palaearctischen Faunengebietes, p. 384. Seitz, 1907, Die Gross-Schmetterlinge der Erde, **2**: 22, pl. 5c. — Monti Aurunci, Campagna, Italien.

melilochsenheimeri Stauder, 1929, Ent. Z., **43**: 32. Tremewan, 1961, Bull. Brit. Mus. (nat. Hist.) Ent., **10** (7): 290, pl. 55, fig. 12.

melilocampaniae Stauder, 1929, Ent. Z., **43**: 79. Tremewan, 1961, Bull. Brit. Mus. (nat. Hist.) Ent., **10** (7): 290, pl. 55, fig. 9.

calabrochsenheimeri Stauder, 1929, Ent. Z., **43**: 132. Tremewan, 1961, Bull. Brit. Mus. (nat. Hist.) Ent., **10** (7): 290, pl. 55, fig. 11.

hybridophila Stauder, 1929, Ent. Z., **43**: 133. Tremewan, 1961, Bull. Brit. Mus. (nat. Hist.) Ent., **10** (7): 290, pl. 55, fig. 10.

ab. **dubia** Staudinger, 1861, in Staudinger & Wocke, Catalog der Lepidopteren Europa's und der angrenzenden Länder, p. 21. Seitz, 1907, Die Gross-Schmetterlinge der Erde, **2**: 22.

melilodubia Stauder, 1929, Ent. Z., **43**: 80. Tremewan, 1961, Bull. Brit. Mus. (nat. Hist.) Ent., **10** (7): 292, pl. 55, fig. 14.

ab. **carnioligiussana** Stauder, 1929, Ent. Z., **43**: 80. Tremewan, 1961, Bull. Brit. Mus. (nat. Hist.) Ent., **10** (7): 291, pl. 55, fig. 13.

ssp. **microchsenheimeri** Verity, 1921, Ent. Rec., **33**: 114. Reiss, — Mainarde Mts., Villalatina, Italy, 500 m.

1930, in Seitz, Die Gross-Schmetterlinge der Erde, Supplement, **2**: 34, pl. 3k.

ssp. **pulcherrimastoechadis** Verity, 1921, Ent. Rec., **33**: 109. — Reggio Emilia, Borzeno, Italy.
italpulcherrima Rocci, 1937, Redia, **22**: 134.
ab. **cuneata** Costantini, 1916, Atti Soc. Nat. Mat. Modena, **49**: 18.

ssp. **montivaga** Verity, 1916, Boll. Soc. ent. ital., **47**: 73; 1921, Ent. Rec., **33**: 124. Reiss, 1930, in Seitz, Die Gross-Schmetterlinge der Erde, Supplement, **2**: 35, pl. 3m. — Monti Sibillini, Bolognola, Italia, 1200 m.

ssp. **etrusca** Verity, 1921, Ent. Rec., **33**: 125. Reiss, 1930, in Seitz, Die Gross-Schmetterlinge der Erde, Supplement, **2**: 35. — Pian di Mugnone, Firenze, Italy, 300 m.
ab. **loniceraeformis** Verity, 1916, Boll. Soc. ent. ital., **47**: 74. Reiss, 1930, in Seitz, Die Gross-Schmetterlinge der Erde, Supplement, **2**: 35.
ab. **latiorelimbata** Verity, 1921, Ent. Rec., **33**: 147.
f. t. **microetrusca** Verity, 1946, Redia, **31**: 68. — S. Giusto presso Greve in Chianti, Firenze, Italia, 500 m. (luglio e agosto).

ssp. **aterrima** Verity, 1921, Ent. Rec., **33**: 128. Reiss, 1930, in Seitz, Die Gross-Schmetterlinge der Erde, Supplement, **2**: 35. — Mt. Prato Fiorito, Fegana Valley, North Tuscany, Italy, 600-1000 m.
ab. **latissimelimbata** Verity, 1921, Ent. Rec., **33**: 147.
ab. **magnamacula** Verity, 1921, Ent. Rec., **33**: 148.
ab. **radiis** Verity, 1921, Ent. Rec., **33**: 148.
ab. **radiiszonata** Verity, 1921, Ent. Rec., **33**: 148.
ab. **guttata** Verity, 1921, Ent. Rec., **33**: 149.
ab. **macula** Verity, 1921, Ent. Rec., **33**: 147.
ab. **bimacula** Verity, 1921, Ent. Rec., **33**: 147.

ssp. **oraria** Verity, 1921, Ent. Rec., **33**: 126. Reiss, 1930, in Seitz, Die Gross-Schmetterlinge der Erde, Supplement, **2**: 35. — Antignano near Leghorn; Forte dei Marmi near Viareggio, Italy.

ssp. **gigantea** Rocci, 1913, Soc. ent., **28**: 56; 1914, Atti Soc. ligust. Sci. nat. geogr., **25**: 220. Verity, 1921, Ent. Rec., **33**: 128. Rocci, 1926, Mem. Soc. ent. ital., **4**: 175, fig. 4a. Reiss, 1930, in Seitz, Die Gross-Schmetterlinge der Erde, Supplement, **2**: 34, pl. 3l. — Riviera di Genoa, Liguria, Italia.
ab. **cuprea** Rocci, 1914, Atti Soc. ligust. Sci. nat. geogr., **25**: 219; 1926, Mem. Soc. ent. ital., **4**: 168. Reiss, 1930, in Seitz, Die Gross-Schmetterlinge der Erde, Supplement, **2**: 35.
ab. **violacea** Rocci, 1914, Atti Soc. ligust. Sci. nat. geogr., **25**: 219; 1926, Mem. Soc. ent. ital., **4**: 167. Reiss, 1930, in Seitz,

Die Gross-Schmetterlinge der Erde, Supplement, **2**: 35.

ab. **transalpinoides** Rocci, 1914, Atti Soc. ligust. Sci. nat. geogr., **24**: 114.

transalpinoides Rocci, 1914, Soc. ent. **29**: 41.

ab. **falcata** Rocci, 1921, Atti Soc. ligust. Sci. nat. geogr., **32**: 35; 1926, Mem. Soc. ent. ital., **4**: 167.

ab. **anomala** Rocci, 1914, Atti Soc. ligust. Sci. nat. geogr., **25**: 220; 1926, Mem. Soc. ent. ital., **4**: 168.

ab. **judicariaeformis** Rocci, 1926, Mem. Soc. ent. ital., **4**: 168. Reiss, 1930, in Seitz, Die Gross-Schmetterlinge der Erde, Supplement, **2**: 35.

judicariae Burgeff, 1926, Mitt. münch. ent. Ges., **16**: 69.

ab. **incompleta** Rocci, 1914, Atti Soc. ligust. Sci. nat. geogr., **25**: 220; 1926, Mem. Soc. ent. ital., **4**: 167.

ab. **praecox** Rocci, 1927, Boll. Soc. ent. ital., **59**: 12.

ab. **undecimaculata** Rocci, 1913, Soc. ent., **28**: 56; 1926, Mem. Soc. ent. ital., **4**: 167. Reiss, 1930, in Seitz, Die Gross-Schmetterlinge der Erde, Supplement, **2**: 35.

ab. **impar** Rocci, 1913, Soc. ent., **28**: 56; 1926, Mem. Soc. ent. ital., **4**: 167. Reiss, 1930, in Seitz, Die Gross-Schmetterlinge der Erde, Supplement, **2**: 35.

ab. **septemaculata** Rocci, 1914, Atti Soc. ligust. Sci. nat. geogr., **24**: 114; 1926, Mem. Soc. ent. ital., **4**: 167. Reiss, 1930, in Seitz, Die Gross-Schmetterlinge der Erde, Supplement, **2**: 35.

septemaculata Rocci, 1914, Soc. ent., **29**: 41.

ab. **biconjuncta** Rocci, 1914, Atti Soc. ligust. Sci. nat. geogr., **24**: 114; 1926, Mem. Soc. ent. ital., **4**: 166.

biconjuncta Rocci, 1914, Soc. ent., **29**: 41.

ab. **triconjuncta** Rocci, 1914, Atti Soc. ligust. Sci. nat. geogr., **24**: 114; 1926, Mem. Soc. ent. ital., **4**: 167. Reiss, 1930, in Seitz, Die Gross-Schmetterlinge der Erde, Supplement, **2**: 35.

triconjuncta Rocci, 1914, Soc. ent., **29**: 41.

ab. **plusnotata** Rocci, 1921, Atti Soc. ligust. Sci. nat. geogr., **32**: 35; 1926, Mem. Soc. ent. ital., **4**: 167. Reiss, 1930, in Seitz, Die Gross-Schmetterlinge der Erde, Supplement, **2**: 35.

ab. **parviguttata** Rocci, 1914, Atti Soc. ligust. Sci. nat. geogr., **24**: 114; 1926, Mem. Soc. ent. ital., **4**: 166. Reiss, 1930, in Seitz, Die Gross-Schmetterlinge der Erde, Supplement, **2**: 35.

parviguttata Rocci, 1914, Soc. ent., **29**: 41.

ab. **reducta** Rocci, 1913, Soc. ent., **28**: 56; 1926, Mem. Soc. ent. ital., **4**: 166. Reiss, 1930, in Seitz, Die Gross-Schmetterlinge der Erde, Supplement, **2**: 35.

quadripuncta Rocci, 1914, Atti Soc. ligust. Sci. nat. geogr., **24**: 114; 1926, Mem. Soc. ent. ital., **4**: 166.

quadripuncta Rocci, 1914, Soc. ent., **29**: 41.

ab. **tripunctata** Rocci, 1921, Atti Soc. ligust. Sci. nat. geogr.,

32: 35; 1926, Mem. Soc. ent. ital., **4**: 166. Reiss, 1930, in Seitz, Die Gross-Schmetterlinge der Erde, Supplement, **2**: 35.

ab. **lavanduloides** Rocci, 1921, Atti Soc. ligust. Sci. nat. geogr., **32**: 35; 1926, Mem. Soc. ent. ital., **4**: 166. Reiss, 1930, in Seitz, Die Gross-Schmetterlinge der Erde, Supplement, **2**:35.

ab. **zonata** Rocci, 1914, Atti Soc. ligust. Sci. nat. geogr., **24**: 114; 1926, Mem. Soc. ent. ital., **4**: 166, figs. 7a, 7b. Reiss, 1930, in Seitz, Die Gross-Schmetterlinge der Erde, Supplement, **2**: 35.

zonata Rocci, 1914, Soc. ent., **29**: 41.

ab. **biguttata** Rocci, 1914, Atti Soc. ligust. Sci. nat. geogr., **24**: 114; 1926, Mem. Soc. ent. ital., **4**: 166, fig. 5b. Reiss, 1930, in Seitz, Die Gross-Schmetterlinge der Erde, Supplement, **2**: 35.

biguttata Rocci, 1914, Soc. ent., **29**: 41.

ab. **nigra** Dziurzyński, 1908, Berl. ent. Z., **53**: 7.

seminigrata Rocci, 1914, Atti Soc. ligust. Sci. nat. geogr., **24**: 114; 1926, Mem. Soc. ent. ital., **4**: 166, fig. 5a.

seminigrata Rocci, 1914, Soc. ent., **29**: 41.

ab. **nigrata** Dziurzyński, 1908, Berl. ent. Z., **53**: 28. Seitz, 1912, Die Gross-Schmetterlinge der Erde, **2**: 442. Burgeff, 1914, Mitt. münch. ent. Ges., **5**: 61, pl. 3, fig. 117. Rocci, 1926, Mem. Soc. ent. ital., **4**: 165, fig. 5c.

nigerrima Reiss, 1913, Soc. ent., **28**: 76; 1930, in Seitz, Die Gross-Schmetterlinge der Erde, Supplement, **2**: 35, pl. 3m.

ab. **amplomaculata** Rocci, 1926, Mem. Soc. ent. ital., **4**: 167. Reiss, 1930, in Seitz, Die Gross-Schmetterlinge der Erde, Supplement, **2**: 35.

ab. **costimaculata** Rocci, 1926, Mem. Soc. ent. ital., **4**: 167.

ab. **pseudoetrusca** Rocci, 1926, Mem. Soc. ent. ital., **4**: 165.

ab. **pseudoraria** Rocci, 1926, Mem. Soc. ent. ital., **4**: 165.

ab. **pseudomedicaginis** Rocci, 1926, Mem. Soc. ent. ital., **4**: 165.

ab. **pseudoliguris** Rocci, 1926, Mem. Soc. ent. ital., **4** : 164, figs. 3c, 3d.

ab. **pseudoduponcheli** Rocci, 1926, Mem. Soc. ent. ital., **4**: 174.

ab. **infecta** Rocci, 1926, Mem. Soc. ent. ital., **4**: 168.

ab. **imperfecta** Rocci, 1926, Mem. Soc. ent. ital., **4**: 168.

ab. **basalis** Rocci, 1914, Atti Soc. ligust. Sci. nat. geogr., **24**: 114; 1926, Mem. Soc. ent. ital., **4**: 166.

basalis Rocci, 1914, Soc. ent., **29**: 41.

ab. **mediounita** Rocci, 1913, Soc. ent., **28**: 56; 1926, Mem. Soc. ent. ital., **4**: 167. Reiss, 1930, in Seitz, Die Gross-Schmetterlinge der Erde, Supplement, **2**: 35.

ab. **confluens** Rocci, 1914, Atti Soc. ligust. Sci. nat. geogr., **24**: 114; 1926, Mem. Soc. ent. ital., **4**: 166.

confluens Rocci, 1914, Soc. ent., **29**: 41.

ab. **bongerti** Reiss, 1914, Int. ent. Z., **8**: 158; 1930, in Seitz, Die Gross-Schmetterlinge der Erde, Supplement, **2**: 35, pl. 3l.
ab. **rosea** Rocci, 1926, Mem. Soc. ent. ital., **4**: 168.
ab. **intermedia** Reiss, 1913, Int. ent. Z., **7**: 113.
aurantiaca Rocci, 1926, Mem. Soc. ent. ital., **4**: 168.
ab. **zlatoroga** Reiss, 1913, Int. ent. Z., **7**: 113; 1930, in Seitz, Die Gross-Schmetterlinge der Erde, Supplement, **2**: 35, pl. 3m.
citrina Rocci, 1914, Atti Soc. ligust. Sci. nat. geogr., **24**: 114; 1926, Mem. Soc. ent. ital., **4**: 168.
citrina Rocci, 1914, Soc. ent., **29**: 41.
ab. **fuscoguttata** Rocci, 1914, Atti Soc. ligust. Sci. nat. geogr., **25**: 220; 1926, Mem. Soc. ent. ital., **4**: 166.
f. t. **autumnalis** Reiss, 1914, Int. ent. Z., **8**: 46; 1930, in Seitz, Die Gross-Schmetterlinge der Erde, Supplement, **2**: 35, pl. 3m. — Genova, Ligurien, Italien.
genuensis Rocci, 1914, Atti Soc. ligust. Sci. nat. geogr., **24**: 114; 1926, Mem. Soc. ent. ital., **4**: 168, fig. 4d.
genuensis Rocci, 1914, Soc. ent., **29**: 41.
ab. **sexmaculata** Rocci, 1914, Atti Soc. ligust. Sci. nat. geogr., **25**: 221; 1926, Mem. Soc. ent. ital., **4**: 167, fig. 6c. Reiss, 1930, in Seitz, Die Gross-Schmetterlinge der Erde, Supplement, **2**: 35.
ab. **tenuimarginata** Rocci, 1914, Atti Soc. ligust. Sci. nat. geogr., **24**: 114.
tenuimarginata Rocci, 1914, Soc. ent., **29**: 41.
ssp. **alassica** Dujardin, 1965, Entomops, Nice, no. 2: 47, fig. — Alassio; Ceriana, Liguria, Italie.

ssp. **veneta** Rocci, 1937, Redia, **22**: 136. Holik, 1943, Mitt. münch. ent. Ges., **33**: 319. — Littorale di Venezia ed a Lido, Istria, Italia.
burgeffi Stauder, 1921, Verh. zool.-bot. Ges. Wien, **70**: (178) (infrasubspecific). Tremewan, 1966, Ent. Rec., **78**: 35, pl. 1, fig. 12.
melilofilipendulae Stauder, 1929, Ent. Z., **43**: 31 (infrasubspecific). Tremewan, 1961, Bull. Brit. Mus. (nat. Hist.) Ent., **10** (7): 289, pl. 55, fig. 3.
levrinii Dujardin, 1956, Bull. mens. Soc. linn. Lyon, **25**: 257.
ab. **dubia** Stauder, 1914, Iris, **28**: 16.
pseudodubia Stauder, 1921, Verh. zool.-bot. Ges. Wien, **70**: (178).
ab. **stoechadioides** Stauder, 1914, Iris, **28**: 16. Reiss, 1930, in Seitz, Die Gross-Schmetterlinge der Erde, Supplement, **2**: 34.
ab. **pallescens** Stauder, 1921, Verh. zool.-bot. Ges. Wien, **70**: (177).

f. t. **exigua** Stauder, 1921, Verh. zool.-bot. Ges. Wien, **70**: (177). — Görz; Triest; Istrien.

postlevrinii Dujardin, 1965, Entomops, Nice, no. 2: 47, fig.

ssp. **restituta** Rocci, 1937, Redia, **22**: 137. Holik, 1943, Mitt. münch. ent. Ges., **33**: 318. — Chiapovano, 700 m.; Ternova della Selva, Tolmino, Julisch-Venetien.

ab. **achillfilipendulae** Stauder, 1929, Ent. Z., **43**: 31. Tremewan, 1961, Bull. Brit. Mus. (nat. Hist.) Ent., **10** (7): 289, pl. 55, fig. 2.

ssp. **zarana** Burgeff, 1926, Mitt. münch. ent. Ges., **16**: 67. Reiss, 1930, in Seitz, Die Gross-Schmetterlinge der Erde, Supplement, **2**: 34, pl. 3k. Holik, 1943, Mitt. münch. ent. Ges., **33**: 322. — Zara, Dalmatien.

ab. **incompleta** Holik, 1943, Mitt. münch. ent. Ges., **33**: 322.

ssp. **illyrica** Holik, 1943, Mitt. münch. ent. Ges., **33**: 323. — Vucija bara, südliche Herzegowina.

ab. **pseudomanni** Schawerda, 1916, Verh. zool.-bot. Ges. Wien, **66**: 247. Reiss, 1930, in Seitz, Die Gross-Schmetterlinge der Erde, Supplement, **2**: 34. Holik, 1943, Mitt. münch. ent. Ges., **33**: 325.

ab. **quinquemaculata** Holik, 1943, Mitt. münch. ent. Ges., **33**: 326.

ssp. **balcanirosea** Holik, 1943, Mitt. münch. ent. Ges., **33**: 327. Daniel, 1957, Acta Mus. maced. Sci. nat., **4**: 214; 1964, Prirod. Muz. Scopje, no. 2: 17. — Kara-Orman, Ochrid, Petrina-Planina, Serbisch-Mazedonien.

ssp. **sharensis** Daniel, 1957, Acta Mus. maced. Sci. nat., **4**: 215; 1964, Prirod. Muz. Scopje, no. 2: 17. — Vratnica, Shar-planina, Mazedonien, 900 m.

ssp. **praeochsenheimeri** Verity, 1939, Ent. Rec., **51**: (19). Daniel, 1958, Fragmenta Balcanica, **2**: 42. — Mt. Olympus, Greece.

ssp. **athonis** Koch, 1942, Iris, **56**: 92. — Berg Athos, Halbinsel Chalkidikae, Griechenland.

ssp. **stoechadis** Borkhausen, 1793, Rheinisches Magazin, **1**: 628. Esper, 1797, Die Schmetterlinge, Supplement, **2** (2): 18, 19, pl. 41, fig. 3 (as *lavandulae* Esper [partim]). Ochsenheimer, 1808, Die Schmetterlinge von Europa, **2**: 83. Freyer, 1842, Neuere Beiträge zur Schmetterlingskunde, **4**: 138, pl. 368, figs. 1-4. Oberthür, 1896, Études d'Entomologie, **20**: 46, pl. 7, fig. 110. Spuler, 1906, in Hofmann, Die Schmetterlinge Europas, **2**: 159, pl. 77, fig. 18. Seitz, 1907, Die Gross-Schmetterlinge der Erde, **2**: 22. Oberthür, 1910, Études de Lépidoptérologie comparée, **4**: 527, 536. Verity, 1921, Ent. Rec., **33**: 127. Forster & Wohlfahrt, 1958, Die Schmetterlinge Mitteleuropas, **3**: 97, pl. 10, figs. 43, 44. — Piemont, Italien.

lavandulae Hübner, 1790, Beiträge Geschichte der Schmetterlinge, **2**: 69, pl. 3, fig. O (preoccupied by **lavandulae** Esper, 1783); 1796, Sammlung europäischer Schmetterlinge, **2**: 17, pl. 4, fig. 24; 1806, ibidem, Der Ziefer, p. 79.

medicaginis Hübner, 1796, Sammlung europäischer Schmetterlinge, **2**: 16, pl. 4, fig. 20; 1806, ibidem, Der Ziefer, p. 82.

myrmeca Heydenreich, 1851, Lepidopterorum Europaeorum Catalogus Methodicus, p. 22 (nomen nudum).

ab. **judicariae** Calberla, 1895, Iris, **8**: 218. Spuler, 1906, in Hofmann, Die Schmetterlinge Europas, **2**: 159. Seitz, 1907, Die Gross-Schmetterlinge der Erde, **2**: 22.

ab. **roseopicta** Turati, 1919, Atti Soc. ital. Sci. nat., **58**: 173.

ab. **oberthueri** Dziurzyński, 1908, Berl. ent. Z., **53**: 5, 12, 29 (as *oberthüri*). Seitz, 1912, Die Gross-Schmetterlinge der Erde, **2**: 442.

ab. **flava** Burgeff, 1926, in Strand, Lepid. Cat., **33**: 56. Reiss, 1930, in Seitz, Die Gross-Schmetterlinge der Erde, Supplement, **2**: 34.

ssp. **caeruleochsenheimeri** Verity, 1930, Mem. Soc. ent. ital., **9**: 23. Reiss, 1933, in Seitz, Die Gross-Schmetterlinge der Erde, Supplement, **2**: 276. — Vanzone, Valle Anzasca, Italia, 700 m.

ab. **biconjuncta** Verity, 1916, Boll. Soc. ent. ital., **47**: 74. Reiss, 1930, in Seitz, Die Gross-Schmetterlinge der Erde, Supplement, **2**: 34.

ssp. **frigidoliguris** Rocci, 1937, Redia, **22**: 135. — Mte. Calisio, Trento, Italia, 800 m.

ssp. **eridanea** Rocci, 1941, Boll. Ist. Ent. Univ. Bologna, **13**: 121, pl. 2, figs. 26-34, pl. 3, figs. 1-5. — Collina di Torino (Maddalena), Piedmont, Italia, 600-700 m.

ssp. **liguris** Rocci, 1925, Boll. Soc. ent. ital., **57**: 97; 1926, Mem. Soc. ent. ital., **4**: 158, fig. 2a. Reiss, 1930, in Seitz, Die Gross-Schmetterlinge der Erde, Supplement, **2**: 35. — Alta Val Bisagno, Appennino ligure, Italia, 300-600 m.

ab. **pseudoetrusca** Rocci, 1926, Mem. Soc. ent. ital., **4**: 165.

ab. **pseudoraria** Rocci, 1926, Mem. Soc. ent. ital., **4**: 165.

ab. **pseudomedicaginis** Rocci, 1926, Mem. Soc. ent. ital., **4**: 165, 174.

ab. **pseudoduponcheli** Rocci, 1926, Mem. Soc. ent. ital., **4**: 174.

ssp. **microstoechadis** Rocci, 1925, Boll. Soc. ent. ital., **57**: 97; 1926, Mem. Soc. ent. ital., **4**: 172, fig. 8a. Reiss, 1930, in Seitz, Die Gross-Schmetterlinge der Erde, Supplement, **2**: 35. — Alta Val Scrivia, Monte Maggio, Appennino ligure, Italia, 700-1500 m.

ssp. **hyperstoechadis** Rocci, 1937, Redia, **22**: 135. — Mte. Sumbra; Palasaccio, 500-1400 m.; Palazzuolo di Romagna, Italia.

ssp. **subliguris** Rocci, 1937, Redia, **22**: 135. — Strada della Futa; Sassello, Torriglia, 700 m., Italia.

ssp. **microfrigida** Rocci, 1937, Redia, **22**: 136. — Mte. Bondone; Mte. Gazza, 1200 m., Trento, Italia.

ssp. **abdita** Rocci, 1941, Boll. Ist. Ent. Univ. Bologna, **13**: 122. — Forno Alpia Graie, Piedmont, Italia, 1300 m.

ssp. **segusina** Rocci, 1941, Boll. Ist. Ent. Univ. Bologna, **13**: 123, pl. 3, figs. 6-11. — Valle di Susa, Meana, Piedmont, Italia, 800-1000 m.

ssp. **microsegusina** Rocci, 1941, Boll. Ist. Ent. Univ. Bologna, **13**: 123, pl. 3, figs. 12, 14. — Claviere, Cesana, 1300-1900 m.; Sestrière, Alpi Cozie, 1600 m., Italia.

ssp. **duponcheli** Verity, 1921, Ent. Rec., **33**: 124. Duponchel, 1835, in Godart & Duponchel, Histoire naturelle des Lépidoptères ou Papillons de France, Supplement, **2**: 73, pl. 6, figs. 5, 6 (as *medicaginis* Hübner). Oberthür, 1909, Études de Lépidoptérologie comparée, **3**, pl. 28, fig. 158 (as *medicaginis* Hübner); 1910, ibidem, **4**: 537 (as *medicaginis* Hübner). Reiss, 1930, in Seitz, Die Gross-Schmetterlinge der Erde, Supplement, **2**: 34. — Nice, Alpes-Maritimes, France.

ab. **pseudostoechadis** Burgeff, 1926, in Strand, Lepid. Cat., **33**: 56. Oberthür, 1909, Études de Lépidoptérologie comparée, **3**, pl. 28, fig. 157 (as *medicaginis* Hübner). Reiss, 1930, in Seitz, Die Gross-Schmetterlinge der Erde, Supplement, **2**: 34.

ab. **flava** Oberthür, 1910, Études de Lépidoptérologie comparée, **4**: 536. Reiss, 1930, in Seitz, Die Gross-Schmetterlinge der Erde, Supplement, **2**: 34. Tremewan, 1961, Bull. Brit. Mus. (nat. Hist.) Ent., **10** (7): 290, pl. 55, fig. 4.

ab. **coffaea** Dujardin, 1956, Bull. mens. Soc. linn. Lyon, **25**: 257.

f. t. **microduponcheli** Verity, 1946, Redia, **31**: 67. — Vence, Alpes-Maritimes, France, 300 m.

ssp. **claraduponcheli** Dujardin, 1956, Bull. mens. Soc. linn. Lyon, **25**: 256. — La Colmiane (Valdeblore), Alpes-Maritimes, France, 1500 m.

ssp. **callimorpha** Dujardin, 1965, Entomops, Nice, no. 2: 42, fig. — Étangs de Villepey, Fréjus; Collobrières; Bormes; Roquebrune; la Môle; La Garde; St. Tropez; Hyères; Fréjus les Hoirs; La Nartelle; Forêt de Janas; Six-Fours; Villepey; La Tour de Mare; La Foux, Var, France.

f. t. **septembrina** Dujardin, 1965, Entomops, Nice, no. 2: 43, fig. — Étangs de Villepey, Fréjus, Var, France (forma autumnalis).

ssp. **siagnica** Dujardin, 1965, Entomops, Nice, no. 2: 43, fig. — Mandelieu (à l'Ouest de Cannes), Alpes-Maritimes, France.

ssp. **pentasema** Dujardin, 1965, Entomops, Nice, no. 2: 46, fig. — Esteng (dans le haut Var), Alpes-Maritimes, France, 1500 m.

ssp. **willaumei** Dujardin, 1965, Entomops, Nice, no. 2: 46, fig. — Auron, Alpes-Maritimes, France, 1500-1600 m.

ssp. **anceps** Oberthür, 1910, Études de Lépidoptérologie comparée, **4**: 551. Verity, 1921, Ent. Rec., **33**: 109. Reiss, 1930, in Seitz, Die Gross-Schmetterlinge der Erde, Supplement, **2**: 34. — St. Baume, Hyères, Var, France méridionale.

Tremewan, 1961, Bull. Brit. Mus. (nat. Hist.) Ent., **10** (7): 290, pl. 55, fig. 7.

quinquemacula Bethune-Baker, 1922, Ent. Rec., **34**: 74. Tremewan, 1961, Bull. Brit. Mus. (nat. Hist.) Ent., **10** (7): 290, pl. 55, fig. 8.

ssp. **oberthueriana** Burgeff, 1926, Mitt. münch. ent. Ges., **16**: 67 (as *oberthüriana*), (nomen novum for *alpina* Boisduval sensu Oberthür). Oberthür, 1910, Études de Lépidoptérologie comparée, **4**: 525 (as *alpina* Boisduval). Reiss, 1930, in Seitz, Die Gross-Schmetterlinge der Erde, Supplement, **2**: 34. Dufay, 1966, Bull. mens. Soc. linn. Lyon, **35**: 72. Digne, Basses-Alpes, Frankreich.

ab. **angelicaeformis** Verity, 1921, Ent. Rec., **33**: 109. Reiss, 1930, in Seitz, Die Gross-Schmetterlinge der Erde, Supplement, **2**: 34.

ssp. **maior** Esper, 1797, Die Schmetterlinge, Supplement, **2** (2): 19, pl. 41, fig. 4. Südfrankreich [Montpellier].

ochsenheimeri Zeller, 1847, Isis von oken, p. 303. Spuler, 1906, in Hofmann, Die Schmetterlinge Europas, **2**: 160, pl. 77, fig. 19b. Seitz, 1907, Die Gross-Schmetterlinge der Erde, **2**: 23, pl. 5g. Oberthür, 1910, Études de Lépidoptérologie comparée, **4**: 547. Querci, 1912, in Oberthür, Études de Lépidoptérologie comparée, **6**: 146, 166. Verity, 1921, Ent. Rec., **33**: 112. Burgeff, 1926, Mitt. münch. ent. Ges., **16**: 66. Reiss, 1930, in Seitz, Die Gross-Schmetterlinge der Erde, Supplement, **2**: 34; 1942, Ent. Z., **56**: 11; 1950, Jber. naturf. Ges. Graubünden, **82**: 116, fig. 25. Forster & Wohlfahrt, 1958, Die Schmetterlinge Mitteleuropas, **3**: 97, pl. 10, figs. 41, 42.

meridionalis Vorbrodt, 1913, in Vorbrodt & Müller-Rutz, Die Schmetterlinge der Schweiz, **2**: 268.

sexmaculata Vorbrodt, 1913, in Vorbrodt & Müller-Rutz, Die Schmetterlinge der Schweiz, **2**: 267.

sexmaculata Vorbrodt, 1913, in Vorbrodt & Müller-Rutz, Die Schmetterlinge der Schweiz, **2**: 267 (as *angelicae* Ochsenheimer ab.).

ab. **semicingulata** Tremewan, 1965, Ent. Rec., **77**: 90.

ab. **stoechadioides** Reiss, 1950, Jber. naturf. Ges. Graubünden, **82**: 117, fig. 26.

charon Boisduval, 1834, Icones historique des Lépidoptères nouveaux ou peu connus, **2**: 61, pl. 54, fig. 9 (preoccupied by **charon** Hübner, 1796, ssp. of **viciae** Denis & Schiffermüller). Tremewan, 1961, Bull. Brit. Mus. (nat. Hist.) Ent., **10** (7): 288, pl. 54, fig. 36.

ab. **translonicerae** Stauder, 1929, Ent. Z., **43**: 80. Tremewan, 1961, Bull. Brit. Mus. (nat. Hist.) Ent., **10** (7): 289, pl. 55, fig. 1.

ab. **latelimbata** Verity, 1921, Ent. Rec., **33**: 147.
ab. **rubra** Dziurzyński, 1908, Berl. ent. Z., **53**: 5, 29. Seitz, 1912, Die Gross-Schmetterlinge der Erde, **2**: 442.
ab. **apicalimaculata** Vorbrodt, 1913, in Vorbrodt & Müller-Rutz, Die Schmetterlinge der Schweiz, **2**: 270.
ab. **basiconfluens** Vorbrodt, 1913, in Vorbrodt & Müller-Rutz, Die Schmetterlinge der Schweiz, **2**: 269, fig. 11.
ab. **medioconfluens** Vorbrodt, 1913, in Vorbrodt & Müller-Rutz, Die Schmetterlinge der Schweiz, **2**: 269, fig. 12.
ab. **apicaliconfluens** Vorbrodt, 1913, in Vorbrodt & Müller-Rutz, Die Schmetterlinge der Schweiz, **2**: 269, figs. 13, 14.
ab. **basimedioconfluens** Vorbrodt, 1913, in Vorbrodt & Müller-Rutz, Die Schmetterlinge der Schweiz, **2**: 269, fig. 15.
ab. **medioapicaliconfluens** Vorbrodt, 1913, in Vorbrodt & Müller-Rutz, Die Schmetterlinge der Schweiz, **2**: 269, fig. 16.
ab. **trimaculata** Vorbrodt, 1913, in Vorbrodt & Müller-Rutz, Die Schmetterlinge der Schweiz, **2**: 269, fig. 17.
ab. **costaliconfluens** Vorbrodt, 1913, in Vorbrodt & Müller-Rutz, Die Schmetterlinge der Schweiz, **2**: 269.
ab. **costalielongata** Vorbrodt, 1913, in Vorbrodt & Müller-Rutz, Die Schmetterlinge der Schweiz, **2**: 268.
ab. **apicalielongata** Vorbrodt, 1913, in Vorbrodt & Müller-Rutz, Die Schmetterlinge der Schweiz, **2**: 268.
ab. **apicejuncta** Verity, 1916, Boll. Soc. ent. ital., **47**: 74.
ab. **parallela** Vorbrodt, 1913, in Vorbrodt & Müller-Rutz, Die Schmetterlinge der Schweiz, **2**: 269.
ab. **confusa** Vorbrodt, 1921, Mitt. schweiz. ent. Ges., **13**: 204.
ab. **incarnata** Vorbrodt, 1921, Mitt. schweiz. ent. Ges., **13**: 204.
f. t. **autumnalis** Burgeff, 1921, Mitt. münch. ent. Ges., **11**: 52. Reiss, 1930, in Seitz, Die Gross-Schmetterlinge der Erde, Supplement, **2**: 34. — Umgebung von Montpellier, Südfrankreich.

ssp. **frigidochsenheimeri** Verity, 1930, Mem. Soc. ent. ital., **9**: 23. Reiss, 1933, in Seitz, Die Gross-Schmetterlinge der Erde, Supplement, **2**: 276. — Sappada, Alpi carniche, Italia, 1300-1400 m.
ab. **oblongamacula** Verity, 1930, Mem. Soc. ent. ital., **9**: 24. Reiss, 1933, in Seitz, Die Gross-Schmetterlinge der Erde, Supplement, **2**: 276.
ssp. **hyperfrigida** Rocci, 1937, Redia, **22**: 136. — Sappada, Alpi carniche, Italia, 1800-2000 m.

ssp. **veldenensis** Reiss, 1942, Ent. Z., **56**: 12. — Velden am Wörthersee, Österreich, 500 m.

ssp. **paulula** Verity, 1921, Ent. Rec., **33**: 89. Reiss, 1930, in Seitz, Die Gross-Schmetterlinge der Erde, Supplement, **2**: 34. — Stelvio Pass (Stilfser Joch), Alps.

ssp. **mannii** Herrich-Schäffer, 1852, Systematische Bearbeitung der Schmetterlinge von Europa, **6**: 44; 1851, ibidem, **2**, pl. 16, figs. 109, 110 (non-binominal). Spuler, 1906, in Hofmann, Die Schmetterlinge Europas, **2**: 160 (partim). Seitz, 1907, Die Gross-Schmetterlinge der Erde, **2**: 23, pl. 5f. Reiss, 1942, Ent. Z., **56**: 9. — Grossglockner, Kärnten, Österreich.

ab. **parvimaculata** Reiss, 1964, Coridon, (A) **6**: 10.

ab. **quinquemaculata** Burgeff, 1926, in Strand, Lepid. Cat., **33**: 53. Reiss, 1930, in Seitz, Die Gross-Schmetterlinge der Erde, Supplement, **2**: 33.

ab. **basimediounita** Reiss, 1964, Coridon, (A) **6**: 10.

ab. **cytisi** Reiss, 1942, Ent. Z., **56**: 10.

ssp. **thomanni** Reiss, 1950, Jber. naturf. Ges. Graubünden, **82**: 113, fig. 23. — Pontresina (Heutal), Ober Engadin, Schweiz, 2200-2400 m.

ab. **ornata** Reiss, 1950, Jber. naturf. Ges. Graubünden, **82**: 114, fig. 24.

ab. **stoechadina** Burgeff, 1926, Mitt. münch. ent. Ges., **16**: 66. Reiss, 1930, in Seitz, Die Gross-Schmetterlinge der Erde, Supplement, **2**: 33.

ab. **confluens** Reiss, 1950, Jber. naturf. Ges. Graubünden, **82**: 114.

ssp. **altarhaetica** Reiss, 1950, Jber. naturf. Ges. Graubünden, **82**: 114. — Bergün-Tuors, Albulatal, Schweiz, 1500 m.

ab. **pseudothomanni** Reiss, 1950, Jber. naturf. Ges. Graubünden, **82**: 115.

ab. **quinquemaculata** Vorbrodt, 1913, in Vorbrodt & Müller-Rutz, Die Schmetterlinge der Schweiz, **2**: 270. Reiss, 1930, in Seitz, Die Gross-Schmetterlinge der Erde, Supplement, **2**: 33.

ab. **herta** Reiss, 1950, Jber. naturf. Ges. Graubünden, **82**: 115.

ssp. **vitrea** Burgeff, 1926, Mitt. münch. ent. Ges., **16**: 66. Reiss, 1930, in Seitz, Die Gross-Schmetterlinge der Erde, Supplement, **2**: 34, pl. 3k. — Stalden, Saas-Fee, Walliser Alpen, Schweiz.

ab. **carnea** Lacreuze, 1945, Mitt. schweiz. ent. Ges., **19**: 255.

ssp. **wiegeli** Reiss, 1953, Z. wien. ent. Ges., **38**: 267, pl. 18, figs. A7, A8. — Col. Pradat, Ladinia, Ostdolomiten, Norditalien, 1700-2000 m.

ab. **parvimaculata** Reiss, 1953, Z. wien. ent. Ges., **38**: 267, pl. 18, fig. C8.

ab. **mediounita** Reiss, 1953, Z. wien. ent. Ges., **38**: 267, pl. 18, fig. B7.

ab. **cytisi** Reiss, 1953, Z. wien. ent. Ges., **38**: 267, pl. 18, fig. C7.

ab. **apiceconjuncta** Reiss, 1953, Z. wien. ent. Ges., **38**: 267, pl. 18, fig. B8.

ssp. **submanni** Reiss, 1942, Ent. Z., **56**: 10.

Gries am Brenner, Österreich, 1000-1200 m.

ssp. **subalpivolans** Reiss, 1942, Ent. Z., **56**: 8 (nomen novum for *subalpina* Reiss).
subalpina Reiss, 1925, Int. ent. Z., **19**: 230 (preoccupied by **subalpina** Calberla, 1895, ssp. of **romeo** Duponchel).
subalpina Reiss, 1926, Die Zygaenen Deutschlands, p. 21.

Oberaudorf, Bayr. Allgäu, Süddeutschland.

ssp. **richteri** Reiss, 1942, Ent. Z., **56**: 11.
ab. **parvimaculata** Reiss, 1964, Coridon, (A) **6**: 10.

St. Leonhard, Pitztal, Tirol, Österreich, 1400 m.

ssp. **kochelensis** Reiss, 1942, Ent. Z., **56**: 7; 1950, Jber. naturf. Ges. Graubünden, **82**: 116.
ab. **flava** Reiss, 1964, Coridon, (A) **6**: 10.
ab. **pallida** Reiss, 1964, Coridon, (A) **6**: 10.

Kochel, Bayr. Alpen, Süddeutschland, 700-800 m.

ssp. **pulchrior** Verity, 1921, Ent. Rec., **33**: 90. Reiss, 1930, in Seitz, Die Gross-Schmetterlinge der Erde, Supplement, **2**: 33. Holik, 1932, Iris, **46**: 123; 1939, Sborn. ent. Odd. nár. Mus. Praze, **17**: 43. Reiss, 1942, Ent. Z., **56**: 6. Forster & Wohlfahrt, 1958, Die Schmetterlinge Mitteleuropas, **3**: 97, pl. 10, figs. 34, 38, 39.
paupercula Verity, 1921, Ent. Rec., **33**: 90.
austriahungarica Reiss, 1922, Int. ent. Z., **16**: 77; 1926, Die Zygaenen Deutschlands, p. 21. Holik, 1929, Int. ent. Z., **23**: 2 (footnote 7).
ab. **cingulata** Tremewan, 1965, Ent. Rec., **77**: 89; 1966, ibidem, **78**: 35, pl. 1, fig. 13.
ab. **klosi** Skala, 1929, Ent. Z., **42**: 319.
ab. **holingeri** Povolný & Gregor, 1946, Folia ent., Brno, **9**, Supplement, 12: 35, pl. 2, fig. 7b.
ab. **latoconfluens** Kelecsényi, 1887, Ent. Z., **1**: 21.
confluens Dziurzyński, 1906, Ent. Z., **19**: 185 (nomen nudum). Reiss, 1926, Die Zygaenen Deutschlands, p. 22.
conflua Derenne, 1931, Lambillionea, Supplement, **31**: 158.
ab. **medioconfluens** Holik, 1943, Mitt. münch. ent. Ges., **33**: 333.
ab. **apicaliconfluens** Holik, 1943, Mitt. münch. ent. Ges., **33**: 333.

Klosterneuburg, Vienna, Austria.

f.t. **autumnalis** Holik, 1943, Mitt. münch. ent. Ges., **33**: 333.

Lovrin, nordwestlich von Temesvar, Transsylvanien (September).

ssp. **fatracola** Reiss, 1942, Ent. Z., **56**: 9. Povolný, 1945, Folia

Arvaer Magura, Fatra Gebirge, Böhmen, 450 m.

ent., Brno, **8**: 79. Gregor & Povolný, 1955, Sborn. ent. Odd. nár. Mus. Praze, **30**: 261, 272, pl. 5, fig. 4.

ssp. **lublinensis** Holik, 1939, Ann. Mus. zool. Polon., **12**: 82, pl. 3, figs. 91-95. — Tomaszów Lubelski, Polen.

ab. **analiconfluens** Holik, 1939, Ann. Mus. zool. Polon., **12**: 73, 78, pl. 3, fig. 90.

ab. **confluenta** Dąbrowski, 1965, Acta zool. cracov., **10**: 129, fig. 131e; pl. 9, fig. 2 (transitional).

ab. **sandeciensis** Klemensiewicz, 1912, Spraw. Kom. fizyogr. Krajow, **46**: 19.

apicaliconfluens Holik, 1939, Ann. Mus. zool. Polon., **12**: 82.

ssp. **germanica** Reiss, 1922, Int. ent. Z., **16**: 76. Esper, 1779, Die Schmetterlinge, **2**, pl. 16, figs. a-e; 1781, ibidem, **2**: 138. Hübner, 1796, Sammlung europäischer Schmetterlinge, **2**: 15, pl. 5, fig. 31; 1806, ibidem, Der Ziefer, p. 80. Spuler, 1906, in Hofmann, Die Schmetterlinge Europas, **2**: 160, pl. 77, fig. 19. Reiss, 1926, Die Zygaenen Deutschlands, p. 20, 21, pl. 2, fig.; 1930, in Seitz, Die Gross-Schmetterlinge der Erde, Supplement, **2**: 33, pl. 3i; 1937, in Schneider, Jh. Ver. vaterl. Naturk. Württemb., **93**: 127; 1942, Ent. Z., **56**: 6; 1949, Entomon, **1**: 172. Haaf, 1952, Veröff. zool. Staatssamml. Münch., **2**: 152, 154, 157, pl. 12 (as *filipendulae* Linné). Bergmann, 1953, Die Grossschmetterlinge Mitteldeutschlands, **3**: 46. Heuser & Jöst, 1959, Mitt. Pollichia, (3) **6**: 126. — Bad-Cannstatt, Umgebung von Stuttgart; Weilderstadt (Württemberg); Würzburg, Bayern; Süddeutschland und Mitteldeutschland.

ab. **pseudoalpina** Reiss, 1922, Int. ent. Z., **16**: 83; 1926, Die Zygaenen Deutschlands, p. 22; 1930, in Seitz, Die Gross-Schmetterlinge der Erde, Supplement, **2**: 33.

ab. **quinquemaculata** Reiss, 1964, Coridon, (A) **6**: 11.

ab. **basimaculata** Vorbrodt, 1913, in Vorbrodt & Müller-Rutz, Die Schmetterlinge der Schweiz, **2**: 270.

ab. **cytisi** Hübner, 1796, Sammlung europäischer Schmetterlinge, **2**: 15; 1796, ibidem, **2**, pl. 4, fig. 26 (as *cythisi*); 1806, ibidem, Der Ziefer, p. 81. Spuler, 1906, in Hofmann, Die Schmetterlinge Europas, **2**: 160, pl. 77, fig. 19a. Seitz, 1907, Die Gross-Schmetterlinge der Erde, **2**: 22. Reiss, 1926, Die Zygaenen Deutschlands, p. 22.

ab. **costalielongata** Reiss, 1964, Coridon, (A) **6**: 10.

ab. **trivittata** Reiss, 1964, Coridon, (A) **6**: 10, fig. 8.

ab. **rubescens** Ziegler, 1911, Int. ent. Z., **5**: 139.

ab. **polygalae** Esper, 1783, Die Schmetterlinge, **2**: 222, pl. 34, fig. 3. Meigen, 1830, Systematische Beschreibung der Europäischen Schmetterlinge, **2**: 77, pl. 57, fig. 9. Burgeff, 1914, Mitt. münch. ent. Ges., **5**: 42, 61, pl. 4, figs. 143, 144. Reiss, 1926, Die Zygaenen Deutschlands, p. 22; 1930, in Seitz, Die Gross-Schmetterlinge der Erde, Supplement, **2**: 33.

marginata Burgeff, 1906, Ent. Z., **20**: 162.

ab. **fereflava** Reiss, 1964, Coridon, (A) **6**: 11.	
ab. **flava** Reiss, 1964, Coridon, (A) **6**: 11. Schweitzer, 1913, Iris, **27**: 101, pl. 4, fig. 15.	
ab. **pallida** Reiss, 1922, Int. ent. Z., **16**: 83; 1926, Die Zygaenen Deutschlands, p. 22, pl. 1, fig.; 1930, in Seitz, Die Gross-Schmetterlinge der Erde, Supplement, **2**: 33, pl. 3i.	
ab. **brunneola** Reiss, 1964, Coridon, (A) **6**: 11.	
ssp. **fischeri** Le Charles, 1949, Rev. franç. Lépid., **12**: 178.	Nonnenbruch; Cernay; Mulhouse, Haut-Rhin, France.
ssp. **escheburgica** Reiss, 1942, Ent. Z., **56**: 4.	Bergedorf, Escheburg, Norddeutschland, 100 m.
ssp. **wieterensis** Reiss, 1942, Ent. Z., **56**: 7. ab. **brunneola** Reiss, 1942, Ent. Z., **56**: 7.	Wieter, Weserbergland, Deutschland, 250-300 m.
ssp. **pseudopulchrior** Reiss, 1942, Ent. Z., **56**: 6.	Maxau, Durlach in Baden (Berghausen und Michelsberg), Südwestdeutschland.
ssp. **calxensis** Le Charles, 1949, Rev. franç. Lépid., **12**: 179. Tremewan, 1961, Bull. Brit. Mus. (nat. Hist.) Ent., **10**(7): 309, pl. 57, fig. 27.	Peyreleau, Aveyron, France.
ssp. **armoricensis** Le Charles, 1949, Rev. franç. Lépid., **12**: 179.	Cancale et Saint-Lunaire, Ille-et-Vilaine, France.
ssp. **pulcherrima** Verity, 1921, Ent. Rec., **33**: 90. Reiss, 1930, in Seitz, Die Gross-Schmetterlinge der Erde, Supplement, **2**: 33, pl. 3k. Tremewan, 1963, Ent. Rec., **75**: 254. ab. **miltosa** Gelin & Lucas, 1912, Catalogue des Lépidoptères observés dans l'ouest de la France, (1): 217. ab. **totanigra** Le Charles, 1949, Rev. franç. Lépid., **12**: 179.	Dompierre-sur-Mer, Charente-Inférieure, France.
ssp. **breillati** Lucas, 1958, Bull. mens. Soc. linn. Lyon, **27**: 67. ab. **subconfluens** Pionneau, 1939, Échange, **55**: 27. ab. **confluens** École Bordelaise, 1933, Acta Soc. linn. Bordeaux, **85**: 139.	La Boucau, Bayonne, Basses-Pyrénées, France.
f.t. **aestivalis** Lucas, 1958, Bull. mens. Soc. linn. Lyon, **27**: 68.	La Boucau, Bayonne, Basses-Pyrénées, France.

ssp. **centrogalliae** Le Charles, 1949, Rev. franç. Lépid., **12**: 178. Dupont, 1900, Bull. Soc. Sci. nat. Elbeuf, **18**: 69. Gadeau de Kerville, 1928, Bull. Soc. ent. Fr., p. 71. Lecaillon, 1928, Bull. Soc. Hist. nat. Toulouse, **57**: 343.

Lardy; Saclas; Orgemont; Seine-et-Oise, France.

ab. **unitella** de Crombrugghe, 1911, Rev. Soc. ent. namur., **11**: 104.

ssp. **torgnica** Reiss, 1942, Ent. Z., **56**: 7. Derenne, 1934, Amat. Papillons, **7**: 150.

Torgny, Belgien.

jottrandi Dufrane, 1949, Lambillionea, **49**: 7.

ab. **communimacula** Selys-Longchamps, 1882, Bull. Soc. ent. Belg., p. cxiv. Seitz, 1907, Die Gross-Schmetterlinge der Erde, **2**: 23.

ab. **bipunctata** Selys-Longchamps, 1882, Bull. Soc. ent. Belg., p. cxiv. Seitz, 1907, Die Gross-Schmetterlinge der Erde, **2**: 23.

ab. **bimacula** Houyez, 1923, Rev. Soc. ent. namur., **23**: 58.

f. loc. **dormali** Lambillion, 1909, Rev. Soc. ent. namur., **9**: 71. Derenne, 1934, Amat. Papillons, **7**: 151.

Les coteaux arides de la citadelle de Namur, Belgique.

ab. **unita** Lambillion, 1909, Rev. Soc. ent. namur., **9**: 71.

ssp. **limmenica** Reiss, 1942, Ent. Z., **56**: 4. Lempke, 1961, Tijdschr. Ent., **104**: 159, fig. 27, pl. 12, figs. 1-8.

Dünen längs der Nordsee, Limmen, Holland.

ab. **albomaculata** Lempke, 1961, Tijdschr. Ent., **104**: 160.

ssp. **nederlandica** Reiss, 1942, Ent. Z., **56**: 5. Lempke, 1961, Tijdschr. Ent., **104**: 159, fig. 27, pl. 12, figs. 9-12.

Nunspeet, Holland.

ssp. **osterodica** Reiss, 1942, Ent. Z., **56**: 5.

Osterode, Ost-Preussen, 100-200 m.

ssp. **stettina** Burgeff, 1926, Mitt. münch. ent. Ges., **16**: 65 (nomen novum for *stettinensis* Reiss). Reiss, 1930, in Seitz, Die Gross-Schmetterlinge der Erde, Supplement, **2**: 33, pl. 3i; 1942, Ent. Z., **56**: 3.

Umgebung von Höckendorf, Stettin, Pommern, Norddeutschland.

stettinensis Reiss, 1922, Int. ent. Z., **16**: 76 (preoccupied by **stettinensis** Reiss, 1922, ssp. of **lonicerae** Scheven); 1926, Die Zygaenen Deutschlands, p. 20, 21, pl. 2, fig.

ab. **pseudotutti** Guhn, 1932, Ent. Jb., **41**: 93.

ab. **sarothamni** Guhn, 1932, Ent. Jb., **41**: 92.

ab. **purpuraloides** Guhn, 1932, Ent. Jb., **41**: 92.

ab. **flava** Reiss, 1964, Coridon, (A) **6**: 11.

ab. **chrysanthemi** Borkhausen, 1789, Naturgeschichte der Europäischen Schmetterlinge nach systematischer Ordnung, **2**: 166, pl. 1, fig. 1. Esper, 1789, Die Schmetterlinge, Supplement, **2**(2): 1, pl. 37, fig. 1. Borkhausen, 1793, Rheinisches Magazin, **1**: 647. Hübner, 1796, Sammlung europäischer Schmetterlinge, **2**: 16, pl. 3, fig. 17; 1806, ibidem, Der Ziefer,

p. 80. Seitz, 1907, Die Gross-Schmetterlinge der Erde, **2**: 22. Dziurzyński, 1908, Berl. ent. Z., **53**: 31, pl. 1, fig. 7. Reiss, 1930, in Seitz, Die Gross-Schmetterlinge der Erde, Supplement, **2**: 33. Alberti, 1955, Ent. Z., **65**: 89-91.

ssp. **anglicola** Tremewan, 1960, Ent. Gaz., **11**: 189. South, 1908, The Moths of the British Isles, **2**: 340, pl. 147, figs. 4,5. Ford, 1955, Moths, pl. 8, fig. 10. South, 1961, The Moths of the British Isles, **2**: 334, pl. 133, figs. 3,4. Tremewan, 1961, Bull. Brit. Mus. (nat. Hist.) Ent., **10**(7): 284, pl. 54, fig. 19; 1961, Coridon, (A) **1**: 3, pl. C1, fig. 10. Hyde, 1964, Animals, **3** (6): 162-164, fig. 1 (text fig. as *lonicerae* Scheven). Tremewan & Manley, 1964, Ent. Rec., **76**: 149-153. Baynes, 1964, A Revised Catalogue of Irish Macrolepidoptera (Butterflies and Moths), p. 88. — Tring, Hertfordshire, England (southern England to Scotland, Ireland).

ab. **minor** Tutt, 1899, A Natural History of the British Lepidoptera, **1**: 509. Tremewan, 1961, Bull. Brit. Mus. (nat. Hist.) Ent., **10**(7): 287, pl. 54, fig. 35.

ab. **spoliata** Cockayne, 1954, Ent. Rec., **66**: 68. Barrett, 1895, The Lepidoptera of the British Islands, **2**, pl. 60, fig. 1h. Tremewan, 1961, Bull. Brit. Mus. (nat. Hist.) Ent., **10** (7): 285, pl. 54, fig. 24.

ab. **nigrolimbata** Cockayne, 1954, Ent. Rec., **66**: 68, pl. 2, fig. 9. Tremewan, 1961, Bull. Brit. Mus. (nat. Hist.) Ent., **10** (7): 285, pl. 54, fig. 27.

ab. **trivittata** Tutt, 1899, A Natural History of the British Lepidoptera, **1**: 509. Seitz, 1907, Die Gross-Schmetterlinge der Erde, **2**: 23. Tremewan, 1961, Bull. Brit. Mus. (nat. Hist.) Ent., **10**(7): 286, pl. 54, fig. 29.

ab. **confluens** Oberthür, 1896, Études d'Entomologie, **20**: 45, pl. 8, fig. 132; 1910, Études de Lépidoptérologie comparée, **4**: 559. Tremewan, 1961, Bull. Brit. Mus. (nat. Hist.) Ent., **10**(7): 284, pl. 54, fig. 21.

ab. **proconfluens** Tutt, 1899, A Natural History of the British Lepidoptera, **1**: 512. Tremewan, 1961, Bull. Brit. Mus. (nat. Hist.) Ent., **10**(7): 284, pl. 54, fig. 22.

ab. **quinquejuncta** Tutt, 1899, A Natural History of the British Lepidoptera, **1**: 512. Barrett, 1895, The Lepidoptera of the British Islands, **2**: 134, pl. 60, fig. 1g. Tremewan, 1961, Bull. Brit. Mus. (nat. Hist.) Ent., **10**(7): 285, pl. 54, fig. 23.

ab. **conjuncta** Tutt, 1899, A Natural History of the British Lepidoptera, **1**: 510. Seitz, 1907, Die Gross-Schmetterlinge der Erde, **2**: 23. Tremewan, 1961, Bull. Brit. Mus. (nat. Hist.) Ent., **10**(7): 285; 1961, Coridon, (A)**1**: 3, pl. C1, fig. 11.

ab. **miniata** Tutt, 1899, A Natural History of the British Lepidoptera, **1**: 510. Tremewan, 1961, Bull. Brit. Mus. (nat. Hist.) Ent., **10**(7): 286.

ab. **grisescens** Oberthür, 1896, Études d'Entomologie, **20**: 45, pl. 8, fig. 135. Barrett, 1895, The Lepidoptera of the British Islands, **2**, pl. 60, fig. 1f. Seitz, 1912, Die Gross-Schmetterlinge der Erde, **2**: 442. Tremewan, 1961, Bull. Brit. Mus. (nat. Hist.) Ent., **10**(7): 284, pl. 54, fig. 20; 1961, Coridon, (A) **1**: 4, pl. C2, fig. 1.

ab. **intermedia** Tutt, 1899, A Natural History of the British Lepidoptera, **1**: 510. Barrett, 1895, The Lepidoptera of the British Islands, **2**: 134, pl. 60, fig. 1e. Seitz, 1912, Die Gross-Schmetterlinge der Erde, **2**: 442. Ford, 1955, Moths, pl. 8, fig. 11. Tremewan, 1961, Bull. Brit. Mus. (nat. Hist.) Ent., **10**(7): 286; 1961, Coridon, (A) **1**: 4, pl. C2, fig. 4 (as *aurantia* Tutt).

ab. **aurantia** Tutt, 1899, A Natural History of the British Lepidoptera, **1**: 510. Seitz, 1912, Die Gross-Schmetterlinge der Erde, **2**: 442. Tremewan, 1961, Bull. Brit. Mus. (nat. Hist.) Ent., **10**(7): 286, pl. 54, fig. 28; 1961, Coridon, (A) **1**: 4, pl. C2, fig. 3 (as *intermedia* Tutt).

ab. **flava** Robson, 1884, Young. Nat., **5**: 236. Barrett, 1895, The Lepidoptera of the British Islands, **2**: 134, pl. 60, figs. 1c, 1d. South, 1908, The Moths of the British Isles, **2**: 340, pl. 148, fig. 6. Reiss, 1926, Die Zygaenen Deutschlands, pl. 1, fig.; 1930, in Seitz, Die Gross-Schmetterlinge der Erde, Supplement, **2**: 33, pl. 3i. Ford, 1955, Moths, pl. 8, fig. 11. South, 1961, The Moths of the British Isles, **2**: 334, pl. 130, fig. 6. Tremewan, 1961, Bull. Brit. Mus. (nat. Hist.) Ent., **10**(7): 286, pl. 54, fig. 30; 1961, Coridon, (A) **1**: 4, pl. C2, fig. 2.

cerinus Robson, 1886, Young Nat., **7**: 192.

lutescens Cockerell, 1887, Entomologist, **20**: 151. Bairstow, 1877, Ent. mon. Mag., **14**: 67. Priest, 1879, Entomologist, **12**: 225.

citrinus Webb, 1891, Ent. Rec., **1**: 331.

flava Oberthür, 1896, Études d'Entomologie, **20**: 43, pl. 8, fig. 133. Tremewan, 1961, Bull. Brit. Mus. (nat. Hist.) Ent., **10**(7): 286, pl. 54, fig. 31.

ab. **griseorosea** Cockayne, 1954, Ent. Rec., **66**: 68. Tremewan, 1961, Bull. Brit. Mus. (nat. Hist.) Ent., **10**(7): 285, pl. 54, fig. 26.

ab. **brunnescens** Cockayne, 1940, Ent. Rec., **52**: 91. Tremewan, 1961, Bull. Brit. Mus. (nat. Hist.) Ent., **10**(7): 285, pl. 54, fig. 25.

f.t. **stephensi** Dupont, 1900, Bull. Soc. Sci. nat. Elbeuf, **28**: 77 (nomen novum for *hippocrepidis* Hübner sensu Stephens). Stephens, 1828, Illustrations of British Entomology, Haustellata, **1**: 109 (as *hippocrepidis* Hübner). Tremewan, 1958, Ent. Gaz., **9**: 193; 1960, ibidem, **11**: 189; 1961, Bull. Brit. Mus.

Coombe-wood; Darenth-wood, Kent [North Downs, Surrey, England].

(nat. Hist.) Ent., **10**(7): 287, pl. 54, fig. 34; 1961, Coridon, (A) **1**: 3, pl. C2, fig. 6.

tutti Rebel, 1901, in Staudinger & Rebel, Catalog der Lepidopteren des Palaearctischen Faunengebietes, p. 384. Spuler, 1906, in Hofmann, Die Schmetterlinge Europas, **2**: 160. Seitz, 1907, Die Gross-Schmetterlinge der Erde, **2**: 23.

ab. **lutescens** Tutt, 1899, A Natural History of the British Lepidoptera, **1**: 533. Stephens, 1828, Illustrations of British Entomology, Haustellata, **1**: 109. Wood, 1833, Index Entomologicus, p. 11, pl. 4, fig. 6a.

flava Burgeff, 1926, in Strand, Lepid. Cat., **33**:53.

f. loc. **degenerata** Tremewan, 1958, Ent. Gaz., **9**: 192. Tutt, 1899, A Natural History of the British Lepidoptera, **1**: 532 (partim). Tremewan, 1961, Bull. Brit. Mus. (nat. Hist.) Ent., **10**(7): 287, pl. 54, fig. 32. — Chattenden, Kent, England.

ab. **pallida** Tutt, 1899, A Natural History of the British Lepidoptera, **1**: 533. Tremewan, 1961, Bull. Brit. Mus. (nat. Hist.) Ent., **10**(7): 288, pl. 54, fig. 33.

ssp. (?ab.) **lismorica** Reiss, 1931, Int. ent. Z., **25**: 345; 1931, ibidem, **25**: 359, fig.; 1933, in Seitz, Die Gross-Schmetterlinge der Erde, Supplement, **2**: 276. Tremewan, 1958, Ent. Gaz., **9**: 190; 1960, ibidem, **11**: 190. — Insel Lismore, Schottland.

ssp. **filipendulae** Linné, 1758, Systema Naturae, ed. X, p. 494; 1746, Fauna Suecica, p. 156. Ochsenheimer, 1808, Die Schmetterlinge von Europa, **2**: 54. Boisduval, 1828, Essai sur une Monographie des Zygénides, p. 59, pl. 4, fig. 1. Nolcken, 1867, Lepidopterologische Fauna von Estland, Livland und Kurland, **1**: 99. Oberthür, 1910, Études de Lépidoptérologie comparée, **4**: 553. Reiss, 1922, Int. ent. Z., **16**: 76; 1926, Die Zygaenen Deutschlands, p. 19, 21; 1930, in Seitz, Die Gross-Schmetterlinge der Erde, Supplement, **2**: 33. Nordström & Wahlgren, 1941, Svenska Fjärilar, p. 327, pl. 46, fig. 7. Alberti, 1958, Mitt. zool. Mus. Berl., **34**: 325. — Suecia [Uppsala; Wisby; Slite; Thorsburg; Kristineborg; Stockholm].

aries Retzius, 1783, Genera et Species Insectorum, p. 35.

ssp. **arctica** Schneider, 1880, Tromsö Mus. Aarsh, **3**: 85. Aurivillius, 1889, Nordens Fjärilar, p. 53 (as *mannii* Herrich-Schäffer). Reiss, 1930, in Seitz, Die Gross-Schmetterlinge der Erde, Supplement, **2**: 33. — Bejern, 67° N, Lappland.

hybr. **angloitalica** Tutt, 1906, A Natural History of the British Lepidoptera, **5**: 37 (**filipendulae anglicola** Tremewan ♂ X **filipendulae maior** Esper ♀). Holik, 1933, Iris, **47**: 17.

hybr. **italoanglica** Tutt, 1906, A Natural History of the British Lepidoptera, **5**: 37 (**filipendulae maior** Esper ♂ X **filipendulae anglicola** Tremewan ♀). Holik, 1933, Iris, **47**: 17.

hybr. **bavarica** Burgeff, 1914, Mitt. münch. ent. Ges., **5**: 61 (**filipendulae kochelensis** Reiss ♂ X **filipendulae gigantea** Rocci

♀). Reiss, 1930, in Seitz, Die Gross-Schmetterlinge der Erde, Supplement, **2**: 34. Holik, 1933, Iris, **47**: 18.

hybr. **intermedia** Tutt, 1906, A Natural History of the British Lepidoptera, **5**: 36 (**filipendulae anglicola** Tremewan ♂ X **lonicerae transferens** Verity ♀). Holik, 1933, Iris, **47**: 16. Alberti, 1939, Ent. Z., **53**: 173 (E.R. 393), fig. 1 (**filipendulae** Linné X **lonicerae** Scheven). Cockayne & Darlow, 1941, Ent. Rec., **53**: 113-114, pl. 6, figs. 1-3 (**Z. filipendulae** Linné X **Z. lonicerae** Scheven).

Biology.

Barrett, 1895, The Lepidoptera of the British Islands, **2**: 133, pl. 60, figs. 1i, j. Blaschke, 1914, Die Raupen Europas mit ihren Futterpflanzen, pl. 6, fig. 15. Boisduval, Rambur & Graslin, 1832, Collection iconographique et historique des Chenilles, pl. 1, figs. 1-3, pl. 2, fig. 6. Buckler, 1886, The Larvae of the British Butterflies and Moths, **2**: 97, pl. 19, figs. 4, 4a-4e. Burgeff, 1910, Z. wiss. InsektBiol., **6**: 43, 97; 1912, ibidem, **8**: 184, 197, 198; 1921, Mitt. münch. ent. Ges., **11**: 50; 1950, Portug. acta biol., (A) Goldschmidt: 663-728; 1951, Biol. Zbl., **70**: 1-23; 1965, Nachr. Akad. Wiss. Göttingen, **2**, mat.-phys. Kl., no. 1: 7, figs. 2c, fig. 3 (**filipendulae germanica** Reiss ♂ X **trifolii barcelonensis** Reiss f. loc. **saleria** Burgeff ♀); 1965, ibidem, no. 14: 201, figs. Briggs, 1871, Trans. ent. Soc. Lond., p. 435. Carpenter, 1937, J. Soc. Brit. Ent., **1**: 176. Döring, 1955, Zur Morphologie der Schmetterlingseier, p. 119, pl. 18, fig. 258. Dorfmeister, 1854, Verh. zool.-bot. Ver. Wien, **4**: 479. Dujardin, 1965, Entomops, Nice, no. 2: 42. Esper, 1779, Die Schmetterlinge, **2**, pl. 16, figs. a-c; 1781, ibidem, **2**: 138. Ford, 1955, Moths, pl. 24, figs. 5, 6. Füge, 1907, Ent. Z., **20**: 302. Guhn, 1932, Ent. Jb., **41**: 92. Hafner, 1911, Ent. Z., **25**: 209. Holik, 1937, Lambillionea, **37**: 15-24, 32-45, 80-91; 1938, ibidem, **38**: 51-58, 79-88, 95-102; 1938, Ent. Rdsch., **55**: 320-323, 331-333; 1946, Rev. franç. Lépid., **10**: 250-261, 273-280; 1953, Ent. Z., **63**: 26. Hrubý, 1964, Prodromus Lepidopter Slovenska, p. 479. Hübner, 1806, Geschichte europäischer Schmetterlinge, pl. [75], figs. 1c, d. Jones, Parsons & Rothschild, 1962, Nature, **193**: 52-53. Kiefer, 1933, Int. ent. Z., **27**: 252-256; 1934, ibidem, **27**: 521-524. Koch, 1955, Wir Bestimmen Schmetterlinge, **2**: 60, 61, pl. 1, fig. 14, pl. 15, fig. 14. Lane, 1962, Ent. Gaz., **13**: 11, fig. A. Millière, 1869/74, Iconographie et Description de Chenilles et Lépidoptères, **3**: 60, pl. 107, figs. 1-6. Moore, 1892, Ent. Rec., **3**: 37. Oberthür, 1910, Études de Lépidoptérologie comparée, **4**: 538. Przibram, 1909, Experimental-Zoologie, **2**: 119, pl. 9, fig. 8. Ochsenheimer, 1808, Die Schmetterlinge von Europa, **2**: 54. Pazsiczky, 1917, Rovart. Lapok, **24**: 88 (**filipendulae** ♂ X **ephialtes** ♀). Reiss, 1926, Die Zygaenen Deutschlands, p. 7, 23; 1930, in Seitz, Die Gross-Schmetterlinge der Erde, Supplement, **2**: 36; 1958, Z. wien. ent. Ges., **43**: 161. Roüast, 1883, Catalogue des Chenilles européennes connues, p. 22. Sarlet, 1964, Mém. Soc. r. ent. Belg., **29**: 9, fig. 1b. Seitz, 1907, Die Gross-Schmetterlinge der Erde, **2**: 22, 23. Sepp, 1762, Nederlandsche Insecten, **1**: 89, pl. 22, figs. 1-9. Seppänen, 1954, Suomen Suurperhostoukkien Ravintokasvit, p. 257. South, 1961, The Moths of the British Isles, **2**, pl. 132, fig. 4. Spuler, 1910, in Hofmann, Die Raupen der Schmetterlinge Europas, pl. 10, fig. 1, pl. 50, fig. 23. Tremewan & Manley, 1964, Ent. Rec., **76**: 149-153. Tutt, 1899, A Natural History of the British Lepidoptera, **1**: 520, 534. Webb, 1896, Ent. Rec., **7**: 255. Zeller, 1847, Isis von oken, p. 303.

SECTION 10

trifolii Esper

Distribution: Algeria, Morocco, Portugal, Spain, central Europe, Great Britain, Sicily, Balkans.

ssp. **australis** Oberthür, 1910, Études de Lépidoptérologie comparée, **4**: 491. Holl, 1912, Bull. Soc. Hist. nat. Afr. N., **4**: 116. Reiss, 1930, in Seitz, Die Gross-Schmetterlinge der Erde, Supplement, **2**: 36. Przegendza, 1932, Ent. Z., **46**: 116, fig. 40 (as *syracusia* Zeller). Tremewan, 1961, Bull. Brit. Mus. (nat. Hist.) Ent., **10** (7): 298, pl. 56, fig. 12. — Lambèze, Algérie.

australis Heydenreich, 1851, Lepidopterorum Europaeorum Catalogus Methodicus, p. 21 (nomen nudum).

australis Lederer, 1852, Verh. zool.-bot. Ver. Wien, **2**: 71 (nomen nudum).

ab. **ruficostata** Holl, 1912, Bull. Soc. Hist. nat. Afr. N., **4**: 117. Reiss, 1930, in Seitz, Die Gross-Schmetterlinge der Erde, Supplement, **2**: 36. Tremewan, 1961, Bull. Brit. Mus. (nat. Hist.) Ent., **10**(7): 299, pl. 56, fig. 14.

ab. **pseudocaerulescens** Burgeff, 1914, Mitt. münch. ent. Ges., **5**: 62. Reiss, 1930, in Seitz, Die Gross-Schmetterlinge der Erde, Supplement, **2**: 36.

ab. **aurorina** Oberthür, 1910, Études de Lépidoptérologie comparée, **4**: 493. Tremewan, 1961, Bull. Brit. Mus. (nat. Hist.) Ent., **10**(7): 298, pl. 56, fig. 13.

ssp. **magnaustralis** Verity, 1926, Ent. Rec., **38**: 23; 1925, ibidem, **37**: 117, pl. 8, figs. 20, 21. Reiss, 1930, in Seitz, Die Gross-Schmetterlinge der Erde, Supplement, **2**: 37. — Khenchela, Algeria.

ssp. **seriziati** Oberthür, 1876, Études d'Entomologie, **1**: 33; — Collo, Algérie.

1878, ibidem, **3**: 41, pl. 5, fig. 7; 1888, ibidem, **12**: 26; 1890, ibidem, **13**: 21, pl. 8, figs. 71-73. Seitz, 1907, Die Gross-Schmetterlinge der Erde, **2**: 21, pl. 4k. Oberthür, 1910, Études de Lépidoptérologie comparée, **4**: 506. Verity, 1925, Ent. Rec., **37**: 117, pl. 8, fig. 10. Reiss, 1944, Z. wien. ent. Ges., **29**: 68, pl. 37, figs. 60, 61. Tremewan, 1961, Bull. Brit. Mus. (nat. Hist.) Ent., **10** (7): 299, pl. 56, fig. 15.

ab. **rusicadica** Stauder, 1914, Z. wiss. InsektBiol., **10**: 173, pl. 1, fig. 14. Reiss, 1930, in Seitz, Die Gross-Schmetterlinge der Erde, Supplement, **2**: 37.

ab. **rubra** Dziurzyński, 1910, Int. ent. Z., **4**: 194.

ab. **confluens** Dziurzyński, 1910, Int. ent. Z., **4**: 194.

ab. **nigra** Dziurzyński, 1906, Ent. Z., **19**: 185. Seitz, 1907, Die Gross-Schmetterlinge der Erde, **2**: 21, pl. 4k.

ab. **pseudoaustralis** Burgeff, 1914, Mitt. münch. ent. Ges., **5**: 62. Reiss, 1930, in Seitz, Die Gross-Schmetterlinge der Erde, Supplement, **2**: 37.

ssp. **lucida** Reiss, 1944, Z. wien. ent. Ges., **29**: 71, pl. 37, figs. 62, 63. Rothschild, 1933, Novit. zool., **38**: 324 (as *seriziati* Oberthür). — Ketama, Rif, Marokko, 1500 m.

ab. **pseudoseriziati** Reiss, 1944, Z. wien. ent. Ges., **29**: 71.

ab. **pseudoaustralis** Reiss, 1944, Z. wien. ent. Ges., **29**: 71.

ab. **rusicadica** Reiss, 1944, Z. wien. ent. Ges., **29**: 71.

ab. **helia** Marten, 1944, Z. wien. ent. Ges., **29**: 198.

ab. **corax** Reiss, 1944, Z. wien. ent. Ges., **29**: 71.

ssp. **diffusemarginata** Rothschild, 1933, Novit. Zool., **38**: 324. Reisser, 1933, Eos, Madr., **9**: 283 (as *seriziati* Oberthür). Reiss, 1941, Z. wien. EntVer., **26**: 290, pl. 31, figs. 13, 14; 1944, Z. wien. ent. Ges., **29**: 69. Tremewan, 1961, Bull. Brit. Mus. (nat. Hist.) Ent., **10** (7): 299, pl. 56, fig. 16. — Hauta Kasdir, Rif Mts., Morocco, 1750 m.

ab. **tenuelimbata** Reiss, 1941, Z. wien. EntVer., **26**: 291, pl. 31, fig. 15.

ab. **pseudoseriziati** Reiss, 1944, Z. wien. ent. Ges., **29**: 70.

ab. **rusicadica** Reiss, 1944, Z. wien. ent. Ges., **29**: 70.

ab. **nigra** Reiss, 1944, Z. wien. ent. Ges., **29**: 70.

ssp. **caerulescens** Burgeff, 1914, Mitt. münch. ent. Ges., **5**: 62. Rambur, 1866, Catalogue systématique des Lépidoptères de l'Andalousie, p. 134, pl. 1, fig. 5. Reiss, 1930, in Seitz, Die Gross-Schmetterlinge der Erde, Supplement, **2**: 37, pl. 4a; 1936, Ent. Rdsch., **54**: 90, pl. 2, figs. Tremewan, 1961, Ent. Rec., **73**: 7. Reiss & Tremewan, 1964, Ent. Rec., **76**: 135. Schmidt-Koehl, 1965, Ent. Z., **75**: 281 (as *australis* Oberthür). — Sierra de Alfacar, Granada, Südspanien.

caerulescens Oberthür, 1910, Études de Lépidoptérologie comparée, **4**: 493 (infrasubspecific). Tremewan, 1961, Bull. Brit. Mus. (nat. Hist.) Ent., **10** (7): 298, pl. 56, fig. 11. Reiss & Tremewan, 1964, Ent. Rec., **76**: 135.

ab. **pseudoaustralis** Reiss, 1929, Int. ent. Z., **22**: 357; 1930, in Seitz, Die Gross-Schmetterlinge der Erde, Supplement, **2**: 37.	
ssp. **tenuelimbata** Romei, 1927, Ent. Rec., **39**: 108. Reiss, 1930, in Seitz, Die Gross-Schmetterlinge der Erde, Supplement, **2**: 37 (as forma). *augustiniana* Fernández, 1929, Mem. Soc. esp. Hist. nat., **15**: 600, figs. 12, 13. Agenjo, 1948, Eos, Madr., **24**: 399.	Jerez del Marquesada, Sierra Nevada, Spain, 3600 ft.
ssp. **altivolans** Reiss, 1936, Ent. Rdsch., **54**: 90, pl. 2, fig.	Pt. del Lobo, Sierra Nevada, Südspanien, 1800-2000 m.
ssp. **espunnica** Reiss, 1936, Ent. Rdsch., **54**: 90, pl. 2, figs.	Sierra de Espuña, Murcia, Südostspanien.
ssp. **lusitaniaemixta** Verity, 1930, Mem. Soc. ent. ital., **9**: 25. Reiss, 1933, in Seitz, Die Gross-Schmetterlinge der Erde, Supplement, **2**: 277 (as *lonicerae* Scheven ssp.); 1936, Ent. Rdsch., **54**: 90.	Serra da Estrela, Portugal, 800-1500 m.
ab. **omniconfluens** Monteiro, 1956, Brotéria, **25**: 50.	
ssp. **hibera** Verity, 1926, Ent. Rec., **38**: 9; 1925, ibidem, **37**: 117, pl. 8, figs. 1, 2. Reiss, 1930, in Seitz, Die Gross-Schmetterlinge der Erde, Supplement, **2**: 36, pl. 4a.	Oviedo, Prov. Oviedo, Asturia, Spain.
ab. **hiberuncula** Verity, 1926, Ent. Rec., **38**: 10; 1925, ibidem, **37**: 117, pl. 8, figs. 3-5. Reiss, 1930, in Seitz, Die Gross-Schmetterlinge der Erde, Supplement, **2**: 36.	
ssp. **pajini** Tremewan, 1963, Ent. Rec., **75**: 8, pl. 1, figs. 18, 19; 1963, ibidem, **75**: 254. Tremewan & Manley, 1965, Ent. Rec., **77**: 11.	Riano, Leon, 3500 ft.; Puerto de Pandatrave, Leon, 5000 ft.; Puerto de San Glorio, Santander, 5100 ft.; Spain.
ssp. **guadarramica** Reiss, 1936, Ent. Rdsch., **54**: 89, pl. 2, figs. Tremewan, 1963, Ent. Rec., **75**: 254.	El Escorial, Kastilien, Spanien.
ssp. **laincalvo** Agenjo, 1948, Eos, Madr., **24**: 397. Tremewan, 1963, Ent. Rec., **75**: 254. Tremewan & Manley, 1965, Ent. Rec., **77**: 11.	Estépar, Burgos, España, 810 m.
ssp. **noguerensis** Reiss, 1936, Ent. Rdsch., **54**: 89, pl. 2, fig. ab. **rosella** Verity, 1926, Ent. Rec., **38**: 24.	Sierra Alta de Albarracin, Teruel, Spanien, 1750 m.
ssp. **barcelonensis** Reiss, 1922, Int. ent. Z., **15**: 175. Verity, 1925,	Küste von Barcelona, Spanien.

Ent. Rec., **37**: 117, pl. 8, fig. 22 (as *syracusia* Zeller). Reiss, 1929, Int. ent. Z., **22**: 356; 1930, in Seitz, Die Gross-Schmetterlinge der Erde, Supplement, **2**: 36.

f. t. **intricata** de Sagarra, 1925, Butll. Inst. catal. Hist. nat., **25**: 272. Reiss, 1930, in Seitz, Die Gross-Schmetterlinge der Erde, Supplement, **2**: 36, pl. 3o. — Llobregat, Catalonia, España.

intricata Verity, 1926, Ent. Rec., **38**: 23.

ab. **depravata** de Sagarra, 1925, Butll. Inst. catal. Hist. nat., **25**: 273. Reiss, 1930, in Seitz, Die Gross-Schmetterlinge der Erde, Supplement, **2**: 36, pl. 3o.

clorinda Bethune-Baker, 1926, Ent. Rec., **38**: 84. Reiss, 1929, Int. ent. Z., **22**: 357. Grosvenor & Hewer, 1931, Ent. Rec., **43**: 23. Tremewan, 1961, Bull. Brit. Mus. (nat. Hist.) Ent., **10** (7): 298, pl. 56, fig. 17.

f. loc. **saleria** Burgeff, 1965, Nachr. Akad. Wiss. Göttingen, **2**, math.-phys. Kl., no. 1: 7, figs. 2a, 2b, fig. 3. — El Saler, Litoral von Valencia, Spanien.

ssp. **duponcheliana** Oberthür, 1910, Études de Lépidoptérologie comparée, **4**: 495, 663; 1896, Études d'Entomologie, **20**: 46, pl. 8, fig. 150. Foulquier, 1918, in Oberthür, Études de Lépidoptérologie comparée, **16**: 264. Reiss, 1930, in Seitz, Die Gross-Schmetterlinge der Erde, Supplement, **2**: 36. Tremewan, 1961, Bull. Brit. Mus. (nat. Hist.) Ent., **10** (7): 297, pl. 56, fig. 7. — Vernet-les-Bains, Pyrénées-Orientales, France.

ssp. **dumezi** Lucas, 1958, Bull. mens. Soc. linn. Lyon, **27**: 68. Verity, 1925, Ent. Rec., **37**: 117, pl. 8, figs. 42, 43 (as *duponcheliana* Oberthür). — Bords de l'Hérault, St. Guilhem le Désert, France.

f. t. **duponcheliella** Verity, 1926, Ent. Rec., **38**: 59; 1925, ibidem, **37**: 117, pl. 8, figs. 44, 45. Reiss, 1930, in Seitz, Die Gross-Schmetterlinge der Erde, Supplement, **2**: 36. — Rognac near Marseilles, France.

aestivalis Lucas, 1958, Bull. mens. Soc. linn. Lyon, **27**: 68.

ssp. **olbiana** Oberthür, 1910, Études de Lépidoptérologie comparée, **4**: 496; 1909, ibidem, **3**, pl. 28, fig. 168. Seitz, 1912, Die Gross-Schmetterlinge der Erde, **2**: 442. Verity, 1925, Ent. Rec., **37**: 117, pl. 8, figs. 46, 47. Reiss, 1930, in Seitz, Die Gross-Schmetterlinge der Erde, Supplement, **2**: 36, pl. 3o. Tremewan, 1961, Bull. Brit. Mus. (nat. Hist.) Ent., **10** (7): 297, pl. 56, fig. 6. — Le Ceinturon, Hyères, Var, France méridionale.

ssp. **syracusia** Zeller, 1847, Isis von oken, p. 301. Freyer, 1852, Neuere Beiträge zur Schmetterlingskunde, **6**: 39, pl. 506, figs. 3, 4. Herrich-Schäffer, 1852, Systematische Bearbeitung der Schmetterlinge von Europa, **6**: 44. Spuler, 1906, in Hofmann, Die Schmetterlinge Europas, **2**: 159, pl. 77, fig. 16b. Seitz, — Syracus, Sizilien.

1907, Die Gross-Schmetterlinge der Erde, **2**: 21, pl. 4k. Oberthür, 1910, Études de Lépidoptérologie comparée, **4**: 494. Reiss, 1930, in Seitz, Die Gross-Schmetterlinge der Erde, Supplement, **2**: 36. Tremewan, 1961, Bull. Brit. Mus. (nat. Hist.) Ent., **10** (7): 297, pl. 56, fig. 8.

paraustralis Verity, 1926, Ent. Rec., **38**: 11, 24; 1925, ibidem, **37**: 117, pl. 8, figs. 18, 19. Reiss, 1930, in Seitz, Die Gross-Schmetterlinge der Erde, Supplement, **2**: 36.

siciliae Verity, 1926, Ent. Rec., **38**: 22; 1925, ibidem, **37**: 117, pl. 8, figs. 13, 14.

ab. **incarnata** Turati, 1908, Nat. sicil., **20**: 15, pl. 1, fig. 17. Seitz, 1912, Die Gross-Schmetterlinge der Erde, **2**: 442.

f. t. **secundogenita** Verity, 1926, Ent. Rec., **38**: 23; 1925, ibidem, **37**: 117, pl. 8, fig. 23. Reiss, 1930, in Seitz, Die Gross-Schmetterlinge der Erde, Supplement, **2**: 36. Catania, Plaja, Sicily.

ssp. **trinacria** Verity, 1917, Bull. Soc. ent. Fr., p. 224; 1925, Ent. Rec., **37**: 117, pl. 8, figs. 11, 12. Reiss, 1930, in Seitz, Die Gross-Schmetterlinge der Erde, Supplement, **2**: 36 (as ab.). Tremewan, 1961, Bull. Brit. Mus. (nat. Hist.) Ent., **10** (7): 297, pl. 56, fig. 9. Lupo, Palermo, Sicily.

ab. **punctonotata** Verity, 1926, Ent. Rec., **38**: 12. Tremewan, 1966, Ent. Rec., **78**: 35, pl. 1, fig. 14.

ab. **kruegeri** Ragusa, 1924, Boll. Lab. Zool. Portici, **18**: 88 (as *krügeri*). Reiss, 1930, in Seitz, Die Gross-Schmetterlinge der Erde, Supplement, **2**: 36 (as *krügeri*). Tremewan, 1961, Bull. Brit. Mus. (nat. Hist.) Ent., **10** (7): 297, pl. 56, fig. 10.

ssp. **trifolii** Esper, 1783, Die Schmetterlinge, **2**: 223, pl. 34, figs. 4, 5. Hübner, [1796]-[24th December 1799], Sammlung europäischer Schmetterlinge, **2**, pl. 17, fig. 79; 1806, ibidem, Der Ziefer, p. 80; [1808]-[20th June 1813], ibidem, **2**, pl. 29, figs. 134, 135. Ochsenheimer, 1808, Die Schmetterlinge von Europa, **2**: 47. Boisduval, 1828, Essai sur une Monographie des Zygénides, p. 54, pl. 3, fig. 7; 1834, Icones historique des Lépidoptères nouveaux ou peu connus, **2**: 59, pl. 54, fig. 8. Freyer, 1839, Neuere Beiträge zur Schmetterlingskunde, **3**: 14, pl. 200, fig. 4. Herrich-Schäffer, 1846, Systematische Bearbeitung der Schmetterlinge von Europa, **2**: 37. Spuler, 1906, in Hofmann, Die Schmetterlinge Europas, **2**: 158, pl. 75, fig. 51a, pl. 77, fig. 16. Seitz, 1907, Die Gross-Schmetterlinge der Erde, **2**: 21, pl. 4i. Oberthür, 1910, Études de Lépidoptérologie comparée, **4**: 489. Reiss, 1926, Die Zygaenen Deutschlands, p. 14, pl. 2, fig.; 1930, in Seitz, Die Gross-Schmetterlinge der Erde, Supplement, **2**: 36. Derenne, 1934, Amat. Papillons, **7**: 153. Reiss, 1937, in Schneider, Jh. Ver. vaterl. Naturk. Württemb., **93**: 128; 1949, Entomon, **1**: 173. Haaf, 1952, Veröff. zool. Staatssamml. Münch., **2**: 152, 154, Bei Frankfurt am Main, Deutschland.

157, pl. 12. Bergmann, 1953, Die Grossschmetterlinge Mitteldeutschlands, **3**: 48. Gregor & Povolný, 1955, Sborn. ent. Odd. nár. Mus. Praze, **30**: 262, 272, pl. 5, fig. 2. Forster & Wohlfahrt, 1958, Die Schmetterlinge Mitteleuropas, **3**: 98, pl. 11, figs. 1, 6, 11. Alberti, 1958, Mitt. zool. Mus. Berl., **34**: 326. Heuser & Jöst, 1959, Mitt. Pollichia, (3) **6**: 126. Weber, 1964, Mitt. ent. Ges. Basel, (N.F.), **14**: 41.

pratorum de Villers, 1789, Caroli Linnaei Entomologia, **2**: 114. Ernst, 1782, Papillons d'Europe, **3**: 51, pl. 97, figs. 136a-e.

ruficincta Tutt, 1908, Ent. Rec., **20**: 248. Tremewan, 1965, Ent. Rec., **77**: 88; 1966, ibidem, **78**: 35, pl. 1, fig. 15.

medioconfluens Vorbrodt, 1913, in Vorbrodt & Müller-Rutz, Die Schmetterlinge der Schweiz, **2**: 262.

ab. **sexmaculata** Vorbrodt, 1913, in Vorbrodt & Müller-Rutz, Die Schmetterlinge der Schweiz, **2**: 263.

sexmaculata Burgeff, 1914, Mitt. münch. ent. Ges., **5**: 62. Reiss, 1930, in Seitz, Die Gross-Schmetterlinge der Erde, Supplement, **2**: 36.

ab. **orobi** Hübner, [1808]-[20th June 1813], Sammlung europäischer Schmetterlinge, **2**, pl. 29, fig. 133. Seitz, 1907, Die Gross-Schmetterlinge der Erde, **2**: 21, pl. 4i. Reiss, 1926, Die Zygaenen Deutschlands, p. 15.

ab. **pauperrima** Vorbrodt, 1913, in Vorbrodt & Müller-Rutz, Die Schmetterlinge der Schweiz, **2**: 263, fig. 43. Reiss, 1930, in Seitz, Die Gross-Schmetterlinge der Erde, Supplement, **2**: 36.

ab. **basimaculata** Vorbrodt, 1913, in Vorbrodt & Müller-Rutz, Die Schmetterlinge der Schweiz, **2**: 262, fig. 21.

ab. **basimedioconfluens** Vorbrodt, 1913, in Vorbrodt & Müller-Rutz, Die Schmetterlinge der Schweiz, **2**: 262.

tripuncta Guhn, 1932, Ent. Jb., **41**: 91. Reiss, 1933, in Seitz, Die Gross-Schmetterlinge der Erde, Supplement, **2**: 277.

ab. **doris** Meigen, 1830, Systematische Beschreibung der Europäischen Schmetterlinge, **2**: 84, pl. 58, fig. 4.

basalis Selys-Longchamps, 1872, Ann. Soc. ent. Belg., **15**: lix. Seitz, 1907, Die Gross-Schmetterlinge der Erde, **2**: 21.

ab. **glycirrhizae** Hübner, [13th March 1814]-[31st December 1817], Sammlung europäischer Schmetterlinge, **2**, pl. 30, fig. 138. Freyer, 1836, Neuere Beiträge zur Schmetterlingskunde, **2**: 116, pl. 164, fig. 3. Seitz, 1907, Die Gross-Schmetterlinge der Erde, **2**: 21. Dziurzyński, 1908, Berl. ent. Z., **53**: 24, pl. 1, fig. 3. Reiss, 1926, Die Zygaenen Deutschlands, p. 15.

apicalimaculata Vorbrodt, 1913, in Vorbrodt & Müller-Rutz, Die Schmetterlinge der Schweiz, **2**: 262, fig. 22.

ab. **trivittata** Speyer, 1877, Stettin. ent. Ztg., **38**: 42. Seitz,

1907, Die Gross-Schmetterlinge der Erde, **2**: 21. Reiss, 1926, Die Zygaenen Deutschlands, p. 15.

ab. **minoides** Selys-Longchamps, 1837, Catalogue des Lépidoptères ou Papillons de la Belgique, p. 23. Spuler, 1906, in Hofmann, Die Schmetterlinge Europas, **2**: 158, pl. 75, fig. 51b, pl. 77, fig. 16a. Seitz, 1907, Die Gross-Schmetterlinge der Erde, **2**: 21, pl. 4i. Reiss, 1926, Die Zygaenen Deutschlands, p. 15.

confluens Staudinger, 1871, in Staudinger & Wocke, Catalog der Lepidopteren des Palaearctischen Faunengebiets, p. 47. Bergmann, 1953, Die Grossschmetterlinge Mitteldeutschlands, **3**, pl. 67, figs. A2, A3.

omniconfluens Vorbrodt, 1913, in Vorbrodt & Müller-Rutz, Die Schmetterlinge der Schweiz, **2**: 263, figs. 23, 26.

ab. **rubescens** Burgeff, 1906, Ent. Z., **20**: 162, fig. 5. Reiss, 1930, in Seitz, Die Gross-Schmetterlinge der Erde, Supplement, **2**: 36. Bergmann, 1953, Die Grossschmetterlinge Mitteldeutschlands, **3**, pl. 67, fig. A4.

ab. **marginata** Burgeff, 1906, Ent. Z., **20**: 162, fig. 7.

ab. **pallens** Vorbrodt, 1921, Mitt. schweiz. ent. Ges., **13**: 204.

ab. **carnea** Dziurzyński, 1908, Berl. ent. Z., **53**: 6, 13.

flavescens Ziegler, 1911, Int. ent. Z., **5**: 139.

incarnata Vorbrodt, 1921, Mitt. schweiz. ent. Ges., **13**: 204.

incarnata Guhn, 1932, Ent. Jb., **41**: 91.

ab. **flava** Reiss, 1964, Coridon, (A) **6**: 11.

ab. **orobiflava** Reiss, 1964, Coridon, (A) **6**: 11.

ab. **minoidesflava** Reiss, 1964, Coridon, (A) **6**: 11.

ab. **albomaculata** Dziurzyński, 1910, Int. ent. Z., **4**: 194.

ab. **candida** Burgeff, 1914, Mitt. münch. ent. Ges., **5**: 61, pl. 2, fig. 182. Reiss, 1926, Die Zygaenen Deutschlands, p. 15; 1930, in Seitz, Die Gross-Schmetterlinge der Erde, Supplement, **2**: 36.

albomaculata Locher, 1916, Ent. Z., **30**: 76.

albomaculata Locher, 1919, Ent. Z., **33**: 31.

ab. **wagneri** Dziurzyński, 1909, Jber. wien. ent. Ver., **19**: 135, pl. 1, fig. 5; 1910, Int. ent. Z., **4**: 194. Alberti, 1955, Ent. Z., **65**: 89-91.

minoidesbrunnea Dziurzyński, 1909, Berl. ent. Z., **53**: 250.

f. t. **gracilis** Fuchs, 1880, Stettin. ent. Ztg., **41**: 118. Spuler, 1906, in Hofmann, Die Schmetterlinge Europas, **2**: 159. Seitz, 1907, Die Gross-Schmetterlinge der Erde, **2**: 21. Burgeff, 1914, Mitt. münch. ent. Ges., **5**: 62. Reiss, 1926, Die Zygaenen Deutschlands, p. 14; 1930, in Seitz, Die Gross-Schmetterlinge der Erde, Supplement, **2**: 36. Heuser & Jöst, 1959, Mitt. Pollichia, (3) **6**: 127.

Bornich, Mittleres Rheintal, Westdeutschland.

ssp. **debilia** Holik, 1939, Ann. Mus. zool. Polon., **12**: 90.

Jastarnia, Halbinsel Hel, Polen.

ssp. **orientalis** Hormuzaki, 1902, Soc. ent., **17**: 139. Spuler, 1906, in Hofmann, Die Schmetterlinge Europas, **2**: 159. Seitz, 1907, Die Gross-Schmetterlinge der Erde, **2**: 21. Holik & Sheljuzhko, 1958, Mitt. münch. ent. Ges., **48**: 225. — Alpines Plateaux der Lutschina, Bukowina.

ssp. **abnobae** Koch, 1941, Mitt. münch. ent. Ges., **31**: 555. — Gutach im Mittelschwarzwald, Westdeutschland.

ssp. **vogesiaca** Przegendza, 1932, Ent. Z., **46**: 116, figs. 33-36. Reiss, 1933, in Seitz, Die Gross-Schmetterlinge der Erde, Supplement, **2**: 277. — Le Mont; Lubine; Vogesen, Frankreich.

ssp. **muspratti** Tremewan, 1961, Ent. Rec., **73**: 199; 1962, ibidem, **74**: 128, pl. 2, fig. 13. — Le Lac, St. Jean-de-Luz, Basses-Pyrénées, France.

f. loc. **pusilla** Oberthür, 1910, Études de Lépidoptérologie comparée, **4**: 498. Tremewan, 1962, Ent. Rec., **74**: 129, pl. 2, fig. 14. — Auch, près Lectoure, Gers, France.

ssp. **aquitania** Le Charles, 1946, Bull. Soc. ent. Fr., **51**: 82, pl. 2, figs. — Vallée du Dropt, Mesterrieux, Gironde, France.

ab. **fremonti** Frémont, 1930, P.V. Soc. linn. Bordeaux, **81**: 132.

f. t. **estivalis** Le Charles, 1930/35, in Lhomme, Catalogue des Lépidoptères de France et de Belgique, **1**: 693. — Vallée du Dropt, Mesterrieux, Gironde, France.

aestivalis École Bordelaise, 1933, Act. Soc. linn. Bordeaux, **85**: 139.

ssp. **palustris** Oberthür, 1896, Études d'Entomologie, **20**: 45. Seitz, 1907, Die Gross-Schmetterlinge der Erde, **2**: 21. Oberthür, 1909, Études de Lépidoptérologie comparée, 3, pl. 28, fig. 165; 1910, ibidem, **4**: 500. Verity, 1925, Ent. Rec., **37**: 117, pl. 8, fig. 6. Reiss, 1930, in Seitz, Die Gross-Schmetterlinge der Erde, Supplement, **2**: 36. Tremewan, 1958, Ent. Gaz., **9**: 185; 1961, Bull. Brit. Mus. (nat. Hist.) Ent., **10** (7): 295, pl. 55, fig. 27, pl. 56, fig. 1. — Environs de Rennes, Ille-et-Vilaine, France.

ab. **sexmaculata** Oberthür, 1896, Études d'Entomologie, **20**: 48, pl. 8, fig. 151; 1910, Études de Lépidoptérologie comparée, **4**: 500, 503. Seitz, 1912, Die Gross-Schmetterlinge der Erde, **2**: 442. Tremewan, 1961, Bull. Brit. Mus. (nat. Hist.) Ent., **10** (7): 296, pl. 56, fig. 2.

ab. **confluens** Oberthür, 1896, Études d'Entomologie, **20**: 45, pl. 8, fig. 153; 1909, Études de Lépidoptérologie comparée, **3**, pl. 28, figs. 166, 167; 1910, ibidem, **4**: 503. Verity, 1925, Ent. Rec., **37**: 117, pl. 8, fig. 7. Reiss, 1930, in Seitz, Die Gross-Schmetterlinge der Erde, Supplement, **2**: 36. Tremewan, 1961, Bull. Brit. Mus. (nat. Hist.) Ent., **10** (7): 296, pl. 56, fig. 3.

ab. **confluenssexmaculata** Oberthür, 1896, Études d'Entomologie, **20**: 48, pl. 8, fig. 152; 1910, Études de Lépidoptérologie comparée, **4**: 500. Tremewan, 1961, Bull. Brit. Mus. (nat. Hist.) Ent., **10** (7): 296, pl. 56, fig. 4.

ab. **nigricans** Oberthür, 1907, Bull. Soc. ent. Fr., p. 220; 1910, Études de Lépidoptérologie comparée, **4**: 504. Seitz, 1912, Die Gross-Schmetterlinge der Erde, **2**: 442. Tremewan, 1961, Bull. Brit. Mus. (nat. Hist.) Ent., **10** (7): 296, pl. 56, fig. 5.

ssp. **carueli** Le Charles, 1960, Bull. Soc. ent. Fr., **65**: 103.

Fleury-la-Rivière, Marne, France.

ssp. **subsyracusia** Verity, 1926, Ent. Rec., **38**: 25. Reiss, 1930, in Seitz, Die Gross-Schmetterlinge der Erde, Supplement, **2**: 36.

Plouharnel, Brittany, France; Channel Islands.

ssp. **decreta** Verity, 1926, Ent. Rec., **38**: 57. Tutt, 1899, A Natural History of the British Lepidoptera, **1**: 480, 499 (as *palustris* Oberthür [partim]). South, 1908, The Moths of the British Isles, **2**: 337, pl. 146, figs. 7, 8. Verity, 1925, Ent. Rec., **37**: 117, pl. 8, figs. 34-36. Reiss, 1930, in Seitz, Die Gross-Schmetterlinge der Erde, Supplement, **2**: 36. Ford, 1955, Moths, p. 126, pl. 29, fig. 5. Tremewan, 1960, Ent. Gaz., **11**: 192; 1961, Coridon, (A) **1**: 4, pl. C3, figs. 1, 2. South, 1961, The Moths of the British Isles, **2**: 331, 332, pl. 129, figs. 7, 8.

Chailey Marsh, Sussex, England [Marshes in England and Wales, end of June to early August].

major Tutt, 1897, Ent. Rec., **9**: 88 (preoccupied by **maior** Esper, 1797, ssp. of **filipendulae** Linné). Tremewan, 1961, Bull. Brit. Mus. (nat. Hist.) Ent., **10** (7): 295, pl. 55, fig. 24.

ab. **longicornibus** Verity, 1926, Ent. Rec., **38**: 58; 1925, ibidem, **25**: 117, pl. 8, figs. 40, 41. Reiss, 1930, in Seitz, Die Gross-Schmetterlinge der Erde, Supplement, **2**: 36.

ab. **confluens** Higgs, 1890, Ent. Rec., **1**: 12. Barrett, 1895, The Lepidoptera of the British Islands, **2**, pl. 59, fig. 2b.

ab. **carnea** Cockayne, 1942, Ent. Rec., **54**: 35. Tremewan, 1961, Bull. Brit. Mus. (nat. Hist.) Ent., **10** (7): 295, pl. 55, fig. 25.

ab. **semilutescens** Higgs, 1890, Ent. Rec., **1**: 12. Barrett, 1895, The Lepidoptera of the British Islands, **2**, pl. 59, fig. 2a. Seitz, 1912, Die Gross-Schmetterlinge der Erde, **2**: 442.

ab. **lutescens** Higgs, 1890, Ent. Rec., **1**: 12.

ab. **obscura** Tutt, 1899, A Natural History of the British Lepidoptera, **1**: 487. Seitz, 1907, Die Gross-Schmetterlinge der Erde, **2**: 21. Oberthür, 1909, Études de Lépidoptérologie comparée, **3**, pl. 28, fig. 164; 1910, ibidem, **4**: 499. Alberti,

1955, Ent. Z., **65**: 89-91. Tremewan, 1961, Bull. Brit. Mus. (nat. Hist.) Ent., **10** (7): 295, pl. 55, fig. 26.

ab. **daimon** Porritt, 1911, Ent. mon. Mag., **47**: 203, pl. 3, fig. 1. Woodforde, 1929, Trans. ent. Soc. Lond., **76**: 530, pl. 22, fig. 9. Tremewan, 1961, Coridon, (A) **1**: 5, pl. C3, fig. 4 (as *nigricans* Oberthür).

f. loc. **ytenensis** Briggs, 1888, Young Nat., **9**: 82; 1888, ibidem, **9**: 108, 153, 188. Tremewan, 1960, Ent. Gaz., **11**: 192.

New Forest, Hampshire, England.

ssp. **palustrella** Verity, 1926, Ent. Rec., **38**: 11; 1925, ibidem, **37**: 117, pl. 8, figs. 8, 9. Ford, 1955, Moths, p. 126, 141, pl. 29, fig. 6. Tremewan, 1960, Ent. Gaz., **11**: 192; 1961, Coridon, (A) **1**: 4, pl. C2, figs. 8, 9. Hyde, 1964, Animals, **3** (6): 162-164, fig. 2 (aberration).

minor Tutt, 1899, A Natural History of the British Lepidoptera, **1**: 480 (preoccupied by **minor** Erschoff, 1874, ssp. of **cocandica** Erschoff).

Surrey, England [North Downs; South Downs; Salisbury Plain, etc., chalk downs in southern England, end of May and June].

ab. **orichalca** Tutt, 1899, A Natural History of the British Lepidoptera, **1**: 484. Tremewan, 1961, Bull. Brit. Mus. (nat. Hist.) Ent., **10** (7): 294.

ab. **caerulea** Tutt, 1899, A Natural History of the British Lepidoptera, **1**: 484. Tremewan, 1961, Bull. Brit. Mus. (nat. Hist.) Ent., **10** (7): 294.

ab. **pygmaea** Cockayne, 1954, Ent. Rec., **66**: 67, pl. 2, fig. 7. Tremewan, 1961, Bull. Brit. Mus. (nat. Hist.) Ent., **10** (7): 293, pl. 55, fig. 17.

ab. **obsoleta** Tutt, 1899, A Natural History of the British Lepidoptera, **1**: 485. Christy, 1896, Entomologist, **29**: 341, fig. 1. Reiss, 1930, in Seitz, Die Gross-Schmetterlinge der Erde, Supplement, **2**: 36. Tremewan, 1961, Bull. Brit. Mus. (nat. Hist.) Ent., **10** (7): 294, pl. 55, fig. 23.

ab. **extrema** Tutt, 1899, A Natural History of the British Lepidoptera, **1**: 485. Christy, 1895, Entomologist, **28**: 215; 1896, ibidem, **29**: 341, fig. 2. Seitz, 1912, Die Gross-Schmetterlinge der Erde, **2**: 442. Tremewan, 1961, Bull. Brit. Mus. (nat. Hist.) Ent., **10** (7): 294, pl. 55, fig. 22.

ab. **intermedia** Tutt, 1899, A Natural History of the British Lepidoptera, **1**: 487. Seitz, 1907, Die Gross-Schmetterlinge der Erde, **2**: 21. Tremewan, 1961, Bull. Brit. Mus. (nat. Hist.) Ent., **10** (7): 295.

ab. **lutescens** Cockerell, 1887, Entomologist, **20**: 152. Wellman, 1878, Entomologist, **11**: 102. Christy, 1895, Entomologist, **28**: 214. Seitz, 1907, Die Gross-Schmetterlinge der Erde, **2**: 21. South, 1908, The Moths of the British Isles, **2**: 337, pl. 148, fig. 5. Oberthür, 1910, Études de Lépidoptérologie comparée, **4**: 499. Tremewan, 1961, Coridon, (A) **1**: 5, pl. C2, fig. 10.

South, 1961, The Moths of the British Isles, **2**: 332, pl. 130, fig. 5.

ab. **lutescensbasalis** Tutt, 1899, A Natural History of the British Lepidoptera, **1**: 488. Tremewan, 1961, Bull. Brit. Mus. (nat. Hist.) Ent., **10** (7): 294.

ab. **lutescensglycirrhizae** Tutt, 1899, A Natural History of the British Lepidoptera, **1**: 488. Tremewan, 1961, Bull. Brit. Mus. (nat. Hist.) Ent., **10** (7): 295.

ab. **lutescensconfluens** Tutt, 1899, A Natural History of the British Lepidoptera, **1**: 488. Oberthür, 1909, Études de Lépidoptérologie comparée, **3**, pl. 28, fig. 163 (as *confluens* Tutt); 1910, ibidem, **4**: 499. Tremewan, 1961, Bull. Brit. Mus. (nat. Hist.) Ent., **10** (7): 294; 1961, Coridon (A) **1**: 5, pl. C1, fig. 11.

hybr. **fletcheri** Tutt, 1906, A Natural History of the British Lepidoptera, **5**: 36 (**trifolii decreta** Verity ♂ X **lonicerae transferens** Verity ♀). Holik, 1933, Iris, **47**: 15.

hybr. **grosvenori** Verity, 1926, Ent. Rec., **38**: 58 (**trifolii decreta** Verity X **filipendulae anglicola** Tremewan); 1925, ibidem, **37**: 117, pl. 8, fig. 39.

hybr. **escheri** Standfuss, 1896, Handbuch der paläarktischen Gross-Schmetterlinge, Aufl., **2**: 55, pl. 3, fig. 5 (**trifolii trifolii** Esper ♂ X **filipendulae germanica** Reiss ♀). Holik, 1933, Iris, **47**: 14.

Biology

Abeille, 1909, Mem. Soc. linn. Provence, **1**: 11. Austaut, 1878, Petites Nouv. Ent., **2**: 243. Bigger, 1962, Ent. Gaz., **13**: 55-67, figs. Blaschke, 1914, Die Raupen Europas mit ihren Futterpflanzen, pl. 6, fig. 13. Briggs, 1871, Trans. ent. Soc. Lond., p. 437, 439. Buckler, 1886, The Larvae of the British Butterflies and Moths, **2**: 94, pl. 19, figs. 2, 2a-2d. Burgeff, 1912, Z. wiss. InsektBiol., **8**: 124, 197; 1913, Ent. Z., **27**: 188, 189; 1950, Portug. acta biol., (A) Goldschmidt: 663-728. Chrétien, 1916, Ann. Soc. ent. Fr., **85**: 403. Döring, 1955, Zur Morphologie der Schmetterlingseier, p. 132, pl. 18, fig. 254. Foulquier, 1918, in Oberthür, Études de Lépidoptérologie comparée, **16**: 264. Grosvenor & Hewer, 1931, Ent. Rec., **43**: 23-28 (*clorinda*). Gühn, 1932, Ent. Jb., **41**: 90. Hepp, 1927, Ent. Z., **41**: 288. Holik, 1937, Lambillionea, **37**: 15-24, 32-45, 80-91; 1938, ibidem, **38**: 51-58, 79-88, 95-102; 1938, Ent. Rdsch., **55**: 320-323, 331-333; 1946, Rev. franç. Lépid., **10**: 250-261, 273-280; 1953, Ent. Z., **63**: 27; 1959, Bull. Soc. ent. Mulhouse, p. 17-25, fig. Hrubý, 1964, Prodromus Lepidopter Slovenska, p. 479. Hyde, 1964, Animals, **3** (6): 162-164, fig. 4. Jones, Parsons & Rothschild, 1962, Nature, **193**: 52-53. Kiefer, 1933, Int. ent. Z., **27**: 252-256; 1934, ibidem, **27**:

521-524. Koch, 1955, Wir Bestimmen Schmetterlinge, **2**: 62, 63, pl. 1, figs. 15, 15a. Lane & Rothschild, 1959, Ent. mon. Mag., **95**: 93-94. De Lattin, 1959, Bombus, **2**: 74. Oberthür, 1910, Études de Lépidoptérologie comparée, **4**: 663; 1917, ibidem, **13**, pl. 618, fig. 3538. Ochsenheimer, 1808, Die Schmetterlinge von Europa, **2**: 47. Reiss, 1926, Die Zygaenen Deutschlands, p. 7, 15; 1930, in Seitz, Die Gross-Schmetterlinge der Erde, Supplement, **2**: 37; 1958, Z. wien. ent. Ges., **43**: 161. Roüast, 1883, Catalogue des Chenilles européennes connues, p. 22. Sarlet, 1964, Mém. Soc. r. ent. Belg., **29**: 7, fig. 1a. Seitz, 1907, Die Gross-Schmetterlinge der Erde, **2**: 21. South, 1961, The Moths of the British Isles, **2**, pl. 132, figs. 3, 3a. Spuler, 1910, in Hofmann, Die Raupen der Schmetterlinge Europas, pl. 9, figs. 24a, b. Tutt, 1899, A Natural History of the British Lepidoptera, **1**: 482, 491, 502. Warnecke, 1948, Bombus, [no. 47]: 202. Zeller, 1847, Isis von oken, p. 301. Wiegel, 1965, Mitt. münch. ent. Ges., **55**: 160.

lonicerae Scheven

Distribution: Asia Minor, Armenia, Transcaucasia, Caucasus, Ukraine, Macedonia, Spain, Italy (excluding Sicily), France, Germany east to Siberia, north to Denmark and Scandinavia (including British Isles).

ssp. **natolica** Reiss, 1929, Int. ent. Z., **23**: 152; 1930, ibidem, **24**: 251; 1930, in Seitz, Die Gross-Schmetterlinge der Erde, Supplement, **2**: 38, pl. 4c; 1935, Int. ent. Z., **29**: 221. Holik, 1937, Festschrift zum 60. Geburstage von Professor Dr. Embrik Strand, **3**: 429. Holik & Sheljuzhko, 1958, Mitt. münch. ent. Ges., **48**: 219.

Sultan Dagh, Ak-Schehir, Kleinasien, 1700 m.

ssp. **sarykamyshensis** Holik & Sheljuzhko, 1958, Mitt. münch. ent. Ges., **48**: 218.

Sarykamysh, West-armenien, Kleinasien.

ssp. **nachitshevanica** Holik, 1937, Festschrift zum 60. Geburtstage von Professor Dr. Embrik Strand, **3**: 429, pl. 17, fig. 4. Holik & Sheljuzhko, 1958, Mitt. münch. ent. Ges., **48**: 216.

ab. **brunescens** Holik & Sheljuzhko, 1958, Mitt. münch. ent. Ges., **48**: 217.

Dorf Inakljü, Alagëz mont., Armenisches Bergland, 2000 m.

ssp. **achalcea** Burgeff, 1926, Mitt. münch. ent. Ges., **16**: 70. Reiss, 1930, in Seitz, Die Gross-Schmetterlinge der Erde, Supplement, **2**: 38, pl. 4c. Holik, 1937, Festschrift zum 60. Geburtstage von Professor Dr. Embrik Strand, **3**: 427, pl. 17, fig. 3. Koch, 1939, Mitt. münch. ent. Ges., **29**: 413. Holik & Sheljuzhko, 1958, Mitt. münch. ent. Ges., **48**: 214.

Achalzich, Georgien, Transkaukasien.

ssp. **abbastumana** Reiss, 1922, Int. ent. Z., **15**: 176; 1930, in Seitz, Die Gross-Schmetterlinge der Erde, Supplement, **2**: 38,

Tiflis und Abastuman

pl. 4c. Holik, 1937, Festschrift zum 60. Geburtstage von Professor Dr. Embrik Strand, **3**: 427. Koch, 1939, Mitt. münch. ent. Ges., **29**: 414. Holik & Sheljuzhko, 1958, Mitt. münch. ent. Ges., **48**: 214.	in Georgien, Trans-kaukasien.
ssp. **abchasica** Holik & Sheljuzhko, 1958, Mitt. münch. ent. Ges., **48**: 213.	Suchum, Abchasien, Trans-kaukasien.
ssp. **kindermanni** Oberthür, 1910, Études de Lépidoptérologie comparée, **4**: 544. Freyer, 1841, Stettin. ent. Ztg., **2**: 56 (as *stoechadis* Borkhausen). Reiss, 1930, in Seitz, Die Gross-Schmetterlinge der Erde, Supplement, **2**: 38. Holik, 1937, Festschrift zum 60. Geburtstage von Professor Dr. Embrik Strand, **3**: 420. Holik & Sheljuzhko, 1958, Mitt. münch. ent. Ges., **48**: 208. Tremewan, 1961, Bull. Brit. Mus. (nat. Hist.) Ent., **10** (7): 303, pl. 57, fig. 6.	Caucase [Terek Gebiet, Dagestan, Ciskaukasien].
ssp. **centricaucasica** Holik, 1939, Ann. Mus. zool. Polon., **13**: 253, pl. 23, figs. 14, 15. Holik & Sheljuzhko, 1958, Mitt. münch. ent. Ges., **48**: 212.	Bilagi-Don, Nord-Ossetien, Ciskaukasien, 1800 m.
ssp. **kubanensis** Holik & Sheljuzhko, 1958, Mitt. münch. ent. Ges., **48**: 211. Burgeff, 1926, Mitt. münch. ent. Ges., **16**: 71 (as *kindermanni* Oberthür).	Teberda Gebiet; Kuban Gebiet, Ciskaukasien.
ssp. **uzjana** Holik, 1939, Rev. franç. Lépid., **9**: 280, pl. 7, figs. 7-9. Holik & Sheljuzhko, 1958, Mitt. münch. ent. Ges., **48**: 206.	50 km. südöstlich von Uzjan, Bashkiren Gebiet, Ural-Westhang.
ssp. **kalkanensis** Reiss, 1932, Ent. Rdsch., **49**: 166, pl.1, figs.; 1933, in Seitz, Die Gross-Schmetterlinge der Erde, Supplement, **2**: 277. Holik, 1937, Festschrift zum 60. Geburtstage von Professor Dr. Embrik Strand, **3**: 423. Holik & Sheljuzhko, 1958, Mitt. münch. ent. Ges., **48**: 207.	Mukantash bei Kalkanova, 900 m.; Kalkansee, 820 m.; Urgunerwald, 830 m.; Ural-Osthang.
ab. **citrina** Reiss, 1932, Ent. Rdsch., **49**: 167; 1933, in Seitz, Die Gross-Schmetterlinge der Erde, Supplement, **2**: 277.	
ssp. **ukraina** Przegendza, 1932, Ent. Z., **46**: 117, figs. 37, 38. Holik & Reiss, 1932, in Holik, Iris, **46**: 123, pl. 1, figs. 20-22. Reiss, 1933, in Seitz, Die Gross-Schmetterlinge der Erde, Supplement, **2**: 277. Holik & Sheljuzhko, 1958, Mitt. münch. ent. Ges., **48**: 203.	Bei Kijev, Ukraine.
ab. **apicalidilatata** Sheljuzhko, 1941, Acta Mus. zool. Kijev, **1**: 77, 95, 101.	
ab. **melanitica** Zhicharew, 1928, Mitt. forstl. Versuchsw. Ukr., **9**: 259, 327.	

ssp. **thurneri** Holik, 1943, Mitt. münch. ent. Ges., **33**: 341. Daniel, 1958, Fragmenta Balcanica, **2**: 43; 1964, Prirod. Muz. Scopje, no. 2: 18.
Kara Orman, Serbisch Mazedonien, 700 m.

ssp. **nobilis** Navás, 1924, Arx. Inst. Cienc. Barcelona, **4** (10): 37.
Corcó, Barcelona, España.

ssp. **intermixta** Verity, 1925, Ent. Rec., **37**: 76, 117, pl. 8, figs. 50-53; 1926, ibidem, **38**: 59. Burgeff, 1926, Mitt. münch. ent. Ges., **16**: 67. Reiss, 1930, in Seitz, Die Gross-Schmetterlinge der Erde, Supplement, **2**: 38, pl. 4b, c; 1936, Ent. Rdsch., **54**: 91.
Orihuela, near Albarracin, Teruel, Spain, 1700 m.

ssp. **leonensis** Tremewan, 1961, Ent. Rec., **73**: 8; 1963, ibidem, **75**: 9, pl. 1, figs. 20, 21.
Riano, Leon, Spain, 3650 ft.

ssp. **astilbonta** Dujardin, 1965, Entomops, Nice, no. 2: 51, fig. Tremewan, 1961, Ent. Rec., **73**: 200 (as *major* Frey).
Estaing, Hautes-Pyrénées, 1200 m.; Vallée du Lys, Haute-Garonne; Plateau d'Agnos, 300 m., Basses-Pyrénées, France.

ab. **confluens** Oberthür, 1896, Études d'Entomologie, **20**: 49, pl. 8, fig. 147; 1910, Études de Lépidoptérologie comparée, **4**: 508. Tremewan, 1961, Bull. Brit. Mus. (nat. Hist.) Ent., **10** (7): 302, pl. 57, fig. 2.

confluens Oberthür, 1911, Études de Lépidoptérologie comparée, **5** (1): 197, pl. 63, fig. 586. Tremewan, 1961, Bull. Brit. Mus. (nat. Hist.) Ent., **10** (7): 302, pl. 57, fig. 3.

ssp. **silana** Burgeff, 1914, Mitt. münch. ent. Ges., **5**: 63, pl. 3, figs. 112-114. Verity, 1926, Ent. Rec., **38**: 71. Reiss, 1930, in Seitz, Die Gross-Schmetterlinge der Erde, Supplement, **2**: 38, pl. 4b.
Sila, Calabrien, Süditalien.

ssp. **herthae** Stauder, 1920, Soc. ent., **35**: 23. Verity, 1926, Ent. Rec., **38**: 72. Tremewan, 1966, Ent. Rec., **78**: 36, pl. 1, fig. 16.
Mt. Martinello, Calabrien, Süditalien.

dimorphica Verity, 1926, Ent. Rec., **38**: 60; 1925, ibidem, **37**: 117, pl. 8, figs. 54, 55. Tremewan, 1966, Ent. Rec., **78**: 36.

ab. **minima** Turati, 1923, Boll. Soc. ent. ital., **55**: 120. Verity, 1926, Ent. Rec., **38**: 72.

ab. **brevicornibus** Verity, 1926, Ent. Rec., **38**: 61. Reiss, 1930, in Seitz, Die Gross-Schmetterlinge der Erde, Supplement, **2**: 38.

ssp. **vivax** Verity, 1920, Boll. Lab. Zool. Portici, **14**: 38; 1925, Ent. Rec., **37**: 117, pl. 8, figs. 75, 76; 1926, ibidem, **38**: 71. Reiss, 1930, in Seitz, Die Gross-Schmetterlinge der Erde, Supplement, **2**: 38.
Presso Atina, Mainarde Mts., Prov. Caserta, Italia, 500 m.

ab. **posticeoobscurata** Verity, 1920, Boll. Lab. Zool. Portici, **14**: 39; 1925, Ent. Rec., **37**: 117, pl. 8, fig. 77. Reiss, 1930, in Seitz, Die Gross-Schmetterlinge der Erde, Supplement, **2**: 38.

ssp. **pauper** Verity, 1926, Ent. Rec., **38**: 69; 1925, ibidem, **37**:
Fergna Valley,

117, pl. 8, figs. 61, 62. Reiss, 1930, in Seitz, Die Gross-Schmetterlinge der Erde, Supplement, **2**: 37, pl. 4b.	Sibillini Mts., Marche, Italy, 1400-1700 m.
ab. **centralitaliae** Verity, 1926, Ent. Rec., **38**: 70. Reiss, 1930, in Seitz, Die Gross-Schmetterlinge der Erde, Supplement, **2**: 37.	
ssp. **pauperetincta** Verity, 1926, Ent. Rec., **38**: 70; 1925, ibidem, **37**: 117, pl. 8, figs. 68, 69. Reiss, 1930, in Seitz, Die Gross-Schmetterlinge der Erde, Supplement, **2**: 38.	Bolognola, Sibillini Mts., Italy, 1200 m.
f. t. **autumnalis** Verity, 1916, Boll. Soc. ent. ital., **47**: 73; 1925, Ent. Rec., **37**: 117, pl. 8, fig. 67; 1926, ibidem, **38**: 70. Reiss, 1930, in Seitz, Die Gross-Schmetterlinge der Erde, Supplement, **2**: 38.	Bolognola, Sibillini Mts., Italia, 1200 m.
ssp. **etruriae** Verity, 1926, Ent. Rec., **38**: 70; 1925, ibidem, **37**: 117, pl. 8, figs. 70, 71. Reiss, 1930, in Seitz, Die Gross-Schmetterlinge der Erde, Supplement, **2**: 37.	Mt. Senario, Florence, Italy, 800 m.
ssp. **apenninica** Rocci, 1921, Atti Soc. ligust. Sci. nat. geogr., **32**: 34. Verity, 1925, Ent. Rec., **37**: 117, pl. 8, fig. 58; 1926, ibidem, **38**: 69. Reiss, 1930, in Seitz, Die Gross-Schmetterlinge der Erde, Supplement, **2**: 37.	Calestano; Casaselvatica; Grontone; Appennino ligure, Italia.
ssp. **divulgata** Rocci, 1941, Boll. Ist. Ent. Univ. Bologna, **13**: 124, pl. 3, figs. 16-18.	Collina di Torino, 700 m.; Valle di Susa, 800 m.; Piedmont, Italia.
ssp. **martinensis** Reiss, 1929, Int. ent. Z., **22**: 358; 1930, in Seitz, Die Gross-Schmetterlinge der Erde, Supplement, **2**: 37.	St. Martin-Vésubie, Alpes-Maritimes, Frankreich.
ssp. **stoechadimima** Dujardin, 1956, Bull. mens. Soc. linn. Lyon, **25**: 256.	St. Barnabé, haute Plaine des Rochers, massif du Cheiron, Alpes-Maritimes, France, 950-1000 m.
ssp. **microdoxa** Dujardin, 1965, Entomops, Nice, no. 2: 53, fig.	Sainte Baume (région du Plan d'Aups), Var, France Sud, 700 m.
ssp. **oenoda** Dujardin, 1965, Entomops, Nice, no. 2: 52, fig. Dufay, 1966, Bull. mens. Soc. linn. Lyon, **35**: 72.	Col de Fontbelle, 1350 m.; Cousson, 1500 m.; Méailles;
ab. **burrasi** Tremewan, 1960, Ent. Rec., **72**: 209.	
ab. **nigerrima** Curtis, 1934, Ent. Rec., **46**: 37. Tremewan,	

1961, Bull. Brit. Mus. (nat. Hist.) Ent., **10** (7): 302, pl. 57, fig. 4.

Digne; Larche, 1600 m.; Allos, 1400 m.; env. de Fours, 1700 m.; Colmars, 1300-1500 m., Basses-Alpes; Les Achards; Chalet Puy St. Vincent; Ailefroide; Pelvoux; Monetier-les-Bains; Céüze, 1800 m.; Col du Lautaret, 1900 m., Hautes-Alpes, France.

ssp. **rechei** Dujardin, 1965, Entomops, Nice, no. 2: 53, fig.

ab. **asymetrica** Oberthür, 1911, Études de Lépidoptérologie comparée, **5** (1): 197, pl. 63, fig. 587. Tremewan, 1961, Bull. Brit. Mus. (nat. Hist.) Ent., **10** (7): 299, pl. 56, fig. 18.

ab. **flava** Oberthür, 1896, Études d'Entomologie, **20**: 43, pl. 8, fig. 148. Tremewan, 1961, Bull. Brit. Mus. (nat. Hist.) Ent., **10** (7): 300, pl. 56, fig. 19.

Meyrueis, Lozère; Mont Aigoual, Cévennes; Massif Central, France Sud.

ssp. **alpiumgigas** Verity, 1926, Ent. Rec., **38**: 73 (nomen novum for *major* Frey); 1925, ibidem, **37**: 117, pl. 8, fig. 87. Rocci, 1941, Boll. Ist. Ent. Univ. Bologna, **13**: 124, pl. 3, figs. 13, 15.

St. Nicolas im Visptal, Schweiz.

major Frey, 1880, Die Lepidopteren der Schweiz, p. 67 (preoccupied by **maior** Esper, 1797, ssp. of **filipendulae** Linné). Christ, 1880, Mitt. schweiz. ent. Ges., **6**: 43. Perlini, 1905, Forme di Lepidotteri esclusivamente Italiane, p. 53, pl. 1, fig. 16. Spuler, 1906, in Hofmann, Die Schmetterlinge Europas, **2**: 159. Seitz, 1907, Die Gross-Schmetterlinge der Erde, **2**: 21, pl. 5a. Reiss, 1930, in Seitz, Die Gross-Schmetterlinge der Erde, Supplement, **2**: 37; 1950, Jber. naturf. Ges. Graubünden, **82**: 118, fig. 27. Tremewan, 1961, Bull. Brit. Mus. (nat. Hist.) Ent., **10** (7): 301, pl. 56, fig. 27.

freyi Le Charles, 1930/35, in Lhomme, Catalogue des Lépidoptères de France et de Belgique, **1**: 695.

ab. **pfaehleri** Vorbrodt & Müller-Rutz, 1917, Mitt. schweiz. ent. Ges., **12**: 498 (as *pfähleri*).

ab. **semidiaphana** Stauder, 1929, Ent. Z., **43**: 80. Tremewan, 1961, Bull. Brit. Mus. (nat. Hist.) Ent., **10** (7): 303, pl. 57, fig. 5.

ab. **basiconfluens** Vorbrodt, 1913, in Vorbrodt & Müller-Rutz, Die Schmetterlinge der Schweiz, **2**: 265.

ab. **medioconfluens** Vorbrodt, 1913, in Vorbrodt & Müller-Rutz, Die Schmetterlinge der Schweiz, **2**: 265.

ab. **basimedioconfluens** Vorbrodt, 1913, in Vorbrodt & Müller-Rutz, Die Schmetterlinge der Schweiz, **2**: 265.

ab. **analiconfluens** Vorbrodt, 1913, in Vorbrodt & Müller-Rutz, Die Schmetterlinge der Schweiz, **2**: 265.

ab. **costalielongata** Vorbrodt, 1913, in Vorbrodt & Müller-Rutz, Die Schmetterlinge der Schweiz, **2**: 265.

ab. **parallela** Vorbrodt, 1914, in Vorbrodt & Müller-Rutz, Die Schmetterlinge der Schweiz, Nachtrag, **2**: 648.

ab. **incendium** Oberthür, 1909, Études de Lépidoptérologie comparée, **3**, pl. 22, fig. 105; 1910, ibidem, **4**: 513. Seitz, 1912, Die Gross-Schmetterlinge der Erde, **2**: 442. Burgeff, 1914, Mitt. münch. ent. Ges., **5**: 63, pl. 3, fig. 103. Tremewan, 1961, Bull. Brit. Mus. (nat. Hist.) Ent., **10** (7): 302, pl. 57, fig. 1.

crassimaculata Vorbrodt, 1913, in Vorbrodt & Müller-Rutz, Die Schmetterlinge der Schweiz, **2**: 264.

rubrosuffusa Verity, 1926, Ent. Rec., **38**: 74.

ssp. **magismaculata** Verity, 1926, Ent. Rec., **38**: 72; 1925, ibidem, **37**: 117, pl. 8, figs. 78, 79. Reiss, 1950, Jber. naturf. Ges. Graubünden, **82**: 118. — Geneva, Switzerland; Mayrhofen, Tyrol.

ab. **pseudomajor** Reiss, 1950, Jber. naturf. Ges. Graubünden, **82**: 119.

ssp. **glacieimagismaculata** Verity, 1930, Mem. Soc. ent. ital., **9**: 25. — Sappada, Alpi carniche, Italia.

ssp. **glaciei** Verity, 1926, Ent. Rec., **38**: 73; 1925, ibidem, **37**: 117, pl. 8, fig. 86. Reiss, 1930, in Seitz, Die Gross-Schmetterlinge der Erde, Supplement, **2**: 37. — Formazza Valley, 1400 m., Lepontine Alps, between the Rhone Valley and the Canton Tessin, Switzerland.

ab. **alpiumnana** Verity, 1926, Ent. Rec., **38**: 73; 1925, ibidem, **37**: 117, pl. 8, figs. 88-90. Reiss, 1930, in Seitz, Die Gross-Schmetterlinge der Erde, Supplement, **2**: 37.

ssp. **lonicerae** Scheven, 1777, Der Naturforscher, Halle, **10**: 97. Schäffer, 1766, Icones Insectorum circa Ratisbonam indigenorum, **1**, pl. 16, figs. 6, 7. Esper, 1780, Die Schmetterlinge, **2**, pl. 24, figs. 1a, b; 1781, ibidem, **2**: 183. Hübner, 1796, Sammlung europäischer Schmetterlinge, **2**: 16, pl. 2, fig. 7; 1806, ibidem, Der Ziefer, p. 80. Ochsenheimer, 1808, Die Schmetterlinge von Europa, **2**: 49. Boisduval, 1828, Essai sur une Monographie des Zygénides, p. 56, pl. 3, fig. 8. Freyer, 1845, Neuere Beiträge zur Schmetterlingskunde, **5**: 108, pl. 446. Herrich-Schäffer, 1846, Systematische Bearbeitung der Schmetterlinge von Europa, **2**: 36. Speyer & Speyer, 1858, Die geographische Verbreitung der Schmetterlinge Deutsch- — Regensburg, Bayern [Mittel- und Süddeutschland].

lands und der Schweiz, p. 347. Spuler, 1906, in Hofmann, Die Schmetterlinge Europas, **2**: 159, pl. 75, fig. 53, pl. 77, fig. 17. Seitz, 1907, Die Gross-Schmetterlinge der Erde, **2**: 21, pl. 5a. Oberthür, 1910, Études de Lépidoptérologie comparée, **4**: 507; 1911, ibidem, **5** (1): 197, pl. 63, figs. 584, 585. Verity, 1925, Ent. Rec., **37**: 117, pl. 8, figs. 80, 81. Reiss, 1926, Die Zygaenen Deutschlands, p. 15, pl. 2, fig. Burgeff, 1926, Mitt. münch. ent. Ges., **16**: 70. Derenne, 1935, Amat. Papillons, **7**: 172. Reiss, 1937, in Schneider, Jh. Ver. vaterl. Naturk. Württemb., **93**: 128. Koch, 1944, Mitt. münch. ent. Ges., **34**: 66. Reiss, 1949, Entomon, **1**: 173. Haaf, 1952, Veröff. zool. Staatssamml. Münch., **2**: 152, 154, 157, pl. 12. Bergmann, 1953, Die Grossschmetterlinge Mitteldeutschlands, **3**: 50, pl. 69, figs. B1, B2, B4. Gregor & Povolný, 1955, Sborn. ent. Odd. nár. Mus. Praze, **30**: 261, 272, pl. 5, fig. 1. Forster & Wohlfahrt, 1958, Die Schmetterlinge Mitteleuropas, **3**: 98, pl. 11, figs. 2, 7, 12. Alberti, 1958, Mitt. zool. Mus. Berl., **34**: 326. Heuser & Jöst, 1959, Mitt. Pollichia, (3) **6**: 127.

graminis de Villers, 1789, Caroli Linnaei Entomologia, **2**: 115. Ernst, 1782, Papillons d'Europe, **3**: 55, pl. 98, figs. 138a-d.

aspasia Meigen, 1830, Systematische Beschreibung der Europäischen Schmetterlinge, **2**: 83; 1832, ibidem, Nachtrag, **3**: 267, pl. 125, fig. 8.

ussuriensis Reiss, 1929, Int. ent. Z., **22**: 357; 1930, in Seitz, Die Gross-Schmetterlinge der Erde, Supplement, **2**: 38, pl. 4c. Holik, 1935, Ent. Anz., **15**: 87; 1935, Int. ent. Z., **29**: 197; 1942, Ent. Z., **55**: 236. Holik & Sheljuzhko, 1958, Mitt. münch. ent. Ges., **48**: 222. [?Ewgieniewka, Ussuri].

ab. **sphingiformis** Dąbrowski, 1965, Acta zool. cracov., **10**: 138, pl. 9, fig. 4.

ab. **diaphana** Burgeff, 1906, Ent. Z., **20**: 163, fig. 10. Reiss & Tremewan, 1964, Ent. Rec., **76**: 135.

translucens Burgeff, 1926, Mitt. münch. ent. Ges., **16**: 70. Reiss, 1930, in Seitz, Die Gross-Schmetterlinge der Erde, Supplement, **2**: 37. Reiss & Tremewan, 1964, Ent. Rec., **76**: 135.

ab. **parvimaculata** Holik, 1939, Ann. Mus. zool. Polon., **12**: 97, pl. 3, fig. 107.

ab. **kratochvili** Povolný & Gregor, 1946, Folia ent., Brno, **9**, Supplement, 12: 41, 96, pl. 2, fig. 9a.

ab. **sexmaculata** Dziurzyński, 1910, Int. ent. Z., **4**: 194. Reiss, 1926, Die Zygaenen Deutschlands, p. 17.

ab. **centripuncta** Reiss, 1964, Coridon, (A) **6**: 12.

ab. **apicalimaculata** Vorbrodt, 1913, in Vorbrodt & Müller-Rutz, Die Schmetterlinge der Schweiz, **2**: 265.

ab. **apicalielongata** Vorbrodt & Müller-Rutz, 1917, Mitt. schweiz. ent. Ges., **12**: 498.

apicalielongata Vorbrodt, 1914, in Vorbrodt & Müller-Rutz, Die Schmetterlinge der Schweiz, Nachtrag, **2**: 648 (nomen nudum).

ab. **medioconfluens** Reiss, 1964, Coridon, (A) **6**: 12.

ab. **bercei** Sand, 1879, Catalogue raisonné des Lépidoptères du Berry & de l'Auvergne, p. 23. Spuler, 1906, in Hofmann, Die Schmetterlinge Europas, **2**: 159. Oberthür, 1910, Études de Lépidoptérologie comparée, **4**: 508; 1911, ibidem, **5** (1): 197, pl. 53, figs. 588, 589. Reiss, 1930, in Seitz, Die Gross-Schmetterlinge der Erde, Supplement, **2**: 37.

confluens Dziurzyński, 1906, Ent. Z., **19**: 185; 1907, Jber. wien. ent. Ver., **17**: 84, pl. 2, fig. 3.

confluens Burgeff, 1906, Ent. Z., **20**: 162, fig. 6.

ab. **herzi** Slastshevsky, 1911, Horae Soc. ent. Ross., **40**: 128. Holik, 1939, Ann. Mus. zool. Polon., **12**: 97, pl. 3, fig. 106.

ab. **omniconfluens** Vorbrodt, 1913, in Vorbrodt & Müller-Rutz, Die Schmetterlinge der Schweiz, **2**: 265, figs. 26, 27.

ab. **rubescens** Burgeff, 1906, Ent. Z., **20**: 162, fig. 4. Seitz, 1907, Die Gross-Schmetterlinge der Erde, **2**: 21. Reiss, 1926, Die Zygaenen Deutschlands, p. 17; 1930, in Seitz, Die Gross-Schmetterlinge der Erde, Supplement, **2**: 37.

ab. **marginata** Burgeff, 1906, Ent. Z., **20**: 162, fig. 8.

ab. **carnea** Spuler, 1906, in Hofmann, Die Schmetterlinge Europas, **2**: 159. Seitz, 1907, Die Gross-Schmetterlinge der Erde, **2**: 21. Reiss, 1926, Die Zygaenen Deutschlands, p. 16.

ab. **semicitrina** Niepelt, 1924, Int. ent. Z., **18**: 189.

ab. **citrina** Speyer, 1887, Stettin. ent. Ztg., **48**: 334. Spuler, 1906, in Hofmann, Die Schmetterlinge Europas, **2**: 159. Seitz, 1907, Die Gross-Schmetterlinge der Erde, **2**: 21. Dziurzyński, 1908, Berl. ent. Z., **53**: 26, pl. 1, fig. 4.

lutea Nickerl, 1897, Verzeichnis der Insekten Böhmen's, **5**: 8 (nomen nudum).

ab. **hades** Metschl, 1925, Int. ent. Z., **19**: 27. Reiss, 1930, in Seitz, Die Gross-Schmetterlinge der Erde, Supplement, **2**: 37; 1935, Int. ent. Z., **28**: 542. Alberti, 1955, Ent. Z., **65**: 89-91.

ssp. **minuens** Verity, 1926, Ent. Rec., **38**: 72; 1925, ibidem, **37**: 117, pl. 8, figs. 82, 83. — **Brandlberg, Gammersdorf, Germany.**

ssp. **deludens** Koch, 1944, Mitt. münch. ent. Ges., **34**: 72. Tremewan, 1961, Bull. Brit. Mus. (nat. Hist.) Ent., **10** (7): 309, pl. 57, fig. 30. — **Memmingen, Eisenburg, Süddeutschland.**

ssp. **praeacuta** Burgeff, 1926, Mitt. münch. ent. Ges., **16**: 70. Reiss, 1930, in Seitz, Die Gross-Schmetterlinge der Erde, Supplement, **2**: 37.
ab. **oblonga** Guhn, 1932, Ent. Jb., **41**: 92.

Berlin; Brandenburg an der Havel; Tangermünde; Norddeutschland.

ssp. **stettinensis** Reiss, 1922, Int. ent. Z., **16**: 66; 1926, Die Zygaenen Deutschlands, p. 16, pl. 2, fig. (as *linnei* Reiss); 1930, in Seitz, Die Gross-Schmetterlinge der Erde, Supplement, **2**: 37, pl. 4a, b.
ab. **citrina** Reiss, 1964, Coridon, (A) **6**: 12.

Umgebung von Stettin (Höckendorf), Pommern, Norddeutschland.

ssp. **transferens** Verity, 1926, Ent. Rec., **38**: 59. Barrett, 1895, The Lepidoptera of the British Islands, **2**: 130, pl. 59, fig. 3. South, 1908, The Moths of the British Isles, **2**: 339, pl. 147, figs. 1, 2. Verity, 1925, Ent. Rec., **37**: 117, pl. 8, figs. 48, 49. Reiss, 1930, in Seitz, Die Gross-Schmetterlinge der Erde, Supplement, **2**: 36 (as *trifolii* Esper forma). Tremewan, 1960, Ent. Gaz., **11**: 190; 1961, Coridon, (A) **1**: 6, pl. C3, figs. 6, 7. South, 1961, The Moths of the British Isles, **2**: 333, pl. 133, figs. 1, 2.

Tring, Hertfordshire, England [southern England to Northumberland].

brittaniae Verity, 1926, Ent. Rec., **38**: 61; 1925, ibidem, **37**: 117, pl. 8, figs. 56, 57.
misera Verity, 1926, Ent. Rec., **38**: 73; 1925, ibidem, **37**: 117, pl. 8, figs. 84, 85.
ab. **minor** Tutt, 1899, A Natural History of the British Lepidoptera, **1**: 467. Tremewan, 1961, Bull. Brit. Mus. (nat. Hist.) Ent., **10** (7): 300, pl. 56, fig. 22.
minutissima Verity, 1926, Ent. Rec., **38**: 73.
ab. **miniata** Tutt, 1899, A Natural History of the British Lepidoptera, **1**: 467. Tremewan, 1961, Bull. Brit. Mus. (nat. Hist.) Ent., **10** (7): 300, pl. 57, fig. 31.
ab. **eboracae** Prest, 1883, Entomologist, **16**: 274. Seitz, 1907, Die Gross-Schmetterlinge der Erde, **2**: 21. Oberthür, 1910, Études de Lépidoptérologie comparée, **4**: 513. Tremewan, 1961, Bull. Brit. Mus. (nat. Hist.) Ent., **10** (7): 301, pl. 56, fig. 23.
ab. **grisescens** Cockayne, 1954, Ent. Rec., **66**: 67. Barrett, 1895, The Lepidoptera of the British Islands, **2**: 131, pl. 59, fig. 3c. Tremewan, 1961, Bull. Brit. Mus. (nat. Hist.) Ent., **10** (7): 301, pl. 56, fig. 24.
ab. **cuneata** Tutt, 1899, A Natural History of the British Lepidoptera, **1**: 468. Seitz, 1912, Die Gross-Schmetterlinge der Erde, **2**: 442. Tremewan, 1961, Bull. Brit. Mus. (nat. Hist.) Ent., **10** (7): 300, pl. 56, fig. 20.
ab. **centripuncta** Tutt, 1899, A Natural History of the British Lepidoptera, **1**: 468. Seitz, 1912, Die Gross-Schmetterlinge

der Erde, **2**: 442. Tremewan, 1961, Bull. Brit. Mus. (nat. Hist.) Ent., **10** (7): 300, pl. 56, fig. 21.

ab. **trivittata** Tutt, 1899, A Natural History of the British Lepidoptera, **1**: 468. Barrett, 1895, The Lepidoptera of the British Islands, **2**, pl. 59, fig. 3a. Reiss, 1930, in Seitz, Die Gross-Schmetterlinge der Erde, Supplement, **2**: 37. Tremewan, 1961, Bull. Brit. Mus. (nat. Hist.) Ent., **10** (7): 300.

ab. **semilutescens** Hewett, 1890, Ent. Rec., **1**: 60. Seitz, 1912, Die Gross-Schmetterlinge der Erde, **2**: 442.

ab. **lutescens** Hewett, 1890, Ent. Rec., **1**: 60. Seitz, 1912, Die Gross-Schmetterlinge der Erde, **2**: 442.

f. loc. **latomarginata** Tutt, 1899, A Natural History of the British Lepidoptera, **1**: 468. Seitz, 1912, Die Gross-Schmetterlinge der Erde, **2**: 442. Tremewan, 1960, Ent. Gaz., **11**: 191; 1961, Bull. Brit. Mus. (nat. Hist.) Ent., **10** (7): 301, pl. 56, fig. 25; 1961, Coridon, (A) **1**: 6, pl. C3, fig. 10. — Filey, coast of Yorkshire, England.

ssp. **jocelynae** Tremewan, 1962, Ent. Gaz., **13**: 10; 1962, Ent. Rec., **74**: 129, pl. 2, fig. 15. Tremewan & Manley, 1964, Ent. Rec., **76**: 149-153. Richardson, 1965, Ent. Rec., **77**: 16. — Isle of Skye, Inner Hebrides, Scotland.

ssp. **insularis** Tremewan, 1960, Ent. Gaz., **11**: 191; 1961, Bull. Brit. Mus. (nat. Hist.) Ent., **10** (7): 301, pl. 56, fig. 26; 1961, Coridon, (A) **1**: 6, pl. C3, fig. 11. Baynes, 1964, A Revised Catalogue of Irish Macrolepidoptera (Butterflies and Moths), p. 88 (as *transferens* Verity). — Mullinures, Co. Armagh, Ireland.

ssp. **linnei** Reiss, 1922, Int. ent. Z., **16**: 66; 1930, in Seitz, Die Gross-Schmetterlinge der Erde, Supplement, **2**: 37, pl. 4a. Nordström & Wahlgren, 1941, Svenska Fjärilar, p. 327, pl. 46, fig. 6. Hoffmeyer, 1960, De Danske Spindere, ed. 2, p. 214, pl. 16, figs. 12, 13, pl. 24, fig. 7. — Slite, Insel Gotland; Umgebung von Stockholm, Schweden.

ab. **chalybea** Aurivillius, 1888/91, Nordens Fjärilar, p. 53. Seitz, 1907, Die Gross-Schmetterlinge der Erde, **2**: 21. Alberti, 1955, Ent. Z., **65**: 89-91.

ab. **hanseni** Tremewan, 1960, Ent. Rec., **72**: 209. Hoffmeyer, 1960, De Danske Spindere, ed. 2, p. 214, fig.

ssp. **kareliae** Burgeff, 1926, Mitt. münch. ent. Ges., **16**: 69. Reiss, 1930, in Seitz, Die Gross-Schmetterlinge der Erde, Supplement, **2**: 37. Holik & Sheljuzhko, 1958, Mitt. münch. ent. Ges., **48**: 199. — Karelien.

hybr. **worthingi** Tutt, 1906, A Natural History of the British Lepidoptera, **5**: 36 (**lonicerae transferens** Verity ♂ X **trifolii decreta** Verity ♀). Holik, 1933, Iris, **47**: 15.

hybr. **secunda** Tutt, 1906, A Natural History of the British Lepidoptera, **5**: 36 (**lonicerae transferens** Verity ♂ X hybr. **fletcheri** Tutt ♀). Holik, 1933, Iris, **47**: 15.

hybr. **complexa** Tutt, 1906, A Natural History of the British

Lepidoptera, **5**: 37 (hybr. **worthingi** Tutt ♂ X hybr. **fletcheri** Tutt ♀). Holik, 1933, Iris, **47**: 15.

hybr. **complicata** Tutt, 1906, A Natural History of the British Lepidoptera, **5**: 37 (**lonicerae transferens** Verity ♂ X hybr. **complexa** Tutt ♀). Holik, 1933, Iris, **47**: 15.

hybr. **confusa** Tutt, 1906, A Natural History of the British Lepidoptera, **5**: 37 (hybr. **complexa** Tutt ♂ X **trifolii decreta** Verity ♀). Holik, 1933, Iris, **47**: 15.

hybr. **inversa** Tutt, 1906, A Natural History of the British Lepidoptera, **5**: 36 (**lonicerae transferens** Verity ♂ X **filipendulae anglicola** Tremewan ♀). Holik, 1933, Iris, **47**: 16. Haaf, 1952, Veröff. zool. Staatssamml. Münch., **2**: 157, pl. 12 (**lonicerae** Esper ♂ X **filipendulae** Linné ♀).

hybr. **burgeffi** Przegendza, 1926, Ent. Z., **40**: 296, 341, fig. 66 (**lonicerae lonicerae** Scheven ♂ X **ephialtes peucedani** Esper ♀). Reiss, 1933, in Seitz, Die Gross-Schmetterlinge der Erde, Supplement, **2**: 277, pl. 16m. Holik, 1933, Iris, **47**: 16.

peucedanoloniceroides Przegendza, 1933, Ent. Z., **47**: 27.

Biology

Barrett, 1895, The Lepidoptera of the British Islands, **2**: 130, pl. 59, fig. 3d. Blaschke, 1914, Die Raupen Europas mit ihren Futterpflanzen, pl. 6, fig. 14. Boisduval, Rambur & Graslin, 1832, Collection iconographique et Historique des Chenilles, pl. 2, figs. 4, 5. Briggs, 1871, Trans. ent. Soc. Lond., p. 438. Buckler, 1886, The Larvae of the British Butterflies and Moths, **2**: 18, pl. 19, figs. 3, 3a-c. Burgeff, 1912, Z. wiss. InsektBiol., **8**: 125, 197, 198; 1950, Portug. acta biol., (A) Goldschmidt: 663-728; 1951, Biol. Zbl., **70**: 1-23. Carpenter, 1937, J. Soc. Brit. Ent., **1**: 178. Döring, 1955, Zur Morphologie der Schmetterlingseier, p. 120, pl. 18, fig. 259. Dorfmeister, 1854, Verh. zool.-bot. Ver. Wien, **4**: 478. Esper, 1793, Die Schmetterlinge, Supplement, **2** (2): 12, pl. 39, figs. 9-14. Freyer, 1845, Neuere Beiträge zur Schmetterlingskunde, **5**: 108, pl. 446. Grabe, 1924, Int. ent. Z., **18**: 117. Gühn, 1932, Ent. Jb., **41**: 92. Holik, 1937, Lambillionea, **37**: 15-24, 32-45, 80-91; 1938, ibidem, **38**: 51-58, 79-88, 95-102; 1938, Ent. Rdsch., **55**: 320-323, 331-333; 1946, Rev. franç. Lépid., **10**: 250-261, 273-280; 1953, Ent. Z., **63**: 28; 1959, Bull. Soc. ent. Mulhouse, p. 17-25, fig. Hrubý, 1964, Prodromus Lepidopter Slovenska, p. 480. Hübner, 1806, Geschichte europäischer Schmetterlinge, pl. [75], figs. 1a, b. Jones, Parsons & Rothschild, 1962, Nature, **193**: 52-53. Kiefer, 1933, Int. ent. Z., **27**: 252-256; 1934, ibidem, **27**: 521-524. Koch, 1955, Wir Bestimmen Schmetterlinge, **2**: 62, 63, pl. 1, fig. 16, pl. 15, fig. 16. Lane, 1962, Ent. Gaz., **13**: 11, fig. B. Lane & Rothschild, 1961, Entomologist, **94**: 79-81. De Lattin, 1959, Bombus, **2**: 74. Ochsenheimer, 1808, Die Schmetterlinge von Europa, **2**: 49. Reiss, 1926, Die Zygaenen Deutschlands, p. 7, 17; 1930, in Seitz, Die Gross-Schmetterlinge der Erde, Supplement, **2**: 38; 1958, Z. wien. ent. Ges., **43**: 161. Roüast, 1883, Catalogue des Chenilles européennes connues, p. 22. Sarlet, 1964, Mém. Soc. r. ent. Belg., **29**: 8. Seitz, 1907, Die Gross-Schmetterlinge der Erde, **2**: 22. Seppänen, 1954, Suomen Suurperhostoukkien Ravintokasvit, p. 256. South, 1961, The Moths of the British Isles, **2**, pl. 132, figs. 2, 2a. Spuler, 1910, in Hofmann, Die Raupen der Schmetterlinge Europas, pl. 9, figs. 25a, b. Tremewan, 1965, Ent. Gaz., **16**: 87. Tremewan & Manley, 1964, Ent. Rec., **76**: 149-153. Tutt, 1899, A Natural History of the British Lepidoptera, **1**: 472.

Fossil from the Miocene Period
(Fossil aus dem Miozän)

Zygaena miocaenica Reiss, 1936, Ent. Rdsch., **53**: 554-556, pl. 7, figs. 1-4. Thomann, 1950, in Reiss, Jber. naturf. Ges. Graubünden, **82**: 96, 97, fig.

Mittlere Schicht des Randecker Maars, Schwäbische Alb, Württemberg, Süddeutschland (In bituminösem öligem Stinkschiefer (Dysodil) eingebettet).

ADDENDA

aurata Blachier

(p. 19, after ssp. **blachieri** Rothschild)

ssp. **oukaimedeina** Wiegel, 1965, Mitt. münch. ent. Ges., **55**: 116, 122, pl. 4, figs. 1, 2, 4-6, 8, 9, 11-18, 20-24. — Tarigt (Jebel Oukaimeden), 2700-3000 m.; Asif n'Ait Irèn, 2700-2850 m.; Djebel-Anngour-Gebiet, Hoher Atlas, Marokko.

ab. **apicaliconfluens** Wiegel, 1965, Mitt. münch. ent. Ges., **55**: 121, pl. 4, figs. 3, 19.

ab. **interrupta** Wiegel, 1965, Mitt. münch. ent. Ges., **55**: 121, pl. 4, figs. 7, 10.

sarpedon Hübner

(p. 23, after ssp. **confluenta** Reiss ab. **confluens** Reiss)

ab. **toticonfluens** Reiss, 1966, Ent. Rec., **78**: 138, pl. 5, figs. 1, 2.

(p. 23, after f. loc. **kampfi** Marten)

ssp. **altetica** Reiss, 1966, Ent. Rec., **78**: 138, pl. 5, figs. 3, 4. — El Altet, 7 km. ssw. Alicante, Prov. Alicante, Spain, 50 m.

ab. **pseudohispanica** Reiss, 1966, Ent. Rec., **78**: 139.

ssp. **benidormica** Reiss, 1966, Ent. Rec., **78**: 139, pl. 5, figs. 5, 6. — Benidorm, Prov. Alicante, Spain, 130 m.

ab. **pseudohispanica** Reiss, 1966, Ent. Rec., **78**: 140.

ab. **vitrea** Reiss, 1966, Ent. Rec., **78**: 140.

ab. **rubrior** Reiss, 1966, Ent. Rec., **78**: 140, pl. 5, figs. 7, 8.

ab. **nigrata** Reiss, 1966, Ent. Rec., **78**: 140.

(p. 24, after ssp. **tipula** Marten)

ab. **quinquemaculata** Reiss, 1966, Ent. Rec., **78**: 141.

ab. **pseudozapateri** Reiss, 1966, Ent. Rec., **78**: 142.

ab. **vitrea** Reiss, 1966, Ent. Rec., **78**: 141.

ab. **nigrata** Reiss, 1966, Ent. Rec., **78**: 141.

johannae Le Cerf

(p. 55, after ssp. **johannae** Le Cerf ab. **flava** Schwingenschuss)

ssp. **charlottae** Wiegel, 1965, Mitt. münch. ent. Ges., **55**: 125, 132, pl. 5, figs. 1-8, 11-13, 17-28. — Tarigt, 2700-2900 m.; Asif n'Ait Irèn, 2700-2800 m.; Tizi n'ou Addi, 2900 m.; Djebel-Anngour-Gebiet, Hoher Atlas, Marokko.

ab. **latestrigata** Wiegel, 1965, Mitt. münch. ent. Ges., **55**: 131, pl. 5, fig. 10.

ab. **interrupta** Wiegel, 1965, Mitt. münch. ent. Ges., **55**: 131, pl. 5, fig. 9.

ab. **albescens** Wiegel, 1965, Mitt. münch. ent. Ges., **55**: 131, pl. 5, fig. 16.

ab. **aurantiaca** Wiegel, 1965, Mitt. münch. ent. Ges., **55**: 131, pl. 5, fig. 14.
ab. **flava** Wiegel, 1965, Mitt. münch. ent. Ges., **55**: 131, pl. 5, fig. 15.

maroccana Rothschild

(p. 92, after ssp. **maroccana** Rothschild)

ssp. **irhris** Wiegel, 1965, Mitt. münch. ent. Ges., **55**: 134, 157, figs. 1a-1d, 2, 3, pl. 6, figs. 1-32. — Irhris/Tadmamt, 1700-1800 m.; vic. Tadmamt, 1800 m.; Azrou-Agaiouar-Gebiet, 1800 m.; Djebel-Toubkal-Gruppe, Hoher Atlas, Marokko.

loti Denis & Schiffermüller

(p. 129, after ssp. **bitorquata** Ménétriés)

exulantis Kolenati, 1846, Meletemata Entomologica, **5**: 94 (nomen nudum).

(p. 139, after ssp. **janthina** Boisduval)

atrapilosa Bruand, 1850, Mém. Soc. emul. Doubs, (1) **3** (3): 60; 1850, Catalogue systématique et synonymique des Microlépidoptères du Département du Doubs, Addendum général, Diurni, p. 94.

(p. 156, before **theryi** de Joannis)

problematica Naumann

Distribution: Southern Turkey.

problematica Naumann, 1966, Z. wien. ent. Ges., **51**: 10, figs. 1-3, pl. 1, figs. A1-A4, B, C, D. — Vic. Namrun, Weidegeb., 1000-1200 m., Bolkar (Bozoglan) Daglari; Cilicischen Taurus, Südturkei.

nevadensis Rambur

(p. 218, after ssp. **picos** Agenjo)

ssp. **panticosica** Reiss, 1966, Ent. Rec., **78**: 142, pl. 5, figs. 11, 12. Banos de Panticosa, 1600 m.; Panticosa, ca. 1300 m.; Huesca, northern Spain.

trifolii Esper

(p. 256, after ssp. **seriziati** Oberthür ab. **pseudoaustralis** Burgeff)

ssp. **tizeragis** Wiegel, 1965, Mitt. münch. ent. Ges., **55**: 158, 166, figs. 4a-4d, 5, 6, pl. 7, figs. 1, 2, 4-9, 13-20, 22, 23. Tizerag, 2700 m.; Ait Slimane, 2600 m.; Tarigt (Jebel Oukaimeden), 2700-2800 m.; Djebel-Anngour-Gebiet, Hoher Atlas, Marokko.

ab. **rusicadica** Wiegel, 1965, Mitt. münch. ent. Ges., **55**: 165, pl. 7, figs. 21, 24.

ab. **basalis** Wiegel, 1965, Mitt. münch. ent. Ges., **55**: 165, pl. 7, fig. 12.

ab. **glycirhizae** Wiegel, 1965, Mitt. münch. ent. Ges., **55**: 166.

ab. **intermedia** Wiegel, 1965, Mitt. münch. ent. Ges., **55**: 166, pl. 7, figs. 10, 11.

ab. **lutescens** Wiegel, 1965, Mitt. münch. ent. Ges., **55**: 166, 175, pl. 7, fig. 3 (as *lutenscens* (partim)).

INDEX

(Synonyms, preoccupied names, etc., are placed *in italics*)